Biology Now
with Physiology
Second High School Edition

ANNE HOUTMAN
ROSE-HULMAN INSTITUTE OF TECHNOLOGY

MEGAN SCUDELLARI
SCIENCE JOURNALIST

CINDY MALONE
CALIFORNIA STATE UNIVERSITY, NORTHRIDGE

REBECCA BREWER
TROY HIGH SCHOOL, MICHIGAN

W. W. NORTON
NEW YORK · LONDON

W. W. Norton & Company has been independent since its founding in 1923, when William Warder Norton and Mary D. Herter Norton first published lectures delivered at the People's Institute, the adult education division of New York City's Cooper Union. The firm soon expanded its program beyond the Institute, publishing books by celebrated academics from America and abroad. By midcentury, the two major pillars of Norton's publishing program—trade books and college texts— were firmly established. In the 1950s, the Norton family transferred control of the company to its employees, and today— with a staff of four hundred and a comparable number of trade, college, and professional titles published each year— W. W. Norton & Company stands as the largest and oldest publishing house owned wholly by its employees.

Editor: Betsy Twitchell
High School Director of Publishing: Jenna Bookin Barry
College Assistant Editor: Taylere Peterson
High School Editorial Assistant: Layne Broadwater
College Editorial Assistant: Danny Vargo
Developmental Editor: Andrew Sobel
College Project Editor: Christine D'Antonio
High School Project Editor: Melissa Atkin
Manuscript Editor: Stephanie Hiebert
Managing Editor, College: Marian Johnson
Managing Editor, College Digital Media: Kim Yi
Production Manager: Ashley Horna, Eric Pier-Hocking
Media Editor: Kate Brayton
Associate Media Editor: Cailin Barrett-Bressack
Media Project Editor: Jesse Newkirk
Media Editorial Assistant: Gina Forsythe
College Marketing Manager, Biology: Todd Pearson
High School Marketing Manager: Christina Magoulis
Design Director: Hope Miller Goodell
Photo Editor: Ted Szczepanski
Photo Researcher: Fay Torresyap
Permissions Manager: Megan Schindel
Permissions Clearer: Elizabeth Trammell
Composition by MPS North America LLC
Illustrations by Dragonfly Media Group
Manufacturing: TC Transcontinental Interglobe

Permission to use copyrighted material is included in the backmatter of this book.

References to the Next Generation Science Standards are included solely for informational purposes and to aid teacher instruction. NEXT GENERATION SCIENCE STANDARDS is a registered trademark of Achieve. Neither Achieve nor the lead states and partners that developed the Next Generation Science Standards were involved in the production of this product, and do not endorse it.

ISBN 978-0-393-63151-7

W. W. Norton & Company, Inc., 500 Fifth Avenue, New York, NY 10110-0017
wwnorton.com

W. W. Norton & Company Ltd., 15 Carlisle Street, London W1D 3BS

2 3 4 5 6 7 8 9 0

Brief Contents

Contents

 INTRODUCTION

CHAPTER 6: Cell Division 97

Toxic Plastic

UNIT 2: GENETICS

CHAPTER 7: Patterns of Inheritance 117

Dog Days of Science

UNIT 4: BIODIVERSITY

UNIT 5: ECOLOGY

About the Authors

ANNE HOUTMAN is Provost and Vice President of Academic Affairs at Rose-Hulman Institute of Technology, where she is also a full professor of biology. Anne has over 20 years of experience teaching nonmajors biology at a variety of private and public institutions, which gives her a broad perspective of the education landscape. She is strongly committed to evidence-based, experiential education and has been an active participant in the national dialogue on STEM (science, technology, engineering, and math) education for over 20 years. Anne's research interests are in the ecology and evolution of hummingbirds. She grew up in Hawaii, received her doctorate in zoology from the University of Oxford, and conducted postdoctoral research at the University of Toronto.

MEGAN SCUDELLARI is an award-winning freelance science writer and journalist based in Boston, Massachusetts, specializing in the life sciences. She has contributed to *Newsweek, Scientific American, Discover, Nature,* and *Technology Review,* among others, and she was a health columnist for the *Boston Globe.* For five years she worked as a correspondent and later as a contributing editor for *The Scientist* magazine. In 2013, she was awarded the prestigious Evert Clark/Seth Payne Award in recognition of outstanding reporting and writing in science. She has also received accolades for investigative reporting on traumatic brain injury and a feature story on prosthetics bestowed with a sense of touch. Megan received an MS from the Graduate Program in Science Writing at the Massachusetts Institute of Technology and worked as an educator at the Museum of Science, Boston.

CINDY MALONE began her scientific career wearing hip waders in a swamp behind her home in Illinois. She earned her BS in biology at Illinois State University and her PhD in microbiology and immunology at UCLA. She continued her postdoctoral work at UCLA in molecular genetics. She is currently a distinguished educator and a professor at California State University, Northridge, where she is the director of the CSUN-UCLA Stem Cell Scientist Training Program funded by the California Institute for Regenerative Medicine. Her research is aimed at training undergraduates and master's degree candidates to understand how genes are regulated through genetic and epigenetic mechanisms that alter gene expression. She has been teaching nonmajors biology for almost 20 years and has won teaching, mentorship, and curriculum enhancement awards at CSUN.

REBECCA BREWER teaches Advanced Placement and ninth-grade biology at Troy High School in southeastern Michigan. As an enthusiastic educator with more than 18 years of experience, Rebecca hopes her constructivist approach to instruction—which emphasizes student-led learning—inspires a passion for biological concepts. Outside of school, Rebecca trains other AP biology teachers in using electrophoresis equipment. She is also the director of Michigan's Outstanding Biology Teacher Award program and a former honoree. Rebecca earned a bachelor's degree in biology, with minors in chemistry and French, along with a teaching certification, from Michigan State University. After spending nearly a year in France, she returned to Michigan to intern at a Title 1 school and complete her master's degree, also at Michigan State, in curriculum and teaching.

Preface

A good biology class can improve the quality of students' lives. Biology is a part of so many decisions that students will need to make as individuals and as members of society. It helps parents to see the value of vaccinating a child, because they will understand what viruses are and how the immune system works. It helps homeowners in Texas, Florida, and Puerto Rico as they decide how to respond to the ongoing cleanup from recent hurricanes because they understand how an ecosystem functions. It helps students make more informed decisions about their own nutrition because they understand the effects of fat, cholesterol, and vitamins and minerals on our health. The examples are endless. Making informed decisions on these real-world issues requires students to be comfortable with scientific concepts and the process of scientific discovery.

How do we instill that capability in students? The last decade has seen an explosion of research on how students learn best. In a nutshell, they learn best when they see the relevance of a subject to their lives, when they are actively engaged in their learning, and when they are given opportunities to practice critical thinking.

In addition, most biology teachers would agree that our goal is to introduce students to both the key concepts of biology (for example, cells, DNA, evolution) and the tools to think critically about biological issues. Many would add that they want their students to leave the class with an appreciation for the value of science to society and an ability to distinguish between science and the nonscience or pseudoscience that bombards them on a daily basis. The introduction of the Next Generation Science Standards (NGSS) has given us a framework further emphasizing that strong teaching should strive to encourage less rote memorization and provide more opportunities for students to develop explanations and design solutions supported by evidence-based arguments.

How can a textbook help combine the ways students learn best with these goals? When students read them, traditional textbooks are adept at teaching key concepts, and they have recently begun to emphasize the relevance of biology to students' lives. But students may be intimidated by the length of chapters and the amount of difficult text, and they often cannot see the connections between the story and the science. More important, textbooks have not been successful at helping students become active learners and critical thinkers, and most fail to emphasize the process of science or how to assess scientific claims. It was our goal to make *Biology Now* relevant and interactive, to emphase the process of science in short chapters that students *want* to read, while still covering the essential content found in other high school biology textbooks, and to have that storyline unfold in such a way that it helps students connect the sequence of concepts of the Disciplinary Core Ideas.

Following the model of the first edition, each chapter in our book covers a current news story about people *doing* science, reported firsthand by Megan, an experienced journalist who specializes in reporting scientific findings in a compelling and accurate way, and fleshed out with a concise introduction to the science by Anne and Cindy and supplemented with additional content by Rebecca. For this second edition we decided to add content requested by our first-edition adopters. Specifically, we've added a full unit—comprising two new chapters and two revised chapters—on the amazing diversity of life on planet Earth. Not only was more substantial coverage of this topic a common request in feedback about the first edition, but it is also essential material for students, for it is partly through an appreciation of the diversity of life that students develop a personal relationship with the natural world.

Finally, we are thrilled for our book to be part of the online-assessment revolution! The second edition is accompanied by two excellent online homework platforms: a formative system called InQuizitive and a summative system called Smartwork5. We no longer worry that our students aren't seeing the forest for the trees when they read the textbook.

We sincerely hope you enjoy the fruits of our long labors.

Anne Houtman
Megan Scudellari
Cindy Malone
Rebecca Brewer

What's New in the Second Edition?

- New chapter stories on current, fun, and unexpected topics like the Zika virus outbreak, the human microbiome, and the discovery of a CRISPR gene editing technology. New stories include:

Chapter 5: How Cells Work—Rock Eaters
Unusual electricity-"eating" microbes could someday provide a new way to store and produce energy as "bacterial batteries."

Chapter 9: What Genes Are—Pigs to the Rescue
CRISPR is perhaps one of the most exciting discoveries of the last century. Chapter 9 describes one application of the CRISPR genome editing technology: creating organs for transplant . . . in pigs.

Chapter 15: Bacteria and Archaea—Navel Gazing
A team at North Carolina State University leads a citizen science project to sequence the human belly button microbiome and gets some surprising results.

Chapter 16: Plants, Fungi, and Protists—The Dirt on Black-Market Plants
Poaching is illegal, and trafficking of tropical plants such as orchids threatens their survival. A group of scientists is tracking illegal plants from the United States to their source.

Chapter 19: Growth of Populations—Zika-Busting Mosquitoes
The spread of Zika throughout the Americas quickly became a health crisis. Genetically modifying mosquitoes is one of the ways that scientists are attempting to control Zika's spread.

- A new unit on biodiversity, which significantly expands coverage of the vast diversity of life on Earth, with two completely new chapters and two significantly revised chapters. Instructors who wish to continue teaching a brief introduction to biodiversity can do so with the "overview" chapter (Chapter 14). But for those wishing to spend time exploring life on Earth, Chapters 15, 16, and 17 provide thorough science coverage and lively stories.
- New, earlier placement of the chapter on applying science to making critical choices. The "capstone" final chapter in the second edition is now Chapter 2: Evaluating Scientific Claims. Introducing the concept of scientifically literate evaluation of scientific claims early in the book gives students the maximum amount of time to benefit from that skill.
- The introduction of two new end-of-chapter question types:
  - Challenge Yourself, which encourages students to think critically about the chapter's important biological concepts
  - Thinking Multidimensionally, which offers students opportunities to tackle questions using multiple facets of the NGSS Disciplinary Core Ideas, Crosscutting Concepts, and Science and Engineering Practices.
- New animation, interactive, and visually based questions in Smartwork5 and InQuizitive that promote critical thinking, interaction with data, and engagement with biology in the real world.
- New resources in the *Ultimate High School Instructor's Guide*, which will be accessible in print and through the online Interactive Instructor's Guide platform, providing instructors with the ability to easily search and sort for active learning resources by topic, objective, and type of resource.

The perfect balance of science and story

Every chapter is structured around a story about people doing science that motivates students to read and stimulates their curiosity about biological concepts.

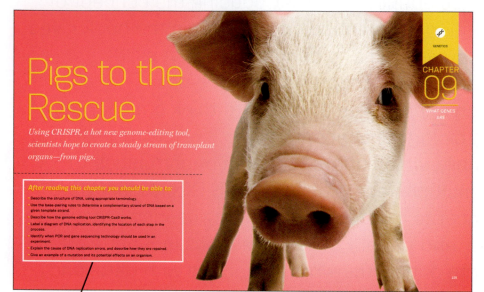

Dynamic chapter-opening spreads inspired by each chapter's story draw students in to the material.

"After reading this chapter you should be able to" sections introduce learning outcomes that preview the concepts presented in each chapter.

Cast-of-character bios

highlight the scientists, researchers, and professors at the center of each story.

GORDON LARK

A geneticist at the University of Utah in Salt Lake City, Gordon Lark initiated the Georgie Project in 1996 to study the genetics of Portuguese water dogs. The national research project has led to valuable knowledge about the genetic basis of health and disease in humans and dogs.

J. G. M. "HANS" THEWISSEN

Paleontologist and embryologist J. G. M. "Hans" Thewissen is a professor and whale expert at Northeast Ohio Medical University in the Department of Anatomy and Neurobiology. He and his lab study ancestral whale fossils and modern whale species.

LISA COOPER

Lisa Cooper is an assistant professor at Northeast Ohio Medical University in the Department of Anatomy and Neurobiology. She earned her PhD in Thewissen's lab.

MICHAEL HELLBERG AND CARLOS PRADA

Michael Hellberg (right) is an evolutionary biologist at Louisiana State University who studies how species evolve in marine environments. Carlos Prada (left) was a graduate student in Hellberg's lab, and is now a postdoctoral researcher at Penn State studying how organisms cope with changes in the environment.

XU XING

Xu Xing is a paleontologist at the Chinese Academy of Sciences in Beijing. He has discovered more than 60 species of dinosaurs and specializes in feathered dinosaurs and the origins of flight.

An inquiry-based approach that builds science skills—asking questions, thinking visually, and interpreting data.

Most **figures** in the book are accompanied by three questions that promote understanding and encourage engagement with the visual content. Answers are provided at the back of the book, making the questions a useful self-study tool.

Q1: Is a zygote haploid or diploid?

Q2: Which cellular process creates a baby from a zygote?

Q3: If a mother or father was exposed to BPA prior to conceiving a child, how might that explain potential birth defects in the fetus?

DRIVING QUESTION

How can a model of the primitive Earth provide new insights into why life is carbon-based?

- HS-LS1.C: Organization for Matter and Energy Flow in Organisms
- HS-PS1.A: Structure and Properties of Matter
- HS-PS1.B: Chemical Reactions

Every chapter-opening spread references the Next Generation Science Standards relevant to the chapter, accompanied by a **Driving Question** that presents a mystery to solve, such as "What evidence is there that modern-day whales evolved from a terrestrial ancestor?" (Chapter 11). Teacher prompts to support the Driving Questions are available in the *Ultimate Guide*.

Engaging, data-driven **infographics** appear in every chapter. Topics range from global renewable energy consumption (Chapter 5) to genetic diseases affecting Americans (Chapter 8) and many more. The infographics expose students to scientific data in an engaging way.

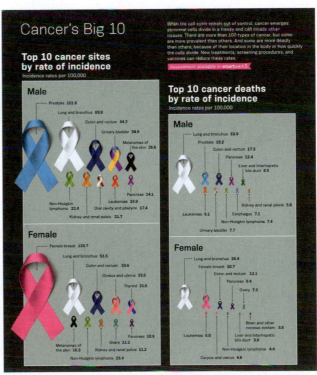

Extensive end-of-chapter review ensures that students see the forest for the trees.

Reviewing the Science

identifies each chapter's key science concepts, providing students with a guide for studying.

End-of-chapter questions

follow Bloom's taxonomy, moving from synthesis (Try Something New), to critical thinking (Challenge Yourself), to application (Leveling Up and Thinking Multidimensionally).

Leveling Up questions,

based on questions the authors use in their classrooms, prompt students to relate biology concepts to their own lives. The questions focus on one of the following themes: "Doing science," "Is it science?," "Life choices," "Looking at data," "What do *you* think?," and "*Write Now* biology."

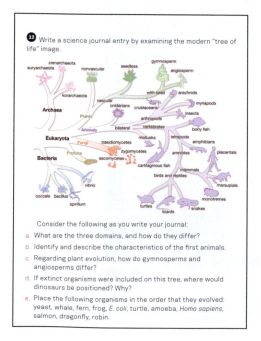

Thinking Multidimensionally questions

allow students to assess their comprehension of two to three facets of the Next Generation Science Standards.

Powerful resources for teaching and assessment

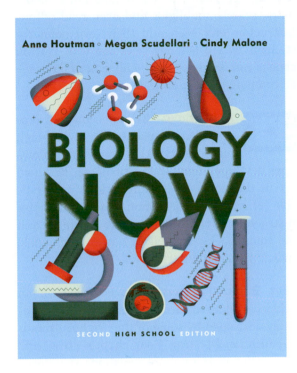

The **Interactive Instructor's Guide** is a searchable database of all the valuable teaching and active learning resources available in the *Ultimate Guide*. Instructors can easily filter by chapter, phrase, topic, or learning objective to find activities with downloadable handouts, streaming video with discussion questions, animations with discussion questions, lecture PowerPoints, and more.

Laura Zapanta, University of Pittsburgh
Tiffany Randall, John Tyler Community College
Rebecca Brewer, Troy High School
A.J. Prawitz, St. James High School
Tina Wnuk, St. James High School

The **Ultimate Guide** provides resources for implementing active learning into the classroom, including sample lecture outlines, coverage of road blocks, multimedia suggestions, animation discussion questions, prompts to accompany Driving Questions, Thinking Multidimensionally tags, and activities.

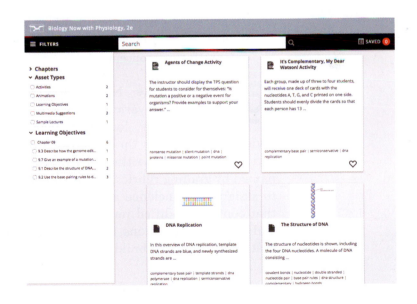

Other presentation tools for instructors

InQuizitive InQuizitive is Norton's easy-to-use adaptive-learning and quizzing tool that improves student understanding of important learning objectives. Students receive personalized quiz questions on the topics they need the most help with. When instructors assign InQuizitive, students come better prepared to lectures and exams. The second-edition course includes new animation questions, story-based questions, and critical-thinking questions.

Smartwork5 Smartwork5 delivers engaging, interactive online homework to students, helping instructors and students reach their teaching and learning goals. The second edition features:

- New infographic questions, which promote interaction with data and engagement with biology in the real world, while making this popular visual feature of the text an assignable activity.
- New story-based questions, which help students to learn and understand the science behind the stories in the text.
- New critical-thinking questions, which prompt students to think critically about important concepts in biology.
- New animation questions, which engage students with the book-specific animations covering biology concepts.

Coursepacks Norton's free coursepacks offer a variety of concept-based opportunities for assessment and review. The Leveling Up questions from the text are available as writing activities, accompanied by grading rubrics, making them easy to assign. Also included are reading quizzes that contain modified images from the text and animation questions, infographic quizzes that help students build skills in reading charts and graphs, and flashcards for student self-study of key terms.

Ebook Norton ebooks give students and instructors an enhanced reading experience at a lower cost than the print textbook. Students are able to have an active reading experience and can take notes, bookmark, search, highlight, and even read offline. Instructors can even add their own notes for students to see as they read the text. Norton ebooks can be viewed on—and synced among—all computers and mobile devices.

Animations Key concepts and processes are explained clearly through high-quality, ADA-compliant animations developed from the meticulously designed art in the book. These animations are available for lecture presentation in the Interactive Instructor's Guide, PowerPoint outlines, and the coursepacks, as well as within our ebook, InQuizitive, and Smartwork5.

Test Bank The test bank is based on an evidence-centered design that was collaboratively developed by some of the brightest minds in educational testing. Each chapter's test bank now includes 75 or more questions structured around the learning objectives from the textbook. Questions are further classified by text section and difficulty, and are provided in multiple-choice, fill-in-the-blank, and short-answer form. New infographic questions in every chapter help test student interpretation of charts and graphs.

Art Files All art and photos from the book are available, in presentation-ready resolution, as both JPEGs and PowerPoints for instructor use.

Lecture Slides Comprehensively revised by book author Cindy Malone, complete lecture PowerPoints thoroughly cover chapter concepts and include images and clicker questions to encourage student engagement.

Correlations to the Next Generation Science Standards

The following correlation maps *Biology Now, Second High School Edition* to the four areas of the Next Generation Science Standards*: Life Science; Physical Science; Engineering, Technology and Applications of Science; and Earth and Space Science. Since the high school biology course grows out of content covered in middle school, we also include a final section on where our textbook revisits important middle school Disciplinary Core Ideas, as well.

Life Science

LS1.A: Structure and Function

Systems of specialized cells within organisms help them perform the essential functions of life.	"The Characteristics of Living Organisms," Ch. 1, p. 6; Figure 16.2, Ch. 16, p. 287;
All cells contain genetic information in the form of DNA molecules. Genes are regions in the DNA that contain the instructions that code for the formation of proteins, which carry out most of the work of cells.	"Life is Cellular," Ch. 4, pp. 60–62; Ch. 10, pp. 174–175; Chapter 15 intro, Ch. 15, pp. 270–273; "Merry Microbes," Ch. 15, pp. 273–276
Multicellular organisms have a hierarchical structural organization, in which any one system is made up of numerous parts and is itself a component of the next level.	Figure 1.9, Ch. 1, p. 14; "Two Cells are Better Than One," Ch. 16, p. 289
Feedback mechanisms maintain a living system's internal conditions within certain limits and mediate behaviors, allowing it to remain alive and functional even as external conditions change within some range. Feedback mechanisms can encourage (through positive feedback) or discourage (negative feedback) what is going on inside the living system.	"The Characteristics of Living Organisms," Ch.1, p. 6; "Through the Barrier," Ch. 4, pp. 64–67

LS1.B: Growth and Development of Organisms

In multicellular organisms individual cells grow and then divide via a process called mitosis, thereby allowing the organism to grow. The organism begins as a single cell (fertilized egg) that divides successively to produce many cells, with each parent cell passing identical genetic material (two variants of each chromosome pair) to both daughter cells. Cellular division and differentiation produce and maintain a complex organism, composed of systems of tissues and organs that work together to meet the needs of the whole organism.	"Divide and Conquer," Ch. 6, pp. 99–101; "Trade Secret," Ch. 6, pp. 101–103; Figure 6.4, Ch. 6, p. 102–103

LS1.C: Organization for Matter and Energy Flow in Organisms

The process of photosynthesis converts light energy to stored chemical energy by converting carbon dioxide plus water into sugars plus released oxygen.	"Energy for Life," Ch. 5, pp. 81–82; "Into the Light, Part 1," Ch. 5, pp. 84–85; Figure 5.7, Ch. 5, p. 86
The sugar molecules thus formed contain carbon, hydrogen, and oxygen: their hydrocarbon backbones are used to make amino acids and other carbon-based molecules that can be assembled into larger molecules (such as proteins or DNA), used for example to form new cells.	Figure 3.1, Ch. 3, p. 50; "Energy for Life," Ch. 5, pp. 81–82
As matter and energy flow through different organizational levels of living systems, chemical elements are recombined in different ways to form different products.	"Chemistry of Life," Ch. 3, pp. 40–43

*Next Generation Science Standards is a registered trademark of Achieve. Neither Achieve nor the lead states and partners that developed the Next Generation Science Standards were involved in the production of this product, and do not endorse it.

As a result of these chemical reactions, energy is transferred from one system of interacting molecules to another. Cellular respiration is a chemical process in which the bonds of food molecules and oxygen molecules are broken and new compounds are formed that can transport energy to muscles. Cellular respiration also releases the energy needed to maintain body temperature despite ongoing energy transfer to the surrounding environment	"Energy for Life," Ch. 5, pp. 81–82; "Into the Light, Part 2," Ch. 5, pp. 88–90; Figure 5.11, Ch. 5, p. 89

LS2.A: Interdependent Relationships in Ecosystems

Ecosystems have carrying capacities, which are limits to the numbers of organisms and populations they can support. These limits result from such factors as the availability of living and nonliving resources and from such challenges such as predation, competition, and disease. Organisms would have the capacity to produce populations of great size were it not for the fact that environments and resources are finite. This fundamental tension affects the abundance (number of individuals) of species in any given ecosystem.	"A Key Loss," Ch. 20, pp. 361–363; Figure 20.5, Ch. 20, p. 362; "A Second Ripple Effect," Ch. 20, pp. 363–367; "Cause and Effect," Ch. 20, p. 366; "Going Green," Ch. 21, pp. 378–380; "Reaching Capacity," Ch. 19, pp. 346–347; "Seeking Change," Ch. 19, pp. 347–350

LS2.B: Cycles of Matter and Energy Transfer in Ecosystems

Photosynthesis and cellular respiration (including anaerobic processes) provide most of the energy for life processes.	Energy for Life, Ch. 5, pp. 81–82; "A Key Loss," Ch. 20, pp. 361–363; Figure 21.5, Ch. 21, p. 382
Plants or algae form the lowest level of the food web. At each link upward in a food web, only a small fraction of the matter consumed at the lower level is transferred upward, to produce growth and release energy in cellular respiration at the higher level. Given this inefficiency, there are generally fewer organisms at higher levels of a food web. Some matter reacts to release energy for life functions, some matter is stored in newly made structures, and much is discarded. The chemical elements that make up the molecules of organisms pass through food webs and into and out of the atmosphere and soil, and they are combined and recombined in different ways. At each link in an ecosystem, matter and energy are conserved.	"Bottom of the Pyramid," Ch. 21, pp. 380–382; Figure 21.5, Ch. 21, p. 382; Figure 20.5, Ch. 20, p. 362; "A Key Loss," Ch. 20, pp. 361–363
Photosynthesis and cellular respiration are important components of the carbon cycle, in which carbon is exchanged among the biosphere, atmosphere, oceans, and geosphere through chemical, physical, geological, and biological processes.	Figure 18.14, Ch. 18, p. 336; "The Carbon Games," Ch. 18, pp. 335–337; "Waiting and Watching," Ch. 18, pp. 337–338

LS2.C: Ecosystem Dynamics, Functioning, and Resilience

A complex set of interactions within an ecosystem can keep its numbers and types of organisms relatively constant over long periods of time under stable conditions. If a modest biological or physical disturbance to an ecosystem occurs, it may return to its more or less original status (i.e., the ecosystem is resilient), as opposed to becoming a very different ecosystem. Extreme fluctuations in conditions or the size of any population, however, can challenge the functioning of ecosystems in terms of resources and habitat availability.	"Going Green," Ch. 21, pp. 378–380; "A Key Loss," Ch. 20, pp. 361–363; Figure 20.5, Ch. 20, p. 362; "A Second Ripple Effect," Ch. 20, pp. 363–367; "Cause and Effect," Ch. 20, p. 366; "Hot and Dry," Ch. 18, pp. 325–326; "A Warmer World," Ch. 18, pp. 326–329
Moreover, anthropogenic changes (induced by human activity) in the environment—including habitat destruction, pollution, introduction of invasive species, overexploitation, and climate change—can disrupt an ecosystem and threaten the survival of some species.	"A Warmer World," Ch. 18, pp. 326–329, "Forest Devastation," Ch. 18, p. 331; "How Big Is Your Ecological Footprint?" Ch. 18, pp. 332–333; "A Warmer World," Ch. 18, pp. 326–329; "Forest Devastation," Ch. 18, p. 331

LS2.D: Social Interactions and Group Behavior

Group behavior has evolved because membership can increase the chances of survival for individuals and their genetic relatives.	Figure 20.12, Ch. 20, p. 370; "Safety in Numbers and Colors," Ch. 20, pp. 368–371

LS3.A: Inheritance of Traits

Each chromosome consists of a single very long DNA molecule, and each gene on the chromosome is a particular segment of that DNA . The instructions for forming species' characteristics are carried in DNA . All cells in an organism have the same genetic content, but the genes used (expressed) by the cell may be regulated in different ways. Not all DNA codes for a protein; some segments of DNA are involved in regulatory or structural functions, and some have no as-yet known function.	"Getting to the Genes," Ch. 7, pp. 118–119; Figure 7.2, Ch. 7, p. 119; "Two-Step Dance, Translation: RNA to Protein," Ch. 10, pp. 179–183; Figure 10.9, Ch. 10, p. 182

LS3.B: Variation of Traits

In sexual reproduction, chromosomes can sometimes swap sections during the process of meiosis (cell division), thereby creating new genetic combinations and thus more genetic variation. Although DNA replication is tightly regulated and remarkably accurate, errors do occur and result in mutations, which are also a source of genetic variation. Environmental factors can also cause mutations in genes, and viable mutations are inherited.	"Unequal Division," Ch. 6, pp. 104–108; Figure 6.8, Ch. 6, p. 107; "Shuffling the DNA," Ch. 6, pp. 108; Figure 6.9, Ch.6, p. 109; Figure 6.10, Ch. 6, p. 110; "A Mysterious Malady," Ch. 8, p. 136; "Painful Pedigree," Ch. 8, pp. 137–139; "Tweaking Gene Expression," Ch.10, pp. 183–184; Figure 10.10, Ch. 10, p. 183
Environmental factors also affect expression of traits, and hence affect the probability of occurrences of traits in a population. Thus the variation and distribution of traits observed depends on both genetic and environmental factors.	"Most Chronic Diseases are Complex Traits," Ch. 7, p. 127; "It's Complicated," Ch. 7, p. 126–128; Figure 7.10, Ch. 7, p. 129

LS4.A: Evidence of Common Ancestry and Diversity

Genetic information, like the fossil record, provides evidence of evolution. DNA sequences vary among species, but there are many overlaps; in fact, the ongoing branching that produces multiple lines of descent can be inferred by comparing the DNA sequences of different organisms. Such information is also derivable from the similarities and differences in amino acid sequences and from anatomical and embryological evidence.	"Clues in the Code," Ch. 11, pp. 201–203, Figure 11.11, Ch. 11, p. 201; Figure 15.2, Ch. 15, p. 270; "Navel Gazing" Intro, Ch. 15, pp. 270–273; Figure 17.3, Ch. 17, p. 304; Figure 17.7, Ch. 17, p. 308; "Hereditary Heirlooms," Ch. 17, p. 309

LS4.B: Natural Selection

Natural selection occurs only if there is both (1) variation in the genetic information between organisms in a population and (2) variation in the expression of that genetic information—that is, trait variation—that leads to differences in performance among individuals.	"Artificial to Natural," Ch. 11, pp. 193–195; Figure 11.4, Ch. 11, p. 195; Figure 12.2, Ch. 12, p. 213; "Rising Resistance," Ch. 12, pp. 215–218
The traits that positively affect survival are more likely to be reproduced, and thus are more common in the population.	"Sex and Selection," Ch. 12, p. 223

LS4.C: Adaptation

Evolution is a consequence of the interaction of four factors: (1) the potential for a species to increase in number, (2) the genetic variation of individuals in a species due to mutation and sexual reproduction, (3) competition for an environment's limited supply of the resources that individuals need in order to survive and reproduce, and (4) the ensuing proliferation of those organisms that are better able to survive and reproduce in that environment.	"Animal Kingdom," Ch. 17, pp. 302–306; "Rise of the Apes," Ch.17, pp. 308–311
Natural selection leads to adaptation, that is, to a population dominated by organisms that are anatomically, behaviorally, and physiologically well suited to survive and reproduce in a specific environment. That is, the differential survival and reproduction of organisms in a population that have an advantageous heritable trait leads to an increase in the proportion of individuals in future generations that have the trait and to a decrease in the proportion of individuals that do not.	"Leaping Lizards," Ch. 13, pp. 232–234; "Get a Backbone!" Ch. 17, p. 305
Adaptation also means that the distribution of traits in a population can change when conditions change.	"Caribbean Corals," Ch. 13, pp. 237–239
Changes in the physical environment, whether naturally occurring or human induced, have thus contributed to the expansion of some species, the emergence of new distinct species as populations diverge under different conditions, and the decline–and sometimes the extinction–of some species.	"What Makes a Species?" Ch. 13, pp. 234–237;"Tussling with Trees," Ch. 14, pp. 261–265; Figure 14.15, Ch. 14, p. 263
Species become extinct because they can no longer survive and reproduce in their altered environment. If members cannot adjust to change that is too fast or drastic, the opportunity for the species' evolution is lost.	"What Makes a Species?" Ch. 13, pp. 234–237; "Tussling with Trees," Ch. 14, pp. 261–265

LS4.D: Biodiversity and Humans

Biodiversity is increased by the formation of new species (speciation) and decreased by the loss of species (extinction).	"Tussling with Trees," Ch. 14, pp. 261–265; Figure 14.15, Ch. 14, p. 263; "The Precious 1 Percent," Ch. 21, pp. 383–386
Humans depend on the living world for the resources and other benefits provided by biodiversity. But human activity is also having adverse impacts on biodiversity through overpopulation, overexploitation, habitat destruction, pollution, introduction of invasive species, and climate change. Thus sustaining biodiversity so that ecosystem functioning and productivity are maintained is essential to supporting and enhancing life on Earth. Sustaining biodiversity also aids humanity by preserving landscapes of recreational or inspirational value.	"Heating Up," Ch. 21, p. 389

Physical Science

PS1.A: Structure and Properties of Matter

Each atom has a charged substructure consisting of a nucleus, which is made of protons and neutrons, surrounded by electrons.	"Ingredients for Life" Intro, Ch. 3, pp. 40–43

PS1.B: Chemical Reactions

Chemical processes, their rates, and whether or not energy is stored or released can be understood in terms of the collisions of molecules and the rearrangements of atoms into new molecules, with consequent changes in the sum of all bond energies in the set of molecules that are matched by changes in kinetic energy.	"Ingredients for Life" Intro, Ch. 3, pp. 40–43

PS3.D: Energy in Chemical Processes and Everyday Life

The main way that solar energy is captured and stored on Earth is through the complex chemical process known as photosynthesis.	"Energy for Life," Ch. 5, pp. 81–82; "Into the Light, Part 1," Ch. 5, pp. 84–85; Figure 5.7, Ch. 5, p. 86

Engineering, Technology, and Applications of Science

ETS1.A: Defining and Delimiting Engineering Problems

Humanity faces major global challenges today, such as the need for supplies of clean water and food or for energy sources that minimize pollution, which can be addressed through engineering. These global challenges also may have manifestations in local communities.	Ch. 5, pp. 80–93; Ch. 10, pp. 174–186; "How Big is Your Ecological Footprint?" Ch. 18, p. 332–333; "Population Control," Ch. 19, pp. 343–345

ETS1.B: Developing Possible Solutions

When evaluating solutions it is important to take into account a range of constraints including cost, safety, reliability and aesthetics and to consider social, cultural and environmental impacts.	Figure 1.5, p. 10; Ch. 5, pp. 80–93; "Deadly with One Allele," Ch. 8, pp. 146–147; "How Big is Your Ecological Footprint?" Ch. 18, p. 332–333; "Seeking Change," Ch. 19, pp. 347–350

ETS1.C: Optimizing the Design Solution

Criteria may need to be broken down into simpler ones that can be approached systematically, and decisions about the priority of certain criteria over others (tradeoffs) may be needed.	Figure 1.6, p. 11

Earth and Space Science

ESS2.D: Weather and Climate

The foundation for Earth's global climate systems is the electromagnetic radiation from the sun, as well as its reflection, absorption, storage, and redistribution among the atmosphere, ocean, and land systems, and this energy's re-radiation into space.	Ch. 18

ESS3.D: Global Climate Change

Though the magnitudes of human impacts are greater than they have ever been, so too are human abilities to model, predict, and manage current and future impacts.	Ch. 18

MS-LS1.A: Structure and Function

All living things are made up of cells, which is the smallest unit that can be said to be alive. An organism may consist of one single cell (unicellular) or many different numbers and types of cells (multicellular).	"The Characteristics of Living Organisms," Ch. 1, p. 6; Ch. 4, p. 60
Within cells, special structures are responsible for particular functions, and the cell membrane forms the boundary that controls what enters and leaves the cell.	"What's in a Cell?" Ch. 4, pp. 68–72
In multicellular organisms, the body is a system of multiple interacting subsystems. These subsystems are groups of cells that work together to form tissues and organs that are specialized for particular body functions.	Figure 1.9, Ch. 1, p. 14

MS-LS1.B: Growth and Development of Organisms

Animals engage in characteristic behaviors that increase the odds of reproduction.	"Sex and Selection," Ch. 12, p. 223
Plants reproduce in a variety of ways, sometimes depending on animal behavior and specialized features for reproduction.	"Searching for Flowers," Ch. 16, pp. 292–294
Genetic factors as well as local conditions affect the growth of the adult plant.	"The Dirt on Black-Market Plants," Chapter 16, pp. 285–297
Organisms reproduce, either sexually or asexually, and transfer their genetic information to their offspring.	"Divide and Conquer," Ch. 6, pp. 99–101; "Trade Secret," Ch. 6, pp. 101–103; Figure 6.4, Ch. 6, p. 102–103; "Unequal Division," Ch. 6, pp. 104–108; Figure 6.8, Ch. 6, p. 107

MS-LS1.C: Organization for Matter and Energy Flow in Organisms

Plants, algae (including phytoplankton), and many microorganisms use the energy from light to make sugars (food) from carbon dioxide from the atmosphere and water through the process of photosynthesis, which also releases oxygen. These sugars can be used immediately or stored for growth or later use.	"Energy for Life," Ch. 5, pp. 81–82; Figure 5.2, Ch. 5, p. 82

MS-LS1.D: Information Processing

Each sense receptor responds to different inputs (electromagnetic, mechanical, chemical), transmitting them as signals that travel along nerve cells to the brain. The signals are then processed in the brain, resulting in immediate behaviors or memories.	"Brain-Body Connection," Ch. 25, pp. 457–459

MS-LS2.A: Interdependent Relationships in Ecosystems

Organisms, and populations of organisms, are dependent on their environmental interactions both with other living things and with nonliving factors.	"Of Wolves and Trees," Chapter 20, pp. 358–373

In any ecosystem, organisms and populations with similar requirements for food, water, oxygen, or other resources may compete with each other for limited resources, access to which consequently constrains their growth and reproduction.	"Safety in Numbers and Colors," Ch. 20, pp. 368–371
Growth of organisms and population increases are limited by access to resources.	"Safety in Numbers and Colors," Ch. 20, pp. 368–371
Similarly, predatory interactions may reduce the number of organisms or eliminate whole populations of organisms. Mutually beneficial interactions, in contrast, may become so interdependent that each organism requires the other for survival. Although the species involved in these competitive, predatory, and mutually beneficial interactions vary across ecosystems, the patterns of interactions of organisms with their environments, both living and nonliving, are shared.	"A Key Loss," Ch. 20, pp. 361–363

MS-LS2.B: Cycle of Matter and Energy Transfer in Ecosystems

Food webs are models that demonstrate how matter and energy is transferred between producers, consumers, and decomposers, as the three groups interact within an ecosystem. Transfers of matter into and out of the physical environment occur at every level. Decomposers recycle nutrients from dead plant or animal matter back to the soil in terrestrial environments or to the water in aquatic environments. The atoms that make up the organisms in an ecosystem are cycled repeatedly between the living and nonliving parts of the ecosystem.	"Food Webs Show How Energy Moves Through a Community," Ch. 20, p. 362

MS-LS2.C: Ecosystem Dynamics, Functioning, and Resilience

Ecosystems are dynamic in nature; their characteristics can vary over time. Disruptions to any physical or biological component of an ecosystem can lead to shifts in all its populations.	"Going Green," Ch. 21, pp. 379–380; "A Key Loss," Ch. 20, pp. 361–363
Biodiversity describes the variety of species found in Earth's terrestrial and oceanic ecosystems. The completeness or integrity of an ecosystem's biodiversity is often used as a measure of its health.	"Dinosaurs and Domains," Ch. 14, pp. 250–253

MS-LS3.A: Inheritance of Traits

Within individual organisms, food moves through a series of chemical reactions in which it is broken down and rearranged to form new molecules, to support growth, or to release energy.	"Getting to the Genes," Ch. 7, pp. 118–119;
Plants, algae (including phytoplankton), and many microorganisms use the energy from light to make sugars (food) from carbon dioxide from the atmosphere and water through the process of photosynthesis, which also releases oxygen. These sugars can be used immediately or stored for growth or later use.	Ch. 10, pp. 174–175
Genes are located in the chromosomes of cells, with each chromosome pair containing two variants of each of many distinct genes. Each distinct gene chiefly controls the production of specific proteins, which in turn affects the traits of the individual. Changes (mutations) to genes can result in changes to proteins, which can affect the structures and functions of the organism and thereby change traits.	"How Genes Work," Ch. 10, pp. 174–175

Variations of inherited traits between parent and offspring arise from genetic differences that result from the subset of chromosomes (and therefore genes) inherited.	"Crisscrossing Plants," Ch. 7, pp. 120–123

MS-LS3.B: Variation of Traits

In sexually reproducing organisms, each parent contributes half of the genes acquired (at random) by the offspring. Individuals have two of each chromosome and hence two alleles of each gene, one acquired from each parent. These versions may be identical or may differ from each other.	"Getting to the Genes, Ch. 7, pp. 118–119; Figure 7.2, Ch. 7, p. 119
In addition to variations that arise from sexual reproduction, genetic information can be altered because of mutations. Though rare, mutations may result in changes to the structure and function of proteins. Some changes are beneficial, others harmful, and some neutral to the organism.	"Getting to the Genes, Ch. 7, pp. 118–119; "Two-Step Dance, Translation: RNA to Protein," Ch. 10, pp. 179–183

MS-LS4.A: Evidence of Common Ancestry and Diversity

Anatomical similarities and differences between various organisms living today and between them and organisms in the fossil record, enable the reconstruction of evolutionary history and the inference of lines of evolutionary descent.	"The Ultimate Family Tree," Ch. 11, pp. 200–201
Comparison of the embryological development of different species also reveals similarities that show relationships not evident in the fully-formed anatomy.	"The Ultimate Family Tree," Ch. 11, pp. 200–201; "Clues in the Code," Ch. 11, pp. 201–203
The collection of fossils and their placement in chronological order (e.g., through the location of the sedimentary layers in which they are found or through radioactive dating) is known as the fossil record. It documents the existence, diversity, extinction, and change of many life forms throughout the history of life on Earth.	"Dinosaurs and Domains," Ch. 14, pp. 250–253; "The History of Life on Earth," Ch. 14, pp. 256–260; "Fossil Secrets," Ch. 11, pp. 195–200

MS-LS4.B: Natural Selection

Natural selection leads to the predominance of certain traits in a population, and the suppression of others.	"Artifice to Natural," Ch. 11, pp. 193–195
In *artificial* selection, humans have the capacity to influence certain characteristics of organisms by selective breeding. One can choose desired parental traits determined by genes, which are then passed on to offspring.	"Artifice to Natural," Ch. 11, pp. 193–195

MS-LS4.C: Adaptation

Adaptation by natural selection acting over generations is one important process by which species change over time in response to changes in environmental conditions. Traits that support successful survival and reproduction in the new environment become more common; those that do not become less common. Thus, the distribution of traits in a population changes.	"Leaping Lizards," Ch. 13, pp. 232–234; "Sex and Selection," Ch. 12, p. 223

MS-LS4.D: Biodiversity and Humans

Changes in biodiversity can influence humans' resources, such as food, energy, and medicines, as well as ecosystem services that humans rely on—for example, water purification and recycling.	"Amazon on Fire," Chapter 18, pp. 324–338

Reviewers

Anne Artz, Preuss School, UC San Diego

Allan Ayella, McPherson College

Davy Barrett, Stratford High School

Erin Baumgartner, Western Oregon University

Joydeep Bhattacharjee, University of Louisiana, Monroe

Rebecca Brewer, Troy High School

Victoria Can, Columbia College Chicago

Lisa Carloye, Washington State University, Pullman

Michelle Cawthorn, Georgia Southern University

Jennifer Chesnut, Ottumwa High School

Craig Clifford, Northeastern State University

Beth Collins, Iowa Central Community College

Julie Constable, California State University, Fresno

Gregory A. Dahlem, Northern Kentucky University

Danielle M. DuCharme, Waubonsee Community College

Robert Ewy, SUNY Potsdam

Clayton Faivor, Ellsworth Community School

Michael Fleming, California State University, Stanislaus

Kathy Gallucci, Elon University

Kris Gates, Pikes Peak Community College

Heather Giebink, Pennsylvania State University

Candace Glendening, University of Redlands

Sherri D. Graves, Sacramento City College

Cathy Gunther, University of Missouri

Robert Heun, University School of Milwaukee

Meshagae Hunte-Brown, Drexel University

Douglas P. Jensen, Converse College

Ragupathy Kannan, University of Arkansas–Fort Smith

Julia Khodor, Bridgewater State University

Jennifer Kloock, Garces Memorial High School

Karen L. Koster, University of South Dakota

Joshua Kreitzer, Karl G. Maeser Preparatory Academy

Dana Robert Kurpius, Elgin Community College

Phi Le, Sacred Heart Cathedral Preparatory

Joanne Manaster, University of Illinois

Mark Manteuffel, St. Louis Community College

Jill Maroo, University of Northern Iowa

Tsitsi McPherson, SUNY Oneonta

John S. Mead, St. Mark's School of Texas

Kiran Misra, Edinboro University of Pennsylvania

Jeanelle Morgan, University of North Georgia

Lori Nicholas, New York University

Fran Norflus, Clayton State University

Christopher J. Osovitz, University of South Florida

Christopher Parker, Texas Wesleyan University

Brian K. Paulson, California University of Pennsylvania

Carolina Perez-Heydrich, Meredith College

Thomas J. Peri, Notre Dame Preparatory School

Kelly Norton Pipes, Wilkes Early College High School

Gordon Plague, SUNY Potsdam

A.J. Prawitz, John F. Hodge High School

Benjamin Predmore, University of South Florida

Jodie Ramsey, Highland High School

Logan Randolph, Polk State College

Debra A. Rinne, Seminole State College of Florida

Michael L. Rutledge, Middle Tennessee State University

Celine Santiago Bass, Kaplan University

Jennifer Scheidler, Sycamore High School

Steve Schwendemann, Iowa Central Community College

Sonja Stampfler, Kellogg Community College

Karen Cruse Suder, The Summit Country Day School

Jennifer Sunderman Broo, Saint Ursula Academy

J. D. Swanson, Salve Regina University

Heidi Tarus, Colby Community College

Larchinee Turner, Central Carolina Technical College

Ron Vanderveer, Eastern Florida State College

Calli A. Versagli, Saint Mary's College

Mark E. Walvoord, University of Oklahoma

Lisa Weasel, Portland State University

Derek Weber, Raritan Valley Community College

Danielle Werts, Golden Valley High School

Elizabeth Wright, Athenian School

Steve Yuza, Neosho County Community College

Biology Now

with Physiology

Caves of Death

Scientists scramble to identify a mysterious scourge decimating bat populations.

After reading this chapter you should be able to:

- Caption a diagram of the scientific method, identifying each step in the process.
- Develop a hypothesis from a given observation and suggest one or more predictions based on that hypothesis.
- Design an experiment using appropriate variables, treatments, and controls.
- Give specific examples of a scientific fact and a scientific theory.
- Create a graphic showing the levels of biological organization.
- Determine whether something is living or nonliving based on the characteristics of living things.

CHAPTER 01

THE NATURE OF SCIENCE

DRIVING QUESTION

How can the process of science determine what may be killing bat populations?

- HS-LS1.A: Structure and Function
- HS-ETS1.C: Optimizing the Design Solution

Every spring for 30 years, Alan Hicks laced up his hiking boots, packed his camera, and set out to count bats in caves in upstate New York. A biologist with the New York State Department of Environmental Conservation, Hicks leads one of the few efforts in the country to collect annual data on bat populations. Since 1980, he had never missed the annual cave trip—until March 17, 2007.

"That day, of all days in my entire career, I stayed at my desk," recalls Hicks, who had remained behind to write a report for his supervisor. A couple of hours after his crew left to inspect some local caves, 15 miles from the Albany office, Hicks's cell phone rang.

"Hey, Al. Something weird is going on here," said a nervous voice. "We've got dead bats. Everywhere."

The line went quiet. "What are we talking here?" asked Hicks. "Hundreds of dead bats?"

"No," said the voice. "Thousands."

At first, Hicks conjectured that the bats had died in a flood, which had happened in that particular cave before. But the next day, a young volunteer who had been out with the team told Hicks to check his e-mail. The volunteer had sent him a picture taken the day before of eight little brown bats (*Myotis lucifugus*) hanging from a cave outcropping. Each one had a fuzzy white nose. This was a surprise because little brown bats do not have white noses.

Hicks e-mailed the picture to every bat researcher he knew. The fuzzy white material looked like a fungus, but there was no previous record of a fungus killing bats. As scientist after scientist looked at the picture, they all replied the same way: "What is that?" Hicks resolved to find out what was killing the bats and whether the white fuzz was involved.

Why was Hicks so interested in saving the bats? And why should any of us care, apart from valuing the preservation of all of Earth's creatures? For one thing, bats help us by devouring insects that would otherwise destroy agricultural crops and forests (see "Bug Zappers" on page 15). And mosquitoes, which bats eat, are the world's most deadly animal to humans: through malaria transmissions, mosquitoes kill hundreds of thousands of people each year.

As a biologist, Hicks took a scientific view of the world—logical, striving for objectivity, and valuing evidence over other ways of discovering the truth. **Science** is a body of knowledge about the natural world, but it is much more than just a mountain of data. Science is an evidence-based process for acquiring that knowledge.

- Science deals with the natural world, which can be detected, observed, and measured.
- Science is based on evidence that can be demonstrated through observations and/or experiments.
- Science is subject to independent validation and peer review.
- Science is open to challenge by anyone at any time on the basis of evidence.
- Science is a self-correcting enterprise.

To gather knowledge, Hicks would apply the **scientific method** (**Figure 1.1**). The scientific method is not a set recipe that scientists follow in a rigid manner. Instead, the term is meant to capture the core logic of how science works. Some people prefer to speak of the **process of science** rather than the scientific method. Whatever we call it, the practices that produce scientific knowledge can be applied across a broad range of disciplines—including bat biology.

Keep in mind that, as powerful as the scientific method is, it is restricted to seeking natural causes to explain the workings of our world. There are other areas of inquiry that science cannot address. The scientific method cannot tell us what is morally right or wrong. For example, science can inform us about the differences between humans and other animals, but it cannot identify the morally correct way to act on that information. Science also cannot speak to the existence of God or any other supernatural being. Nor can it tell us what is beautiful or ugly, which poems are most lyrical, or which paintings are most inspiring. So, although science exists comfortably alongside different belief systems—religious, political, and personal—it cannot answer all questions.

ALAN HICKS

Alan Hicks is a retired bat specialist who began the investigation of a mysterious bat illness while working for the New York Department of Environmental Conservation.

Figure 1.1

The scientific method

The scientific method is a logical process that helps us learn more about the natural world. ▶

Q1: What were the original observation and question of the scientists studying the sick bats?

Q2: At what point in the scientific method would a scientist decide on the methods she should use to test her hypothesis?

Q3: How might you explain the scientific method to someone who complains that "scientists are always changing their minds; how can we trust what they say?"

But science is the best way to answer questions about the natural world. The first two steps of the scientific method are to *gather observations* and *form a hypothesis*. Hicks didn't waste a moment of time before applying the scientific method to the question of the white fuzz. Bats were dying. "Bats are part of the planet and vital members of the ecosystem," says Hicks. "They play an important role in the environment in which we live."

Bat Crazy

On March 18, the day after the first dead bats were discovered, Hicks entered the cave to make observations—a key part of the scientific process. An **observation** is a description, measurement, or record of any object or phenomenon. Hicks's team observed that the sick bats had not only white noses, but also depleted fat reserves, meaning that the bats did not have enough stored energy to get through the winter. The bats also had white fuzz on their wings with scarred and dying wing tissue, and they were behaving

abnormally, waking up early from hibernation and leaving the cave when it was still too cold outside to hunt.

Hicks's team also observed that the illness cut across species—many different types of bats were getting sick—and the bats exhibited a high rate of death: in some cases, up to 97 percent of infected bats died. Hicks and others began to call the illness white-nose syndrome (WNS). They still didn't know what caused the syndrome, but its characteristics led them to the assumption that the cause was a living organism (see "The Characteristics of Living Organisms" on page 6).

"For the first few years, we were just sleuthing," says Paul Cryan, a research biologist with the U.S. Geological Survey (USGS), and one of the scientists who received the original e-mailed picture from Hicks. From that first picture, Cryan was involved in trying to pinpoint the cause. "We were trying to understand something that had never happened before in a group of animals that was poorly understood."

In the caves, Hicks began collecting dead bats and sending them to laboratories around the nation. In those labs, technicians scraped samples from the bats' noses and wings, rubbed the samples into petri dishes (shallow glass or plastic plates containing a nutrient solution used to grow microorganisms), and watched to see whether the white fuzz would grow. Time after time, many different types of bacteria and fungi grew on the dishes, speckling them with dots of different-colored colonies, but none of the samples were unusual. Nothing special or dangerous appeared to be present on the bats.

One researcher, a young microbiologist named David Blehert, decided to try something different. Blehert worked at the USGS National Wildlife Health Center in Madison, Wisconsin. In December 2007, Hicks called Blehert. Blehert listened carefully as Hicks described how WNS was spreading. "He said, 'We have a major problem on our hands,'" recalls Blehert. "It turns out he was 100 percent right."

Hicks described to Blehert the conditions under which the bats lived during hibernation—caves in upstate New York, where the temperature was often between 30°F and 50°F. Blehert realized that most of the laboratories, including his, were trying to grow the samples

The Characteristics of Living Organisms

All living things share certain features that characterize life.

1. *They are composed of one or more cells.* The **cell** is the smallest and most basic unit of life; all organisms are made of one or more cells. Larger organisms are made up of many different kinds of specialized cells and are known as *multicellular organisms*.

2. *They reproduce autonomously using DNA.* All living organisms are able to **reproduce**, to make new individuals like themselves. **DNA** is the genetic material that transfers information from parents to offspring. A segment of DNA that codes for a distinct genetic characteristic is called a *gene*. Life, no matter how simple or how complex, uses this inherited genetic code to direct the structure, function, and behavior of every cell.

3. *They obtain energy from the environment to support metabolism.* All organisms need **energy** to survive. Organisms use a wide variety of methods to capture this energy from their environment. The capture, storage, and use of energy by living organisms is known as *metabolism*.

4. *They sense the environment and respond to it.* Living organisms **sense** many aspects of their external environment, from the direction of sunlight to the presence of food and mates. All organisms gather information about the environment by sensing it, and then respond appropriately.

5. *They maintain a constant internal environment.* Living organisms sense and respond to not only the external environment, but also their internal conditions. All organisms maintain constant internal conditions—a process known as **homeostasis**.

6. *They can evolve as groups.* **Evolution** is a change in the genetic characteristics of a group of organisms over generations. When a characteristic becomes more or less common across generations, evolution has occurred within the group.

	Rock	Virus	Fungus	Plant	Animal
Composed of one or more cells	✗	✗	✔	✔	✔
Autonomously reproduce themselves	✗	✗	✔	✔	✔
Obtain energy from their environment	✗	✗	✔	✔	✔
Sense their environment and respond to it	✗	✗	✔	✔	✔
Maintain a constant internal environment (homeostasis)	✗	✗	✔	✔	✔
Can evolve as groups	✗	✔	✔	✔	✔
Living	✗	?	✔	✔	✔

from the bats at room temperature—a method conducive to the growth of many fungi. But in the caves, any living thing would have to grow at cold temperatures, so Blehert and his technicians took samples from dead bats, put them on petri dishes, and placed the dishes in the fridge.

At the same time, Melissa Behr, an animal disease specialist at the New York State Health Department, accompanied Hicks on a trip to a local cave (**Figure 1.2**). Behr swabbed a sample of the white fuzz directly from a bat in the cave, immediately spread it onto a glass slide, and looked at it under a microscope. A unique fungus was on the plate. The fungus was visible in little white fuzzy patches of cells, and up close, the individual spores of the fungus appeared crescent-shaped—different from all the other "normal" microbes growing on the bats' skin, and different from any fungus known to the researchers.

But Behr's single observation wasn't enough evidence to convince anyone that the strange-looking fungus was the cause of WNS. To be of use in science, an observation must be repeatable, preferably by multiple techniques. Independent observers should be able to see or detect the same object or phenomenon, at least some of the time.

In this case, Blehert was able to reproduce Behr's results by an independent technique. After letting his plates sit in the fridge for a few weeks, Blehert removed them and observed white patches of the same strange, crescent-shaped fungal spores. "OK, we now have in laboratory culture what Melissa captured when she collected white material in the caves," thought Blehert. "We've got it."

Prove Me Wrong

In science, just as in everyday life, observations lead to questions, and questions lead to potential explanations. For example, if you flip on a light switch but the light does not turn on, you wonder why, and then you look for an explanation: Is the lamp plugged in? Has the lightbulb burned out? You then identify one of these explanations as the most likely hypothesis for why the light did not turn on.

Figure 1.2

Preparing to enter the bat cave
Scientists suit up to collect more observations on the infected bats and the environmental conditions in the bats' roosting cave.

Q1: Which step(s) in the scientific method does this photograph illustrate?

Q2: What types of environmental data might the researchers have collected?

Q3: Why do you think the researchers are wearing protective gear?

A scientific **hypothesis** (plural "hypotheses") is an informed, logical, and plausible explanation for observations of the natural world. From the start, Hicks hypothesized that a new, cold-loving fungus was the primary cause of death in the bats. After observing the unique crescent-shaped fungal spores, Behr and Blehert agreed with this hypothesis. "It was the simplest

DAVID BLEHERT

A microbiologist and branch chief of the Wildlife Disease Diagnostic Laboratories at the National Wildlife Health Center, David Blehert studies a variety of fungal and bacterial pathogens that are harmful to bats, humans, and other species.

solution," says Blehert. "We had bats with a white fungus that nobody had ever seen before growing on them, so that was the most likely thing that was doing it."

But other scientists disagreed. A fungus itself is rarely deadly to a mammal; more often, a fungus causes an annoying, but not lethal, skin infection or is a secondary response after an animal gets sick from a viral or bacterial infection. So scientists proposed other hypotheses for the cause of WNS. Some suggested the fungus was a secondary effect of an underlying condition, such as a viral infection. Others hypothesized that an environmental contaminant, such as a pesticide, was the cause of death. "There were so many different hypotheses," says Cryan. "But that's what is beautiful about the scientific process. You observe as much as you can, and from those observations you can form multiple hypotheses. Science doesn't proceed by just landing on the right hypothesis the first time."

One of the joys, and challenges, of the scientific method is that after scientists suggest competing hypotheses, they then test their own hypotheses against those of others. A scientific hypothesis must be constructed in such a way that it is potentially **falsifiable**, or refutable. In other words, it must make predictions that can be clearly determined to be true or false, right or wrong (**Figure 1.3**). A well-constructed hypothesis is precise enough to make predictions that can be expressed as "if . . . then" statements.

For example, *if* WNS is caused by a transmissible fungus, *then* healthy bats that hibernate in contact with affected bats should develop the condition. *If* the fungus is secondary to an underlying condition, *then* the infection will occur in bats only after the primary underlying condition is present. *If* an environmental contaminant is the cause, *then* bats with WNS symptoms will have elevated levels of that contaminant in their blood or on their skin.

In each "if . . . then" case, it is possible to design tests able to demonstrate that a prediction is right or wrong. Although predictions can be shown to be true or false, the same is not true of hypotheses. Hypotheses can be *supported*, but no amount of testing can *prove* a hypothesis is correct with complete certainty (**Figure 1.4**).

The reason a hypothesis cannot be proved is that there might be another factor, unmeasured or unobserved, that explains why the prediction is true. For example, consider the first prediction stated in the previous paragraph—that healthy bats hibernating in contact with affected bats will develop WNS. If this is true, the reason might be that the healthy bats were infected by a fungus from their neighbor, supporting the hypothesis that the disease is caused by a transmissible fungus. Alternatively, related bats may tend to hibernate together in the same cave, and the disease, or at least vulnerability to the disease, might be genetically based. The hypothesis that the disease is fungal is *supported* but not *proved* by the correctness of this prediction.

Blehert set out to test the hypothesis that he, Behr, and Hicks had put forward—that a unique, cold-loving fungus was the primary cause of death in the bats. One can test a hypothesis through observational studies or experimental studies. Blehert's first study was observational. Observational studies can be purely **descriptive**— reporting information (**data**) about what is found in nature. Observational studies can also be **analytical**—looking for (analyzing)

1 **Observations and questions:** Bats are observed with white noses. What is causing the white fuzz? These bats are dying at higher rates than bats without white noses. Why?

3 **Predictions:** *If* the white noses are caused by a transmissible fungus, *then* healthy bats that hibernate in contact with affected bats should develop the condition. *If* the white noses are caused by a deadly fungus, *then* healthy bats inoculated with the fungus should develop white noses and die at higher rates.

2 **Hypothesis:** Bats with white noses are infected with a fungus, and this fungus is causing death.

Figure 1.3

From observation to hypothesis to testable prediction

Scientific tests prove Lucky Strike milder than any other principal brand!

These scientific tests, confirmed by independent consulting laboratory, prove Lucky Strike mildest of 6 major brands tested!

MARLENE DIETRICH *says:*

"I smoke a smooth cigarette—Lucky Strike!"

Let your own taste and throat be the judge! For the rich taste of fine tobacco—for smoothness and mildness . . .

THERE'S NEVER A ROUGH PUFF IN A LUCKY!

L.S./M.F.T. — Lucky Strike Means Fine Tobacco

So round, so firm, so fully packed—so free and easy on the draw

Figure 1.4

Hypotheses are supported or not supported, but never proved

Although the claim of scientifically confirmed mildness in this vintage advertisement for cigarettes seems ridiculous, "science" is still used to sell products today. Most Americans see thousands of advertisements every day, and many of these make "scientific" claims that are exaggerated or inaccurate.

Q1: State the hypothesis that this advertisement is claiming was scientifically tested.

Q2: State a prediction that comes from this hypothesis. Is it testable? Why or why not?

Q3: Explain in your own words why the hypothesis cannot be "proved."

patterns in the data and addressing how or why those patterns came to exist. The tools of **statistics**—a branch of mathematics that can quantify the reliability of data—help scientists determine how well those patterns support a hypothesis. Observational studies usually rely on both descriptive and analytical methods to test predictions made by a hypothesis.

In 2009, Blehert, Behr, and Hicks published a scientific paper in which they described the results from inspecting 117 dead bats. They identified microscopic damage caused by a specific kind of fungus in 105 of the bats, and isolated and identified the fungus from a subset of 10 of them. It was a type of cold-loving fungus belonging to a group of fungi called *Geomyces*. They named this new species *Geomyces destructans*.

Their observational study revealed a correlation between white fungus on the noses of bats and bat illness and death. Observational studies suggest possible causes for a phenomenon, but they do not establish a cause-effect relationship. To demonstrate that the fungus was actually causing the illness—and not just correlated with it—Blehert designed and conducted an experiment. Testing scientific hypotheses often involves both observational and experimental approaches (**Figure 1.5**).

Catching the Culprit

An **experiment** is a repeatable manipulation of one or more aspects of the natural world. Blehert's experiment was to take healthy bats into his laboratory and expose them to the fungus. Like analytical observational studies, experimental studies use statistics to determine whether the experimental results support or refute the hypothesis being tested.

In studying nature, whether through observations, experiments, or both, scientists focus on **variables**, characteristics of any object or individual organism that can change. In a scientific experiment, a researcher typically manipulates

MELISSA BEHR

Melissa Behr, formerly with the New York Department of Health, is now a doctor of veterinary medicine at South Dakota State University. She conducts research on the pathology and biology of bats and teaches at the UW veterinary school.

Descriptive: Looking for physical evidence of WNS

Analytical: Measuring the weight of each bat

Experimental: Injecting a fungicide to determine whether it will protect a bat from WNS

One research team; three approaches

Figure 1.5

Testing hypotheses using multiple approaches

Scientists set up an underground laboratory in Tennessee's New Mammoth Cave to test hypotheses about white-nose syndrome (WNS) using descriptive, analytical, and experimental approaches.

Q1: Give a possible hypothesis that could be tested by weighing the bats.

Q2: State the hypothesis being tested in the photo on the bottom right.

Q3: Explain in your own words why an experimental study is the only way to show a cause-effect relationship.

a single variable, known as the **independent variable**. In this case, Blehert's independent variable was fungal exposure. Some bats were exposed; others were not. A **dependent variable** is any variable that responds, or could potentially respond, to changes in the independent variable. Blehert's dependent variable was any sign of WNS on the healthy bats.

If we think of the independent variable as the cause, then the dependent variable is the effect. In the most basic experimental design, a researcher manipulates a single independent variable and tracks how that manipulation changes the value

of a dependent variable. Blehert manipulated his independent variable—exposing some bats to the fungus but not others—and then tracked his dependent variable, whether the bats showed symptoms of WNS.

Blehert made sure his experiment was a controlled experiment. A **controlled experiment** measures the value of the dependent variable for two groups of subjects that are comparable in all respects, except that one group is exposed to a change in the independent variable and the other group is not. In this case, healthy bats were either exposed to the fungus

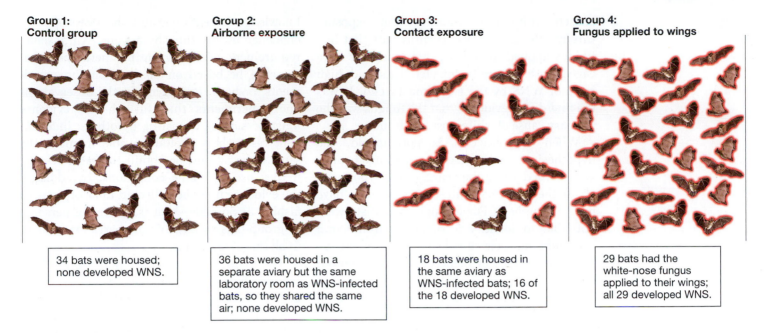

Group 1:
Control group

34 bats were housed;
none developed WNS.

Group 2:
Airborne exposure

36 bats were housed in a
separate aviary but the same
laboratory room as WNS-infected
bats, so they shared the same
air; none developed WNS.

Group 3:
Contact exposure

18 bats were housed in
the same aviary as
WNS-infected bats; 16 of
the 18 developed WNS.

Group 4:
Fungus applied to wings

29 bats had the
white-nose fungus
applied to their wings;
all 29 developed WNS.

Figure 1.6

Blehert's experimental design

Blehert and his colleagues captured 117 healthy bats and brought them into the laboratory. They divided the bats into control and treatment groups and observed them for 102 days.

Q1: Which is the control group in this experiment, and what are the three treatment groups?

Q2: What is the hypothesis being tested in this experiment?

Q3: In one or two sentences, state the conclusions you can draw from the experiment. Was the hypothesis supported? Why or why not?

or not exposed. Typically, a researcher obtains a sufficiently large sample of study subjects and assigns them randomly to two groups. Randomization helps ensure that the two groups are comparable to start with.

One group, the **control group**, is maintained under a standard set of conditions with no change in the independent variable. Blehert had 34 healthy bats in his control group; he kept these bats in the laboratory, under the same conditions as the other bats, except for one difference: he did not expose them to *Geomyces destructans*.

The other group, known as the experimental or **treatment group**, is maintained under the same standard set of conditions as the control group, but the independent variable is manipulated. Blehert exposed 83 healthy bats to the fungus. Of these, 36 were exposed to the

fungus through the air, 18 were put into close contact with naturally infected bats, and 29 had the fungus applied directly to their wings (**Figure 1.6**).

As noted earlier, when scientists test a prediction of a hypothesis and find it upheld, the hypothesis is said to be supported. Scientists can be relatively confident in a supported hypothesis, but they cannot say that the hypothesis has been proved true. Even well-established scientific ideas can be overturned if new evidence against the prevailing view comes to light. Albert Einstein is famously reported to have said, "No amount of experimentation can ever prove me right; a single experiment can prove me wrong."

After watching both the control group and the treatment group, Blehert saw what he had hypothesized: physical exposure to the fungus caused white-nose syndrome, but exposure through the air did not. Healthy bats that had fungus applied directly to their wings or were caged with naturally infected bats had high rates of WNS by the experiment's end. It was the first direct evidence that the fungus was the primary cause of white-nose syndrome.

When a prediction is not upheld, the hypothesis is reexamined and changed, or it is discarded. Over the years, other hypotheses about the cause of white-nose syndrome have not been upheld. For example, scientists were not able to identify a single environmental contaminant at elevated levels in infected bats. A follow-up study by Blehert and others showed that the fungus not only leads to symptoms of WNS in bats, but is sufficient to cause death (**Figure 1.7**).

The story of the bats with white noses is just a single example of the scientific process. One of the greatest strengths of science is that scientific knowledge is tentative and therefore open to challenge at any time by anyone. An absolute requirement of the scientific method is that evidence be based on observations or experiments, or both. Furthermore, the observations and experiments that furnish the evidence must be subject to testing by others; independent researchers should be able to make the same observations, or obtain the same results, if they use the same conditions. In addition, the evidence must be collected in as objective a fashion as possible—that is, as free of personal or group bias as possible. Blehert's experiment fit all these conditions.

The main mechanism for policing personal or group bias and even outright fraud in science is **peer-reviewed publication**. Peer-reviewed publications are found in scientific journals that publish original research after it has passed the scrutiny of experts who have no direct involvement in the research under review. Before Blehert's research was published, it was reviewed by numerous scientists who had not participated in the experiment. If reviewers have concerns during the peer-review process, such as whether the evidence is strong enough to support a hypothesis, they can ask the paper's authors to address those concerns (for example, by gathering additional evidence) and to resubmit the paper. Blehert's paper passed the peer-review process and was published in the scientific journal *Nature* in 2011. At that point, says Blehert, the evidence that *G. destructans* causes WNS was strong enough that "I think we'd convinced most people." (In 2013, scientists renamed *G. destructans* as *Pseudogymnoascus destructans* when the fungus's genus was reclassified.)

But identifying the cause of WNS did not stop the disease from spreading. By March 2008, just a year after Hicks's team found the thousands of dead bats near Albany, more bats were found dead and dying in caves across Vermont, Massachusetts, and Connecticut. Within a year, the disease had spread as far as Tennessee and Missouri. The spread of WNS is a fact: bats around the United States are dying. In casual conversation, we typically use the term "fact" to mean a thing that is known to be true. A scientific **fact** is a direct and repeatable observation of any aspect of the natural world.

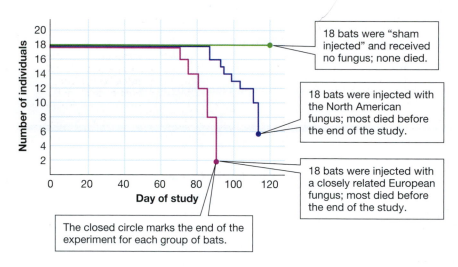

Figure 1.7

Experiments support the hypothesis that the fungus causes WNS in bats

The experiment whose results are plotted here was conducted by some of the same researchers interviewed for this story.

Q1: What is the control group in this experiment, and what are the two treatment groups?

Q2: At day 40, approximately how many individuals were alive in each treatment group? At day 80? At day 100?

Q3: In one or two sentences, state the conclusions you can draw from the experiment. Was the hypothesis supported? Why or why not?

Scientific fact should not be confused with scientific **theory**. Outside of science, people often use the word "theory" to mean an unproven explanation. In science, a theory is a hypothesis, or a group of related hypotheses, that has received substantial confirmation through diverse lines of investigation by independent researchers. Scientific theories have such a high level of certainty that we base our everyday actions on them. For example, the *germ theory of disease*, formally verified by Robert Koch in 1890, is the basis for treating infections and maintaining hygiene in the modern world (**Figure 1.8**).

No End in Sight

According to the U.S. Fish and Wildlife Service, white-nose syndrome has killed more than 6 million bats across the United States since 2007 and shows no signs of slowing. As of this writing, the disease has spread to 29 U.S. states and 5 Canadian provinces. In March 2016, hikers found a bat in Washington State with WNS, and Rhode Island's first case was confirmed two months later. Almost all species of bats that hibernate in these regions have been affected, including little brown bats and

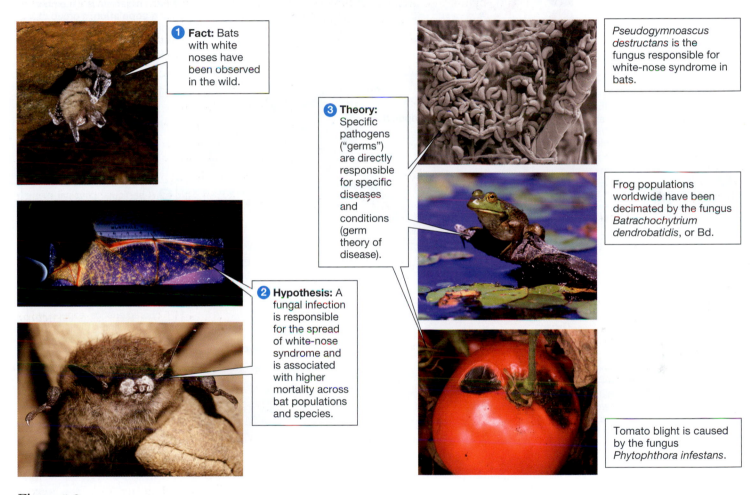

1 Fact: Bats with white noses have been observed in the wild.

3 Theory: Specific pathogens ("germs") are directly responsible for specific diseases and conditions (germ theory of disease).

2 Hypothesis: A fungal infection is responsible for the spread of white-nose syndrome and is associated with higher mortality across bat populations and species.

Pseudogymnoascus destructans is the fungus responsible for white-nose syndrome in bats.

Frog populations worldwide have been decimated by the fungus *Batrachochytrium dendrobatidis*, or Bd.

Tomato blight is caused by the fungus *Phytophthora infestans*.

Figure 1.8

Facts, hypotheses, and theories

It is important to distinguish among facts, hypotheses, and theories when thinking and talking about science.

Q1: Give one *fact* about bats that you learned from this chapter.

Q2: What is another example of evidence for the *germ theory of disease*? (*Hint:* Think about human diseases.)

Q3: Explain in your own words the difference between a fact and a hypothesis, and between a hypothesis and a theory.

endangered Indiana bats, both of which have been particularly hard-hit.

The fungus appears to be related to a type of fungus common in caves in Europe, so a human traveler from Europe most likely carried it across the Atlantic and into the Albany cave, where it infected its first bat in the United States. Researchers continue to explore exactly how the fungus kills the bats. It appears to wake them from hibernation too many times during the winter, so the bats use up their fat reserves too soon and do not survive the months of cold weather. The fungus also eats through bats' delicate wings, which are important not only for flight, but also for maintaining healthy levels of water, oxygen, and carbon dioxide in the bats' bodies. The impact of this fungus on bats is a powerful example of how a microorganism can affect many levels of life, from individual tissues and organs up to whole populations, communities, and even the ecosystem itself (**Figure 1.9**).

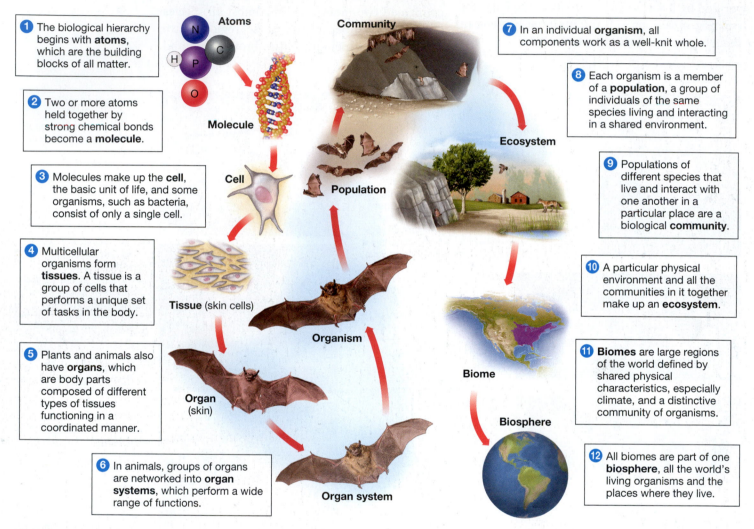

1 The biological hierarchy begins with **atoms**, which are the building blocks of all matter.

2 Two or more atoms held together by strong chemical bonds become a **molecule**.

3 Molecules make up the **cell**, the basic unit of life, and some organisms, such as bacteria, consist of only a single cell.

4 Multicellular organisms form **tissues**. A tissue is a group of cells that performs a unique set of tasks in the body.

5 Plants and animals also have **organs**, which are body parts composed of different types of tissues functioning in a coordinated manner.

6 In animals, groups of organs are networked into **organ systems**, which perform a wide range of functions.

7 In an individual **organism**, all components work as a well-knit whole.

8 Each organism is a member of a **population**, a group of individuals of the same species living and interacting in a shared environment.

9 Populations of different species that live and interact with one another in a particular place are a biological **community**.

10 A particular physical environment and all the communities in it together make up an **ecosystem**.

11 **Biomes** are large regions of the world defined by shared physical characteristics, especially climate, and a distinctive community of organisms.

12 All biomes are part of one **biosphere**, all the world's living organisms and the places where they live.

Atoms, N, H, P, C, O — Molecule — Cell — Tissue (skin cells) — Organism — Organ (skin) — Organ system — Community — Population — Ecosystem — Biome — Biosphere

Figure 1.9

The biological hierarchy

The **biological hierarchy** is a way to visualize the breadth and scope of life, from the smallest structures to the broadest interactions between living and nonliving systems.

Q1: Give examples of other kinds of organs that mammals such as bats have. (*Hint:* Think of the organs in your own body.)

Q2: Are bats in California part of the community of bats in upstate New York if they are of the same species? Why or why not?

Q3: Is the soil in a cave where bats live a part of the bats' population, community, or ecosystem? Explain your reasoning.

Bug Zappers

Bats are skilled natural exterminators, consuming billions upon billions of insects each year, including crop pests and mosquitoes. For this reason, bats play a critical role in agriculture and potentially in human health with the onset of increased mosquito-borne viruses such as West Nile and Zika. The loss of bat populations to white-nose syndrome would have a significant impact on farms, forests, and people around the country.

Assessment available in smartwork**5**

1.3 MILLION

A single colony of **150 big brown bats** (*Eptesicus fuscus*) in Indiana is estimated to eat nearly **1.3 million pest insects each year.**

4–8 GRAMS

A single little brown bat (*Myotis lucifugus*) can consume **4–8 grams of insects each night** during the active season.

660–1,320 METRIC TONS

Extrapolating the diet of a single bat to 1 million bats estimated to have died from WNS, between **660 and 1,320 metric tons of insects** are no longer being consumed each year in WNS-affected areas.

$22.9 BILLION

If bats disappeared entirely from the United States, it would cost the agricultural industry roughly **$22,900,000,000 per year** to save crops by dealing with the insects no longer being eaten by bats.

Today, WNS continues to spread, and scientists have not yet found a way to stop it, though many are trying. Recently, an international team of scientists identified a few populations of bats in Asia and the United States that survived the fungal infection. Evidence from those bats suggests they evolved resistance to the deadly disease. Researchers have also had some luck treating infected bats with a type of bacteria that kills the fungus, but they are still working on ways to deliver the treatment to large groups of bats.

For now, bats continue to die across America every winter. Hicks worries that students, hikers, and tourists will visit caves in the United States, not see any bats, and think that's normal. "In 2006, in one big cave in the Adirondacks, we counted over 185,000 bats. Anywhere you shined a light, there was a bat," says Hicks. "You go in now, and there's not a bat in sight."

REVIEWING THE SCIENCE

- **Science** is both a body of knowledge about the natural world and an evidence-based process for generating that knowledge.

- The **scientific method** represents the core logic of the process by which scientific knowledge is generated. The scientific method requires that we (1) make **observations**, (2) devise a **hypothesis** to explain those observations, (3) generate predictions from that hypothesis, (4) test those predictions, and (5) share the results of the tests for **peer review** by fellow scientists.

- A hypothesis cannot be proved true; it can only be supported or not supported. If the predictions of a hypothesis are not supported, the hypothesis is rejected or modified. If the predictions are upheld, the hypothesis is supported.

- We can test hypotheses by making further observations or by performing **experiments** (controlled, repeated manipulations of nature) that will either uphold the predictions or show them to be incorrect.

- In a scientific experiment, the **independent variable** is manipulated by the investigator. Any variable that can potentially respond to the changes in the independent variable is called a **dependent variable**.

- A scientific **fact** is a direct and repeatable observation of any aspect of the natural world. A scientific **theory** is a major idea that has been supported by many observations and experiments.

- The term **biological hierarchy** refers to the many levels at which life can be studied: atom, molecule, cell, tissue, organ, organ system, organism, population, community, ecosystem, biome, biosphere.

- All living organisms have the following characteristics in common: (1) They are built of **cells**; some are single-celled and some are multicellular. (2) They **reproduce**, using **DNA** to pass genetic information from parent to offspring. (3) They take in **energy** from their environment. (4) They **sense** and respond to their environment. (5) They exhibit **homeostasis**, maintaining constant internal conditions. (6) They can **evolve** as groups.

THE QUESTIONS

Challenge Yourself

1 Describe one observation, one hypothesis, and one experiment from the white-nose syndrome research discussed in this chapter.

2 Which of the following statements is a scientific hypothesis (that is, it makes testable predictions)? (Select only one.)

(a) Even though no one else can see him, the ghost of my dog lives in my backyard.

(b) The Atkins diet helps people lose more weight and keep it off than Weight Watchers does.

(c) People born under the sun sign Aquarius are kinder and cuter than those born under Scorpio.

(d) It is unethical to text while driving.

(e) none of the above

3 Consider an experiment in which subjects are given a pill to test its effectiveness in reducing the duration of a cold. Which of the following is the best way to treat the control group?

(a) Give the control group two pills instead of one.

(b) Do nothing with the control group.

(c) Give the control group a pill that looks like the test pill but does nothing.

(d) Let the control group choose whether or not to take any pills.

(e) Expose the control group to the cold virus.

Try Something New

4 Review the map showing the confirmed and suspected outbreaks of WNS. Is each statement that follows a testable, properly formulated scientific hypothesis for the spread of WNS to bats in Washington State? Explain why or why not. (*Hint:* Think about the criteria for a scientific hypothesis.)

a. Did bats infected with WNS arrive in Washington State via human smuggling for research purposes?

b. The WNS fungus was transported from the Midwest to Washington State via a bat found in Flight 1701's cargo hold.

c. Bats in Washington State were infected by contaminated clothing and equipment from visiting researchers checking for WNS in these colonies.

d. How did bats end up with WNS in Washington State, so far away from all other sites of contamination?

e. A mysterious cloud of fungal particles must have floated over Washington State in early 2016.

5 If a virus was discovered that had the ability to reproduce itself without the help of a cell and its machinery, several experiments could be performed in order to show that it should be categorized as "alive." For each experiment, specify which one of the six criteria for being classified as "alive" would be tested.

_____ a. Place a set number of viral particles in a nutrient broth and check for an increase in viral particles after several days.

_____ b. Place a set number of viral particles in a broth where all the nutrients are corralled to one side of the container and check for viral particle accumulation in the nutrient-rich area of the container.

_____ c. Place a set number of viral particles in a nutrient broth and check for waste products (broken-down nutrients) in the broth after several days.

_____ d. Place a set number of viral particles in a nutrient broth for several months, adding different nutrients daily. Randomly isolate viral particles and sequence their DNA to determine whether there have been changes in genetic characteristics of this population of viral particles compared to the DNA of the original population before culturing.

_____ e. Perform biochemical and microscopic analyses on a number of viral particles to determine whether they have the main components of a living cell, including a plasma membrane, ribosomes, and DNA.

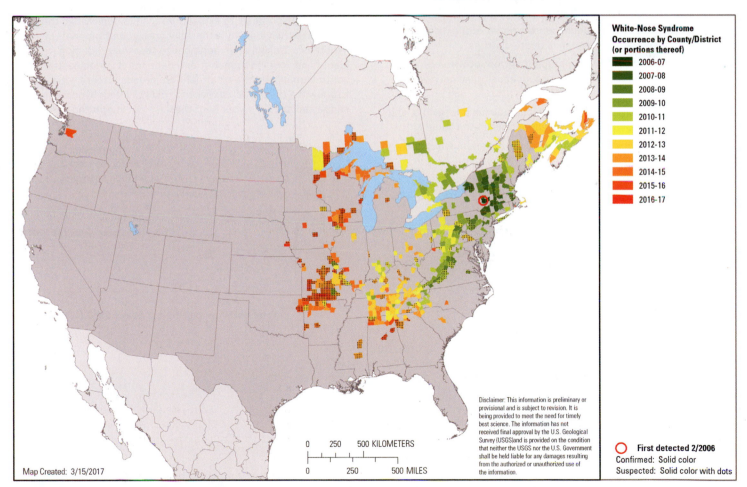

Map depicts the first time WNS is reported suspect or confirmed in a county or district (or portions thereof); each time period in the legend spans a winter bat hibernation period.

Citation: White-nose syndrome occurrence map - by year (2017). Data Last Updated: 3/15/2017. Available at: https://www.whitenosesyndrome.org/resources/map.

6 Mad cow disease, or bovine spongiform encephalopathy, appears to be caused by a novel infectious agent: a protein that replicates by causing related proteins to modify their structure from a harmless shape to a dangerous one. These prions (short for "proteinaceous infectious particles") also appear to be the cause of several other spongiform encephalopathy diseases, such as scrapie in sheep, and kuru and Creutzfeldt-Jakob disease in humans. Which of the following observations or experiments would *not* support the hypothesis that a prion causes spongiform encephalopathy?

(a) The brains of many sheep with scrapie contain prion proteins, but the brains of most sheep without scrapie do not.

(b) There is a high incidence of kuru in populations of people who consume brain tissue from prion-infected animals.

(c) Coyotes that feed on cows with mad cow disease do not subsequently develop spongiform encephalopathy.

(d) When introduced into sheep brain cells in culture, prions cause the normal proteins to change shape into dangerous prion proteins.

(e) When prions are fed to sheep, most of them subsequently develop scrapie, whereas sheep not fed prions do not develop scrapie.

7 Label each of the following statements as an observation, a hypothesis, an experiment, or a result.

_____ a. People who ingest a liquid contaminated with a rotavirus will subsequently experience acute diarrhea.

_____ b. Seventeen out of twenty students given rotavirus-contaminated chocolate bars subsequently experienced acute diarrhea.

_____ c. Ten students were given a liquid containing a rotavirus, and 10 other students were given uncontaminated liquid.

_____ d. People exposed to rotaviruses often experience acute diarrhea.

_____ e. The latest outbreak of acute diarrhea on campus was due to a rotavirus-contaminated elevator button in the dorms.

Leveling Up

8 **What do *you* think?** Insectivorous bats save the agriculture industry billions of dollars each year because they eat insects that damage crops. Other species of bats pollinate some crops. What would you predict about crop yields in areas that are experiencing heavy bat mortality from WNS? Does your prediction affect your level of concern about WNS? Why or why not?

9 **Doing science** Although you are an expert on white-nose syndrome in bats, you have been asked to contribute your scientific expertise to understanding the fungal infection that is decimating frog populations worldwide: *Batrachochytrium dendrobatidis*, a.k.a. Bd.

a. Read the May 2013 *National Geographic* article on how Bd has been spread around the world: http://news.nationalgeographic .com/news/2013/13/130515-chytrid-fungus-origin-african -clawed-frog-science.

b. The graph provided here, based on data from a scientific paper published in 2013, shows that frog species with higher average body temperatures tend to have lower Bd infection rates. It also shows that, compared to males of the same species, females tend to have higher average body temperatures and lower Bd infection rates. Is this observational study descriptive or analytical? Explain your answer.

c. Propose a hypothesis to explain the results depicted in the graph, and then identify a testable prediction from that hypothesis. Design an experiment to test the prediction, identifying your independent and dependent variables, the control group, and the treatment conditions. Create a graph to show the results you expect to find (1) if your hypothesis is supported and (2) if your hypothesis is not supported.

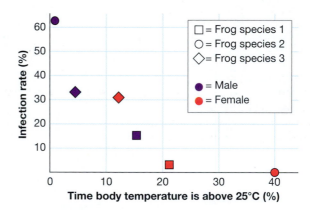

Thinking Multidimensionally

10 Ringworm isn't actually a worm. It's a fungus that infects the skin leaving an open, red scaly patch.

The image above shows mushrooms growing on a lawn. According to medieval folklore, the resulting "fairy rings" formed from dancing elves and fairies.

Compare the patterns observed in the two images, and investigate the underlying cause of these two phenomena.

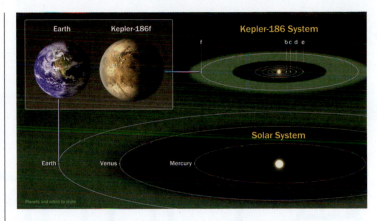

11 You've been recruited by NASA to travel to Kepler-186f, an Earth-sized exoplanet where liquid water might pool on the surface. In your search for extraterrestrial life, you come across an unusual substance that you suspect is alive. It moves when touched, gets bigger, forms smaller pieces, and can take in tiny objects surrounding it. Apply the characteristics of life to defend or refute if this new discovery constitutes life.

For more, visit digital.wwnorton.com/bionow2 for access to:

A Critical Choice

A mother peels back the controversy to decide whether to vaccinate her children.

After reading this chapter you should be able to:

- Evaluate a scientific claim, using the process described in the chapter.

- Explain the importance of scientific literacy for making informed decisions.

- Distinguish between secondary and primary literature, and explain the role of peer review in the latter.

- Compare and contrast basic and applied research, and give an example of each.

- Distinguish between correlation and causation, and give a clear example of each.

- Determine whether a scientific claim is based on real science or pseudoscience.

CHAPTER 02

EVALUATING
SCIENTIFIC
CLAIMS

DRIVING QUESTION

Why is vaccine safety being questioned?

- HS-ETS1.C: Optimizing the Design Solution

In 2009, Anna Eaton discovered she was pregnant. Several of her friends were also expecting or had recently had children, so the friends' afternoon chats were soon about all things baby: nursery colors, pediatricians, car seats. Every once in a while, however, one of the mothers broached a subject far more contentious than strollers and baby carriers. One of their hot topics of discussion became vaccination.

Vaccination is the injection of material—typically an inactivated and harmless infectious organism or parts of such an organism (e.g., a single protein)—that stimulates the immune system to protect against future exposure to that pathogen (**Figure 2.1**). When the body's immune system is exposed to a vaccine, it recognizes the inactivated organism (or its parts) as an invader and mounts an attack against it. Upon later contact with the infectious organism, a vaccinated individual's immune system remembers the inactivated organism from the vaccine and is already armed and bristling. The inactivated organism from the vaccine cannot cause disease because it was made harmless by the process of creating the vaccine. The vaccine itself cannot cause disease because it is not a functional infectious organism and therefore cannot replicate.

Over the past 200 years, scientists have developed vaccines to protect against dozens of pathogens, starting with the smallpox virus. In England in 1796, Dr. Edward Jenner first vaccinated his gardener's 8-year-old son against smallpox using the cowpox virus, because it was known that individuals who had previously suffered cowpox were immune to smallpox. Jenner's breakthrough ultimately led to a worldwide smallpox vaccination campaign, which was so successful that in 1980 the disease was officially declared eradicated by the World Health Organization. Vaccines have also been developed for the Tic Tac–shaped bacteria that cause the upper respiratory infection diphtheria, the round and highly contagious rotavirus that causes vomiting and diarrhea, the tiny airborne bacteria that cause pertussis (also known as whooping cough), and many more pathogens.

Before vaccines, children died in large numbers from smallpox, diphtheria, whooping cough, polio, and many other diseases (**Figure 2.2**). The infectious pathogens that cause those diseases still exist today in our environment, but because vaccines protect people, we almost never see those infections. Today, the Centers for Disease Control and Prevention (CDC), the public health branch of the U.S. government, recommends that children receive 10 vaccines, given over a total of 24 doses, between birth and 15 months of age. Additional vaccinations are recommended between the ages of 18 months and 18 years, including the annual influenza ("flu") vaccine, also recommended for all adults each year (see "Flu Shot" on page 24).

Eaton, pregnant with her first baby, was a microbiologist by training. She worked at the Cleveland Clinic in Ohio, an academic medical center, before moving on to teach high school biology and chemistry, as well as bioscience manufacturing at a local community college. Some of her friends were also scientists, and one pregnant friend adamantly commented, "Of course we're going to vaccinate. We're scientists."

Vaccine — Vaccine — Antibodies — Disease organisms — Antibodies

A vaccine with a harmless form of a virus (or other organism) is injected under the skin.

The vaccine stimulates the immune system to produce antibodies (in green) that recognize the virus.

When the individual is exposed to the virus after vaccination, the new antibodies are primed to attack the invader.

Figure 2.1

How vaccines work
A vaccine trains the body's immune system to fight infection.

Q1: Describe in one sentence how a vaccine creates immunity to a virus.

Q2: Why is it impossible to become infected with a virus from a vaccine composed of viral proteins?

Q3: Natural immunity occurs without a vaccine, just by exposure to a particular stimulus, like the chicken pox virus. Explain why people don't get chicken pox twice.

ANNA EATON

Anna Eaton was, at the time of her research into vaccines, a high school and community college science teacher in Cleveland Heights, Ohio.

Diphtheria
1920–22: 13,191 (7.5%)
2000–13: 0 (0%)

Pertussis
1922–25: 7,363 (5%)
2000–13: 17 (0.1%)

Smallpox
1600s: 70% Native Americans
1721: 844 (14%) Bostonians
2000–13: 0 (0%) U.S.

Polio
1950–53: 8,050 (9%)
2000–13: 0 (0%)

Tetanus
1922–26: 1,182 (90%)
2000–13: 4 (14%)

Figure 2.2

Vaccines save lives

The numbers in each box represent each disease's average annual number of deaths and percent mortality rate in the United States during prevaccination periods (for example, 1950–53 for polio) and the postvaccination twenty-first century.

Q1: Before vaccinations, which diseases had the highest and lowest mortality rates? What are these mortality rates?

Q2: After vaccinations, which diseases have the highest and lowest mortality rates? What are these mortality rates?

Q3: If there were a sudden outbreak of pertussis at a university where pertussis vaccinations were not required and no one was protected, how many students would die? What is the probability or chance that you would die if infected?

But Eaton was not so confident. She had seen media stories suggesting that vaccines might cause autism spectrum disorders, and when she searched online, she read scary accounts from parents who had had a child vaccinated and then watched as their child developed autism. "I'm on message forums, encountering a hundred different moms with stories," recalls Eaton. "It was frightening."

Stories and anecdotes are powerful. They can change how we feel or think about a subject. But anecdotal evidence is not scientific evidence. Anecdotes are not representative of collected data or collected scientific observations, and therefore they cannot reliably give us the big picture of a subject or phenomenon. Scientific evidence can.

Eaton's scientist friend said there were data backing the safety of vaccines, but Eaton had not seen those data and wasn't about to inject her new baby girl with something she knew nothing about. "Feelings of distrust are in all of us, and possibly the strongest in a new mother," says Eaton. "Even though she was a friend of mine, it was still not enough for me to just go in the office and have my daughter vaccinated."

True or False?

The statement that vaccines cause autism, or the opposite statement that vaccines are safe, is a **scientific claim**, a statement about how the world works that can be tested using the scientific method. We are exposed to scientific claims every day—dozens of them, in fact. They are not all true. In 2004, editors at the magazine *Popular Science* asked one of their journalists to write down every scientific claim he heard in a day and evaluate each one. He recorded a whopping 106 claims and dug into each of them, from Cheerios' claim that it reduces cholesterol (supported by scientific evidence) to the claim that a face cream infused with vitamin A would revitalize skin (not supported). Of the 106 claims, most were bogus.

The majority of scientific claims come from advertisers. Though companies are legally bound to tell the truth, not all of them do. For example, in 2005 an Italian footwear company named Vibram introduced the $100 FiveFingers running shoes, glove-like shoes that mimic the feel of barefoot running (**Figure 2.3**). In a marketing campaign, the company touted the health benefits of these unique shoes, advertising that they could reduce foot injuries and strengthen foot muscles. In 2013, however, two peer-reviewed studies of over a hundred joggers found that the shoes actually increased the likelihood of injury. Soon, Vibram was sued for false advertising,

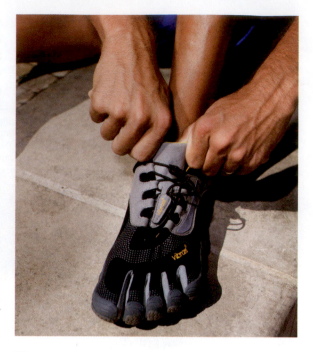

Figure 2.3

Shoes designed to mimic bare feet
FiveFingers running shoes did not live up to the manufacturer's scientific claims.

and in 2014 the company settled that lawsuit for $3.75 million, agreeing to refund FiveFingers shoe purchases and remove all health claims about the shoes from its advertisements.

Scientific claims also come from special-interest groups and organizations that exist to advance certain causes, often for political or religious reasons. These claims include statements about global warming, evolution, and medical care. Again, such claims are often untrue. Therefore, it is important to question the "truth" of a claim when you hear it.

We, the public, are not simply consumers of science and technology. We are participants. By voting on issues that have a scientific underpinning, we shape the course of science and influence which technologies are used, as well as where and how they're used. Although our personal values and political leanings are likely to influence how we vote, the underlying science should also be taken into consideration. **Scientific literacy**, an understanding of the basics of science and the scientific process, enables us to make informed decisions about the world around us and to communicate our knowledge to others. Our hope is that this book will help you become scientifically literate.

Flu Shot

Why do we need a flu shot every year? Why doesn't it protect us for life, like other vaccines do? In fact, each year's flu vaccine *does* protect you for life from a particular strain of the flu virus, but the virus mutates rapidly, and every year's flu strain is likely to be different from the previous year. To the immune system, the new strain looks like a completely new virus. But that's not all! That pesky flu virus can even combine itself with other flu viruses, from animals like birds and pigs, to create deadly flu pandemics like the H1N1 strain, or "swine flu," in 2009.

Each flu season, the CDC uses current flu cases to try to determine what the next season's flu viruses will be. It then creates a vaccine to protect us from the anticipated viruses. Sometimes the experts are extremely accurate and the flu vaccines are highly effective against that season's virus. Sometimes, however, they are not so accurate, and the vaccine is not as protective. It's a viral genetics puzzle and a guessing game, with a little attempt at clairvoyance thrown in. Still, even in years when the vaccine does not provide full coverage for that season, it is likely to protect you from future manifestations of the flu.

Here's the bottom line: The more yearly flu vaccines you receive, the more flu viruses your immune system will be ready to fight, and the less likely you will be to catch the seasonal flu. More important, you will be less likely to succumb to a deadly flu pandemic.

Eaton didn't know the source of the original claim that vaccines cause autism, but she knew that in order to be scientifically literate and make an informed decision for her daughter, she needed to find it. Scientific claims directly affect our lives because we make decisions based on them. Some of these are small decisions: Should I take a multivitamin every morning? How often should I exercise?

Others are larger decisions: Should I vote to support carbon taxes? Does a cell phone cause tumors from radiation? Should I vaccinate my children? The good news is that you can learn to evaluate scientific claims. You can be skeptical about claims and, using critical thinking, make scientifically literate decisions for yourself.

Credentials, Please

Caroline was born in January 2010 (**Figure 2.4**). Eaton had Caroline vaccinated when she was two months old, but then Eaton found herself gripped with fear. "I worried I was doing more harm than good," she says. Because of her rising fears, she did not get Caroline's four-month vaccinations.

Eaton sought out a pediatrician who would discuss her fears about vaccination. Several pediatricians refused to talk with her about a delayed vaccination schedule; some even refused to treat Caroline if she was not vaccinated. Finally, Eaton found a pediatrician who was willing to listen and discuss. After Eaton shared her concerns about vaccine safety, the pediatrician handed Eaton a book, *Vaccines and Your Child*, by a pediatrician named Paul Offit. In the book, Offit laid out how vaccines work, how they are made, and which risks are real and which are false. He also provided documentation, with detailed references to scientific papers, showing that there is no scientific evidence that vaccines cause autism.

Eaton found the book to be well written and informative, but she wondered whether she could trust Offit. She checked his **credentials**—a first step toward assessing the strength of a person's scientific claim. Does the person making a scientific claim have a PhD or MD? Is the degree in the field in which they are making the claim? PhDs in physics do not have training in germ theory, for example, and medical doctors do not have training in atmospheric science.

Figure 2.4

Anna and Caroline Eaton
Anna Eaton had concerns about vaccinations for her first child, Caroline.

Offit had an MD from the University of Maryland and was a professor of vaccinology and pediatrics at the University of Pennsylvania. He was also chief of the Division of Infectious Diseases and the director of the Vaccine Education Center at the Children's Hospital of Philadelphia. In other words, Offit had an advanced degree in an appropriate field and held a job at a reputable university. While good credentials alone do not guarantee that a source is trustworthy, scientists practice for many years in their area of expertise, and their scientific claims tend to be based on that expertise and carefully stated.

In addition to evaluating someone's credentials, it is important to assess whether the person making a scientific claim has an agenda or **bias** (a prejudice or opinion for or against something). Does the person have an ideological, political, or religious belief that will be supported by the scientific claims being made? Does the person have a

PAUL OFFIT

Paul Offit is a pediatrician and chief of the Division of Infectious Diseases at the Children's Hospital of Philadelphia.

conflict of interest? Does he or she stand to make money in any way if others accept those claims?

Doing scientific research almost always requires money, so it is important to take into account where the money comes from. In North America, the vast majority of **basic research** in science is funded by the federal government—that is, by taxpayers. Basic research is intended to expand the fundamental knowledge base of science. In the United States, Capitol Hill appropriates more than $30 billion each year for basic and medical research in the life sciences, including biomedicine and agriculture. Researchers must compete vigorously for the limited funds, and this competition helps ensure that public money goes toward supporting high-quality science. Research funded by the government is normally not considered biased, since the funding comes from taxpayers.

But industries and businesses spend a great deal of money funding science as well, often in areas of **applied research**, in which scientific knowledge is applied to human issues and often commercial applications. In some cases, researchers funded by industry may have a bias in favor of whatever that industry is selling. Funding from industry does not necessarily mean that a scientific claim is incorrect, but the claim should be looked at closely to rule out possible bias.

Offit, Eaton read, had worked with other researchers to develop the RotaTeq vaccine against rotavirus, a virus that can cause death from severe diarrhea. The pharmaceutical company Merck had purchased the vaccine, and Offit had received an unspecified amount of money from that transaction. This financial compensation raised the possibility that Offit might have some bias toward the use of the rotavirus vaccine. Though she had enjoyed Offit's book, Eaton decided she did not want to take his word alone as an answer to her questions about vaccines. "In the end, I decided to just read about it, do a lot of research, and talk to a lot of people," she says.

To the Books

Eaton dove into the secondary literature on vaccines. When investigating a scientific claim, your first stop should be the Internet or the library to get a basic overview of the topic from the **secondary literature**, which summarizes and synthesizes an area of research. Textbooks, review articles, and popular science magazines such as *National Geographic*, *Popular Science*, and *Scientific American* are good secondary sources.

Eaton went to her local library and came home with a stack of books, including a thick vaccine textbook so that she could learn the underlying science of how a vaccine works inside the body. She read about how vaccines stimulate cells of the immune system to protect a person from a virus or bacteria. Once, while she was reading, Eaton's husband walked into the room and looked at her with surprise. "What are you doing?" he asked. "I'm looking for answers!" she replied.

For secondary literature on the Internet, try to visit sites that are affiliated with the government, a university, or a respected institution like a major hospital or museum. Wikipedia often has overview articles that link to science blogs and review articles in science journals. Like Eaton did, it's important to check the credentials of the person or people behind a resource, especially on the Internet. Anonymous sources are not to be trusted.

When evaluating a scientific claim, you may need more detailed information than is available in the secondary literature, especially if you're dealing with a particularly important life decision or if the area of science involved is changing rapidly. In that case, you should next review the **primary literature**, where scientific research is first published (**Figure 2.5**). Primary sources include technical reports, conference proceedings, and dissertations, but the most important primary sources are peer-reviewed scientific journals such as *Science*, *Ecology*, and the *Journal of the American Medical Association* (*JAMA*).

Pulling from references she found in the secondary literature, Eaton compiled a stack of primary literature on vaccines. If her library didn't carry a particular journal, she acquired a copy through interlibrary loan or, in some cases, even e-mailed the journal directly and got articles for free. She found dozens and dozens of papers about vaccines.

Correlation or Causation?

One of the first papers that Eaton came across in her investigation—one that had made huge waves in the media—was published in the *Lancet* in

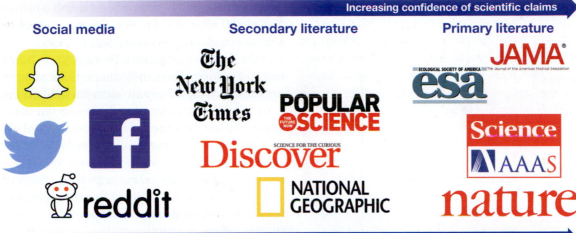

Increasing confidence of scientific claims →

Social media | Secondary literature | Primary literature

JAMA®
ECOLOGICAL SOCIETY OF AMERICA — The Journal of the American Medical Association
esa
The New York Times
POPULAR SCIENCE
Science
SCIENCE FOR THE CURIOUS
Discover
AAAS
reddit
NATIONAL GEOGRAPHIC
nature

Increasingly challenging to read →

Figure 2.5

Scientific claims in the media and literature

It is easy to find and read scientific claims in social media. However, this is not a good source of scientific information. For help in making important life choices, it is important to go to the secondary literature or even the primary scientific literature for accurate and reliable information.

Q1: Why are we less confident of scientific claims made over social media?

Q2: Where would you place a blog in this figure? Would it matter whether or not it was written by a practicing scientist? Explain your reasoning.

Q3: Give an example of when you would rely on secondary literature to evaluate a scientific claim and an example of when you would go to the primary literature. What is the basis of that decision?

1998. The *Lancet* is a well-known peer-reviewed medical journal and therefore a reputable source. The paper was unremarkably titled "Ileal-Lymphoid-Nodular Hyperplasia, Non-specific Colitis, and Pervasive Developmental Disorder in Children." It was a study of 12 children ranging in age from 3 to 10 years old who had experienced a loss of language skills—a symptom of autism spectrum disorders—as well as diarrhea and abdominal pain. Parents of 8 of the 12 children said the onset of symptoms occurred shortly after the child's immunization with the measles, mumps, and rubella (MMR) vaccine.

The authors of the paper, a team of 12 researchers, concluded that more research was needed to study a possible relation between the observed brain dysfunction, bowel problems, and the MMR vaccine. In a press conference when the paper was published, one of the authors, a British doctor named Andrew Wakefield, stated that he believed single vaccines, rather than the MMR triple vaccine, were likely to be "safer" for children. The study and press conference sparked widespread fear among parents that the MMR vaccine could cause autism.

The study was published around the same time that officials began documenting an increase in rates of autism. Since the early 1970s, when organizations started counting the number of people with autism for the first time, the incidence of autism has increased 20- to 30-fold in the United States and other countries. In 2002, the estimate was that one in 150 children aged eight years (the age of peak prevalence) had the disorder. In 2004, that number had risen to one in 125. In 2006, one in 110. By the year 2012, the most recent year with available data, an estimated one in every 68 eight-year-olds had an autism spectrum disorder. The CDC has said that the rise in rates is likely due to heightened disease awareness, more screening

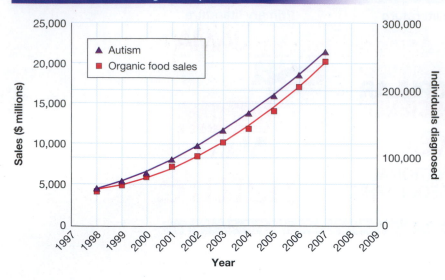

The real cause of increasing autism prevalence?

Figure 2.6

Correlation is not causation: organic food and autism

Reddit user Jasonp55 created a tongue-in-cheek demonstration of why it is important not to assume causation from correlation. Jason used real data on organic food sales and the prevalence of autism from 1998 to 2007. The two are highly correlated, but one does not cause the other.

Q1: How much did organic food sales grow during the period covered in the graph? How much did the incidence of autism grow?

Q2: Why might both organic food sales and autism prevalence have increased during this time period? A Reddit user in the original discussion thread suggested that both might be affected by increasing wealth in the United States. How might increased wealth affect these variables?

Q3: In what way has the vaccine-autism debate been confused by people misinterpreting correlation as causation?

within schools, and a willingness to label the condition. But after Wakefield suggested that the MMR vaccine might be causing autism, the press and other organizations began to report that the rise in autism rates was caused by the increased use of vaccines.

Linking rising rates of autism with increased use of vaccines is a correlation. **Correlation** means that two or more aspects of the natural world behave in an interrelated manner: if one shows a particular value for one aspect, we can predict a value for the other aspect. But correlation does not establish **causation**, in which a change in one aspect *causes* a change in another. Correlations may suggest possible causes for a

phenomenon, but they do not establish a cause-and-effect relationship. For example, there is also a correlation between organic food sales and the increase in autism. From 1998 to 2007, rates of autism increased hand in hand with organic food sales. They are correlated, but there is no scientific evidence that eating organic food causes autism (**Figure 2.6**).

Another correlation that spurred fears about vaccines is that the onset of autism symptoms occurs at about the same age that children receive vaccinations. Most children receive the MMR vaccine at about 15 months old, which is shortly before the first symptoms of autism are often noticed. Parents of children with autism saw their children begin to exhibit symptoms of the illness after their vaccination and therefore directly observed a correlation between the administration of the vaccine and the onset of an autism spectrum disorder.

Yet, again, correlation does not prove causation. Only scientific experiments can demonstrate causation. So, was Wakefield's claim based on good science? Did the MMR vaccine cause autism?

Real or Pseudo?

Unfortunately, sometimes a claim that superficially looks like science is actually **pseudoscience**. Pseudoscience is characterized by scientific-sounding statements, beliefs, or practices that are not actually based on the scientific method. Asking a few simple questions at each step of the progression toward a claim purported to be "scientific" can help you distinguish science from pseudoscience (**Figure 2.7**).

Using these criteria, Eaton analyzed Wakefield's claim. It was quickly obvious that his study did not live up to the standards of good scientific research (**Figure 2.8**). First, the study was small—only 12 children participated—yet the conclusions were grand: Wakefield suggested that all children should stop receiving the MMR vaccine. Sample size is extremely important in observational studies, where small samples may skew data to one extreme. Large sample sizes are more representative of the population, are less likely to be affected by outliers, and provide the power to draw more accurate conclusions.

A second problem with the study was that it was not made up of a random sample of

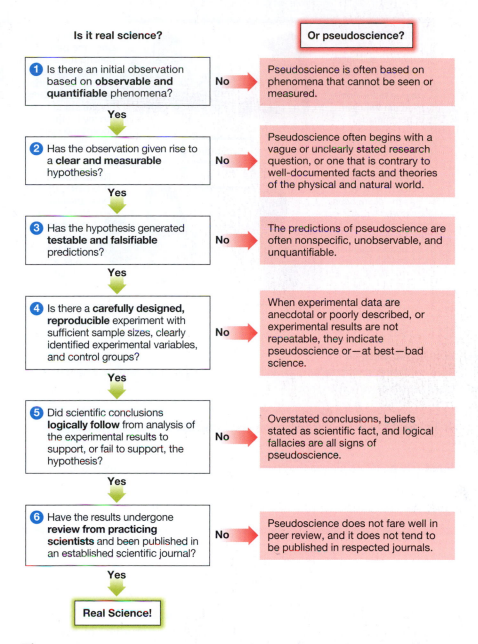

Is it real science?

Or pseudoscience?

1 Is there an initial observation based on **observable and quantifiable** phenomena?

No → Pseudoscience is often based on phenomena that cannot be seen or measured.

Yes ↓

2 Has the observation given rise to a **clear and measurable** hypothesis?

No → Pseudoscience often begins with a vague or unclearly stated research question, or one that is contrary to well-documented facts and theories of the physical and natural world.

Yes ↓

3 Has the hypothesis generated **testable and falsifiable** predictions?

No → The predictions of pseudoscience are often nonspecific, unobservable, and unquantifiable.

Yes ↓

4 Is there a **carefully designed, reproducible** experiment with sufficient sample sizes, clearly identified experimental variables, and control groups?

No → When experimental data are anecdotal or poorly described, or experimental results are not repeatable, they indicate pseudoscience or—at best—bad science.

Yes ↓

5 Did scientific conclusions **logically follow** from analysis of the experimental results to support, or fail to support, the hypothesis?

No → Overstated conclusions, beliefs stated as scientific fact, and logical fallacies are all signs of pseudoscience.

Yes ↓

6 Have the results undergone **review from practicing scientists** and been published in an established scientific journal?

No → Pseudoscience does not fare well in peer review, and it does not tend to be published in respected journals.

Yes ↓

Real Science!

Figure 2.7

Science or pseudoscience?

A series of simple questions based on the scientific method can help you determine whether a "scientific" study is real science or pseudoscience.

Q1: State the hypothesis of the people who believe that vaccines cause autism. What is an alternative hypothesis?

Q2: What part(s) of the figure show(s) where Wakefield's study failed to meet the standards of the scientific method?

Q3: Why does only one arrow point to "real science," whereas multiple arrows point to pseudoscience?

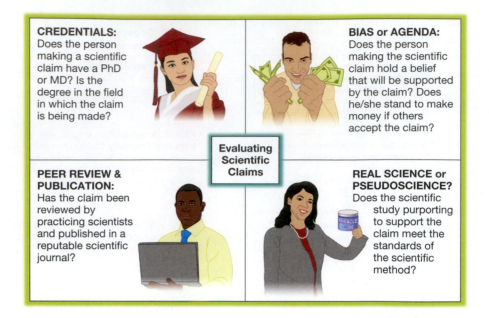

Figure 2.8

Evaluating scientific claims

We are constantly bombarded with scientific claims. A few simple questions will help you evaluate which of these claims are valid and which are not.

Q1: Why is it important to know the education and expertise of a person making a scientific claim?

Q2: List at least five possible biases that people making scientific claims might have.

Q3: Describe a situation in which you might not dismiss the scientific claim of a person who did not have appropriate credentials, or who had a bias toward the claim.

children. Instead, the subjects were picked specifically for their symptoms. Third, the study did not have a control group, such as children who had been vaccinated but did not show signs of autism, or children who had autism but had not been vaccinated. Finally, and most important, the finding could not be repeated. In fact, as of October 2016, at least 110 published papers, studying millions of children, have found no evidence of a link between vaccines and autism. Vaccination does not cause autism, and the two are not even directly correlated. In fact, studies first published in March 2014 provide evidence that autism begins in the womb, before birth.

In one of the most recent papers, researchers at the University of Sydney in Australia, who stated no conflicts of interest, performed a **meta-analysis** (work combining results from different studies) of five *cohort studies*—observational studies of a group of people over a certain period of time—involving 1,256,407 children. The researchers also looked at five *case-control studies*—studies comparing patients with a disease to those without—involving 9,920 children. The data showed no relationship between vaccination and autism, autism and MMR, or autism and thimerosal (a mercury-containing preservative in some vaccines). The work, published in the journal *Vaccine*, included data from the United States, United Kingdom, Japan, and Denmark on vaccines for measles, mumps, diphtheria, tetanus, and pertussis.

Many more peer-reviewed studies with good control groups and sufficient sample sizes, conducted by researchers without a conflict of interest, have shown that the number of vaccines given from birth to 15 months of age under the CDC's recommended vaccination schedule is safe. Researchers have also found that delaying the vaccination schedule increases a child's risk of contracting a disease.

Today it is clear that the paper that started it all, Wakefield's 1998 *Lancet* paper, proposed a claim that is incorrect. All told, it's OK for science to be wrong. There is no expectation that every single one of the thousands of studies published each year will be correct. Yet in the case of the *Lancet* paper, there was bias and wrongdoing. Years after publication of the paper, it came to light that Andrew Wakefield had received large amounts of money as a paid expert for lawyers who were suing vaccine manufacturers. Wakefield had also applied for a patent on a vaccine that would rival the most commonly used MMR vaccine.

Because of these conflicts of interest, in 2010 the General Medical Council, an organization that licenses all medical doctors in the United Kingdom, concluded that Wakefield's conduct had been "irresponsible and dishonest" and revoked his medical license. Eventually, 10 of the paper's 11 other authors retracted their support of the study and its conclusions. Ultimately, in February of 2010, the *Lancet* officially retracted the paper—a very rare action for publishers (**Figure 2.9**). A peer-reviewed paper is retracted (withdrawn as untrue or inaccurate) by a publisher or author when its findings are no longer considered trustworthy because of error, plagiarism, a violation of ethical guidelines, or other scientific misconduct.

Fears versus Facts

Wakefield's paper had not yet been retracted when Eaton was making her decision about whether to vaccinate Caroline. But Eaton followed the guidelines to distinguish science from pseudoscience, and she concluded that Wakefield and his colleagues had looked at too small a sample of children. Therefore, the paper "did not make a lasting impression," recalls Eaton. So she looked beyond that one paper to others. The multitude of papers she found, all

The Lancet, Volume 351, Issue 9103, Pages 637 - 641, 28 February 1998
doi:10.1016/S0140-6736(97)11096-0 (?) Cite or Link Using DOI

< Previous Article | Next Article >

This article was retracted

RETRACTED: Ileal-lymphoid-nodular hyperplasia, non-specific colitis, and pervasive developmental disorder in children

Dr AJ Wakefield FRCS [a] [✉], SH Murch MB [b], A Anthony MB [a], J Linnell PhD [a], DM Casson MRCP [b], M Malik MRCP [b], M Berelowitz FRCPsych [c], AP Dhillon MRCPath [a], MA Thomson FRCP [b], P Harvey FRCP [d], A Valentine FRCR [e], SE Davies MRCPath [a], JA Walker-Smith FRCP [a]

Summary

Background
We investigated a consecutive series of children with chronic enterocolitis and regressive developmental disorder.

Methods
12 children (mean age 6 years [range 3–10], 11 boys) were referred to a paediatric gastroenterology unit with a history of normal development followed by loss of acquired skills, including language, together with diarrhoea and

RETRACTED

Figure 2.9

The paper that precipitated the vaccine-autism scare is debunked and retracted

with controls and larger sample sizes, assured her that vaccination does not cause autism. In the end, her research and scientific literacy led Eaton to fully vaccinate all of her children on the CDC's recommended schedule.

In making her decision, Eaton was aware that vaccines can have side effects. Most are minor, like a sore arm or a low-grade fever. In some very rare cases, however, vaccination can result in a serious reaction. An allergic reaction to the hepatitis B vaccine, for example, is estimated to occur once in 1.1 million doses. A serious allergic reaction to the MMR vaccine is seen less than once in a million doses. The CDC maintains a list of side effects and constantly monitors the safety of vaccines (**Figure 2.10**).

Weighing the benefits of avoiding disease against the small risk of side effects made the choice to vaccinate her children clear, says Eaton. "We [parents] are often so focused on possible effects of the vaccines that we don't think about our kids getting these diseases" that vaccines prevent, she says. The same was not true for your great-grandparents; in the early 1900s, one in five children worldwide died of a vaccine-preventable disease.

When making life choices based on scientific claims, you sometimes need to consider nonscientific aspects of your decision. Your

Figure 2.10

Evaluating vaccines: an ongoing process

Vaccines are continually tested and evaluated for effectiveness, safety, and side effects. The severity of the disease a vaccine works against, as well as how likely a child is to be infected with the disease, is also taken into consideration in determining whether to recommend the vaccine.

Q1: Why do vaccine manufacturers begin with tests on animals or cell lines before moving on to adult human subjects?

Q2: What ongoing testing and reporting are vaccines subjected to?

Q3: What do ACIP, FDA, and CDC stand for, and what is the role of each in evaluating vaccines?

values, ethical stances, and religious beliefs will make some choices more acceptable than others. For example, if you believe it is not ethical to eat meat, the scientific finding that leaner meats are healthier than fattier meats won't matter to you.

It is also important to consider how your choices will affect other people. Choosing not to vaccinate a child, for instance, affects the entire community. Rates of vaccination in the United States have been hovering around 75 percent because of misconceptions about the safety of vaccines, and this low level of immunization puts at risk not only the children who are not vaccinated, but also those who *cannot* be vaccinated, such as infants too young for a vaccine

or people who are genetically unable to respond to a vaccine. When a critical portion of a population is vaccinated, then the spread of disease is contained—a concept known as **herd immunity** (**Figure 2.11**). In other words, vaccinating a large number of people keeps germs out of circulation and protects the vulnerable members of the community.

When parents opt not to vaccinate their children, herd immunity disappears. And because of that loss of herd immunity, infectious diseases of the past, some thought to be eliminated, are roaring back. In August 2013, a megachurch in Texas whose founder had spoken out against vaccines made headlines after 21 members of its congregation contracted

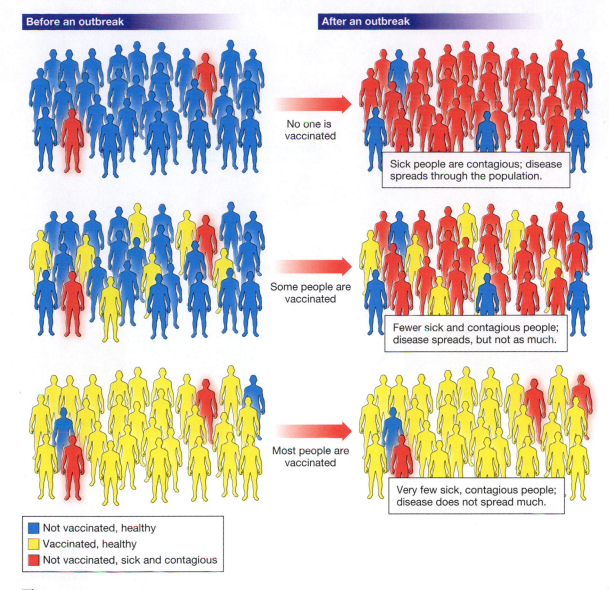

Before an outbreak | **After an outbreak**

No one is vaccinated →

Sick people are contagious; disease spreads through the population.

Some people are vaccinated →

Fewer sick and contagious people; disease spreads, but not as much.

Most people are vaccinated →

Very few sick, contagious people; disease does not spread much.

- ■ Not vaccinated, healthy
- ■ Vaccinated, healthy
- ■ Not vaccinated, sick and contagious

Figure 2.11

Vaccine prevalence and herd immunity

When the majority of the population is not vaccinated, an outbreak of disease will spread further and may cause disease and death in members of the population who are too young to be vaccinated or have a compromised immune system. However, when most of the population is vaccinated against a contagious disease, the disease will spread little during an outbreak; this tighter containment of a disease is the result of what is known as herd immunity. Image modified from the National Institute of Allergy and Infectious Diseases (NIAID).

Q1: What happens to an immunized person when a disease spreads through a population? (*Hint:* In the graphic, follow an immunized individual before and after a disease spreads.)

Q2: Explain why a disease is less likely to spread to vulnerable members of a population if most people are immunized.

Q3: How does vaccination help an individual person? How does it help that person's community?

Safety in Numbers

In 1999, the United Kingdom began a meningitis vaccination program for children up to the age of 18. Meningitis rates rapidly fell off throughout the population—children and adults alike—thanks to both the effectiveness of the vaccine and the protective effects of herd immunity. Models show that without herd immunity, the number of meningitis cases would have rebounded. The U.S. CDC recommends meningitis vaccination up to the age of 23, as outbreaks have been reported on college campuses in recent years.

Assessment available in smartwork5

Meningitis cases in England and Wales

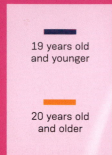

19 years old and younger

20 years old and older

Prevaccination · Postvaccination

Observed cases

Predicted cases (with herd immunity)

Predicted cases (no herd immunity)

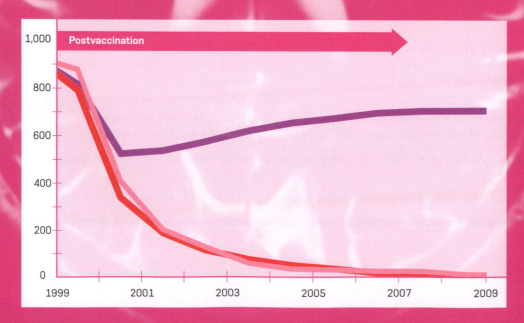

Postvaccination

measles. Sixteen of them were unvaccinated. In 2014, measles cases in the United States hit a 20-year high, with 667 reported cases. And in the 2013–14 flu season, 90 percent of the 100 American children who died of complications from the flu had not received the flu vaccine. Decisions based on scientific claims have serious consequences.

Today, Eaton is thankful that all of her children are fully vaccinated (**Figure 2.12**). She encourages her friends, relatives, and colleagues to foster scientific literacy and use their critical-thinking skills to distinguish science from pseudoscience. Not everyone will want to read a vaccine textbook, Eaton admits with a laugh, but anyone can think critically and follow the process of evaluating scientific claims.

"Look to see what has credible backing and what doesn't. This issue transcends vaccines," says Eaton. "We should all have the tools to become scientifically literate citizens."

Figure 2.12

Anna Eaton and her children

Although Eaton initially had concerns about vaccinations for her first child, Caroline, her research gave her confidence that vaccines are a safe and responsible choice. Caroline and her younger brothers are now all up-to-date on their vaccinations.

REVIEWING THE SCIENCE

- A **scientific claim** is a statement about how the world works that can be tested using the scientific method. To evaluate scientific claims, it is important to look at the **credentials** and **bias** of those making the claim.

- **Scientific literacy** requires a basic understanding of scientific facts and theories, and of the process of science. It is important to be scientifically literate in order to make well-informed life decisions.

- Scientific claims can be found in advertising, in social media, in the popular press, and in scientific publications. The best source for a review or overview of a scientific topic is the **secondary literature**, including the popular press, reputable websites, or review articles in scientific journals.

- The actual experimental or observational results related to a scientific claim are found in the **primary literature**, articles published in scientific journals that have undergone peer review.

- With **correlation**, two or more aspects of the natural world behave in an interrelated manner. With **causation**, a change in one aspect *causes* a change in another.

- **Basic research** explores questions about the natural world and expands the fundamental knowledge base of science. **Applied research** seeks to use the knowledge gained from basic research to address human issues and concerns. It may involve developing commercial applications.

- **Pseudoscience** is characterized by scientific-sounding statements, beliefs, or practices that do not meet the standards of the scientific method.

THE QUESTIONS

Challenge Yourself

1 An example of basic research is

(a) a study of how hummingbirds learn song.

(b) an investigation of how the melting of polar ice caps affects agriculture.

(c) looking at possible genetic contributions to autism spectrum disorder.

(d) designing more effective vaccines for dangerous infectious diseases.

(e) exploring how agricultural waste can be turned into fuel.

2 For each of the following, select the term that best describes the type of literature it represents: primary, secondary, or neither.

_____ a. A research study from Dr. Drake and colleagues on the blood sugar levels of diabetic rats that eat only kale, compared to control diets, is published in a peer-reviewed scientific journal.

_____ b. In the infomercial selling his own health care products, Dr. Horton states, "I believe my personal cure of daily meditation and yoga for diabetes is more effective than any drug I've ever prescribed in 20 years of practicing medicine."

_____ c. Dr. DeBellard insists in her blog that people will lose weight in a healthy manner if they acquire some of her personal parasitic tapeworm.

_____ d. In an article published in the *Annual Review of Nutrition*, a peer-reviewed journal, Dr. Pepper summarizes the last 10 years of basic research on diet and diabetes.

3 Which of the following situations has the greatest potential for biased or inaccurate results in an experimental process?

(a) Ms. Ochoa-Bolton is an outside consultant who is conducting a health and wellness survey for a pharmaceutical company. She does not know the name of the company, nor does she know the name of the drug being tested during the survey.

(b) Ms. Adamian is a research technician surveying a study group on response to a new cold remedy. She knows only the e-mail address of each subject and asks them identical questions by computer.

(c) Dr. Wisidagama is evaluating cancer patients for their responses to a new therapeutic drug. She knows which patients are receiving the placebo and which are receiving the drug.

(d) Dr. Waters is analyzing biopsy samples from rats that have been given either a placebo or an experimental drug believed to reduce inflammation. Each sample is identified by a code number such that she cannot tell which treatment each rat received.

(e) Ms. Nuno is conducting a survey of weight loss regimens as part of her master's degree project. Her online survey is anonymous and asks each participant the same questions.

Try Something New

4 Determine whether each of the following statements is likely to represent real science or pseudoscience. In each case of likely pseudoscience, identify which scientific standard is not met.

_____ a. Dr. Oz says that green coffee beans will burn fat, so you can lose weight without dieting.

_____ b. The *Wall Street Journal* reports that climate scientists have conspired to exaggerate the effects of global climate change.

_____ c. A group of researchers reports at a national scientific conference that they have found a genetic link to autism.

_____ d. Many astrologers agree that people born under the sun sign Aquarius are more intelligent than those born under Scorpio.

_____ e. A study published in the scientific journal *Diabetes* finds that sleeping in a cooler room may increase metabolic rate and insulin sensitivity.

5 The following graphs illustrate the incidence of pertussis (whooping cough) cases in the United States. The first graph organizes the data by year from 1922 to 2012, with the inset showing a zoomed-in view of 1990–2012. The second graph organizes the data by age group from 1990 to 2012. (The 2013 data are not complete in either graph.) DTP, Tdap, and DTaP are different formulations of the vaccine that covers tetanus, diphtheria, and pertussis.

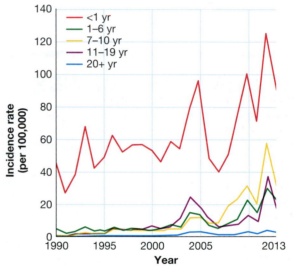

a. Describe what the first graph shows. What do each of the axes represent? What does any point on the line show? What general trend is seen, if all the data shown in the graph are considered?

b. Why do you think there was an increase in cases of pertussis in the first decade of the twenty-first century?

c. Describe what the second graph shows. What do each of the axes represent? What does any point on each of the different lines show? What general trend is seen, if all the lines are considered? What differs among the lines?

d. Compare the incidence of pertussis cases in children under 1 year old and people over the age of 20.

e. Summarize your reflections on reviewing these graphs and the relative risks of pertussis and the pertussis vaccine. What would you recommend to someone trying to decide whether to vaccinate a child?

6 Human papillomavirus (HPV) is the most common sexually transmitted disease in the United States; almost all sexually active people are infected with HPV at some point. HPV can be contagious even when an infected person shows no symptoms, and symptoms may not appear until years after infection. Some but not all strains of HPV can cause genital warts and cancer.

The HPV vaccine is very effective and is recommended for all young people from the ages of 11 to 26 years. What steps would you go through to decide whether to vaccinate yourself or your child with the HPV vaccine? (*Hint:* Use the method for evaluating scientific claims, including assessing real science versus pseudoscience.)

Leveling Up

7 **Life choices** Most American universities require students to have the following immunizations: MMR (measles, mumps, and rubella), varicella (chicken pox), and Tdap (tetanus, diphtheria, and pertussis/whooping cough). Most also recommend or require hepatitis A and B, meningococcal conjugate (meningitis), HPV (human papillomavirus), poliovirus, and the annual flu vaccine.

a. Which of these vaccines have you received?

b. If you have not received one or more of them, why not?

c. Some school districts allow parents to "opt out" of vaccines because of their belief system. Do you think this option should be allowed for university students? Why or why not? What are the possible consequences of allowing people to opt out of vaccinations?

8 *Write Now* **biology: evaluating scientific claims** John Oliver, in a 2014 episode of *Last Week Tonight*, discussed a poll finding that one in four Americans is skeptical of global climate change. He dismissed the poll results and compared it to a poll asking, "Which number is bigger, 15 or 5?" or "Do owls exist?" Why does Oliver feel that what the American public believes about climate change is not relevant? (*Hint:* It could be argued that each of the people polled was making a scientific claim, either in support of or against the scientific consensus on climate change.)

9 **Is it science?** Watch the 2011 movie *Contagion*, which depicts the spread of a fictional virus. While watching the movie, make a list of the scientific concepts that are presented, and note whether you think they represent real science or pseudoscience. Then use your textbook and Internet research to determine whether you were correct or incorrect in each case. If you evaluated any of the concepts incorrectly, reflect on how you came to your initial conclusions.

10 **What do *you* think?** Read Peg Kehret's novel *Small Steps: The Year I Got Polio*, the author's true account of contracting polio as a seventh-grader. Keep a reflective journal while reading the book, documenting your emotional reactions to each chapter in this story. While reading, also note where you learned something new that you did not know previously about the poliovirus, the symptoms of polio, and the treatment options (or lack thereof) in the late 1940s. In your final entry, please reflect on how this novel did or did not change your feelings toward vaccination.

Thinking Multidimensionally

11 Explain how this graph exemplifies that correlation does not imply causation.

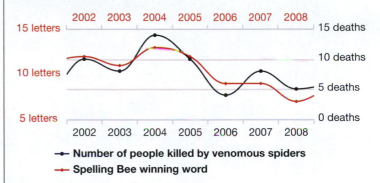

— Number of people killed by venomous spiders
— Spelling Bee winning word

12 Scientific claims permeate American culture. Investigate the popular claims below to determine if they are supported by scientific evidence.

Claim 1: Eating turkey will make you sleepy because it contains lots of tryptophan.

a. What is tryptophan?

b. What hormone is tryptophan used to create?

c. What is the role of this hormone in the body?

d. How do tryptophan levels in turkey compare to other meats and to other foods?

e. What happens to the physiology of the body after consuming any large meal?

Claim 2: To stay hydrated, drink at least 8 glasses of water a day.

a. Where do most of the fluids people consume in a day come from?

b. Are other drinks acceptable substitutes for water?

c. Should a person's weight and lifestyle also factor into recommended amounts of fluid intake? If so, how?

d. Are there limits on how much you should drink each day?

e. Research what happened in 2014 to a high school football player in Georgia who drank 2 gallons of water and 2 gallons of Gatorade, and explain why this occurred.

For more, visit digital.wwnorton.com/bionow2 for access to:

Ingredients for Life

A dusty box, rediscovered after 50 years, sparks a scientific treasure hunt for the origins of life on Earth.

After reading this chapter you should be able to:

- Compare and contrast the types of bonds used to form molecules.
- Diagram the hydrogen bonds that form between water molecules and explain how those bonds produce the unique properties of water.
- Explain the difference between hydrophilic and hydrophobic molecules, and between acidic and basic molecules.
- Predict whether a solution has a high or low concentration of free hydrogen ions by using its pH number.
- Describe the chemical qualities of carbon that make it the basis of life on Earth.
- Create a graphic depicting how each of the four main classes of biomolecules and their subunits are related.

CELLS

CHAPTER
03

CHEMISTRY OF LIFE

DRIVING QUESTION

How can a model of the primitive Earth provide new insights into why life is carbon-based?

- HS-LS1.C: Organization for Matter and Energy Flow in Organisms
- HS-PS1.A: Structure and Properties of Matter
- HS-PS1.B: Chemical Reactions

Jim Cleaves hauled yet another box to the dumpster. The young, dark-haired researcher at UC San Diego had the unenviable task of cleaning out the laboratory of his PhD adviser, the recently retired chemist Stanley Miller. It was 2003. Cleaves tossed box after box—20 years' worth of chemical samples from experiments done by Miller and his former students. As the lab emptied, a small cardboard box high on a shelf caught Cleaves's eye. It was labeled "Electric Discharge Samples."

"That box looked like something I'd really regret throwing out," says Cleaves. So instead of making a trip to the trash, Cleaves gave the box to another former student and friend of Miller's, marine chemist Jeffrey Bada at the Scripps Institution of Oceanography. Once again, the box was relegated to a shelf, where it sat, unopened, for years. Neither Bada nor Cleaves had any idea there was a scientific treasure trove inside, preserved for them by Miller.

In 1952, Miller was a thin, bespectacled graduate student at the University of Chicago, looking for a thesis idea. He approached Harold Urey, a Nobel Prize–winning chemist who, a year and a half earlier, had proposed a radical hypothesis about Earth's early atmosphere and the origins of life. Scientists knew that several key types of **matter** (anything that has mass and occupies a volume of space) existed on the early Earth, but they debated which types were necessary for the emergence of life 3–4 billion years ago. One type of matter is an **element**, a pure substance that has distinct physical and chemical properties, and that cannot be broken down into other substances by ordinary chemical methods. There are 98 natural elements known to us, and another 20 have been created in laboratories.

An **atom** is the smallest unit of an element that retains the element's distinctive properties. Atoms make up all common materials, including this book, the air, and you. Every atom has a dense core called a **nucleus** (plural "nuclei") made up of positively charged **protons** and electrically neutral **neutrons**. A cloud of negatively charged **electrons** surrounds the nucleus (**Figure 3.1**). Electrons have significantly less mass than protons and neutrons have: if an electron weighed as much as a 1-liter bottle of water, a proton or neutron would be as heavy as a car.

The number of protons in an atom's nucleus is called its **atomic number** and is unique to that element. **Isotopes** of an element have the same number of protons but different numbers of neutrons. The sum of the number of protons and the number of neutrons is the **atomic mass number** of an isotope. The atomic mass number is how much mass is in an element or, in other words, how much it weighs. For example, the most common isotope of carbon has 6 protons and 6 neutrons, giving it an atomic mass number of 12, and we call it carbon-12 (^{12}C). The isotope of carbon with 6 protons and 8 neutrons is carbon-14 (^{14}C).

Atoms interact with other atoms via electrons; they can donate electrons, accept electrons, and even share electrons. When two atoms share electrons, they form a **covalent bond**. Atoms linked by covalent bonds form **molecules**. Molecules that include at least one carbon atom are referred to as **organic molecules**. Urey and others suspected that gas molecules in early Earth's atmosphere combined to form the earliest organic molecules, which later assembled into the first living organism.

Urey proposed that Earth's early atmosphere resembled that of other planets in our solar system, including Jupiter, Saturn, and Uranus. He suggested that, like the atmospheres of those planets, our atmosphere was once rich with important **chemical compounds**, molecules that contain atoms from two or more elements. Earth's early atmosphere, he hypothesized, was made up of mainly four compounds: the gases methane (CH_4), ammonia (NH_3), hydrogen (H_2), and water vapor (H_2O).

The young Miller was intrigued by Urey's hypothesis, and he asked Urey if he could perform an experiment to test it, to see whether such a combination of gases could be used to create other simple compounds. Urey dismissed the idea, saying that it was too difficult for a student's thesis and that, if it didn't work, Miller would have nothing to show in order to graduate. But Miller pestered him until at last Urey

STANLEY MILLER

Stanley Miller was an American chemist who designed the first experiment to mimic Earth's early atmosphere and pioneered the study of the origins of life. He died in 2007 at the age of 77.

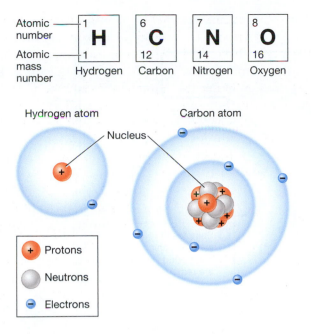

Figure 3.1

Atomic structure

The electrons, protons, and neutrons of these hydrogen and carbon atoms are shown greatly enlarged in relation to the size of the whole atom.

Q1: How many protons, electrons, and neutrons does the hydrogen atom shown here have? What are the atomic number and the atomic mass number of the hydrogen atom?

Q2: What are the atomic number and the atomic mass number of the carbon isotope shown?

Q3: Nitrogen-11 is an isotope of nitrogen that has 7 protons and 4 neutrons. What are the atomic number and atomic mass number of nitrogen-11?

agreed to let Miller work on the project for six months, a year at the most. As it turned out, all Miller needed was a few weeks.

Miller mixed the four gases together in a large glass apparatus. He then zapped the swirling cloud of compounds with a continuous electrical spark to mimic lightning strikes on early Earth, which he surmised would break apart the compounds (**Figure 3.2**). The process of breaking existing chemical bonds and creating new ones is known as a **chemical reaction**. The **reactants** (the gases, in this case) undergo a chemical change and form new molecules, called **products**. Some chemical reactions, like the one Miller devised, require energy in order to proceed. Others release energy.

After the first day of the experiment, a pink liquid pooled in the bottom of the apparatus. By the end of a week, it had turned "deep red and turbid," Miller reported. The color changes, he later discovered, reflected the formation of new products through chemical reactions.

To Miller and Urey's pleased surprise, the resulting red broth contained several biologically significant compounds, including five **amino acids**, small molecules important to life.

HAROLD UREY

Harold Urey was an American chemist who won the Nobel Prize in Chemistry in 1934 for the discovery of an isotope of hydrogen. He was the first to speculate that Earth's early atmosphere was composed of ammonia, methane, hydrogen, and water vapor.

3 An electrical spark continuously fires during the experiment.

1b Gases are added into this chamber of the apparatus.

2 The water becomes vapor and travels to the flask containing the gases.

5 Samples of the water/organic compound "soup" are removed here.

4 The water cools and becomes liquid as it moves through the apparatus, carrying organic compounds created by chemical reactions in the chamber above.

1a Water is poured into the apparatus and heated.

Figure 3.2

Miller's original spark discharge experiment

When Stanley Miller zapped a mixture of methane (CH_4), ammonia (NH_3), hydrogen (H_2), and water vapor (H_2O) with electrical sparks, he produced compounds called amino acids.

Q1: Before the experiment was run, the apparatus was sterilized and then carefully sealed. Why was this an important thing to do?

Q2: Why is inclusion of methane in the gas flasks an essential part of the hypothesis that complex organic molecules were formed in the early atmosphere of Earth? (*Hint:* What makes a molecule organic?)

Q3: Answer this question after reading about Miller's "steam injection" experiments: Where was the steam injected in the experimental apparatus?

There are hundreds of known amino acids, but only 20 are the building blocks for proteins, a major class of molecules found in the cells of living things. Of the five amino acids that Miller isolated, three were part of that group of 20. Thus, from a simple mixture of gases Miller had indeed created some, but not all, of the amino acids used in life on Earth.

At a crowded seminar, Miller presented his results. Urey sat in the front row. Colleagues at the meeting recalled that after the presentation, the famed physicist Enrico Fermi turned to Urey and said, "I understand that you and Miller have demonstrated that this is one path by which life might have originated. Harold, do you think it was the way?" Urey looked at Fermi and replied, "Let me put it this way, Enrico. If God didn't do it this way, he overlooked a good bet!"

Miller and Urey's experiment, which became an instant classic in the scientific community, demonstrated that the basic chemicals for life could arise under natural conditions. Miller continued to perform spark discharge experiments, tweaking the experimental conditions and types of gases in the hope of producing more amino acids. But for an unknown reason, perhaps lack of time, Miller never published or followed up on many of his results—that is, until the mysterious box sitting on Jeffrey Bada's shelf was finally opened.

One Picture, a Thousand Experiments

In 2007, Bada was visiting the University of Texas to give a lecture. His talk was scheduled immediately after a talk by another close friend of Miller's, Antonio Lazcano, a biologist at the School of Sciences at the National Autonomous University of Mexico in Mexico City. The two men agreed to review each other's lecture slides to make sure their talks didn't overlap.

One of Lazcano's slides caught Bada's eye. It was a picture of a small glass vial, labeled as containing a residue from Miller's early experiments. Bada asked Lazcano about it. Lazcano explained that during a visit to his friend, Miller had pulled a cardboard box off a shelf, lifted out the vial, and let Lazcano take a picture. At the time, Miller told Lazcano that it was one of the leftover samples of his spark discharge experiments. Miller had saved them all.

"I was flabbergasted," says Bada. "I'd known Stanley since 1965, and he never once mentioned it." It dawned on Bada that the box might still exist. He called his lab and asked if anyone had seen the box that Cleaves had given him four years earlier. As soon as Bada returned to the lab, he found and opened the box. It was like a scientist's Christmas; the box was full of carefully labeled plastic boxes containing thin glass

Figure 3.3

Electric discharge samples from Miller's experiments

vials, many with films of dried brown, tar-like gunk in the bottom. "It was on the order of 200–300 vials," says Bada (**Figure 3.3**). "It was extracts from experiments throughout the course of his life."

Luckily, Miller kept notebooks detailing the specific contents of each vial. He had performed two other experiments shortly after his original spark discharge work, using variations on the original apparatus. In one, a different method generated the spark. In another, which caught Bada's attention, hot steam was injected directly into the spark chamber.

HENDERSON (JIM) CLEAVES

Jim Cleaves is an organic geochemist at the Carnegie Institution for Science in Washington, DC. As a graduate student of Stanley Miller in 2003, he discovered old vials from Miller's 1950s experiments. Today, Cleaves continues to study how life arose on Earth.

Bada decided to analyze the contents of the vials. He sent the samples to Jason Dworkin, also one of Miller's former students, at NASA's Goddard Space Flight Center in Maryland. The Goddard Center is home to an advanced mass spectrometer, an instrument that measures the weights of tiny amounts of matter as a way to identify them.

The results were stunning. In 1953, Miller had identified just five amino acids, three of which were common constituents of proteins. In 2008, analyzing the same samples with more sophisticated techniques, Bada's team identified 14 amino acids in the original experiment and a whopping 22 in the unpublished steam experiment. Bada was intrigued. Why did the second experimental apparatus result in more kinds of amino acids?

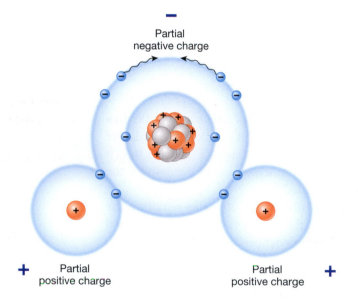

Figure 3.4

Water molecules are polar

The polarity of water molecules is the cause of water's unique properties.

Q1: Where are the covalent bonds in this figure?

Q2: This figure shows a water molecule (H_2O). A hydrogen molecule (H_2) consists of two hydrogen nuclei that share two electrons. Draw a simple diagram of a hydrogen molecule indicating the positions of the two electrons.

Q3: When table salt (sodium chloride, NaCl) dissolves in water, it separates into a sodium ion (Na^+) and a chloride ion (Cl^-). Which portion of a water molecule would attract the sodium ion, and which portion would attract the chloride ion?

The World of Water

The major difference between the two experiments was steam—hot water vapor shooting through the gas chamber. Water is essential for life because of its unique chemical properties, which enable molecules to dissolve and interact in special ways through chemical reactions.

A water molecule is made up of two hydrogen atoms and one oxygen atom held together by shared electrons—that is, by covalent bonds. Electrons moving around an atom's nucleus have different energy levels, so we can think of the electrons as segregated into rings or shells, each of which can contain up to a fixed number of electrons. The innermost shell, closest to the nucleus, can hold up to two electrons. Moving outward, the next two shells can hold up to eight electrons each.

A shell needs to be full to be stable, so in an effort to fill their shells, atoms may form covalent bonds. In a covalent bond, atoms share electrons in their outermost shell of electrons, also called the valence shell. In a water molecule, the oxygen atom uses an electron from each of two hydrogen atoms to fill its outer shell, increasing its count of electrons from six to eight. The hydrogen atoms also benefit from the bond: they fill their outer (and only) shell with the two electrons needed to be stable.

However, the electrons are not shared equally; they spend more time near the oxygen atom than near the hydrogen atoms. Because electrons are negatively charged particles, the oxygen end of a water molecule therefore takes on a slightly negative charge, and the hydrogen ends become slightly positively charged. This lopsided electron sharing means that water is a **polar molecule** (**Figure 3.4**).

In addition to covalent bonds, two other types of **chemical bonds** attach atoms to one another: ionic bonds and hydrogen bonds. Atoms that have lost or gained electrons are called **ions**. Since electrons are negatively charged, an atom that has gained an electron is a negative ion, and an atom that has lost an electron is a positive ion. When a negatively charged ion and a positively charged ion are in the same vicinity, they will chemically attract each other and form an **ionic bond**. Common table salt, sodium chloride (NaCl), is composed of sodium and chlorine held together by ionic bonds. Unlike a covalent bond, no electrons are shared in an ionic bond.

The third means of attaching atoms to one another is called a **hydrogen bond**. Hydrogen bonds are weak electrical attractions between a hydrogen atom with a partial positive charge and a neighboring atom with a partial negative charge. Molecules of water bind to each other through hydrogen bonds because the negatively charged oxygen end of one water molecule weakly attracts one of the positively charged hydrogen ends of another water molecule. A single hydrogen bond is about 20 times weaker than a covalent bond, but water makes up for that lack of strength with sheer quantity. The collective cross-linking of many, many water molecules through hydrogen bonds amounts to a potent force.

The polarity of water molecules and hydrogen bonding explain nearly all of the special properties of water, which were critical in Miller's experiments. The foremost of these was that water was able to break apart the compounds in the flask. As you may have noticed the last time you soaked a dirty dish, water has an incredible ability to dissolve other materials. This is because water molecules form hydrogen bonds with other polar molecules, like sugars or, in Miller's experiment, ammonia. The formation of hydrogen bonds with polar molecules causes those compounds to dissolve in water. Such compounds are said to be **soluble**; that is, they mix completely with the water.

A **solution** is any combination of a **solute** (a dissolved substance, such as sugar) and a **solvent** (the fluid, such as water, into which the solute has dissolved). Water is called the "universal solvent" because it successfully dissolves so many substances. The polar nature of water molecules, however, means that they will *not* interact with uncharged or nonpolar substances, such as fat or oil. Molecules that are soluble in water (such as salt) are called **hydrophilic** ("water-loving"); molecules that don't dissolve well in water (such as oil) are called **hydrophobic** ("water-fearing"). **Figure 3.5** shows these processes in action.

When Bada and his colleagues published their results from reanalyzing Miller's vials, some scientists proposed that Miller's second experiment, in which he shot a jet of steam into the spark, had resulted in more amino acids because the hot water enabled a wider variety of chemical reactions. Whether or not that was the case, water was central to Miller's success.

Oil molecules are hydrophobic. They are excluded from water and tend to clump together.

Olive oil

Vinegar molecules are hydrophilic. They are held in solution by water molecules.

Vinegar

Figure 3.5

Hydrophilic substances dissolve in water, but hydrophobic substances do not

Q1: Describe what will happen to the molecules of olive oil if you shake the bottle and then leave it alone for an hour. What about the molecules of vinegar?

Q2: What would happen if you added another fat to the bottle, such as bacon grease, and shook it?

Q3: Given how sugar behaves when it is mixed into coffee or tea, would you predict that it is hydrophobic or hydrophilic?

JEFFREY BADA

Jeffrey Bada is a chemist at the Scripps Institution of Oceanography at UC San Diego. With Jim Cleaves, he closely analyzed and then duplicated the Miller-Urey experiments. He is also a leading scientist studying organic compounds beyond Earth, including in meteorites and on Mars.

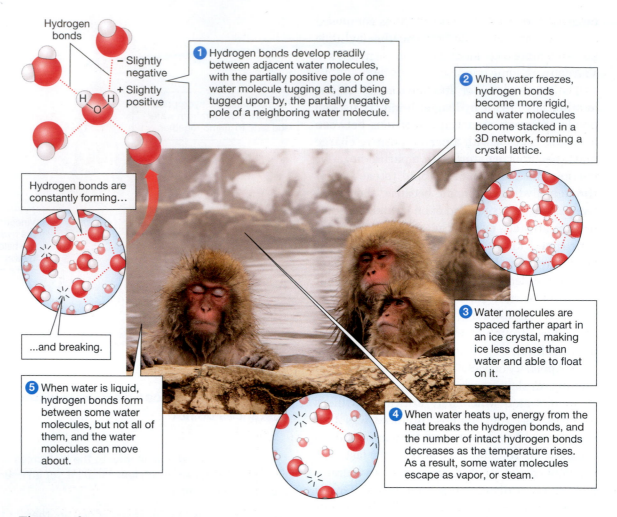

Hydrogen bonds

– Slightly negative

+ Slightly positive

H O H

1 Hydrogen bonds develop readily between adjacent water molecules, with the partially positive pole of one water molecule tugging at, and being tugged upon by, the partially negative pole of a neighboring water molecule.

2 When water freezes, hydrogen bonds become more rigid, and water molecules become stacked in a 3D network, forming a crystal lattice.

Hydrogen bonds are constantly forming...

...and breaking.

3 Water molecules are spaced farther apart in an ice crystal, making ice less dense than water and able to float on it.

5 When water is liquid, hydrogen bonds form between some water molecules, but not all of them, and the water molecules can move about.

4 When water heats up, energy from the heat breaks the hydrogen bonds, and the number of intact hydrogen bonds decreases as the temperature rises. As a result, some water molecules escape as vapor, or steam.

Figure 3.6

Water molecules change state as hydrogen bonds increase or decrease

Japanese snow macaques escape the cold with a daily dip in natural hot springs. Water can be seen here in its liquid, solid, and gas states.

Q1: Identify where in the picture water can be seen in its liquid, solid, and gas states.

Q2: In the gas state, water molecules move too rapidly and are too far apart to form hydrogen bonds. Compare the volumes occupied by an equal number of water molecules in the liquid, solid, and gas states.

Q3: Explain in your own words how ice floats on water.

Water can exist in all three states of matter: liquid, solid, and gas. Hydrogen bonds explain the physical properties of water in these three states. Though water is composed of two elements that are gases at room temperature (hydrogen and oxygen), it forms a liquid at room temperature because hydrogen bonds stick water molecules together, keeping them close. Those hydrogen bonds are constantly forming and breaking in water, creating a nonstop jostling that gives water its liquid form (**Figure 3.6**, left).

When water chills, its molecules cannot move about as vigorously (**Figure 3.6**, right). As water turns into ice at 0°C (32°F), a stable network of hydrogen bonds emerges. The molecules become spaced farther apart, locked into an orderly pattern known as a crystal lattice. That spacing is the reason ice occupies more space than liquid water. Normal ice is 9 percent less dense than liquid water (density is mass divided by volume), which explains why ice floats on water. This property of water is quite unusual—most substances are *more* dense in the solid state than in the liquid state—but it has helped shape life as we know it. If ice did not float on water, then each time a lake or river froze in the winter, the frozen top layer would sink to the bottom and the new, liquid top layer would freeze. This process would repeat until the entire lake or river was frozen solid, no matter the depth. All the aquatic creatures and plants would be killed in the process, and terrestrial plants and animals would no longer have access to the freshwater.

To boil water, as Miller did to inject steam into his experiments, requires the addition of a significant amount of energy to snap the water molecules' network of hydrogen bonds before the molecules can move fast enough to escape as steam (**Figure 3.6**, bottom). That transition from the liquid to the gas state is a phenomenon known as **evaporation**. By the opposite reaction, as water vapor cools, molecules slowly re-form hydrogen bonds and return to the liquid state—a process known as **condensation**. Without evaporation and condensation, both of which are essential parts of the water cycle, life on Earth could not exist (see Figure 18.12 for more details).

Figure 3.7

Miller's experiments were like volcanic eruptions

An erupting volcano releases all of the gases included in Miller's mixtures, and the volcanic ash contains iron and other metals. Steam is produced when the magma (superheated rock) comes in contact with groundwater, and lightning bolts lance through the cloud of gas and ash.

Q1: Suppose you were going to repeat Miller's experiments. How would you decide how much of each gas to include in the chamber?

Q2: Why did the addition of steam to the gases in Miller's second set of experiments increase the yield of amino acids?

Q3: Miller used electrical energy in his experiment. What other forms of energy were present in the early atmosphere of Earth that could have led to the formation of complex molecules?

The Smell of Success

Water was key to the success of Miller's spark discharge experiments, but another compound would turn out to be almost as important. In 2011, three years after reanalyzing the first of Miller's samples, Bada, Cleaves, and a group of scientists teamed up again to analyze another set of vials in the cardboard box.

In 1958, Miller had performed spark discharge experiments that included a new gas: hydrogen sulfide (H_2S). You may know hydrogen sulfide gas as the cause of the awful smell released by rotten eggs, but it is also released by volcanoes and so may have been present in Earth's early atmosphere (**Figure 3.7**). Miller's lab notes show that he performed the experiment and isolated the amino acids, but then never analyzed them. "Why he never analyzed it, I don't know," says Bada. Bada's team finished the job, identifying 22 amino acids in Miller's test tube, including six that contained sulfur. One of those sulfur-containing amino acids, methionine, initiates the construction of all proteins in cells.

Miller's experiments support the hypothesis that volcanoes—a major source of hydrogen sulfide today—coinciding with lightning, which is often focused around volcanoes, may have played a role in making large and varied quantities of biologically crucial molecules, setting the stage for the evolution of life on Earth. But this is not the only hypothesis about the origins of life. There are other ideas, including one that has gained traction in recent years: that the building blocks for life arrived from space.

Although proteins use only 20 amino acids, scientists have identified 90 amino acids in meteorites, suggesting that the first organic materials could have come to Earth from outer space. At NASA's Ames Research Center in California, astrophysicist Scott Sandford and colleagues have been able to make amino acids from gases using conditions like those that exist in interstellar space.

In several experiments, Sandford's team froze different combinations of gases to simulate ices found on comets and in interstellar clouds (where new stars and planets form), and then bombarded those ices with radiation. And what do their icy-hot recipes create? "We make amino acids all the time," says Sandford. In many cases, they identified amino acids similar to those that Miller's experiment created. "The universe seems to be an organic chemist," says Sandford. "It is hardwired to turn simple molecules into more complicated ones. The consequence of that is that it makes a whole host of products, some of which are biologically interesting."

On October 7, 2008, an asteroid entered Earth's atmosphere over Africa, and a NASA satellite photographed its impact (**Figure 3.8**). Initially 6–15 feet in diameter, the asteroid exploded over eastern Africa, and its fragments landed in Sudan. NASA scientists were able to find pieces of the asteroid on the ground because the dark rocks stood out against the light-colored desert sand. These fragments are called the "Almahata Sitta" or "Station Six" meteorites, named after a train station near the location where pieces were recovered.

When the fragments were analyzed, scientists found 19 different amino acids. The extraterrestrial origin of these amino acids is confirmed by a particular characteristic of amino acid molecules: They have two forms, left-handed and right-handed, that are mirror images of each other. All of the amino acids produced by living organisms on Earth are of the left-handed form; right-handed amino acids are produced only in laboratories. The amino acids in the Almahata Sitta meteorites are a mixture of left-handed and right-handed forms. The presence of right-handed amino acids shows that the amino acids in the fragments really came from space, not from terrestrial organic contamination after the fragments landed.

A NASA satellite photographed the impact of the asteroid on October 7, 2008. The yellow arrow traces the path the asteroid followed, and the reddish-orange blob shows the point at which it exploded after entering Earth's atmosphere.

The asteroid broke into fist-sized pieces that fell into the Nubian Desert in Sudan. The dark color of the fragments made them conspicuous on the desert sand, and many of the fragments were recovered by NASA scientists in February 2009.

Figure 3.8

Asteroid 2008 TC3 contains amino acids that formed in outer space

Q1: How did the NASA scientists find the fragments of the meteorite that exploded over eastern Africa?

Q2: What piece of evidence suggests that amino acids found in the meteorite fragments originated in outer space?

Q3: Speculate on the significance of finding extraterrestrial amino acids.

Getting the Right Mix

Miller never stopped trying to make complex molecules from simple ones. Though his 1953 experiment was originally met with fanfare, scientists later began to dispute the usefulness of his experiments. Methane and ammonia, they argued, didn't exist in large amounts on the early Earth. Instead, new evidence suggested that the atmosphere contained nitrogen gas (N_2) and carbon dioxide (CO_2). "Most people agree now [that Miller and Urey] didn't have the right composition," says Sandford. So, in 1983, 30 years after his original experiment, Miller repeated it using nitrogen and carbon dioxide. But instead of a deep red/brown broth, the liquid produced was clear and seemingly barren. The experiment looked like a failure.

In 2007, Bada and Cleaves decided to revisit that experiment to see what had gone wrong. Instead of just reanalyzing samples, they redid the experiment and discovered that the reactions between the new gases were producing chemicals called nitrites, which destroy amino acids. The solution also became acidic because of the presence of nitrous acid (HNO_2). An **acid** is a hydrophilic compound that dissolves in water and loses one or more hydrogen ions (H^+). By donating H^+ ions to water, acids increase the concentration of free H^+ ions in an aqueous solution. H^+ ions are extremely reactive and can disrupt or alter other chemical reactions. The acidic solution, Bada and Cleaves realized, was preventing amino acids from forming.

To counteract that acidity, Bada added a **base**. Acids and bases are chemical opposites. Unlike acids, bases *accept* hydrogen ions from aqueous surroundings. Because a base removes H^+ ions from solution, it has the overall effect of *reducing* the concentration of free H^+ ions in an aqueous solution (strong bases, like strong acids, can be dangerous because they disrupt chemical reactions important to life). Acids react with bases to have an overall neutralizing effect, reducing the concentration of reactive H^+ ions.

Hydrogen ion concentration is commonly expressed on a scale from 0 to 14, where 0 represents an extremely high concentration of free H^+ ions and 14 represents the lowest concentration. This scale, called the **pH scale**, is logarithmic: each pH unit represents a 10-fold increase or decrease in the concentration of hydrogen ions (**Figure 3.9**). Pure water is said to be neutral at pH 7, in the middle of the pH scale. The addition of acids to pure water raises the concentration of free hydrogen ions, making the solution more acidic and pushing the pH below

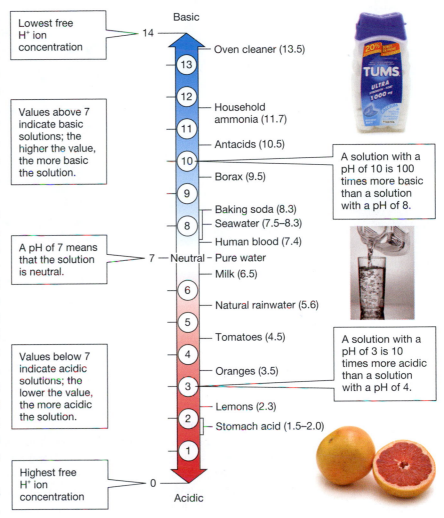

Lowest free H^+ ion concentration

Values above 7 indicate basic solutions; the higher the value, the more basic the solution.

A pH of 7 means that the solution is neutral.

Values below 7 indicate acidic solutions; the lower the value, the more acidic the solution.

Highest free H^+ ion concentration

Basic

14
13 — Oven cleaner (13.5)
12
11 — Household ammonia (11.7)
10 — Antacids (10.5)
9 — Borax (9.5)
8 — Baking soda (8.3)
— Seawater (7.5–8.3)
— Human blood (7.4)
7 — Neutral – Pure water
6 — Milk (6.5)
5 — Natural rainwater (5.6)
4 — Tomatoes (4.5)
3 — Oranges (3.5)
2 — Lemons (2.3)
— Stomach acid (1.5–2.0)
1
0

Acidic

A solution with a pH of 10 is 100 times more basic than a solution with a pH of 8.

A solution with a pH of 3 is 10 times more acidic than a solution with a pH of 4.

Figure 3.9

The pH scale indicates hydrogen ion concentration

Q1: Which has a higher concentration of free hydrogen ions: vinegar, pH 2.8; or milk, pH 6.5?

Q2: What happens to the concentration of free hydrogen ions in your stomach when you drink a glass of milk?

Q3: Black coffee has a pH of 5. Does adding coffee to water (pH 7) increase or decrease the concentration of free hydrogen ions in the liquid?

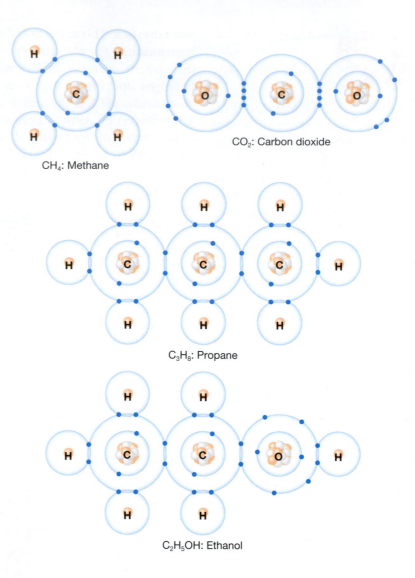

Figure 3.10

Versatile carbon

All organic compounds are built from carbon atoms interacting with atoms of other elements through strong covalent bonds. A carbon atom can bond with up to four other atoms because it has four valence electrons in its outer shell. Recall that atoms must fill their outer shells to be stable and that covalent bonds require the sharing of valence electrons. Count the electrons circling each of the atoms in the molecules shown; all the outer shells are full.

Q1: In methane gas, how many electrons is each hydrogen atom sharing? How many is the carbon atom sharing?

Q2: In carbon dioxide, how many electrons is each oxygen atom sharing? How many is the carbon atom sharing?

Q3: Draw a molecule of formaldehyde (CH_2O). How many electrons is the oxygen atom sharing with the carbon atom? How many is the carbon atom sharing with the oxygen atom and with each hydrogen atom?

the neutral value of 7. Adding a base lowers the concentration of free hydrogen ions in the solution, making the resulting solution more basic and raising the pH above 7.

Bada added a simple base, calcium carbonate, to the experiment in order to raise the pH of his solution. This time, the resulting brew was bursting with amino acids, suggesting that the building blocks of life could indeed have originated on Earth and not been solely delivered to Earth by meteorites or comets. Today, many scientists agree that it is likely that both processes—amino acids arriving from space and originating on Earth—contributed to life as we know it. But no matter where amino acids came from, the logical follow-up question is one that continues to stump scientists to this day: What happened next?

Life's First Steps

"No one really knows how life got started," says Sandford. But researchers do agree on what life requires. If all the water in any living organism were removed, four major classes of large organic molecules, or **biomolecules** (sometimes called "macromolecules"), would remain, all of them critical for living cells: proteins, carbohydrates, nucleic acids, and lipids.

Each of these biologically important molecules is built on a framework of covalently bonded carbon atoms. Carbon is the predominant element in living systems, partly because it can form large molecules that contain thousands of atoms. A single carbon atom can form strong covalent bonds with up to four other atoms (**Figure 3.10**). Carbon atoms can also bond to other carbon atoms, forming long chains, branched molecules, and even rings. No other element is as versatile as carbon in the sheer diversity of complex molecules that can be assembled from it. In fact, while there are only about 4,500 known naturally occurring inorganic molecules (molecules that do not contain a carbon atom) on Earth, the number of known organic molecules is in the range of millions, and the number of those that are not yet uncharacterized is likely several orders of magnitude larger.

Proteins, carbohydrates, and nucleic acids are **polymers**, long strands of repeating units of small molecules called **monomers**. Amino acids are the monomers making up proteins, simple sugars are the monomers in carbohydrates, and nucleotides are the basis of nucleic acid polymers.

All three of these polymers have essential functions for every life-form. **Proteins** are known to be the most numerous and versatile of the biomolecules. Different combinations of the twenty amino acid monomers allow for countless proteins that vary in size and shape, and therefore function (**Figure 3.11**, top left). For example, enzymatic proteins, like polymerases, enable us to copy our DNA (see Chapter 9 for more on DNA replication). Structural proteins give our cells shape. Hormone and receptor proteins, like insulin and its receptor, allow our cells to take up sugars for use as energy. Other equally important protein categories include membrane transport proteins that help move substances into and out of cells, antibody molecules that protect us from disease, storage proteins like LDL and HDL (low-density and high-density lipoproteins, respectively) that carry cholesterol, and venoms and toxins such as the tetanus toxoid.

Carbohydrates are the next-most-versatile biomolecules. They range in size from simple sugar monomers (monosaccharides) and two-monomer sugars (disaccharides) to complex carbohydrates that may contain thousands of monomers (**Figure 3.11**, top right). Simple sugars are the cell's direct fuel to make ATP (adenosine triphosphate), the molecular energy source essential for all cellular work (see Chapter 4 for more on ATP). Other carbohydrates, such as glycogen in animals and starch in plants, are used for energy storage. Three additional complex carbohydrates provide structural support to cells: cellulose, also known as fiber, helps plants to grow tall; chitin forms a hard outer covering to protect organisms without an internal skeleton, such as insects, spiders, and crustaceans; and peptidoglycan is a major component of bacterial cell walls.

The third and most crucial category of polymer, the **nucleic acids**—DNA (deoxyribonucleic acid) and RNA (ribonucleic acid)—form the basis of life itself. Nucleic acids are polymers of nucleotide monomers: DNA is composed of deoxyribonucleotides, and RNA is composed of ribonucleotides (**Figure 3.11**, bottom left). DNA provides living organisms with long-term, stable genetic information storage in a form that is easily copied and passed on to future generations (see Chapter 9). Our genes are DNA. That sounds important, but what about RNA? Without RNA,

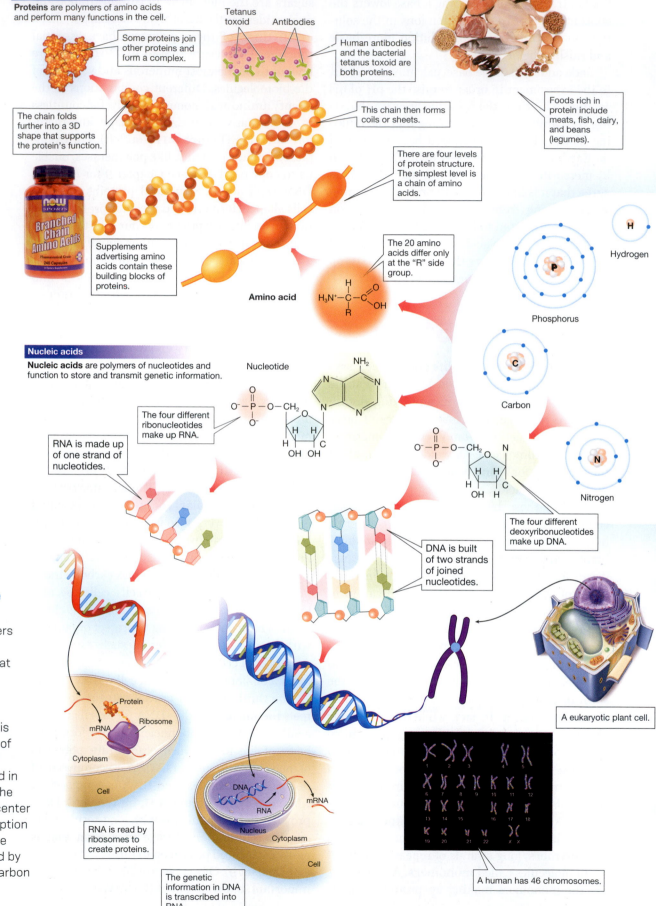

Proteins

Proteins are polymers of amino acids and perform many functions in the cell.

Some proteins join other proteins and form a complex.

Tetanus toxoid   Antibodies

Human antibodies and the bacterial tetanus toxoid are both proteins.

Foods rich in protein include meats, fish, dairy, and beans (legumes).

The chain folds further into a 3D shape that supports the protein's function.

This chain then forms coils or sheets.

There are four levels of protein structure. The simplest level is a chain of amino acids.

Supplements advertising amino acids contain these building blocks of proteins.

Branched Chain Amino Acids

The 20 amino acids differ only at the "R" side group.

Amino acid

H_3N^+ — C — C — OH
H, O, R

H — Hydrogen

P — Phosphorus

Nucleic acids

Nucleic acids are polymers of nucleotides and function to store and transmit genetic information.

Nucleotide

The four different ribonucleotides make up RNA.

RNA is made up of one strand of nucleotides.

C — Carbon

N — Nitrogen

The four different deoxyribonucleotides make up DNA.

DNA is built of two strands of joined nucleotides.

Figure 3.11

Biomolecules are critical for life

A handful of monomers (*mono*, "one") can be assembled into a great variety of polymers (*poly*, "many"). Each of these biologically important molecules is built on a framework of repeating monomers made up of some, and in several cases all, of the atoms shown in the center circle—with the exception of the lipids, which are instead characterized by repeating atoms of carbon and hydrogen.

Protein
Ribosome
mRNA
Cytoplasm
Cell

RNA is read by ribosomes to create proteins.

A eukaryotic plant cell.

DNA
RNA
mRNA
Nucleus
Cytoplasm
Cell

The genetic information in DNA is transcribed into RNA.

A human has 46 chromosomes.

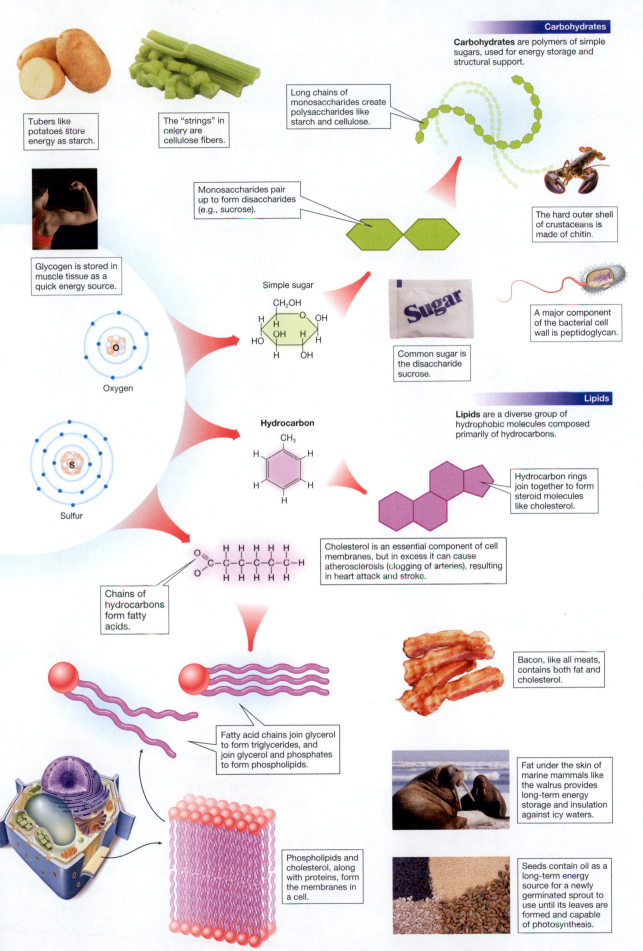

Tubers like potatoes store energy as starch.

The "strings" in celery are cellulose fibers.

Carbohydrates

Carbohydrates are polymers of simple sugars, used for energy storage and structural support.

Long chains of monosaccharides create polysaccharides like starch and cellulose.

Monosaccharides pair up to form disaccharides (e.g., sucrose).

The hard outer shell of crustaceans is made of chitin.

Glycogen is stored in muscle tissue as a quick energy source.

Simple sugar

CH$_2$OH

Oxygen

Sugar

A major component of the bacterial cell wall is peptidoglycan.

Common sugar is the disaccharide sucrose.

Sulfur

Lipids

Lipids are a diverse group of hydrophobic molecules composed primarily of hydrocarbons.

Hydrocarbon

CH$_3$

Hydrocarbon rings join together to form steroid molecules like cholesterol.

Cholesterol is an essential component of cell membranes, but in excess it can cause atherosclerosis (clogging of arteries), resulting in heart attack and stroke.

Chains of hydrocarbons form fatty acids.

Bacon, like all meats, contains both fat and cholesterol.

Fatty acid chains join glycerol to form triglycerides, and join glycerol and phosphates to form phospholipids.

Fat under the skin of marine mammals like the walrus provides long-term energy storage and insulation against icy waters.

Phospholipids and cholesterol, along with proteins, form the membranes in a cell.

Seeds contain oil as a long-term energy source for a newly germinated sprout to use until its leaves are formed and capable of photosynthesis.

Ingredients for Life ▪ 53

What's It All Made Of?

Everything in the universe is composed of matter—from ordinary matter, made of atoms, to dark matter, which may consist of unknown types of particles. Here, we stick with what we know and describe the common elements that compose the world around us.

Assessment available in smartwork5

Earth's Atmosphere

- **78%** Nitrogen
- **21%** Oxygen
- **<1%** Argon
- **<<1%** Other Elements

The Universe

- **75%** Hydrogen
- **23%** Helium
- **2%** Other Elements

The Human Body

- **65%** Oxygen
- **18%** Carbon
- **10%** Hydrogen
- **3%** Nitrogen
- **2%** Calcium
- **1%** Phosphorus
- **1%** Other Elements

Earth's Crust

- **46%** Oxygen
- **28%** Silicon
- **8%** Aluminum
- **5%** Iron
- **4%** Calcium
- **3%** Sodium
- **2%** Potassium
- **2%** Magnesium
- **2%** Other Elements

All proportions are by mass except Earth's atmosphere, which is by volume the information stored in our genes would be stuck there, like a blueprint in a foreign language that no one can decipher. RNA comes in many forms and plays many roles, but its most important job is providing a readable genetic language that enables genes to be expressed as proteins.

A final valuable type of biomolecule is the **lipids**, which are better known as fats, oils, and steroids. Lipids are not polymers, because their structure is not composed of a chain of monomers (**Figure 3.11**, bottom right). Nevertheless, the lipids form a diverse group of biomolecules made up of combinations of hydrocarbons (carbon- and hydrogen-only molecules), fatty acids, or glycerol molecules. For example, triglycerides—composed of three fatty acid molecules linked to a glycerol molecule—provide long-term energy storage in both plants (as oils) and animals (as fats), and serve as insulation against the cold in animals. And phospholipids, which have a hydrophilic half and a hydrophobic half, are the main component of all cellular membranes (see Chapter 4 for more on phospholipids).

Miller, Bada, and other scientists have not yet been able to identify any biomolecules in their prebiotic soups. "It's still very much of an unknown how we go from simple to complex molecules," says Bada. Lipids and carbohydrate-like compounds, however, have been found in meteorites, and Sandford's team at NASA is now attempting to create carbohydrates using its space ice simulations. Other scientists are trying to re-create lipids and nucleic acids in the lab (see Chapter 4).

Fifty More Years

Sadly, Miller died in 2007, never having seen the results of Bada's reanalysis. Today, researchers continue to perform spark discharge experiments with equipment that is essentially the same as in Miller's original design. "We're not done with it," says Jim Cleaves. "There are just so many variations to look at—different combinations of gases, changing the pH, adding metals, et cetera."

"This experiment lives on," agrees Bada. "Not many people can say that an experiment done over 50 years ago is still being investigated today, and finding a wealth of new information." And with better and better instruments to detect the results, who knows what scientists might discover over the next 50 years?

REVIEWING THE SCIENCE

- The physical world is composed of **matter**, which is anything that has mass and occupies space. There are 98 distinct **elements** found naturally on Earth. An **atom** is the smallest unit of an element that maintains its unique properties; it contains positively charged **protons**, uncharged **neutrons**, and negatively charged **electrons**.

- The chemical interactions that cause atoms to associate with each other are known as **chemical bonds**. When an atom loses or gains electrons, it becomes, respectively, a positively or negatively charged **ion**. Ions of opposite charge are held together by **ionic bonds**.

- **Covalent bonds** are formed by the sharing of electrons between atoms. A **molecule** contains at least two atoms that are held together by covalent bonds.

- **Hydrogen bonds** are weak associations between two molecules such that a partially positive hydrogen atom within one molecule is attracted to a partially negative region of the other molecule.

- Partial electrical charges result from the unequal sharing of electrons between atoms, giving rise to **polar molecules**. The polarity of individual water molecules and the hydrogen bonding across water molecules explain nearly all of the special properties of water.

- A **solution** is any combination of a dissolved substance, known as the **solute**, and a fluid into which the solute has dissolved, known as the **solvent**.

- Ions and polar molecules are **hydrophilic**; they readily dissolve in water. Nonpolar molecules cannot associate with water and are therefore **hydrophobic**.

- In **chemical reactions**, bonds between atoms are formed or broken. The participants in a chemical reaction (**reactants**) are modified to give rise to new ions or molecules (**products**).

- The concentration of free hydrogen ions in water is expressed by the **pH scale** and reflects whether a solution is **acidic**, **basic**, or neutral.

- **Chemical compounds** are molecules that contain atoms from at least two different elements. Carbon atoms can link with each other and with other atoms to generate a great diversity of chemical compounds called **organic molecules**. The four main types of large organic molecules, or **biomolecules**, are **proteins**, **carbohydrates**, **nucleic acids**, and **lipids**.

THE QUESTIONS

Challenge Yourself

1 You are asked to determine the classification of an unknown biomolecule. You are told that it can be broken down into only carbon, oxygen, and hydrogen atoms, and nothing else. Using your knowledge of biomolecules and their components, identify which of the following statements about your unknown sample are true. (Select all that apply.)

(a) It is an organic compound.

(b) It contains amino acids.

(c) It contains sugar.

(d) It is a nucleic acid.

(e) It is a carbohydrate.

2 Explain why life on Earth is carbon-based rather than, for example, hydrogen- or oxygen-based.

3 Why did Miller's addition of hydrogen sulfide (H_2S) to the mixture of molecules in the spark discharge experiments increase the number of amino acids he found?

Try Something New

4 In the accompanying figure, a carbon atom resides at each unlabeled corner of the hexagons and wherever a "C" is shown. Each "O" is an oxygen atom, and each "H" is a hydrogen atom. What is this structure?

(a) a fatty acid monomer

(b) two simple sugar molecules linked together

(c) a nucleotide polymer

(d) a hydrocarbon ring polymer

(e) an amino acid

5 Lipoproteins are relatively large, combined clumps of both protein and lipid molecules that circulate in the blood of mammals. They come in two forms, called HDL and LDL, and they act like suitcases to move cholesterol, fatty acid remnants, triglycerides, and phospholipids from one place to another through the bloodstream. (LDL recirculates lipids throughout the body, while HDL takes lipids to the liver to excrete them in feces.) Given that lipids are hydrophobic and proteins can be hydrophilic, which of the following statements is correct?

(a) The lipid portion of LDL does not dissolve in the bloodstream, while the lipid portion of HDL does.

(b) The protein portions of both LDL and HDL can dissolve or interact with the water molecules in the bloodstream.

(c) Neither the protein nor the lipid portions of LDL molecules can interact with water molecules in the bloodstream.

(d) Both the protein and the lipid portions of HDL molecules can interact with water molecules in the bloodstream.

(e) none of the above

6 You have found an unknown cleaning solution in your under-sink cabinet. You have several spring cleaning projects to take care of and would like to use this product if it has the correct pH for the job. You know you need an extremely basic solution to clean your oven, but an acidic solution like vinegar to clean your coffee maker. You borrow a few pH indicator strips from your school's biology lab and test the solution. For each of the following scenarios, select the best answer.

a. The solution tested at a pH of 11; you will be able to clean your (**oven / coffee maker / both / neither**).

b. The solution tested at a pH of 2; you will be able to clean your (**oven / coffee maker / both / neither**).

c. The solution tested at pH 7; you will be able to clean your (**oven / coffee maker / both / neither**).

d. The solution tested at pH 9; you will be able to clean your (**oven / coffee maker / both / neither**).

e. The solution tested at pH 4; you will be able to clean your (**oven / coffee maker / both / neither**).

7 Laundry detergent molecules have a short polar end and a long nonpolar end, enabling them to bind to water (on the polar end) and oils (on the nonpolar end). Suppose you spill some salad dressing on your shirt. Explain how washing your shirt with detergent will help remove the dressing.

Leveling Up

8 **What do _you_ think?** Read the articles "Two Miles Underground, Strange Bacteria Are Found Thriving" in _News at Princeton_ (http://www.princeton.edu/main/news/archive/S16/13/72E53/index.xml?section=newsreleases) and "Martian Underground Could Hold Clues to Life's Origins" (http://www.studentnews.eu/s/3229/68257-Articles/4014087-Martian-underground-could-hold-clues-to-lifes-origins.htm). Then answer the following questions.

a. How are bacteria able to stay alive below Earth's surface or (potentially) below the surface of Mars?

b. How does the discovery of bacteria living 2 miles beneath Earth's surface relate to the importance of the McLaughlin Crater on Mars?

c. What research do you think should next occur to answer the questions brought up by these studies?

d. From what you've read, what do you think is the probability of life on other planets? What would you expect that life to look like?

9 **Is it science?** Life on Earth evolved as carbon-based. Silicon shares many of the chemical properties of carbon, yet it is not a building block of life on Earth. Watch the episode of the TV series _The X-Files_ titled "Firewalker" (http://www.hulu.com/watch/158588), or read through its story line (https://en.wikipedia.org/wiki/Firewalker_%28The_X-Files%29). While doing so, make a list of the scientific concepts presented, and note whether you think they represent real science or pseudoscience. Then use your textbook and Internet research to determine whether you were correct or incorrect in each case. If you evaluated any of the concepts incorrectly, reflect on how you came to your initial conclusions.

Thinking Multidimensionally

10 As a budding gardener, you decide to add several plants to your yard, including a pink hydrangea, which you plant adjacent to a white pine tree. You proudly display your "green thumb" by posting the images on social media.

The next summer, you take additional snapshots of your plants, and your friends are surprised to see that the hydrangeas have changed color!

a. Investigate and report on what caused this effect.

b. Can this color change be reversed, and if so, how?

11 *"I've heard of the theoretical possibility of life based on silicon, but silicon-based life would be of an entirely different order."* Captain James T. Kirk, March 9, 1967

In the first season of *Star Trek*, Kirk, Spock, and McCoy encounter a silicon-based amorphous creature called a Horta on the planet Janus VI.

a. What properties of silicon make it a sci-fi contender for extraterrestrial life-forms?

b. Research the disadvantages to life revolving around silicon instead of carbon, and then write an argument against this possibility in a letter to the crew of the *Enterprise*.

12 Webbed toes create pockets of air that enable the basilisk lizard (also called the Jesus lizard) to literally walk on water. Draw and label the underlying properties of H_2O that also facilitate this remarkable ability.

For more, visit digital.wwnorton.com/bionow2 for access to:

Engineering Life

In 2003, scientists began trying to build an artificial cell from scratch. Today, they're closer than you might think.

After reading this chapter you should be able to:

- Explain cell theory and why it is central to the study of life.
- Describe the differences in the structures of viruses, prokaryotes, and eukaryotes.
- Diagram a plasma membrane, showing how the structure allows some substances in and keeps others out.
- Compare and contrast passive and active transport of materials into and out of cells.
- Differentiate between exocytosis and the three types of endocytosis.
- Describe the role of any given organelle in a eukaryotic cell.
- Identify the main differences between a plant cell and an animal cell.

CHAPTER
04

LIFE IS CELLULAR

DRIVING QUESTION

To create a synthetic cell, which approach is best, and why—start by making the DNA or the plasma membrane?

- HS-LS1.A: Structure and Function
- HS-LS1.C: Organization for Matter and Energy Flow in Organisms
- HS-PS3.D: Energy in Chemical Processes

The sky was still dark when Daniel Gibson hurried into the J. Craig Venter Institute (JCVI) in La Jolla, California. At 5:00 a.m., his footsteps echoed through the empty halls of the building. He reached a laboratory door and slipped inside. There, Gibson peered into a warm incubator, his eyes scanning rows of palm-sized petri dishes. His stomach was in knots. For 3 months the experiment had failed. Would this day—Monday, March 29, 2010—be any different?

Gibson is part of a team at the JCVI with a single, audacious goal: to create life. For more than a decade, this team of scientists and engineers has attempted to build a synthetic, or human-made, **cell**. Cells are the smallest and most basic unit of life—microscopic, self-contained units enclosed by a protective membrane (**Figure 4.1**). The human body is composed of approximately 100 trillion (10^{14}) cells. On that day in 2010, however, the JCVI was trying to synthesize just one cell—a single-celled bacterium.

Gibson's team had sequenced a bacterium's complete genetic information, its **genome**; built a synthetic version of that genome using basic laboratory chemicals; and, finally, replaced the natural DNA of another species of bacterium with the synthetic DNA. **DNA** (deoxyribonucleic acid) is a large and complex molecule that acts as a set of instructions for building an organism, like a blueprint. Almost every cell of every living organism contains DNA. DNA transfers information from parents to offspring, which is why it is essential for reproduction. Life, no matter how simple or how complex, uses this inherited genetic code to direct the structure, function, and behavior of every cell. DNA is made up of many nucleotides held together in a structure called the double helix, a ladderlike assembly twisted along its length into a spiral (see Figure 3.11).

Gibson's boss, the famous geneticist J. Craig Venter, worked for more than 15 years and spent millions of dollars to construct a synthetic DNA helix from chemicals in the laboratory, but Gibson and the team had been unable to get that synthetic DNA to work inside a cell. Every Friday for 3 months, they transplanted the synthetic DNA into a bacterial cell whose own DNA had been removed. The synthetic DNA included a gene to make the cells turn bright blue, so every Monday, Gibson hurried to the incubator and checked the petri dishes for a colony of blue cells. But Monday after Monday, the dishes were barren. "We did the genome transplantation again and again," he recalls, "but nothing was working."

Then, in mid-March, Gibson identified an error in a single gene in the synthetic DNA. A **gene** is a segment of DNA that codes for a distinct genetic characteristic, such as having O-type blood or a dimpled chin. In Gibson's bacterium, the gene with the error was responsible for DNA replication. When it wasn't working, the bacterium couldn't replicate its DNA, and it died. So in late March, Gibson fixed the DNA error, transplanted the genome yet again, and waited.

Salmonella is a single-celled bacterium that is a common cause of food poisoning.

This is one cell of the multicellular plant *Arabidopsis*, which is used extensively in genetic studies.

Yeasts are single-celled, but more complex than bacteria. Some species are critical for making bread and beer, while others are pathogens.

Humans are multicellular animals with many specialized cells, such as these neurons within the central nervous system.

Figure 4.1

An individual organism may consist of a single cell or many cells

All of these photos of cells were taken using electron microscope technology. Color has been added to the images to differentiate structures within the cells.

Life, Rewritten

The JCVI is just one institution among a large group of universities and companies pursuing synthetic biology, a field that aims to design and construct new biological entities with novel and useful functions. Scientists are working to create algae that digest trash and produce energy, microbes that use light and water to create hydrogen gas, and bacteria that produce new kinds of antibiotics to treat infections. With synthetic biology "we can harness what nature has made, but repurpose it," says James Collins, a synthetic biologist at Boston University. "We can reprogram organisms and endow them with novel functions." In addition to useful tools, the pursuit of artificial life sheds light on the very origins of life. "It's going to be a big challenge to create a totally synthetic cell," says Collins, "but it's fundamentally intriguing to explore how life may have arisen on the planet."

To make their own cell, scientists are pushing the boundaries of **cell theory**, one of the unifying principles of biology. Cell theory has two main parts: every living organism is composed of one or more cells, and all cells living today came from a preexisting cell. By trying to engineer a cell in the laboratory, Venter, Gibson, and others are challenging the second part of the definition.

Starting Small

The JCVI's first step toward a synthetic cell was a small one. In 2003, Venter's team flexed its scientific muscles by synthesizing the 11-gene, 5,386-base-pair genome of phiX174, a virus that infects bacteria. A **virus** is a small, infectious agent that can replicate only inside a living cell. Most viruses are little more than stripped-down genetic material wrapped in proteins, yet these pathogens attack and devastate organisms in every kingdom of life, from bacteria to plants and animals. Though the JCVI team successfully created a virus with a synthetic genome, it was not considered the first synthetic life, because scientists debate whether viruses are alive. (For more on this debate, see "Viruses—Living or Not?," page 66.)

Next, Venter and his colleagues moved up to a bacterium, a living organism that consists of a single cell. In 2010, they sequenced and built the genome of a bacterium called *Mycoplasma mycoides* (*M. mycoides*), an organism that can cause the mammary glands of goats to swell. They constructed the genome—a 1.1-million-base-pair DNA sequence—using four necessary ingredients: the nucleotides **adenine** (**A**), **thymine** (**T**), **guanine** (**G**), and **cytosine** (**C**), the building blocks of DNA. A, T, G, and C, organized in different combinations, carry all the instructions for everything a cell does.

The team used a machine to read the nucleotide sequence of the *M. mycoides* genome and then "print out" little bits of that code, creating strands of DNA about 50–80 bases long. They then strung these pieces together using living cells as factories, inserting the short segments into yeasts and *Escherichia coli*—small, single-celled organisms. These organisms interpreted the strands as broken pieces of DNA and stitched them together, creating longer and longer sequences. It was like building the Eiffel Tower from a massive box of Legos, constructing a single support beam at a time. The effort—with many mistakes along the way—took years. "It was very complex," said Venter. "It was a long, involved process."

Congratulations, It's a . . . Cell

Once the DNA sequence of *Mycoplasma mycoides* was complete and intact, it was up to the JCVI team to transfer it into another species and make it work. This was the experiment that almost drove Gibson crazy. The researchers removed all the DNA from a cell of a closely related bacterium, *Mycoplasma capricolum*, and replaced it with the *M. mycoides* synthetic DNA.

After months of trying, on that Monday morning at 5:00 a.m., Gibson cautiously scanned the

J. CRAIG VENTER

J. Craig Venter is an American biologist and founder and CEO of the J. Craig Venter Institute. He led a team to fully sequence and publish the human genome in 2001, and initiated the effort to create the first cell constructed with synthetic DNA.

petri dishes. There, on a single dish, was a group of bright-blue cells—proof that the *M. capricolum* cell had "booted up" the *M. mycoides* DNA and transformed itself into an *M. mycoides* cell (**Figure 4.2**). Gibson was ecstatic. Moments later, he sent a text message to Venter, waking him up. Within the hour, Venter was in the lab with a camera, taking pictures of the tiny blue dollop in the dish. "How does it feel to create life?" Venter asked Gibson. They opened champagne and toasted their success.

Over the following weeks, Gibson repeated the experiment hundreds of times to make sure the blue cells were not an accident or a fluke, confirming that they contained only

Two colonies of *M. mycoides*, transformed from *M. capricolum*.

100 μm

Individual *M. mycoides* cells.

500 nm

Figure 4.2

The first synthetic organism

A colony of bacteria was transformed from *M. capricolum* by the insertion of synthetic DNA of a closely related species of bacterium, *M. mycoides*. The investigators inserted a gene into the synthetic DNA that codes for blue pigment.

Q1: What was the purpose of inserting the gene that codes for blue pigment into the synthetic DNA?

Q2: What part of the transformed bacterium is synthetic?

Q3: Did this experiment create life?

M. mycoides DNA. Every time, the cells with a synthetic genome survived. The team had done it—created the first synthetic cell. "They are living cells," Venter told *The Scientist* magazine when the research was published 2 months later. "The only difference is that they have no natural history. Their parents were the computer."

Once the research was published, the reaction from the academic community, captured in *Nature* magazine, was swift and divided. Some called it a significant advance: "We now have an unprecedented opportunity to learn about life," said Mark Bedau, a professor of philosophy at Reed College in Oregon. Arthur Caplan, a bioethicist at the University of Pennsylvania, said, "Venter's achievement would seem to extinguish the argument that life requires a special force or power to exist. In my view, this makes it one of the most important scientific achievements in the history of mankind."

Others were more hesitant. "Has [Venter] created 'new life'?" asked George Church, a prominent geneticist at Harvard Medical School. "Not really. . . . Printing out a copy of an ancient text isn't the same as understanding the language." Gibson and Venter agree that they did not create life from scratch, but they argue that they did create new life from existing life.

A Different Approach

While Venter and Gibson were making headlines, other scientists were quietly pursuing a different approach to building a cell, working from the bottom up rather than from the top down. One young scientist in California decided to start by building one of the simplest, yet most vital, components of a cell: the layer of molecules that surrounds it.

In the chemistry department at UC San Diego, assistant professor Neal Devaraj was fascinated by the idea of building life and approached synthetic biology from the perspective of chemistry. "Most molecular biologists study what exists," says Devaraj, "but when you're a chemist and constantly make new compounds, you want to engineer something from scratch." So instead of taking a cell apart and determining how

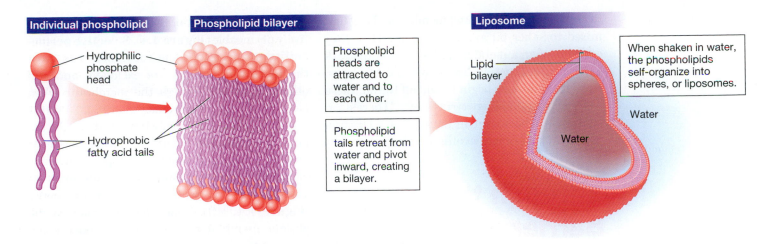

Individual phospholipid

Hydrophilic phosphate head

Hydrophobic fatty acid tails

Phospholipid bilayer

Phospholipid heads are attracted to water and to each other.

Phospholipid tails retreat from water and pivot inward, creating a bilayer.

Liposome

Lipid bilayer

When shaken in water, the phospholipids self-organize into spheres, or liposomes.

Water

Water

Figure 4.3

Liposomes form when phospholipids and water are shaken together

When you shake a mixture of phospholipids and water, the phospholipid bilayers bend and link together to form spheres called liposomes. This simple structure is remarkably similar to the basic structure of a cell.

Q1: Why is it important that the phosphate head of a phospholipid is hydrophilic?

Q2: What essential component of a cell do liposomes lack, and why is that omission important?

Q3: Could the tendency of phospholipid bilayers to spontaneously form spheres have played a role in the origin of life? (*Hint:* Refer to "The Characteristics of Living Organisms" on page 6 of Chapter 1.)

it works, Devaraj decided to try to build a cell artificially, using materials not typically found in nature. "If you want to really understand the principles by which life operates and evolves, the best way to do so is to build a cell from the ground up," he says.

Scientists suspect that one of the first events at the beginning of life on Earth was the formation of a **plasma membrane**, a barrier separating a cell from its external environment. A plasma membrane is made of two layers of **phospholipids**, organic molecules with a water-loving, or *hydrophilic*, head, and a water-fearing, or *hydrophobic*, tail. In water, these molecules form a double layer with heads out and tails in, a barrier that separates the contents of the cell from what lies outside the cell. Thus, the membrane is a **phospholipid bilayer**, a mostly impermeable barrier. When a phospholipid bilayer forms a sphere, or **liposome**, the fluid inside the liposome can have a different composition from the fluid outside (**Figure 4.3**). The ability to maintain an internal environment separate from the external environment is one

of the most critical functions of the plasma membrane of a cell.

Given a container of phospholipids, anyone can make a membrane, says Devaraj. "Making membranes is almost a trivial thing," he says. "You take natural or synthetic phospholipids, add water, and they form membranes." Yet researchers had been unable to form a membrane from scratch—without using preexisting phospholipids. In nature, new phospholipids are created by enzymes embedded in the cell.

Instead of trying to engineer new phospholipids, Devaraj wanted to start with something simpler. He worked with graduate student Itay Budin, who was then at Harvard University and is now a postdoctoral researcher at UC Berkeley,

NE AL DEVARAJ

JNeal Devaraj is a biochemist at UC San Diego who is working to make an artificial cell from the bottom up, starting with the membrane and then building other organelles.

to create a self-assembling membrane. They first mixed together oil and a detergent. Then they added copper, a metal ion, as a catalyst to spark a chemical reaction. With the addition of copper, sturdy membranes begin to bud off the oil; these were self-assembling structures. "There's no equivalent whatsoever in nature," says Devaraj. "Our goal was simply to mimic biology."

Through the Barrier

Devaraj admits that his artificial membrane is far simpler than a real cell's plasma membrane, which is dotted with numerous proteins, including transport proteins. **Transport proteins** are gates, channels, and pumps that allow molecules to move into and out of the cell, making it selectively permeable (**Figure 4.4**). **Selective permeability** means that some substances can cross the membrane, others are excluded, and still others can pass through the membrane when aided by transport proteins.

All movement of substances through the plasma membrane occurs by either active or passive transport. Some transport proteins facilitate **active transport**, the movement of a substance that requires an input of energy (**Figure 4.4**, left). Molecules move across the plasma membrane by active transport when they need to move from a region of *lower* concentration to a region of *higher* concentration. In contrast, **passive transport** is the movement of a substance without the addition of energy (**Figure 4.4**, middle and right). Movement via passive transport is spontaneous.

Figure 4.4

The plasma membrane is a barrier and a gatekeeper
The plasma membrane moves substances in a highly selective fashion, determined in large part by the types of membrane proteins embedded in the phospholipid bilayer. ▶️

Q1: In what ways is the plasma membrane a barrier, and in what ways is it a gatekeeper?

Q2: Why can't ions (such as Na^+) cross the plasma membrane without the help of a transport protein?

Q3: If no energy were available to the cell, what forms of transport would not be able to occur? What forms of transport could occur? (*Hint:* Look ahead at Figures 4.5 and 4.6.)

A primary type of passive transport is **diffusion**, the movement of a substance from a region of *higher* concentration to a region of *lower* concentration (**Figure 4.5**). Water, oxygen, and carbon dioxide usually enter and leave cells by **simple diffusion**: These small, uncharged molecules slip between the large molecules in the phospholipid bilayer without much hindrance.

Water moves in and out of cells (and compartments inside cells) by **osmosis**. Osmosis is a form of simple diffusion because the water molecules are moving from areas of higher concentration to areas of lower concentration (**Figure 4.6**). Osmosis is critical for cellular processes because most cells are at least 70 percent water, and nearly all cellular processes take place in a water-rich environment.

Cells must maintain a stable internal water concentration to function properly, but the concentration of water in most cells changes from moment to moment. For example, when salt molecules move into a cell, the concentration of water in the cell decreases because additional molecules have been added. In response, water

Sugar has just been added to a beaker of water. This beaker is divided by a semipermeable membrane—that is, a membrane with pores large enough to allow water molecules to pass through, but too small for sugar molecules to pass.

Sugar molecule Water molecule Semipermeable membrane Water molecule

Semipermeable membrane

After a period of time, water molecules have moved by osmosis from the right side of the membrane (which had a higher concentration of water) to the left side of the membrane. The concentration of water on the two sides of the membrane is now the same, and the movement of water molecules is now the same in both directions.

At first, the molecules of food coloring are concentrated in one region.

Net movement of the food coloring is from regions of high concentration to regions of low concentration.

Diffusion ceases when food-coloring molecules are evenly distributed. At equilibrium, just as many molecules move into any given region as leave that region, so there is no net change in concentration.

Figure 4.5

Food coloring in water illustrates diffusion

Q1: Is the dye at equilibrium in any of these glasses? Describe how the first glass will look when the dye is at equilibrium with the water.

Q2: Will diffusion mix the molecules of dye evenly through the water, or is it necessary to shake the container to get a uniform mixture?

Q3: Will diffusion mix the dye faster in hot water than in cold water? Why or why not? (*Hint:* Review the discussion of the behavior of water molecules at different temperatures in Chapter 3.)

Figure 4.6

In osmosis, water diffuses across a semipermeable membrane

Osmotic movement of water between a cell and the external environment is critical for maintaining a constant water concentration in the cell, which it needs to function properly.

Q1: What would the second diagram look like if the pores in the semipermeable membrane were too small to allow water molecules to pass through?

Q2: What would the second diagram look like if the pores were large enough to let both water molecules and sugar molecules through?

Q3: The fluid in an IV bag is isotonic to blood. What change would you see in the red blood cells of a patient if a bag of a hypertonic solution was used in error?

molecules immediately move by osmosis—that is, they diffuse—across the plasma membrane into the cell until the concentration of water inside the cell is the same as the concentration on the outside. On the other hand, when salt molecules move out of a cell, the water molecules become more concentrated, and osmotic movement of water out of the cell then restores the concentration of water in the cell to its correct level.

You can imagine, then, that the concentration of solutes within a cell in relation to the concentration *outside* the cell is of critical importance. When cells are surrounded by fluid with the same solute concentration as the cell interior, the extracellular and intracellular environments are said to be **isotonic** to each other (*iso*, "equal"). If the extracellular environment has a higher solute concentration, it is **hypertonic** to the cell interior (*hyper*, "more"). If it has a lower solute concentration than the cell's interior, it is said to be **hypotonic** to it (*hypo*, "less").

The careful balance of concentrations is particularly vital in human blood. Red blood cells are typically in an isotonic solution in our bloodstream, where the solute concentrations inside and outside the cells are the same. Doctors and nurses are taught to administer an IV drip of saline, a solution of salt in water, to dehydrated patients. If a dehydrated patient was administered water instead of a saline solution, the water would dilute the patient's blood, making it hypotonic. Osmosis would then occur, causing water to rapidly diffuse into red blood cells to the point where the cells could burst and die.

Most hydrophobic molecules, even fairly large ones, can pass through the plasma membrane via simple diffusion because they mix readily with the hydrophobic tails that form the core of the phospholipid bilayer. But hydrophilic substances such as sodium ions (Na^+), hydrogen ions (H^+), and larger molecules, including sugars and amino acids, cannot cross the plasma membrane without assistance. These substances move across the plasma membrane by **facilitated diffusion**, a type of passive transport that requires transport proteins (**Figure 4.4**, right).

Devaraj's artificial membrane does not currently contain any transport proteins, he says, so it is impermeable to large hydrophilic molecules. But small molecules such as water can pass through his membrane via simple diffusion.

The plasma membrane also contains **receptor proteins**, which are sites where molecules released by other cells can bind. The binding of a molecule to a receptor protein starts a chain of

Viruses—Living or Not?

Protein coat
RNA
Envelope

You've heard their names: Ebola virus, Zika virus, H5N1, dengue virus. Viruses—microscopic, noncellular infectious particles—are perhaps the smallest biological agents with the greatest impact on human health. Like living organisms, viruses reproduce and evolve, yet they lack some of the key characteristics of life—which is why most scientists today regard viruses as nonliving. For one thing, viruses are not made up of cells. A virus is much simpler than a cell, usually consisting of a small piece of genetic material (for example, DNA) that is wrapped in a protein coat. Some viruses also have an envelope, a lipid layer usually stolen from a cell's plasma membrane, enclosing the central core of genetic material and protein.

Another difference, compared to living organisms, is that viruses lack the many structures within cells that are necessary for critical cellular functions such as homeostasis, autonomous reproduction, and metabolism. To gain these functions, they become "body snatchers": Viruses use their genetic material to make the cells of the organisms they infect do their work for them. They accomplish this feat by invading cells, releasing their genetic material into the cell interior, and "hijacking" the host cell's machinery. Viruses multiply to huge numbers, and viral offspring escape from a host cell either by causing it to burst open or by budding off from the cell, wrapped in a layer of the host cell's plasma membrane.

Uniquely, unlike the case with living organisms, the genetic material that viruses pass from one generation to the next is not always DNA; sometimes it is RNA. Viruses are generally classified by the type of genetic material they possess (type of DNA or RNA molecule), their shape and structure, the type of organism (host) they infect, and the disease they produce. The variant forms of a particular type of virus are called **viral strains** or serotypes. Viruses evolve new strains within a host so quickly that sometimes an antiviral drug or vaccine developed to fight an older strain becomes useless against a new strain.

events inside the cell that causes the cell to do something. For example, the receptors in nerve cells receive molecular signals from other nerve cells that cause the cells to fire. Receptor proteins are key components of a cell's communication system, enabling it to respond appropriately to changes in the organism.

Another Way Through

In addition to transport proteins, there is another way that molecules move into and out of a cell. Sections of the plasma membrane can bulge inward or outward to form packages called **vesicles**. Vesicles move molecules from place to place inside a cell but also transport substances into and out of the cell (**Figure 4.7**).

Cells expel materials in vesicles via **exocytosis** (**Figure 4.7**, top left). The substance to be exported from the cell is packaged into a vesicle, and as the vesicle approaches the plasma membrane, a portion of the vesicle's membrane fuses with the plasma membrane. The inside of the vesicle then opens to the exterior of the cell, discharging its contents.

Endocytosis (**Figure 4.7**, right and lower left) is the opposite of exocytosis. In this process, a section of plasma membrane bulges inward to form a pocket around extracellular fluid,

Exocytosis

Exocytosis is used to eject substances from the cell. Here, a cell ejects waste material into the outside environment.

Endocytosis

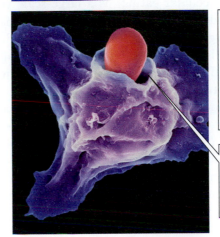

Endocytosis is the reverse of exocytosis, bringing material from the outside of the cell to the inside, enclosed in vesicles. Endocytosis may be nonselective, as shown here, drawing in any and all molecules near the opening of the vesicle.

Here, a white blood cell engulfs a yeast cell through phagocytosis, a form of nonspecific endocytosis.

Receptor-mediated endocytosis

Receptor-mediated endocytosis is a selective process in which only certain molecules bind to receptor proteins. Here, low-density lipoprotein (LDL) particles bind to LDL receptors and are transported to the cell interior.

Cells lining blood vessels have created fluid-filled vesicles through pinocytosis, nonspecific endocytosis of fluid.

Figure 4.7

Substances move into cells by endocytosis and out of cells by exocytosis

Q1: If endocytosis itself is nonspecific, how does receptor-mediated endocytosis bring only certain molecules into a cell?

Q2: What sorts of molecules could be moved by endocytosis or exocytosis, but not by diffusion?

Q3: How does the fluid that enters a cell via pinocytosis differ from the fluid that enters by osmosis?

molecules, or particles. The pocket deepens until the opening in the membrane pinches off and the membrane breaks free as a closed vesicle, now wholly contained within the cell. Endocytosis can be nonspecific or specific. In nonspecific endocytosis, all of the material in the immediate area is surrounded and enclosed; in specific endocytosis, one particular type of molecule is enveloped and imported.

There are three types of endocytosis. **Receptor-mediated endocytosis** (**Figure 4.7**, bottom left) is a form of endocytosis in which receptor proteins embedded in the membrane recognize specific surface characteristics of substances to be incorporated into the cell. For example, our cells use receptor-mediated endocytosis to take up cholesterol-containing packages called low-density lipoprotein (LDL) particles. **Phagocytosis** (**Figure 4.7**, top right), or "cellular eating," is a large-scale version of endocytosis in which particles considerably larger than biomolecules are ingested. Specific cells in the immune system use phagocytosis to ingest an entire bacterium or virus. **Pinocytosis** (**Figure 4.7**, bottom right) is a form of endocytosis that is often described as "cellular drinking" because it involves the capture of fluids. However, the cell does not attempt to collect particular solutions. Pinocytosis is nonspecific: the vesicle budding into the cell contains whatever happened to be dissolved in the fluid when the cell "drank."

There is a long way to go before an artificial plasma membrane will perform processes like endocytosis and exocytosis, says Neal Devaraj. "But just because a research problem is difficult doesn't mean we shouldn't tackle it," he adds.

Prokaryotes versus Eukaryotes

Devaraj's team continues to pursue that ideal. Prior synthetic membranes, their own included, did not have the ability to grow by adding new phospholipids, but in 2015 the team succeeded in designing and synthesizing an artificial membrane that sustained continual growth, just like a living cell. Membranes are important not only because they form the structure of a cell, but because they compartmentalize processes within eukaryotic cells. Depending on the fundamental structure of their cells, all living organisms can be sorted into one of two groups: **prokaryotes** or **eukaryotes** (**Figure 4.8**). *Mycoplasma mycoides* and all other bacteria are prokaryotes, but virtually all the organisms you see every day, including all plants and animals, are eukaryotes.

Eukaryotic cells are larger and more complex than prokaryotic cells: they are roughly ten times wider, with a cell volume about a thousand times greater. Unlike prokaryotic cells, eukaryotic cells have a membrane-enclosed **nucleus** (plural "nuclei") that contains the organism's DNA, and they have a variety of membrane-enclosed subcellular compartments called **organelles**. Through specialization and division of labor, these organelles act like cubicles in a large office, allowing the cell to localize different processes in different places. In contrast, prokaryotic cells are like an open floor plan: they lack a cell nucleus or any membrane-encased organelles.

Though the first fully artificial cell will most likely be a simple prokaryotic cell, synthetic biologists aspire to build a eukaryotic cell. "We're not quite there yet, but it's interesting to think about making complex structures like organelles," says Devaraj. Having achieved a self-assembling and growing artificial membrane, his team is working to create membranes inside preexisting vesicles, mimicking organelles like mitochondria that are made up of membranes within membranes. It is the first step toward creating the variety of organelles inside eukaryotic cells.

What's in a Cell?

The nucleus is the control center of the cell. It contains most of the cell's DNA and may occupy up to 10 percent of the space inside the cell (**Figure 4.9**, top left). Inside the nucleus, long strands of DNA are packaged with proteins into a remarkably small space. The boundary of the nucleus, called the **nuclear envelope**, is made up of two concentric phospholipid bilayers. The nuclear envelope is speckled with thousands of small openings called **nuclear pores**. These pores allow chemical messages to enter and exit the nucleus.

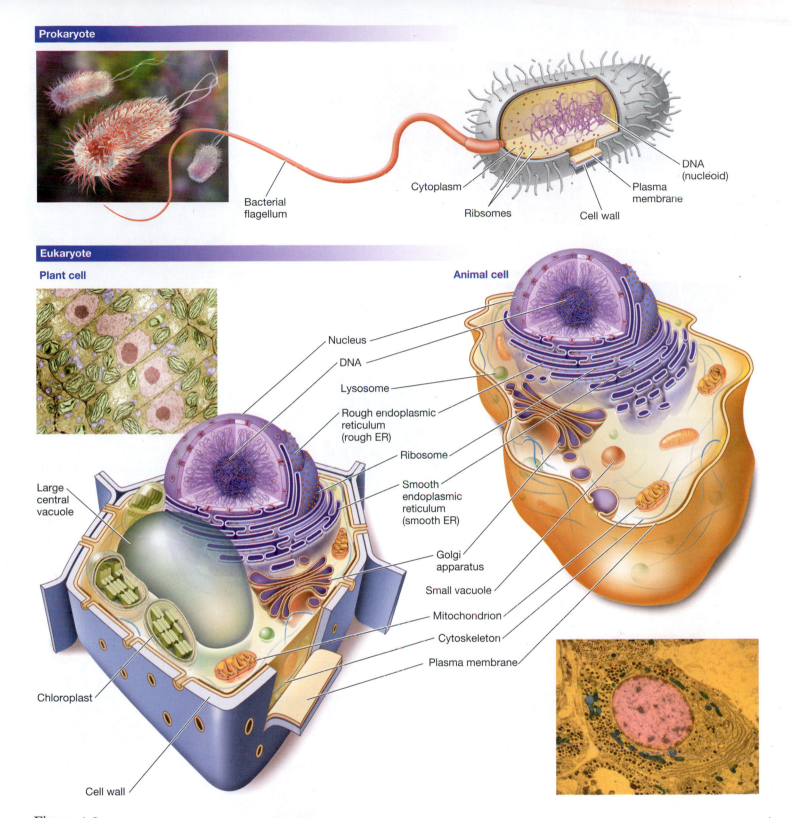

Prokaryote

Bacterial flagellum

Cytoplasm

Ribsomes

Cell wall

Plasma membrane

DNA (nucleoid)

Eukaryote

Plant cell

Animal cell

Nucleus

DNA

Lysosome

Rough endoplasmic reticulum (rough ER)

Ribosome

Smooth endoplasmic reticulum (smooth ER)

Golgi apparatus

Small vacuole

Mitochondrion

Cytoskeleton

Plasma membrane

Large central vacuole

Chloroplast

Cell wall

Figure 4.8

Prokaryotic and eukaryotic cells

A prokaryotic cell, like all cells, contains DNA, cytoplasm, ribosomes, and a plasma membrane. Many prokaryotes also have a cell wall that serves as a kind of exoskeleton. To enable movement, some bacteria possess a flagellum, or several flagella. The components of eukaryotic cells are described in the text and in Figure 4.9. ▶▣

Q1: What structures do prokaryotic and eukaryotic cells have in common?

Q2: What cellular processes occur in both prokaryotic and eukaryotic cells?

Q3: Both plants and animals are eukaryotes, but there are differences in their cellular structure. What are those differences?

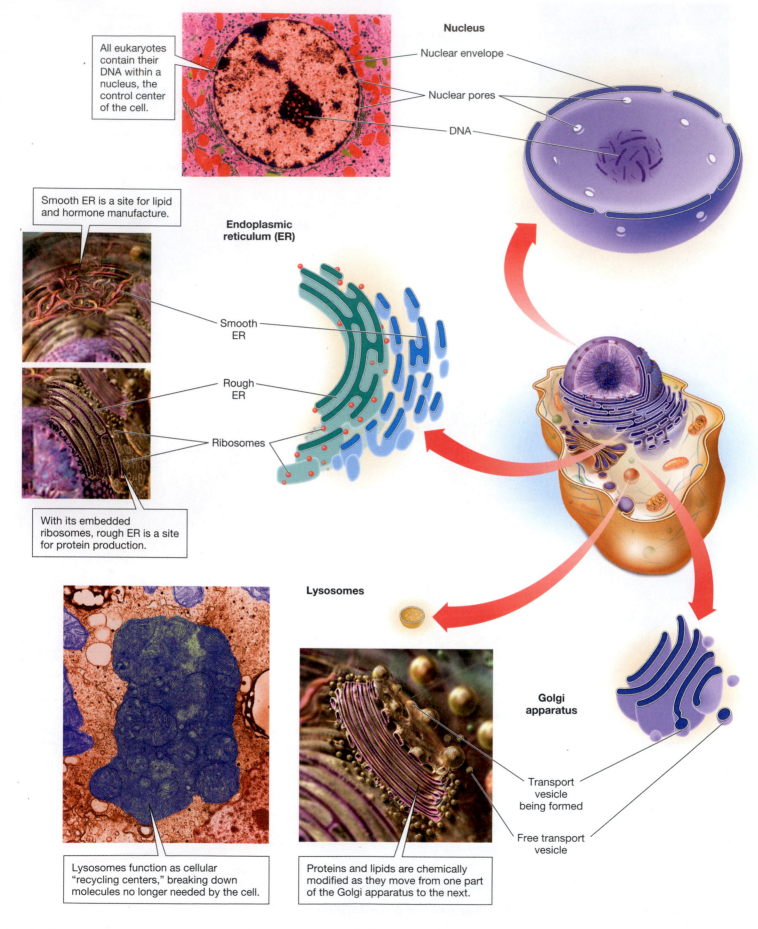

All eukaryotes contain their DNA within a nucleus, the control center of the cell.

Nucleus

Nuclear envelope

Nuclear pores

DNA

Smooth ER is a site for lipid and hormone manufacture.

Endoplasmic reticulum (ER)

Smooth ER

Rough ER

Ribosomes

With its embedded ribosomes, rough ER is a site for protein production.

Lysosomes

Golgi apparatus

Transport vesicle being formed

Free transport vesicle

Lysosomes function as cellular "recycling centers," breaking down molecules no longer needed by the cell.

Proteins and lipids are chemically modified as they move from one part of the Golgi apparatus to the next.

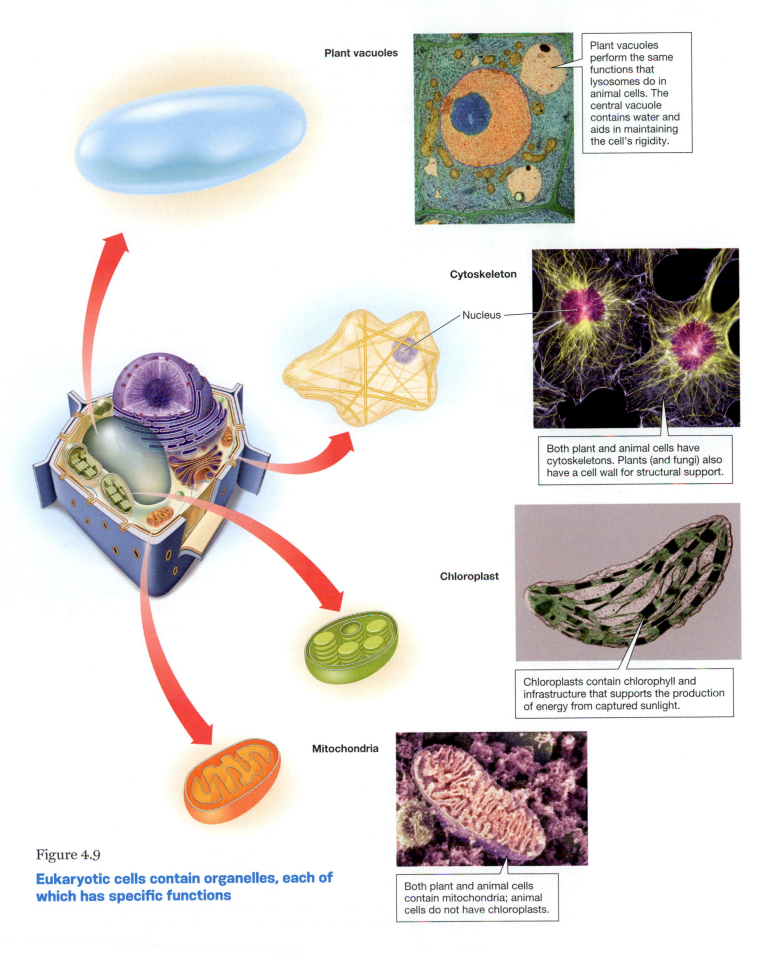

Plant vacuoles

Plant vacuoles perform the same functions that lysosomes do in animal cells. The central vacuole contains water and aids in maintaining the cell's rigidity.

Cytoskeleton

Nucleus

Both plant and animal cells have cytoskeletons. Plants (and fungi) also have a cell wall for structural support.

Chloroplast

Chloroplasts contain chlorophyll and infrastructure that supports the production of energy from captured sunlight.

Mitochondria

Both plant and animal cells contain mitochondria; animal cells do not have chloroplasts.

Figure 4.9

Eukaryotic cells contain organelles, each of which has specific functions

The **endoplasmic reticulum** (**ER**) is an extensive and interconnected network of sacs made of a single membrane that is continuous with the outer membrane of the nuclear envelope (**Figure 4.9**, middle left). The membranes of the ER are classified into two types based on their appearance: smooth and rough. Enzymes associated with the surface of the **smooth ER** manufacture lipids and hormones. In some cell types, smooth-ER membranes also break down toxic compounds. **Ribosomes** embedded in the **rough ER** give it the knobby appearance from which it gets its name. Ribosomes on the rough ER assemble proteins that will be inserted into the cell's plasma membrane or its organelles.

Resembling a pile of flattened balloons, the **Golgi apparatus** is like a post office, packaging and directing proteins and lipids produced by the ER to their final destinations either inside or outside of the cell (**Figure 4.9**, bottom left). Each molecule is first packaged into a transport vesicle. The transport vesicle buds off from the ER membrane and delivers the cargo to its destination by fusing with the membrane of the target compartment. The Golgi apparatus targets these destinations by adding a specific chemical tag to each molecule it receives, like attaching a shipping label to a package.

In animal cells, transport vesicles bring large molecules that will be discarded to **lysosomes**, organelles that act as garbage and recycling centers (**Figure 4.9**, bottom far left). Lysosomes contain a variety of enzymes that degrade biomolecules and release the breakdown products into the cell interior to be discarded or reused. In plant cells, **vacuoles** perform functions similar to those of lysosomes, plus a few additional functions, such as water storage (**Figure 4.9**, top right). Some plant vacuoles stockpile noxious compounds that can deter herbivores from eating plants.

In most eukaryotic cells, the main source of energy is the **mitochondrion** (plural "mitochondria"), a tiny power plant that fuels cellular activities (**Figure 4.9**, bottom right). Mitochondria are made up of double membranes—a smooth external membrane and a folded internal membrane—that form a mazelike interior. Mitochondria use chemical reactions to transform the energy of food molecules into ATP (adenosine triphosphate), the universal cellular fuel, in a process called *cellular respiration*. Mitochondria provide ATP to all eukaryotic cells (both plant and animal), but the cells of plants and some protists have additional organelles called **chloroplasts** that capture energy from sunlight and use it to manufacture food molecules via *photosynthesis* (**Figure 4.9**, middle right). We will explore cellular respiration and photosynthesis in Chapter 5.

A network of protein cylinders and filaments collectively known as the **cytoskeleton** forms the framework of a cell (**Figure 4.9**, middle right). The cytoskeleton organizes the interior of a eukaryotic cell, supports the intracellular movement of organelles such as transport vesicles, and enables whole-cell movement in some cell types. It also gives shape to wall-less cells. Fungi and plants separately evolved cell walls to maintain cell structure, but animals rely only on the cell's cytoskeleton.

Life Goes On

From Venter's complex synthetic genome to Devaraj's simple self-assembling membrane, scientists debate whether the top-down or bottom-up approach will be more successful in the effort to build an artificial cell. But the two can be complementary, says James Collins. "I'm not sure one will win out. I think they bring different things to the table," he notes. Gibson agrees: "I would like to one day be able to combine all of a cell's parts from nonliving components, including the genome, and incubate them, and see if we can get life out of those nonliving components," he says. "It would help us better understand how cells work."

Today, the work on both ends continues. While Devaraj makes his membranes more complex, Gibson and his team (**Figure 4.10**) recently simplified the *Mycoplasma mycoides* genome. With the goal of determining the smallest set of genes needed to maintain life, they broke the *M. mycoides* genome into eight DNA segments and mixed and matched them to see which combinations would produce viable cells. Eventually, they narrowed the genome down to just 473 genes capable of sustaining life. Amazingly, the team could not identify the function of 149 of the 473 genes. "We don't

Sizing Up Life

Cells dramatically range in size, and viruses are even smaller. The amoeba, a single-celled eukaryote that eats smaller unicellular organisms, is visible under a light microscope. But to observe a virus, one would have to use an electron microscope, which relies on a high-voltage beam of electrons to magnify an image.

Assessment available in smartwork**5**

⊢—— 500 μm ——⊣

Amoeba proteus

⊢—— 130 μm ——→

Human egg cell
(shown next to an Amoeba proteus)

⊢—— 30 μm ——→

Skin cell
(shown next to a human egg cell)

⊢—— 8 μm ——→

Red blood cell
(shown next to a skin cell)

⊢—— 3 μm ——→

E. coli bacterium
(shown next to a red blood cell)

⊢—— .22 μm ——→

Measle virus
(shown next to an E. coli bacterium)

⊢—— .13 μm ——→

Influenza virus
(shown next to a measle virus)

⊢—— .03 μm ——→

Rhinovirus
(shown next to an influenza virus)

Figure 4.10

Scientist in Action
Laboratory technician Javier Quinones is setting up the beginning of the sequencing procedure used to sequence the *M. mycoides* genome in the sequencing laboratory at the J. Craig Venter Institute in Rockville, Maryland.

know about a third of essential life, and we're trying to sort that out now," Venter told *Nature* magazine in 2016.

Knowing both the identities and function of a core set of genes necessary for life will make it easier to build other synthetic cells with specific functions, says Gibson. "There's not a single organism in the world where we understand what every gene does," he notes. If they can reach such understanding with those 473 genes in the *M. mycoides* genome, "this would be the first example of that."

REVIEWING THE SCIENCE

- **Cells** are the basic units of all living organisms. **Cell theory** states that all living things are composed of one or more cells, and that all cells living today came from a preexisting cell.

- Every cell is surrounded by a **plasma membrane** that separates the chemical reactions inside the cell from the surrounding environment. The plasma membrane is **selectively permeable** and formed by a **phospholipid bilayer** embedded with proteins that perform a variety of functions.

- In **passive transport**, substances move across the plasma membrane without the direct expenditure of energy. **Active transport** by cells requires energy. **Diffusion** is the passive transport of a substance from a region where it is at a higher concentration to a region where it is at a lower concentration.

- **Osmosis** is the diffusion of water across a selectively permeable membrane. In a **hypotonic** solution, a cell gains water. In a **hypertonic** solution, a cell loses water. In an **isotonic** solution, there is no net movement of water. Cells can actively balance their water content through osmosis.

- Cells export materials by **exocytosis** and import materials by **endocytosis**.

- In **receptor-mediated endocytosis**, **receptor proteins** in the plasma membrane recognize and bind the substance to be brought into the cell.

- **Prokaryotes** are single-celled organisms lacking a nucleus and complex internal compartments. **Eukaryotes** may be single-celled or multicellular, and their cells typically possess many membrane-enclosed compartments.

- By volume, eukaryotic cells can be a thousand times larger than prokaryotic cells. They require internal compartments, or **organelles**, that concentrate and organize cellular chemical reactions.

- The **nucleus** of a eukaryotic cell contains DNA. It is bounded by the **nuclear envelope**, which has **nuclear pores** that allow communication between the nucleus and the cell interior.

- Lipids are made in the **smooth endoplasmic reticulum**. Some proteins are manufactured in the **rough endoplasmic reticulum**.

- The **Golgi apparatus** receives proteins and lipids, sorts them, and directs them to their final destinations.

- In animals, **lysosomes** break down biomolecules such as proteins into simpler compounds that can be used by the cell. Plant **vacuoles** are similar to lysosomes but also store ions and molecules and lend physical support to plant cells.

- **Mitochondria** produce chemical energy for eukaryotic cells in the form of ATP.

- **Chloroplasts** harness the energy of sunlight to make sugars through photosynthesis.

- Eukaryotic cells depend on the **cytoskeleton** for structural support and for the ability to move and change shape. Plants and fungi also have a cell wall that provides structural support.

THE QUESTIONS

Challenge Yourself

1 How does the phospholipid bilayer of a liposome differ from the phospholipid bilayer of the plasma membrane of a cell?

(a) The phospholipid bilayer of a liposome contains only phospholipids, without the proteins that are embedded in the plasma membrane of a cell.

(b) The phospholipid bilayer of a liposome contains two bilayers of phospholipid molecules, whereas the plasma membrane of a cell contains only one.

(c) The phospholipid bilayer of a liposome completely envelops the liposome, whereas the plasma membrane of a cell does not completely envelop the cell.

(d) The phospholipid molecules in the phospholipid bilayer of a liposome are oriented with the lipid ends on the outside of the bilayer and the phosphate groups on the inside.

2 Examine the graph presented here. Using your knowledge of how the processes work, label each curve as representative of either simple diffusion or facilitated diffusion. Explain how you determined which curve was which.

3 Which of the following statements about transport through the plasma membrane of a cell is correct?

(a) Both passive and active transport require the input of energy.

(b) Passive transport does not require the input of energy, and active transport does.

(c) Passive transport does require the input of energy, and active transport does not.

(d) Neither active nor passive transport requires the input of energy.

Try Something New

4 Picture a beaker divided by a semipermeable membrane like the one in Figure 4.6. Imagine that you put 5 grams of sugar in the left side of the beaker and 10 grams of sugar in the right side, and then add water to bring both sides to the same depth. Select the correct terms in the following description of what happens next: The water will move by osmosis to the (**left side / right side**) of the beaker because there are (**more / fewer**) molecules of solute and (**more / fewer**) molecules of water on that side.

5 When the chemical oligomycin is added to cells, the production of ATP is blocked. Which of the following processes would be most affected by oligomycin?

(a) simple diffusion

(b) facilitated diffusion

(c) active transport

(d) All of these processes would be equally affected by oligomycin.

(e) None of these processes would be affected by oligomycin.

6 Use the accompanying image to select the correct terms in the statements below each of the cells shown.

(a) This red blood cell is in a(n) (**hypertonic / hypotonic / isotonic**) solution.

(b) In this environment, normal red blood cells (**gain / lose / neither gain nor lose**) water.

(c) The total solute concentration outside the cell is (**lower than / higher than / equal to**) the total solute concentration inside it.

(d) This red blood cell is in a(n) (**hypertonic / hypotonic / isotonic**) solution.

(e) In this environment, normal red blood cells (**gain / lose / neither gain nor lose**) water.

(f) The total solute concentration outside the cell is (**lower than / higher than / equal to**) the total solute concentration inside it.

(g) This red blood cell is in a(n) (**hypertonic / hypotonic / isotonic**) solution.

(h) In this environment, normal red blood cells (**gain / lose / neither gain nor lose**) water.

(i) The total solute concentration outside the cell is (**lower than / higher than / equal to**) the total solute concentration inside it.

Leveling Up

7 **What do _you_ think?** Viruses display many of the characteristics of living organisms. In particular, they reproduce, creating new virus particles. During reproduction, viruses make copies of their genetic material, and some of the copies contain mutations that are beneficial to the virus. For example, HIV (human immunodeficiency virus), the virus that causes the disease AIDS (acquired immunodeficiency syndrome) mutates so often that its surface proteins change faster than we can develop antiviral drugs. New drug-resistant strains of HIV are appearing constantly. However, viruses can reproduce only after entering a living cell of an organism, because viruses hijack the cell's machinery and use it to produce new viruses. Where does that combination of characteristics place viruses on the scale of nonlife to life? Are viruses living organisms? Nonliving? If neither of those categories fits the properties of viruses, how should they be classified?

8 **What do _you_ think?** Did Venter create "new life" by inserting synthetic DNA copied from the genome of _Mycoplasma mycoides_ into a cell of _Mycoplasma capricolum_ from which the DNA had been removed? Be prepared to support your opinion in a class discussion.

9 **Doing science** Go to the National Science Foundation website (http://www.nsf.gov), and search for and read "Biologists Replicate Key Evolutionary Step in Life on Earth" (Press Release 12-009, January 16, 2012). State the hypothesis that the biologists were testing, describe their experimental design, and explain their results. Was their hypothesis supported? What experiment do you think they should try next? State a hypothesis that you would test using their multicellular yeast samples.

Thinking Multidimensionally

10 A mother with a sick child arrives at your office, and as the attending pediatrician, you need to diagnose and treat the patient. The baby's body appears floppy like a ragdoll, and he is having difficulty breathing and eating. Upon further analysis, you run a genetic test and conclude the child has Pompe disease.

Hypotonia (decreased muscle tone)

A.D.A.M.

Pompe disease affects heart and skeletal muscles and can lead to death. It results when lysosomes fail to break down long polymers of sugar (polysaccharides) into simple sugars like glucose.

a. Without glucose, what process will be hindered in this child?

b. What are the available treatment options?

c. To explain the disease to the mother, use beads and Play-Doh to create a model depicting how lysosomes normally function inside cells and how they fail to work properly in people diagnosed with Pompe disease.

11 After eating a peanut butter and jelly sandwich made from peanuts infected with *Salmonella*, hours later you experience symptoms of food poisoning, namely diarrhea.

a. Use your understanding of endocytosis to show how the bacteria get under the lining of the colon, and demonstrate the principles of osmosis by drawing what is happening on a cellular level inside your digestive tract.

b. Research and report on how peanuts in processing plants can get exposed to this virulent bacterium.

12 "It's alive!" is one of the most classic lines in horror movie history, uttered when the young scientist, Henry Frankenstein, pieced together human body parts, applied electricity, and built his monster.

a. Scientists would argue that Hollywood oversimplified the process to creating life. If you wanted to build a complex organism, what steps need to occur after successfully creating a single-celled eukaryote?

b. Research how cells normally differentiate to create specialized cell types.

c. What are the benefits to being a multicellular, as opposed to a unicellular, organism?

For more, visit digital.wwnorton.com/bionow2 for access to:

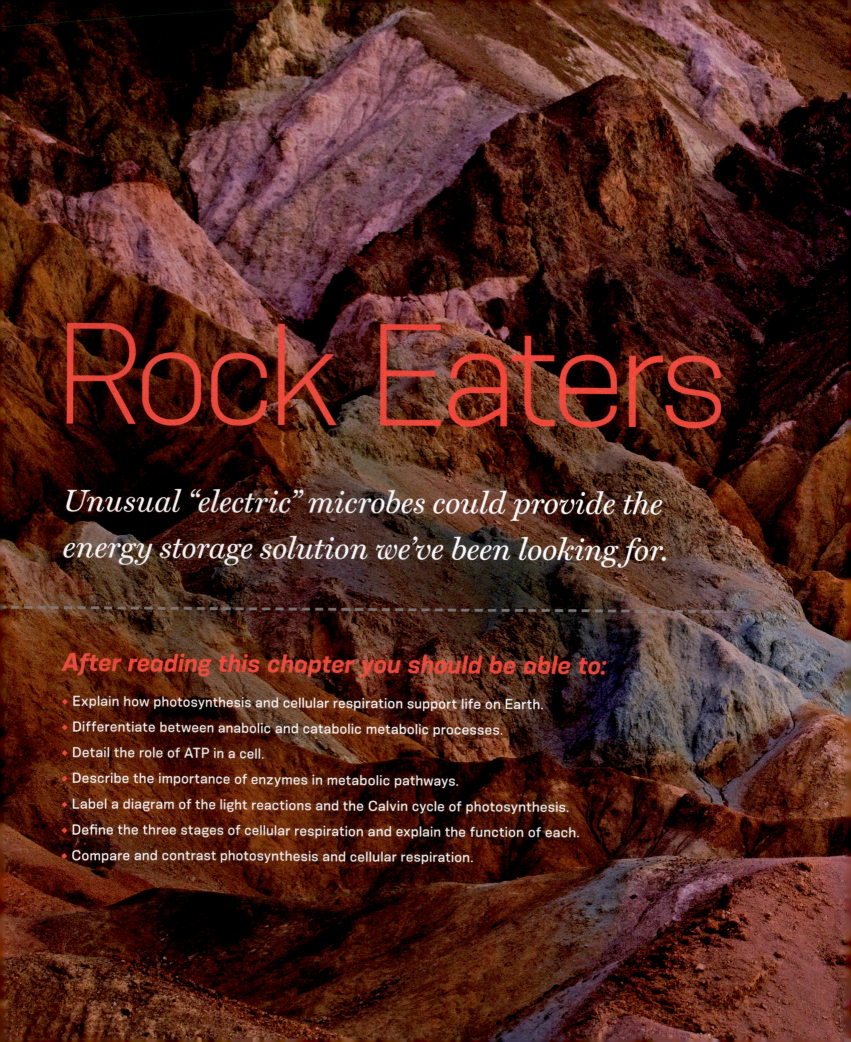

Rock Eaters

Unusual "electric" microbes could provide the energy storage solution we've been looking for.

After reading this chapter you should be able to:

- Explain how photosynthesis and cellular respiration support life on Earth.
- Differentiate between anabolic and catabolic metabolic processes.
- Detail the role of ATP in a cell.
- Describe the importance of enzymes in metabolic pathways.
- Label a diagram of the light reactions and the Calvin cycle of photosynthesis.
- Define the three stages of cellular respiration and explain the function of each.
- Compare and contrast photosynthesis and cellular respiration.

CHAPTER

05

HOW CELLS WORK

DRIVING QUESTION

How can some microbes eat or excrete electricity from rocks and metals and what technologies could benefit from their discovery?

- HS-LS1.C: Organization for Matter and Energy Flow in Organisms
- HS-LS2.B: Cycles of Matter and Energy Transfer in Ecosystems
- HS-ETS1.A: Defining and Delimiting Engineering Problems
- HS-ETS1.B: Developing Possible Solutions

Annie Rowe stands knee-deep in cold, clear salt water. Sweating, she pushes a shovel into the ground and lifts a mound of heavy mud. It's a sunny day on Santa Catalina Island, a rocky retreat off the coast of California, where Rowe has come to collect samples. Once the bucket is full, she sits on a rock and retrieves a mesh filter to sieve out any small invertebrates, such as snails and worms, from the sand. They are not part of the experiment.

A postdoctoral research fellow at the University of Southern California, Rowe pours the filtered sediment into one of her five 10-gallon aquariums. When each aquarium is about two-thirds full, she and a graduate student transport the tanks to a local laboratory on the island. There, they hook each aquarium up to a steady stream of seawater from the harbor.

To the untrained eye, the aquariums appear sterile, empty. In reality, both the water and sand in the tanks are teeming with microscopic organisms, or *microbes*, and Rowe is aiming to find a rare, strange one. Deep in the sand of each aquarium, she buries a metal electrode buzzing with electrical current to attract her quarry: a microbe that eats electricity (**Figure 5.1**).

"I went into microbiology because there's so much diversity in things that microbes can do," says Rowe. As a PhD student, Rowe began studying unusual microbes with the hope of someday putting them to use toward one of the greatest human challenges of our time: developing sources of alternative energy.

Humans rely heavily on fossil fuels such as oil and petroleum to power our cars, trains, planes, and more. Hundreds of millions of years ago, single-celled organisms used carbon dioxide and sunlight to make cell materials, which then became buried and concentrated into oil in Earth's crust. Burning that oil releases that ancient carbon dioxide back into the

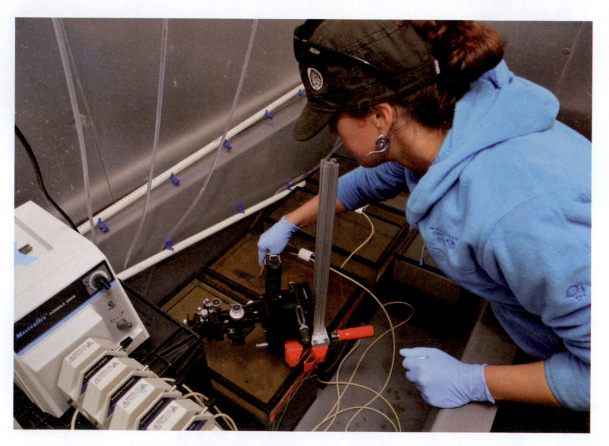

Figure 5.1

Annie Rowe burying electrodes in her sediment aquariums

atmosphere, and a consequence of this increase in carbon dioxide is that the surface of our planet has begun to warm. To prevent further climate change and lessen our dependence on fossil fuels—many of which we pay a premium to import from other countries—scientists and engineers have developed clean, renewable ways to capture energy, such as solar, wind, and water power.

Yet to switch from a fossil fuel–based economy to renewable energy technologies, we need efficient and inexpensive ways to store renewable energy. With the right chemical stabilizers, oil can be left in a barrel for decades, but a direct current of electricity from a wind farm or solar panels needs to be used immediately or fed into the electrical grid. Today's batteries aren't a good storage option; they are bulky, expensive, and not terribly efficient.

Now imagine the ability to siphon electricity from a wind farm or solar panel and transform it into a liquid fuel. That's where electricity-eating microbes come in.

Energy for Life

All living cells require energy. Organisms use energy for growth, reproduction, and defense, and to manufacture the many chemical compounds that make up living cells. They must obtain energy from the living or nonliving components of their environment, and at the very core of making and storing energy is the transfer of *electrons*—subatomic particles with a negative charge. Electrons play an essential role in electricity, magnetism, and many other physical phenomena that shape the world we know. The success of computers, solar cells, cell phones, and other devices is due to our ability to shape and control the flow of electrons.

But long before electrons flowed through computers, they moved through cells. The first law of thermodynamics says that energy cannot be created or destroyed, but it can be changed from one form to another. In other words, cells cannot create energy from nothing, so they must utilize one form of energy and change it to another form. However, organisms can't simply perform the biological equivalent of plugging into an electrical socket and sucking up electricity. That's because the cell membrane is an electrically neutral zone that prevents charged particles from sneaking through, including electrons. So, to move electrons into and out of a cell, living organisms attach electrons to molecules. Plants, for example, smuggle electrons into the cell via water molecules. Humans and other organisms obtain electrons from food (sugars, proteins, fats, etc.). As the Nobel Prize–winning physiologist Albert Szent-Györgyi reportedly said, "Life is nothing but an electron looking for a place to rest."

Cells use and store energy by transferring electrons among molecules via chemical reactions. Thousands of different types of chemical reactions are required to sustain life in even the simplest cell. The term **metabolism** describes all the chemical reactions that occur inside living cells, including those that store or release energy. Most chemical reactions in a cell occur in chains of linked events known as **metabolic pathways**. Metabolic pathways produce key biological molecules in a cell, including important chemical building blocks like amino acids and nucleotides.

Two metabolic pathways drive most of the life around us. The sun is the ultimate source of energy for most living organisms, and in the first process, known as **photosynthesis**, organisms capture energy from the sun and use it to create sugars from carbon dioxide and water (**Figure 5.2**). In this way, photosynthetic organisms such as plants transform light energy into chemical energy stored in the covalent bonds of sugar molecules. These sugar molecules, for example, glucose, fuel the cell's activities, and some are converted to fatty acids to help build cell membranes and to store energy for future needs.

The second important process is **cellular respiration**, a process reciprocal to photosynthesis. During cellular respiration, the cell breaks down sugars into usable energy

ANNETTE ROWE

Annette ("Annie") Rowe is a postdoctoral research associate at the University of Southern California, where she studies microbes that take up electrons from inorganic surfaces.

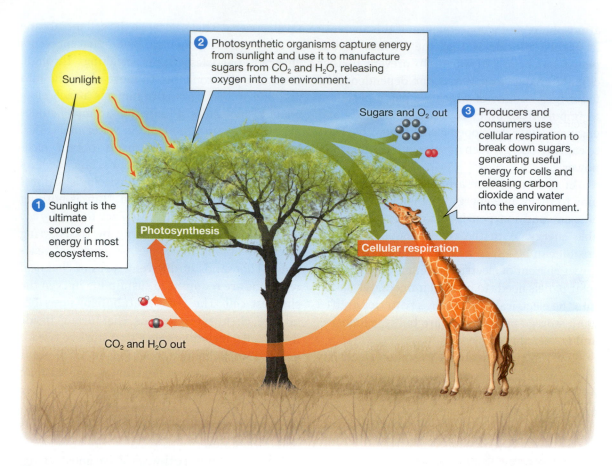

Figure 5.2

From sunlight to usable energy

Photosynthesis transforms sunlight into sugar molecules within the cell, releasing oxygen as a by-product. Without photosynthesis, we would not have access to the sun's energy, and we wouldn't have oxygen in our atmosphere to support life. In a complementary process, cellular respiration breaks down these sugar molecules, allowing organisms to access the energy stored in them.

Q1: Why is photosynthesis called "primary production"?

Q2: How does animal life depend on photosynthesis?

Q3: Explain how photosynthesis and cellular respiration are "complementary" processes.

(**Figure 5.2**). Plants break down the glucose made during photosynthesis, whereas nonphotosynthetic organisms, including animals, eat plants or other animals and break down sugars in that food to release energy. Electrons are transferred back and forth during these two metabolic processes. Metabolism, at its core, is a feat of juggling electrons.

Rowe spent a week on Catalina setting up her tanks and then let them sit for 3 months, untouched. During that time, her instruments began to detect evidence of life: the negative charge flowing through the electrodes in the sand began to steadily increase. Something was stealing the electrons, consuming more and more electricity.

An Unusual Pathway

Scientists in the 1980s first discovered two types of bacteria that expel—breathe, actually—electrons onto metal. Because it was assumed

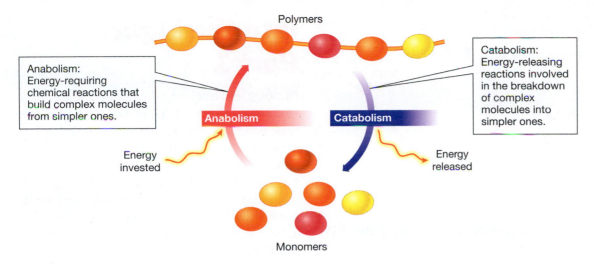

Polymers

Anabolism: Energy-requiring chemical reactions that build complex molecules from simpler ones.

Anabolism

Catabolism

Catabolism: Energy-releasing reactions involved in the breakdown of complex molecules into simpler ones.

Energy invested

Energy released

Monomers

Figure 5.3

Anabolism builds biomolecules; catabolism breaks them down

Metabolic reactions either build or break down molecules. Molecule-building reactions (anabolism) cost energy, and molecule-breakdown reactions (catabolism) release energy.

Q1: What source of energy would plants use for anabolic reactions? Would an animal use the same kind of energy?

Q2: What source of energy would plants release in catabolic reactions? Would an animal release the same kind of energy?

Q3: Create a mnemonic or jingle that helps you remember the difference between anabolism and catabolism.

that electrons couldn't cross the cell membrane, the discovery was a big surprise. For decades, however, the way the electron transfer worked remained a mystery. Then, in 2006, a scientific team identified a trio of proteins that form a bridge in the bacterial cell membrane, allowing the transfer of electrons from the inside to the outside of the cell.

Inspired by these "rock breathers" that expelled electrons onto a mineral surface, scientists began looking for "rock eaters." If nature produced microbes that spit out electrons, why not microbes that directly ingest electrons from minerals?

All living cells have two main types of metabolism: anabolism and catabolism (**Figure 5.3**). **Anabolism** refers to metabolic pathways that create complex molecules from simpler compounds—a process used by all cells to make the basic building blocks of the cell. These processes often require an energy input. Photosynthesis is an example of anabolism where the energy is provided by sunlight: as plants sit in

the noonday sun, the cells in their green leaves capture energy to make glucose from carbon dioxide; a complex molecule is assembled from simpler ones. **Catabolism**, unsurprisingly, refers to the opposite: metabolic pathways that release chemical energy in the process of breaking down complex molecules. Plant cells rely on cellular respiration, a catabolic process, to break down the glucose made by photosynthesis.

All of this metabolic activity requires energy, so cells need **energy carriers** to deliver usable "on-demand" energy. Every living cell uses **ATP** (adenosine triphosphate), a small, energy-rich organic molecule, to carry energy from one part of the cell to another. ATP powers almost all activities in the cell, such as moving molecules and ions in and out of the cell, sending nerve impulses, triggering muscle contractions, and moving organelles around inside the cell. In addition to powering a cell's activities, ATP fuels metabolic reactions and most other enzymatic reactions. If a cell exhausts its ATP supply, it will die.

Figure 5.4

Cells store and deliver energy using ATP

Loading a high-energy phosphate (P) on ADP transforms the otherwise sedate molecule into the live wire that is ATP. But turning ADP and phosphate groups into ATP—the universal energy currency of cells—takes metabolic energy.

Q1: Define ATP in your own words.

Q2: How is ATP involved in anabolism and catabolism? (*Hint:* Review Figure 5.3.)

Q3: Arsenic disrupts ATP production. Why would this characteristic cause it to be a potent poison?

Much of the usable energy in ATP is held in its energy-rich phosphate bonds (**Figure 5.4**). Energy is released when a molecule of ATP loses its terminal phosphate group, breaking into a molecule of **ADP** (<u>a</u>denosine <u>d</u>iphosphate) and a free phosphate. Converting ADP and phosphate back into ATP takes metabolic energy.

ATP is not the only energy carrier that cells rely on; NADPH, NADH, and FADH$_2$ are also energy carriers. Each one is a specialist in terms of the amount of energy it carries and the types of chemical reactions to which it supplies energy and from which it receives energy. ATP, because of its phosphate bonds, carries the most energy.

ALFRED SPORMANN

Alfred Spormann is a professor of civil and environmental engineering, and of chemical engineering, at Stanford University. He investigates novel microbial metabolism, including how microbes transport electrons between themselves and surfaces.

Into the Light, Part 1

In 2009, environmental engineers at Pennsylvania State University identified the first elusive rock eaters by exposing a mix of microbes to a negatively charged electrode (**Figure 5.5**). One of the species, *Methanobacterium palustre*, survived on the electricity, taking in electrons and using them in a metabolic pathway to convert carbon dioxide to methane. One news article summed up the experiment with the headline BUG EATS ELECTRICITY, FARTS BIOGAS. The Penn State engineers proposed, but were unable to prove, that the bacteria ("bugs") were ingesting naked electrons straight from the surface of the electrode.

A few years later, Alfred Spormann launched his own investigation into bacteria that seemed to be "eating" electrons. Spormann, a microbiologist at Stanford University, has long been interested in outliers, microbes with unusual metabolisms. Tall and thin with a throaty German accent, Spormann has spent more than 25 years studying such microbes, from those in the human gut that contribute to irritable bowel syndrome to bacteria that could be used to help decontaminate groundwater. "I'm really interested in exceptions to the rules, and in understanding the plasticity of microbial metabolism," says Spormann. Typically, the microbes he studies are **anaerobic**, growing in places with little or no oxygen. These cells don't need oxygen

Figure 5.5

A negatively charged electrode used to grow "rock eaters"

for metabolism; in fact, some are even poisoned by it.

Oxygen plays an important role in photosynthesis, considered by many to be the most important life process on Earth; it is the way our planet stores energy from the sun and produces oxygen for animals to breathe. In the cells of algae and all plants, photosynthesis takes place inside chloroplasts, organelles that look like green, oval gumballs when viewed under a light microscope (**Figure 5.6**). Chloroplasts contain an extensive network of structures called thylakoids, piled up like stacks of pancakes, that contain enzymes needed for photosynthesis. Also embedded in those membranes is a green pigment, called **chlorophyll**, that is specialized for absorbing light energy.

Photosynthesis takes place in two principal stages: the *light reactions* and the *light-independent reactions*, or the *Calvin cycle* (**Figure 5.7**). During the **light reactions**, chlorophyll molecules absorb energy from sunlight and use that energy to split water (**Figure 5.7**, left). The splitting of water produces oxygen gas (O_2)—the oxygen that we breathe—as a by-product that is released into the atmosphere. More important for photosynthesis, electrons and protons (H^+) from the light reactions are handed over to other molecules via the **electron transport chain**, an elaborate chain of chemical events that ultimately generates ATP and NADPH. The light reactions depend on protein complexes embedded in the chloroplast membrane, including photosystems I and II, and ATP synthase.

In the next stage, the **light-independent reactions**, or **Calvin cycle**—a series of chemical reactions—convert carbon dioxide (CO_2) into sugar, using energy delivered by ATP, and electrons and hydrogen ions donated by NADPH (**Figure 5.7**, right). Enzymes catalyze these reactions at each step; the enzyme needed in the first step—and the most abundant enzyme on the planet—is **rubisco**. This process is also known as **carbon fixation**. By capturing inorganic carbon atoms from CO_2 gas and converting them into glucose, the Calvin cycle reactions make carbon from the nonliving world available to the photosynthetic organisms and eventually to other living organisms, including us.

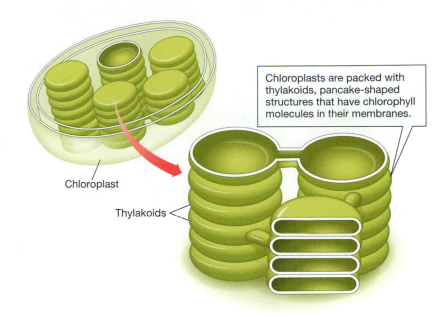

Chloroplasts are packed with thylakoids, pancake-shaped structures that have chlorophyll molecules in their membranes.

Chloroplast

Thylakoids

Figure 5.6

Chloroplasts are the site of photosynthesis in eukaryotes

Most plants and algae have multiple chloroplasts to contain their chlorophyll. Photosynthetic bacteria embed chlorophyll directly into the plasma membrane.

Q1: Is chlorophyll found only within chloroplasts?

Q2: What could be an advantage of concentrating chlorophyll molecules in the membranes of chloroplasts?

Q3: What is the advantage of having multiple chloroplasts per cell?

Catalyzing Reactions

In about 2014, Spormann's team began studying an "electric" microbe that lives in water, *Methanococcus maripaludis*, to see whether it was capable of directly ingesting electrons. It wasn't. The researchers found that the microbe actually excretes an enzyme onto the surface of an electrode to do its dirty work. Nearly all metabolic reactions are facilitated by enzymes. **Enzymes** are biological catalysts—molecules that speed up chemical reactions. Without the action of enzymes—most of which are proteins—metabolism would be extremely slow, and life as we know it could not exist. Enzymes work by positioning **substrates**—molecules that will react to form new products—in

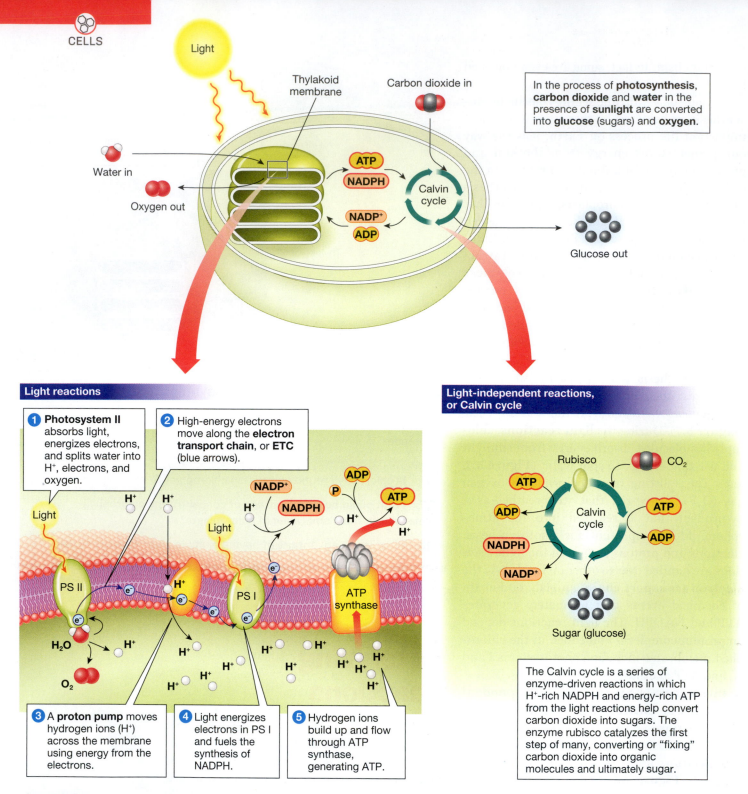

Light

Thylakoid membrane

Carbon dioxide in

In the process of **photosynthesis**, **carbon dioxide** and **water** in the presence of **sunlight** are converted into **glucose** (sugars) and **oxygen**.

Water in

Oxygen out

ATP
NADPH
Calvin cycle
NADP⁺
ADP

Glucose out

Light reactions

1 **Photosystem II** absorbs light, energizes electrons, and splits water into H⁺, electrons, and oxygen.

2 High-energy electrons move along the **electron transport chain**, or **ETC** (blue arrows).

Light

H⁺   H⁺   NADP⁺   P   ADP

Light   NADPH   ATP

H⁺   H⁺

PS II   H⁺   PS I   ATP synthase   H⁺

e⁻   e⁻   e⁻   e⁻   e⁻

H₂O   H⁺   H⁺   H⁺   H⁺   H⁺   H⁺
H⁺   H⁺   H⁺   H⁺
O₂   H⁺   H⁺   H⁺   H⁺

3 A **proton pump** moves hydrogen ions (H⁺) across the membrane using energy from the electrons.

4 Light energizes electrons in PS I and fuels the synthesis of NADPH.

5 Hydrogen ions build up and flow through ATP synthase, generating ATP.

Light-independent reactions, or Calvin cycle

Rubisco   CO₂

ATP

ADP   Calvin cycle   ATP

NADPH   ADP

NADP⁺

Sugar (glucose)

The Calvin cycle is a series of enzyme-driven reactions in which H⁺-rich NADPH and energy-rich ATP from the light reactions help convert carbon dioxide into sugars. The enzyme rubisco catalyzes the first step of many, converting or "fixing" carbon dioxide into organic molecules and ultimately sugar.

Figure 5.7

Photosynthesis occurs in two stages

The light reactions generate energy carriers; the light-independent Calvin cycle reactions create sugars. Rubisco contributes to fixing carbon during the Calvin cycle, but enzymes are critical throughout both stages of photosynthesis. ▶️

Q1: What is the source of the carbon dioxide used for photosynthesis?

Q2: Which products of the light reactions of photosynthesis does the Calvin cycle use?

Q3: What are the two major products of photosynthesis?

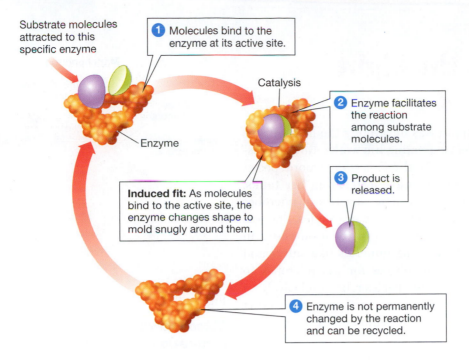

Substrate molecules attracted to this specific enzyme

1 Molecules bind to the enzyme at its active site.

Enzyme

Catalysis

2 Enzyme facilitates the reaction among substrate molecules.

Induced fit: As molecules bind to the active site, the enzyme changes shape to mold snugly around them.

3 Product is released.

4 Enzyme is not permanently changed by the reaction and can be recycled.

Figure 5.8

Enzymes are molecular matchmakers

Enzymes dramatically increase the rate of chemical reactions by positioning molecules so that they more easily form or break chemical bonds.

Q1: Why is it important that enzymes are not permanently altered when they bind with substrate molecules?

Q2: How would a higher temperature or higher salt concentration make it more difficult for an enzyme to function effectively?

Q3: If a cell was unable to produce a particular enzyme necessary for a metabolic pathway, describe how the absence of that enzyme would affect the cell.

an orientation that favors the making or breaking of chemical bonds (**Figure 5.8**).

Each enzyme binds only to a specific substrate or substrates and catalyzes a specific chemical reaction, such as rubisco catalyzing the first step of the Calvin cycle. An enzyme's function is based on its chemical characteristics and the three-dimensional shape of its **active site**—the location within the enzyme where substrates bind. When molecules bind to the active site, the enzyme changes shape—a process called **induced fit**. The enzyme's shape, and therefore its activity, can be affected by temperature, pH, and salt concentration. Because the enzyme's active site is not permanently changed as reactions are catalyzed, an enzyme, like rubisco, is used over and over.

The enzyme that Spormann identified was unusual because it was excreted to the exterior of the cell. Most enzymes work on the inside of cells. A multistep metabolic pathway can proceed rapidly and efficiently because the required enzymes are physically close together and the products of one enzyme-catalyzed chemical reaction serve as the basis for the next reaction in the series, as is especially true for metabolic pathway reactions.

In the case of *M. maripaludis*, the excreted enzyme's role was to grab an electron from the metal, pair it with a proton from water, and create a hydrogen atom—a familiar food for microbes, and easily passed through the cell membrane. "They [*M. maripaludis* bacteria] found a way to produce a compound that is easily metabolized by the cells," says Spormann. And although this particular microbe wasn't eating naked electrons, Spormann has since isolated a microbe that does directly take up electrons. His team has yet to publish the details.

Into the Light, Part 2

As Spormann was unraveling the mechanisms of how electron-eating life works, Rowe's prospecting trip to Catalina uncovered more species of rock eaters (**Figure 5.9**). From her tanks, Rowe identified 30 new varieties of microbes sucking electrons from the electrodes. She was even able to grow a few of them on plates in the laboratory, using minerals like sulfur and iron as electron sources for the microbes to munch. Not all the species relied exclusively on electrons as a food source; that is, they didn't just live off mineral sediments. Like eukaryotic cells, most microbial cells use the traditional route of cellular respiration for obtaining energy, converting sugars into energy through catabolic reactions.

During cellular respiration, the carbon-carbon bonds in glucose molecules are broken,

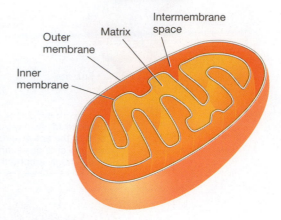

Mitochondrion

Figure 5.10

Mitochondria are the site of cellular respiration in eukaryotes

and each carbon atom is released into the environment in a molecule of CO_2, with water (H_2O) as a by-product. Thus, cellular respiration is the reciprocal process of photosynthesis, as illustrated earlier, in **Figure 5.2**. In most eukaryotes, cellular respiration is a multistep process that occurs in the cell's mitochondria (**Figure 5.10**). Cellular respiration is the main way that animal cells, which do not perform photosynthesis, obtain energy. We digest glucose from food sources; break it down into carbon dioxide, hydrogen, and electrons; and then process those captured electrons for energy.

Simply put, cellular respiration consists of three stages: glycolysis, the Krebs cycle, and oxidative phosphorylation (**Figure 5.11**). The first stage, **glycolysis**, takes place in the cytoplasm of the cell. During glycolysis, sugars (mainly glucose) are split to make a three-carbon compound called pyruvate. This process results in two useful molecules of ATP and two molecules of NADH for each glucose molecule that is split (**Figure 5.11**, top left). In other words, glycolysis converts some of the chemical energy of glucose into the chemical energy of NADH and ATP.

From an evolutionary standpoint, glycolysis was probably the earliest means of producing ATP from food molecules, and it is still the primary means of energy production in many prokaryotes, including some of Rowe's microbes. However, because glucose is only partially broken down through this process, the

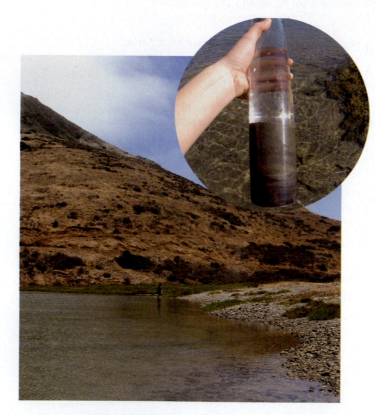

Figure 5.9

Sediment containing rock eaters, collected off the coast of Santa Catalina Island

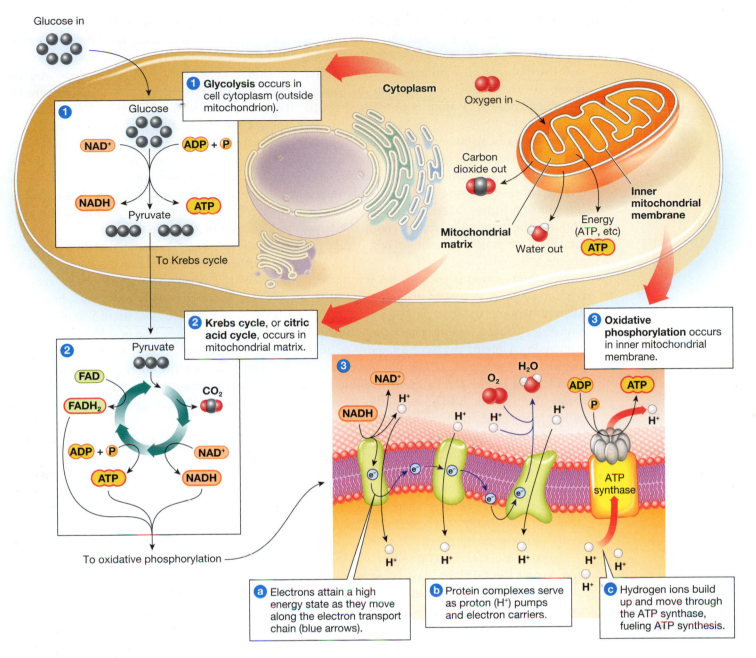

Figure 5.11

Cellular respiration is highly efficient in eukaryotes

In glycolysis, each six-carbon glucose molecule is converted into two molecules of pyruvate. The next two stages of cellular respiration require oxygen. The Krebs cycle (bottom left) releases carbon dioxide and generates high-energy molecules. Oxidative phosphorylation (bottom right), the last stage in cellular respiration, produces more ATP than any other metabolic pathway. ▶

Q1: What are the products of cellular respiration?

Q2: Considering the inputs and products of each process, why is cellular respiration considered the reciprocal process to photosynthesis?

Q3: Which of the three stages of cellular respiration—glycolysis, the Krebs cycle, or oxidative phosphorylation—could organisms have used 4 billion years ago, before photosynthesis by cyanobacteria released oxygen into the atmosphere?

energy yield from glycolysis is small. For most eukaryotes, glycolysis is just the first step in extracting energy from sugars, and the second and third stages of cellular respiration, which occur in the mitochondria, help to generate much more ATP than is possible through glycolysis alone.

Glycolysis is an anaerobic process; that is, it does not require oxygen. Cells experiencing low oxygen levels—such as our muscle cells during intense exercise or cancer cells in the interior of a tumor mass—use glycolysis alone because there is not enough oxygen available for the later stages of cellular respiration to proceed.

But when oxygen is present, eukaryotic cells rely on all three stages of cellular respiration. After pyruvate is made in the cytoplasm during glycolysis, it enters the mitochondria, where it is broken down during the second stage of cellular respiration: a sequence of enzyme-driven reactions known as the **Krebs cycle** or **citric acid cycle**. The carbon backbone of the pyruvate molecule is taken apart, releasing carbon dioxide (**Figure 5.11**, bottom left). The breakdown of carbon backbones by the Krebs cycle produces a large bounty of energy carriers, including ATP, NADH, and $FADH_2$. Essentially, during the Krebs cycle the remaining chemical energy of glucose is completely converted to the chemical energy of these energy carriers.

The largest output of ATP, however, is generated during the third and last stage of cellular respiration: **oxidative phosphorylation**. The electrons and hydrogen atoms are removed from NADH and $FADH_2$ and handed over to molecular O_2 through an electron transport chain, creating water (H_2O). In the process, a large amount of ATP is generated (**Figure 5.11**, bottom right). During oxidative phosphorylation, the chemical energy of NADH and $FADH_2$ is converted into the chemical energy of ATP. In fact, oxidative phosphorylation can generate 15 times as much ATP as can glycolysis alone.

ATP production in mitochondria is crucially dependent on oxygen; that is, the Krebs cycle and oxidative phosphorylation are strictly **aerobic** processes. Highly aerobic tissues, like your muscles, have high concentrations of mitochondria and a rich blood supply to deliver the large amounts of O_2 needed to support their activity. Even so, intense exercise exhausts this oxygen supply, at which point your cells switch to glycolysis, as mentioned already. But how do the cells extract sufficient energy via glycolysis alone?

Under anaerobic conditions, these muscle cells use a fermentation pathway. **Fermentation** begins with glycolysis, followed by a special set of reactions whose only role is to help perpetuate glycolysis. This process enables organisms to generate ATP through glycolysis alone, when aerobic ATP production is constrained by low oxygen levels (**Figure 5.12**). If glycolysis cannot be perpetuated, the cells will run out of ATP and die.

Many species of bacteria that live in places like oxygen-deficient swamps, sewage, or deep layers of soil never use oxygen or are actually poisoned by oxygen. Similar to rock-breathing microbes, many microbes can utilize other electron acceptors, such as iron or sulfur compounds. When these run out, anaerobic organisms use the fermentation pathway to regenerate molecules needed to perpetuate glycolysis, and exclusively generate ATP through glycolysis without ever using oxygen.

Most eukaryotic cells can perform both aerobic and anaerobic ATP production, depending on the oxygen levels in their environment. Because the Krebs cycle and oxidative phosphorylation produce magnitudes more ATP than glycolysis and the fermentation pathway yield, these cells always use oxygen and aerobic respiration when levels permit.

Together, photosynthesis and cellular respiration enable cells to store and utilize energy from the sun. They are two sides of the same coin; the products of one reaction are the ingredients for the other. Photosynthesis, an anabolic process, requires energy (sunlight) and CO_2, and releases O_2 and glucose. Cellular respiration, a catabolic process, requires O_2 and glucose, and releases CO_2 and energy.

Unlike the situation with photosynthesis and cellular respiration—metabolic processes that have been well studied for decades—researchers don't yet know the details of how rock eaters store and use energy. "There isn't really any pathway worked out yet," says Rowe. "Truthfully, it's amazing that it works." She's eager to find out how. Of the 30 species Rowe has isolated, she is currently focused on 8, working to develop tools to identify which enzymes are involved in their metabolism.

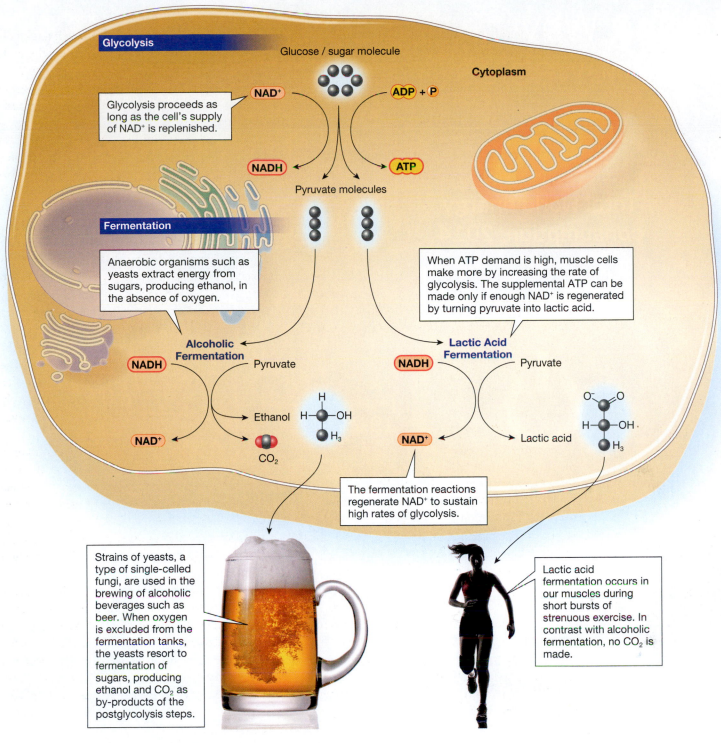

Figure 5.12

Fermentation produces energy in the absence of oxygen

When the oxygen supply is inadequate to support ATP production through cellular respiration, fermentation enables ATP production through glycolysis alone.

Q1: Which product released by fermentation accounts for the bubbles in beer?

Q2: Bakers of yeast breads also rely on fermentation, allowing bread to "rise" before baking. Describe what is occurring with the yeast as the bread rises.

Q3: Explain in your own words why lactic acid builds up in your muscles during strenuous physical activity.

Making Way for Renewables

Microbes that "eat" electricity may someday provide an efficient way to store renewable energy. That's good, because we need such storage technologies as soon as possible. Technologies that produce renewable energy are no longer "alternative"—they're mainstream and compete financially with fossil fuel producers. From solar power to biofuels, the fuel economy is rapidly changing toward more sustainable solutions.

Assessment available in smartwork5

Total world energy consumption, 2014

Total renewables
19.2%

Fossil fuels
78.3%

Nuclear power
2.5%

Traditional biomass
(wood and agricultural fuels)
8.9%

Modern renewables
10.3%

Renewable heat energy from biomass, geothermal, and solar technologies
4.2%

Hydropower
3.9%

Renewable power from wind, biomass, geothermal, and solar technologies
1.4%

Biofuels
0.8%

Bacterial Batteries

Because cell membranes are insulated, researchers long believed that cells could not transport electrons directly across their membranes, but now it appears they can. In 2010, in fact, Rowe's postdoctoral adviser, Mohamed El-Naggar, a pioneer in the rock breather field, found that the tiny hairlike structures called pili (singular "pilus") on some bacteria are conductive (**Figure 5.13**)—just as conductive, in fact, as silicon, the basis of most electronics. One single nano-sized pilus, which researchers refer to as a bacterial "nanowire," can transport about 106 electrons per second—enough to sustain the respiration of a whole cell.

Today, Rowe, Spormann, and others are focusing on how to use these unique bacteria in "bacterial batteries." Energy absorbed during the day from a solar panel currently has to be either instantly used or siphoned off into the electrical grid. But what if it could be passed into a tank of bacteria instead, where the microbes would gobble up the electricity and store it in chemical bonds in the form of methane or another natural gas? In that case the gas could be stored until needed.

Figure 5.13

The rock eater *Methanococcus maripaludis*, with electrically conductive hairlike pili

"If you could convert electrical energy, through microbial processes, to a liquid fuel, that could be a really valuable way of storing energy," says Rowe. This electricity-to-fuel scenario could also be run in reverse, using microbes to digest a fuel source such as raw sewage and convert it into electricity.

There's also the intriguing scenario that someday our electronic devices will be powered by microorganisms that manipulate the flow of electrons. In this rapidly evolving field, the possibilities are endless. And so the hunt for rock eaters continues. "We don't know how many organisms do or don't do it," adds Spormann. "It's entirely possible that rock eating is more widespread than previously thought. We just don't know because it's so new."

REVIEWING THE SCIENCE

- The sun is the source of energy fueling most living organisms. Plants, algae, and some bacteria gain energy from their environment through **photosynthesis**. Most organisms use **cellular respiration** to extract usable energy from sugar molecules. In chemical terms, photosynthesis is the opposite of cellular respiration.

- All the many chemical reactions involved in the capture, storage, and use of energy by living organisms are collectively known as **metabolism**. Energy-releasing breakdown reactions like cellular respiration are **catabolism**; energy-requiring synthesis reactions like photosynthesis are **anabolism**.

- A **metabolic pathway** is a multistep sequence of chemical reactions, with each step catalyzed by a different enzyme.

- **Energy carriers** store energy and deliver it for cellular activities. **ATP** is found in all cells and is the most commonly used energy carrier.

- Photosynthesis takes place in chloroplasts and occurs in two stages. In the **light reactions**, energy is absorbed using pigment molecules that include

chlorophyll as electrons flow along the **electron transport chain**. The light reactions create the energy carriers ATP and NADPH, splitting water molecules and releasing oxygen gas.

- The energy carriers are then used to convert carbon dioxide into sugar molecules during the **light-independent reactions**, or **Calvin cycle**. In the first of the reactions, the enzyme **rubisco** catalyzes the fixation of CO_2.

- **Enzymes** are biological catalysts, usually small proteins, that speed up chemical reactions. An enzyme's function is based on its chemical characteristics and the three-dimensional shape of its **active site**.

- Cellular respiration occurs in three stages: (1) **glycolysis**, which yields small amounts of ATP and NADH; (2) the **Krebs cycle**, which releases carbon dioxide and produces NADH, $FADH_2$, and ATP; and (3) **oxidative phosphorylation**, which generates many molecules of ATP.

- In the absence of oxygen, **fermentation** breaks down the products of glycolysis into alcohol or lactic acid.

THE QUESTIONS

Challenge Yourself

1 The Calvin cycle reactions are sometimes called the "light-independent reactions" or "dark reactions" to contrast them with the light (or light-dependent) reactions. Can the Calvin cycle be sustained in algae that are kept in total darkness for several days? Why or why not?

2 After strenuous exercise, you may notice that your muscles burn and feel sore the next day. Which statement best explains this phenomenon?

(a) Proteins in muscle cells are being digested to provide energy.

(b) Carbon dioxide is building up in muscle cells and changing their pH.

(c) Spontaneous combustion occurs during strenuous exercise, so avoid it at all costs!

(d) ATP is accumulating in muscle cells, causing a burning sensation.

(e) Without adequate oxygen, muscle cells are fermenting pyruvate into lactic acid.

3 The graph presented here depicts the activation energy, or the amount of energy needed for a reaction to proceed, with and without an enzyme.

a. Which reaction requires more energy to proceed—the one with or without an enzyme? How do you know?

b. Is this reaction anabolic or catabolic? How do you know?

Try Something New

4 In 2012, an Illinois man was killed by cyanide poisoning after he won a million dollars in the lottery. Cyanide is a lethal poison because it interferes with the electron transport chain in mitochondria. What effect would cyanide have on cellular respiration?

(a) Glycolysis, the Krebs cycle, and oxidative phosphorylation would all be inhibited.

(b) The Krebs cycle would be inhibited, but oxidative phosphorylation would not.

(c) Oxidative phosphorylation would be inhibited.

(d) Glycolysis, the Krebs cycle, and oxidative phosphorylation would all be stimulated.

5 Plants in the genus *Ephedra* have been harvested for their active substance ephedrine for centuries. Ephedrine is used to reduce the symptoms of bronchitis and asthma, as a stimulant and study aid, and as an appetite suppressant. It is also the main ingredient in the illegal production of methamphetamine. Since 2006, the sale of ephedrine and related substances has been limited and monitored in the United States. One effect of ingesting ephedrine is greatly increased metabolism, which has been known to kill users of ephedrine. How might an increased metabolic rate cause death?

6 You have been transported into the future, where a nanosensor can be inserted into a living cell and subsequently travel into any organelle. The sensor relays information stating that it has lodged itself in a compartment of the mitochondrion, where there is a concentration gradient of hydrogen ions (H^+) and levels of ATP production are high. From what you know about cellular respiration in mitochondria, where is the sensor lodged?

(a) inner mitochondrial membrane

(b) mitochondrial matrix

(c) chloroplast stroma

(d) cell cytoplasm

(e) plasma membrane

Leveling Up

7 **What do *you* think?** What would happen if a virus destroyed all photosynthetic organisms on Earth?

8 *Write Now* **biology: calorie-burning fat** Your friend has e-mailed you a link (see below) to a *New York Times* article on "brown" fat. He has been trying to lose weight and wants to know whether you think it would be a good idea for him to spend more time in the cold, rather than continuing to exercise regularly. He is also interested that the article mentions ephedrine's ability to stimulate brown fat, and he asks if you think he should begin to take ephedrine supplements. Compose a reply to your friend addressing the following points (using one to two paragraphs for each one). [*Note:* You may need to do further reading to answer (b) and (c).]

To research your answer, begin with the following *New York Times* article, published on April 8, 2009: "Calorie-Burning Fat? Studies Say You Have It" (http://www.nytimes.com/2009/04/09/health/research/09fat.html). Consult and reference at least two additional sources in your answer.

a. Explain in detail how brown fat burns calories when someone is chilled.

b. Explain how ephedrine affects metabolism and what its possible side effects are.

c. Contrast (a) and (b) with the effect of exercise on metabolism, both in the short term and in the longer term by increasing muscle mass.

d. In your final paragraph, advise your friend as to whether he should begin spending time in the cold to increase weight loss or take ephedrine supplements, and whether he should continue to exercise regularly. Justify your opinion with data and logic.

9 Is it science? Wouldn't it be amazing if humans could manufacture their own food through photosynthesis like plants do? Use Internet research to determine whether this idea is real science or science fiction. Is there evidence that animals can use the sun's energy directly and/or perform photosynthesis in nature? In the lab? If so, how would that capability affect life as we know it?

Thinking Multidimensionally

10 Watch the ending of the Ironman triathlon footage from 1997 by searching YouTube for "Ironman triathlon 1997."

"Bonking" is the term for what these women experienced as a result of depleted glycogen reserves (a long polymer of glucose) in their muscles. Without glucose available for cellular respiration, ATP production drastically declines and may even come to a halt.

a. You are a sports newscaster who covered this race from start to finish. Now that they've crossed the finish line, what questions do you have for these athletes?

b. Research the role ATP plays in muscle contractions, and describe how ATP powers cellular work.

c. Explain the mechanisms of how both fermentation and muscle fatigue were also likely contributors to this dramatic finish.

11 You have been asked to prepare a dessert for a holiday gathering, so you search Pinterest for recipe ideas and decide to make a Jell-O dessert like the one pictured above. You mix fresh pineapple, watermelon, kiwi, strawberries, and grapes with the Jell-O and then add the toppings. After allowing it to cool for 12 hours, you open the fridge and are shocked to see a slushy dessert that hasn't solidified! What happened? Describe the science behind why this dessert bombed in the style of an online post for the Pinterest Fail website (http://pinterestfail.com).

For more, visit digital.wwnorton.com/bionow2 for access to:

Toxic Plastic

Two ruined experiments expose the health risks of chemicals in everyday products.

After reading this chapter you should be able to:

- Label a figure of the major stages of the cell cycle, and explain the processes that occur during each of these stages.
- Compare and contrast cell division by binary fission, mitosis, and meiosis.
- Distinguish between sister chromatids and homologous chromosomes.
- Diagram, using the appropriate terms, the steps in mitosis and in meiosis.
- Explain the importance of the checkpoints in the cell cycle and the consequences of bypassing those checkpoints.
- Identify the ways in which meiosis and fertilization together produce genetically diverse offspring.

CELLS

CHAPTER
06

CELL DIVISION

DRIVING QUESTION

What evidence is there that some chemicals in plastics can impact cell division?

- HS-LS1.B: Growth and Development of Organisms
- HS-LS3.B: Variation of Traits

It began as a run-of-the-mill experiment. In 1989, biologists Ana Soto and Carlos Sonnenschein at Tufts University in Massachusetts were studying how the hormone estrogen regulates the growth of cells in the female reproductive system. For their research, the duo developed an experimental setup consisting of human breast tumor cells growing in plastic bottles called cell culture flasks (**Figure 6.1**, top). The flasks were filled with a liquid containing a factor that prevented the cells from multiplying. But when estrogen was added to the flasks, the cells grew. When estrogen was absent, they didn't.

One day, suddenly and surprisingly, cells in the flasks began growing even when estrogen hadn't been added. "What had worked for years didn't work anymore," says Sonnenschein. The two scientists immediately stopped their experiments and began searching for the cause. "It smacked of contamination," recalls Soto, as if estrogen had somehow gotten into the flasks. But after weeks of searching, Soto and Sonnenschein still couldn't identify a source of contamination. They became so paranoid that they suspected someone was entering the lab at night and secretly dripping estrogen into their flasks.

Almost 10 years later, in August 1998, geneticist Patricia Hunt at Case Western Reserve University in Ohio stared dumbfounded at another experimental anomaly. Hunt was studying why older women are at increased risk of having children with chromosomal abnormalities, like Down syndrome, in which an individual has 47 **chromosomes**—the tiny, stringlike structures in cells that contain genes—instead of the usual 46 (see Chapter 8 for more on chromosomes). She hypothesized that hormone levels have an impact on that increased risk. To test her hypothesis, Hunt

Figure 6.1

Two research models, two unusual results

Breast tumor cells (top) and mouse oocytes (bottom) grew in unexpected ways, leading scientists to study them more deeply.

raised groups of mice with varying levels of hormones and checked their egg cells for abnormal numbers of chromosomes (**Figure 6.1**, bottom).

The experiment was almost complete when Hunt went in to check on the control mice one last time. A control population is a necessary baseline for comparison against an experimental population; in this case, the control was a group of healthy mice whose hormone levels had not been altered. Using a light microscope, Hunt examined mouse oocytes—precursors to egg cells—at the moment just before the cells

ANA SOTO AND CARLOS SONNENSCHEIN

Ana Soto is a biologist at Tufts University who studies how cell division is affected by sex steroids. Carlos Sonnenschein, also a professor at Tufts University, studies chemicals that disrupt hormone systems in mammals.

undergo a specialized type of cell division that produces the eggs. She was shocked. The cells were a mess, the chromosomes scrambled. A whopping 40 percent of the resulting eggs had chromosomal defects. "The controls were completely bonkers," says Hunt. "One week they were fine, the next week they weren't. That's when we knew something was going on."

Like Soto and Sonnenschein, Hunt scrutinized every method and every piece of lab equipment used in the experiment, looking for the culprit. But as weeks passed, she couldn't figure out what had ruined her experiment. Soto, Sonnenschein, and Hunt didn't know it at the time, but their botched experiments would change the course of their scientific careers forever. The three would spend the next decade identifying, tracking, and investigating a toxic chemical that pervades our environment.

Divide and Conquer

In Soto and Sonnenschein's experiment, the breast cells were multiplying under the wrong circumstances. In Hunt's experiment, the mouse oocytes were not producing egg cells correctly. In both cases, something was interfering with the ability of the cells to divide—disrupting the cell cycle. The **cell cycle** is a sequence of events that make up the life of a typical eukaryotic cell, from the moment of its origin to the time it divides to produce two daughter cells. The time it takes to complete a cell cycle depends on the organism, the type of cell, and the life stage of the organism. Human cells, for example, typically have a 24-hour cell cycle, while Hunt's mouse oocytes can take days to complete a cycle. Some fly embryos, on the other hand, have cell cycles that are only 8 minutes long.

There are two main stages in the cell cycle of eukaryotes: *interphase* and *cell division*—each marked by distinctive cell activities (**Figure 6.2**). **Interphase** is the longest stage of the cell cycle; most cells spend 90 percent or more of their life span in interphase. During this phase, the cell takes in nutrients, manufactures proteins and other substances, expands in size, and conducts special functions depending on the cell type. Neurons in the brain, for example,

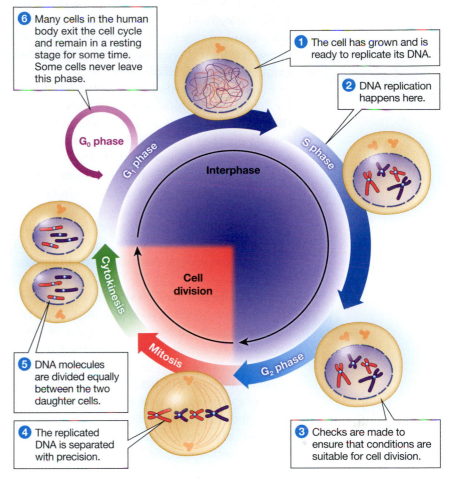

6 Many cells in the human body exit the cell cycle and remain in a resting stage for some time. Some cells never leave this phase.

1 The cell has grown and is ready to replicate its DNA.

2 DNA replication happens here.

3 Checks are made to ensure that conditions are suitable for cell division.

4 The replicated DNA is separated with precision.

5 DNA molecules are divided equally between the two daughter cells.

G_0 phase

G_1 phase

S phase

G_2 phase

Interphase

Cell division

Cytokinesis

Mitosis

Figure 6.2

The cell cycle

The eukaryotic cell cycle consists of two major stages: interphase and cell division.

Q1: When is DNA replicated during the cell cycle?

Q2: When in the cell cycle does DNA separate into the two genetically identical daughter cells?

Q3: If a cell is not destined to separate into daughter cells, what phase does it enter? Is this part of the cell cycle?

transmit electrical impulses, while beta cells in the pancreas release insulin.

Interphase can be divided into three main intervals: G_1, S, and G_2. The **G_1 phase** (for "gap 1") is the first phase in the life of a newborn cell. In cells that are destined to divide, preparations for cell division begin during the **S phase**

("S" stands for "synthesis"). A critical event during the S phase is the copying, or *replication*, of all the cell's DNA molecules, which contain the organism's genetic information. The **G$_2$ phase** (for "gap 2") occurs after the S phase but before the start of cell division.

Early cell biologists bestowed the term "gap" on the G$_1$ and G$_2$ phases because they believed those phases to be less significant periods in the life of a cell than are the S phase and cell division. We now know that the "gap" phases are often periods of growth during which both the size of the cell and its protein content increase. Furthermore, each "gap" phase serves as a checkpoint to prepare the cell for the phase immediately following it, ensuring that the cell cycle does not progress unless all conditions are suitable.

Cell division is the last stage in the cell cycle of an individual cell. As cell division begins, the cell contains twice the usual amount of DNA because of DNA replication during the S phase.

Not all cells complete the cell cycle. Many types of cells—neurons and beta cells, for example— become specialized shortly after entering G$_1$, and they pull out of the cell cycle to enter a nondividing state called the G$_0$ phase. The **G$_0$ phase** can last for a period ranging from a few days to the lifetime of the organism.

Cells begin cell division for two basic reasons: (1) to reproduce and (2) to grow and repair a multicellular organism. Most single-celled organisms use cell division to produce offspring through **asexual reproduction**. Asexual reproduction generates *clones*, offspring that are genetically identical to the parent.

Cell division occurs in all living organisms (eukaryotes and prokaryotes) and involves copying the DNA in the parent cell, then delivering one copy of it to each of two daughter cells. Most prokaryotes carry their genetic material in just one loop of DNA. These cells reproduce through **binary fission**, a type of cell division in which a

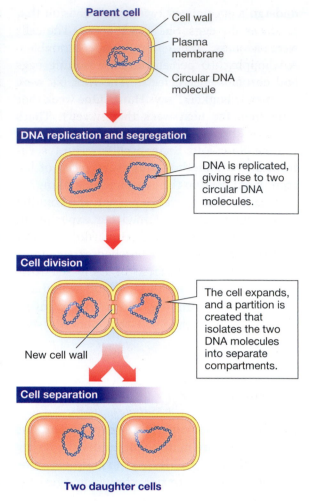

Figure 6.3

Cell division in a prokaryote

Many prokaryotes reproduce asexually in a type of cell division known as binary fission.

Q1: Name one similarity between cell division in prokaryotes and eukaryotes.

Q2: Why is binary fission referred to as "asexual reproduction"?

Q3: Name one difference between cell division in prokaryotes and eukaryotes.

PATRICIA HUNT

Patricia Hunt is a reproductive biologist at Washington State University in Pullman, Washington, who studies meiosis in mammalian cells.

cell simply copies the circular chromosome and each daughter cell receives one copy of the DNA loop—resulting in two cells that are genetically identical (**Figure 6.3**).

Soto, Sonnenschein, and Hunt, however, were studying eukaryotic cells, where cell division is more complicated than binary fission. Eukaryotic DNA forms multiple, distinct linear chromosomes wrapped around proteins and coiled into fibers that have to be unwound, replicated, and equally distributed between the two daughter cells. In addition to the multiple-chromosome problem, eukaryotic DNA lies inside a nucleus, enclosed by a double layer of membranes that make up the nuclear envelope. In most eukaryotes, the nuclear envelope must be disassembled in the dividing cell and then reassembled in each of the daughter cells toward the end of cell division. To make things even more complicated, eukaryotic cells undergo two types of division, depending on cell type: asexual reproduction through *mitosis*, and sexual reproduction—the production of sperm and eggs—through *meiosis*.

Trade Secret

Back in Massachusetts, Soto and Sonnenschein spent 4 months trying to figure out why their experiment had stopped working—how unknown estrogen was getting into their cell culture flasks and causing the cells to divide. By trial and error, they determined that a compound seemed to be shedding from the walls of the plastic tubes in which they stored the liquid being added to the cell culture flasks.

They called the tube manufacturer, who confirmed that an ingredient was added to make the tubes more impact resistant. But the company refused to reveal the identity of the "trade secret" ingredient. Soto and Sonnenschein spent a year purifying the secret ingredient and finally identified a compound called nonylphenol, a chemical used to make detergents and hard plastics. The reason for all their problems became clear: nonylphenol mimics the action of estrogen.

Like estrogen, nonylphenol activates **mitotic division**, a type of cell division that generates two genetically identical daughter cells from a single parent cell in eukaryotes. Mitotic division consists of two steps that occur after interphase: *mitosis* and *cytokinesis*. The first step, **mitosis**, refers to the division of the copied chromosomes in the nucleus. Mitosis is divided into four main

Cancer: Uncontrolled Cell Division

Cancer accounts for almost 600,000 deaths in the United States each year—nearly one in every four deaths. Only heart disease kills more people. Over the course of a lifetime, an American male has a nearly one in two chance of being diagnosed with cancer; American women fare slightly better, with a one in three chance of developing cancer. There are more than 200 different types of cancer, but the big four—lung, prostate, breast, and colon cancers—combine to account for about half of all cancers. More than 15 million Americans alive today have been diagnosed with cancer and are either in remission or undergoing treatment. The National Cancer Institute estimates that the collective price tag for the various forms of cancer is more than $100 billion per year.

Every cancer begins with a single rogue cell that starts dividing without the checkpoints of a normal cell. This runaway cell division rapidly creates a cell mass known as a **tumor**. Tumors that remain confined to one site are **benign**. Because benign tumors can usually be surgically removed, they are generally not a threat to the patient's survival. However, an actively growing benign tumor is like a cancer-in-training. Because these tumor cells are not subject to the monitoring that occurs at checkpoints during the cell cycle of a normal cell, their descendants can become increasingly abnormal—changing shape, increasing in size, and ultimately ceasing normal cell functions. As tumor cells progress toward a cancerous state, they begin secreting substances that cause **angiogenesis**, the formation of new blood vessels. The resulting increase in blood supply to the tumor is important for delivering nutrients to it and whisking waste away from it, allowing the tumor to grow larger.

Most cells in the adult animal body are firmly anchored in one place and will stop dividing if they are detached from their surroundings—a phenomenon known as **anchorage dependence**. But some tumor cells may acquire anchorage *in*dependence, the ability to divide even when released from their attachment sites. When tumor cells gain anchorage independence and start invading other tissues, they are transformed into **cancer cells**, also known as **malignant cells**. Cancer cells may break loose from their attachment sites and enter blood or lymph vessels to emerge in distant locations throughout the body, where they form new tumors. The spread of a disease from one organ to another is known as **metastasis**. Metastasis typically occurs at later stages in cancer development. Once a cancer has metastasized to form tumors in multiple organs, it may be very difficult to fight.

Cancer cells multiply rapidly wherever they establish themselves, overrunning neighboring cells, monopolizing oxygen and nutrients, and starving normal cells in the vicinity. Without restraints on their growth and migration, cancer cells steadily destroy tissues, organs, and organ systems. The normal function of these organs is then seriously impaired, and cancer deaths are ultimately caused by the failure of vital organs.

Figure 6.4

Cell division in a eukaryote

Mitotic cell division is composed of two main stages: mitosis (with four substages) and cytokinesis. ▶

> **Q1:** Do all cells in an organism enter each stage of mitosis at the same time? (*Hint:* See image of onion root tip at far left in the figure.)
>
> **Q2:** What happens between the end of interphase and early prophase that changes the appearance of the chromosomes?
>
> **Q3:** Explain in your own words the role of the mitotic spindle in mitosis.

phases: *prophase, metaphase, anaphase,* and *telophase*. Each phase is defined by easily identifiable events (**Figure 6.4**).

A parent cell sets up for an upcoming mitotic division by replicating its DNA during the S phase of interphase, well before mitosis gets under way. Note that DNA in the nucleus is not tightly packaged during the gap and synthesis phases. This is because the DNA must be accessible for replication and for conducting the business of the cell. Then, as cell division begins, each long, double-stranded DNA molecule is attached to proteins that help pack it for cell division into a chromosome. This packing is

necessary because every DNA molecule is enormously long, even in the simplest eukaryotic cells. When a chromosome is replicated, two identical DNA molecules called **sister chromatids** are produced. These sister chromatids are firmly attached at a central region of the chromosome called the **centromere** and do not separate until metaphase (**Figure 6.5**).

One of the main objectives of mitosis is to separate those sister chromatids, pulling them apart at the centromere and delivering one of each to the opposite ends of the parent cell. Eukaryotic cells have evolved an elaborate choreography to minimize the risk of mistakes during

Mitosis

Anaphase

Telophase

Cytokinesis

Interphase

Cell division

Nuclear envelope forming

Chromosomes decondensing

New chromosomes

3 The sister chromatids are separated and become individual chromosomes. These new chromosomes are moved toward the poles of the cell by the shortening of the mitotic spindle.

4 The new chromosomes reach the poles. The nuclear envelopes re-form, and the chromatin decondenses.

Overlapping with the end of mitosis, the cell physically divides into two daughter cells through cytokinesis.

the equal and symmetrical partitioning of the replicated genetic material. Normally, no daughter cell winds up short a chromosome, nor does it acquire duplicates. Unless an error occurs, each daughter cell inherits the same genetic information that the parent cell possessed in the G_1 phase of its life.

After the replicated DNA has been divided in two, half to each end of the parent cell, the cytoplasm is divided by a process called **cytokinesis** ("cell movement"), like pulling apart a ball of Silly Putty into two halves. Cytokinesis gives rise to two self-contained daughter cells that are clones of each other.

Mitotic division can serve both the eukaryotic organism's need to replace itself (to reproduce) and its need to add new cells to its body. Many multicellular eukaryotes use mitotic division to reproduce asexually, including seaweeds, fungi, and plants, and some animals, such as sponges and flatworms. All multicellular organisms also rely on mitotic division for the growth of tissues and organs and the body as a whole, and for repairing injured tissue and replacing worn-out cells. Mitosis is why children grow taller and why skin closes over a cut.

Good Cells Gone Bad

Cell division is not always a good thing. Runaway cell division can create a tumor (see "Cancer: Uncontrolled Cell Division" on page 101). In a developing organism, it can also cause an organ such as the heart or liver to form incorrectly and not function properly. It is little wonder, then, that the cell cycle is carefully controlled in healthy individuals. The decision to divide a cell is made during the G_1 phase of the cell cycle in response to internal and external signals. In humans, external signals that influence the commitment to divide include hormones and proteins called growth factors. Some hormones and growth factors act like the gas pedal in a car and push a cell toward cell division; others act like a brake and prevent cell division.

After a cell enters the cell cycle, special *cell cycle regulatory proteins* are activated. These proteins "throw the switch" that enables the cell to pass through critical checkpoints and proceed from one phase of the cell cycle to the

One chromosome

During gap phase

Synthesis phase

At the beginning of mitosis

One replicated chromosome

Centromere

Two sister chromatids

Centromere

Each chromosome is replicated before mitosis begins.

One replicated chromosome

Two sister chromatids

As mitosis begins, the chromatin packs even more tightly, and the chromosome becomes easily visible with a light microscope.

Figure 6.5

Chromosomes are copied and condensed in preparation for cell division

Chromosomes spend the majority of the cell cycle unpackaged (left). They are copied (replicated) during synthesis, then tightly packaged, or condensed, during early mitosis.

Q1: Why is it important for a chromosome to be copied before mitosis?

Q2: Are sister chromatids attached at the centromere considered to be one or two chromosomes?

Q3: Why is the chromosome's DNA tightly packed for mitosis and cytokinesis? (*Hint:* Think about what would happen if it were unpackaged, as during interphase.)

next (**Figure 6.6**). For example, upon receiving the appropriate signals, cell cycle regulatory proteins advance a cell from the G_1 phase to the S phase by triggering chromosome replication and other processes associated with it.

Cell cycle regulatory proteins also respond to negative internal or external control signals. Internal signals will pause a cell in the G_1 phase under any of the following conditions: the cell is too small, the nutrient supply is inadequate, or the cell's DNA is damaged. G_2 pauses in the same circumstances, as well as when chromosome duplication in the S phase is incomplete for any reason.

Nonylphenol interferes with the G_0 and G_1 checkpoints. In essence, it gives the cell a green light to enter the cell cycle at a time when the cell would not normally divide. So, when Soto and Sonnenschein realized that nonylphenol enables human breast cells and rat uterine cells to divide inappropriately, they became concerned. If nonylphenol was being used in everyday plastics, it was possible that healthy human cells were exposed to it on a regular basis.

In a 1991 paper detailing their discovery, Soto and Sonnenschein wrote that nonylphenol might be interfering with science experiments like theirs and that, even more important, it could be harmful to humans. "From the very beginning, we realized that this could be a health problem," says Sonnenschein.

Unequal Division

In Ohio, Patricia Hunt was having no success determining why her control mouse eggs divided abnormally, with either too many or too few chromosomes. After searching in vain for months, one day she noticed that something was wrong with

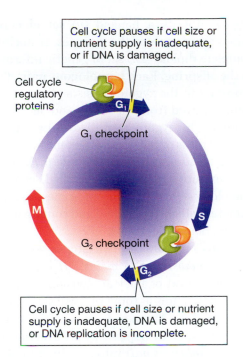

Cell cycle pauses if cell size or nutrient supply is inadequate, or if DNA is damaged.

Cell cycle regulatory proteins

G_1

G_1 checkpoint

M

S

G_2 checkpoint

G_2

Cell cycle pauses if cell size or nutrient supply is inadequate, DNA is damaged, or DNA replication is incomplete.

Figure 6.6

The cell cycle must pass through checkpoints to proceed

Just two of the known cell cycle checkpoints are depicted in this diagram. Checkpoints also operate in both the S phase and the M phase (mitosis).

Q1: What could happen if the cell's checkpoints are disabled?

Q2: What is the advantage of stopping the cell cycle if the cell's DNA is damaged?

Q3: Which part of the cell cycle may have been influenced in Soto and Sonnenschein's breast tumor cell experiments?

the plastic mouse cages: the water bottles were leaking and the plastic cage walls were hazy.

She asked around and found out that, months earlier, a substitute janitor had used detergent with a high pH instead of the normal low-pH detergent to clean the cages and bottles. With some lab work, she identified the chemical oozing from the corroded plastic. "It took essentially one washing with the wrong detergent," says Hunt, "and once it was damaged, [the plastic] started to leach bisphenol A."

Bisphenol A, or BPA, was first synthesized in 1891. A synthetic hormone that, like nonylphenol, mimics estrogen, BPA was tested in the 1930s by clinicians seeking a hormone replacement therapy for women who needed estrogen, but it wasn't as effective as other substitutes. In the 1940s and '50s, the chemical industry found another use for BPA, as a chemical component of a clear, strong plastic. Manufacturers began to incorporate it into eyeglass lenses, water and baby bottles, the linings of food and beverage cans, and more. Unfortunately, as Hunt found out, BPA doesn't necessarily stay in those products. Not all the BPA used to make plastics gets locked into chemical bonds, and what doesn't get locked into bonds can work free—especially if the plastic is heated, such as when a baby bottle is warmed, or if the plastic is exposed to a harsh chemical, as Hunt's mouse cages were. Because of its prevalence in products and its ability to leach out of them, BPA is one of the most common chemicals we are exposed to in everyday life.

To confirm the hypothesis that BPA caused the mouse egg abnormalities, Hunt's team re-created the original event. They intentionally damaged a set of new cages and put healthy female mice in them. They also had the mice drink from damaged water bottles. Later, when they examined the eggs of these mice, they saw the same toxic effects as before: 40 percent of the eggs had abnormal chromosomes. The eggs showed errors in meiosis.

Meiosis is a specialized type of cell division that kicks off **sexual reproduction**, the process by which genetic information from two individuals is combined to produce offspring. Sexual reproduction has two steps: cell division through meiosis, and fertilization. BPA was affecting the first of these steps—meiosis.

We learned earlier that mitosis produces daughter cells with the same number of chromosomes as the parent cell. These non–sex cells are called **somatic cells**. In contrast, meiosis produces **gametes**, daughter cells that have half the chromosome count of the parent cell. The differences between mitosis and meiosis are the reason why the somatic cells of plants and animals have twice as much genetic information as their gametes have. The double set of genetic information possessed by somatic cells is known as the **diploid** set (represented by $2n$), and the single set possessed by gametes is called the **haploid** set (represented by n).

Fertilization, the fusion of two gametes, results in a single cell called the **zygote**. The

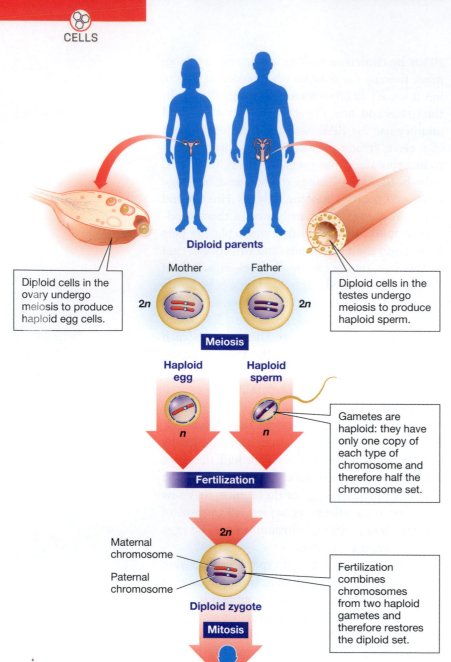

Diploid parents

Mother   Father

Diploid cells in the ovary undergo meiosis to produce haploid egg cells.

Diploid cells in the testes undergo meiosis to produce haploid sperm.

2n   2n

Meiosis

Haploid egg   Haploid sperm

n   n

Gametes are haploid: they have only one copy of each type of chromosome and therefore half the chromosome set.

Fertilization

2n

Maternal chromosome

Paternal chromosome

Diploid zygote

Fertilization combines chromosomes from two haploid gametes and therefore restores the diploid set.

Mitosis

Diploid offspring (2n)

Figure 6.7

Fertilization creates a zygote from the fusion of two gametes

In species with two sexes, female gametes are *eggs* and male gametes are *sperm*. This figure shows only one of the 23 homologous pairs found in human cells.

Q1: Is a zygote haploid or diploid?

Q2: Which cellular process creates a baby from a zygote?

Q3: If a mother or father was exposed to BPA prior to conceiving a child, how might that explain potential birth defects in the fetus?

zygote inherits a haploid (n) set of chromosomes from each of the gametes, restoring the complete diploid ($2n$) set of genetic information to the offspring. Each **homologous pair** of chromosomes in the zygote consists of one chromosome received from the father and one from the mother. The zygote then divides by mitosis to create a mass of cells that will eventually develop into a mature organism (**Figure 6.7**). All cells in a mature organism are diploid, containing homologous pairs of chromosomes, except for gametes, which are haploid and contain only one of each homologous pair.

Meiosis occurs in two stages after interphase—*meiosis I* and *meiosis II*—each involving one round of nuclear division followed by cytokinesis, for a total of two cell divisions (**Figure 6.8**). **Meiosis I** reduces the chromosome number to haploid by *separating one of each homologous pair* into two different daughter cells. Each homologous chromosome lines up with its partner and then separates to the two ends of the cells, and cytokinesis occurs. **Meiosis II** *separates each sister chromatid* into two different daughter cells. This time, the phases of the division cycle are almost exactly like those of mitosis: sister chromatids separate to the two ends of the cells, and cytokinesis occurs, leading to an equal segregation of chromatids into two new daughter cells. In summary, meiosis I produces two haploid cells (n) with one of each pair of duplicated chromosomes, and in meiosis II these two haploid cells separate the sister chromatids and each divide, giving rise to a total of four haploid cells (n) with unduplicated chromosomes. Each haploid gamete now has half of the chromosome set found in the original diploid cell ($2n$) that underwent meiosis.

BPA is toxic because it disrupts the process of meiosis, hindering the ability of the chromosomes to separate into four haploid cells. Hunt realized that if BPA was disrupting meiosis in mice, it could be doing the same in humans. And if a human gamete (either the sperm or the egg) does not contain the correct number of chromosomes, fertilization typically results in a miscarriage.

Hunt was nervous about publishing the results of her experiment. "We knew we were stepping into a landmine," she says. "We knew the paper would get some press, because essentially we were publishing that this chemical—used in

Prophase I **Metaphase I** **Anaphase I** **Telophase I and Cytokinesis I**

1 Each replicated chromosome pairs with its homologue.

2 Homologous chromosome pairs line up at the metaphase plate, aided by the spindle fibers.

3 The paternal and maternal homologous chromosomes separate through the shortening of the spindle fibers.

4 The first cytokinesis takes place, producing two haploid cells.

Meiosis II

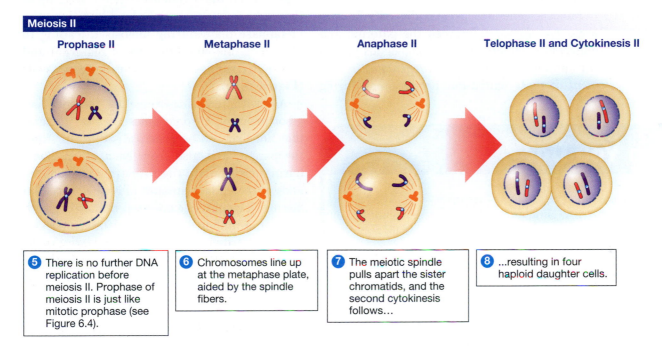

Prophase II **Metaphase II** **Anaphase II** **Telophase II and Cytokinesis II**

5 There is no further DNA replication before meiosis II. Prophase of meiosis II is just like mitotic prophase (see Figure 6.4).

6 Chromosomes line up at the metaphase plate, aided by the spindle fibers.

7 The meiotic spindle pulls apart the sister chromatids, and the second cytokinesis follows...

8 ...resulting in four haploid daughter cells.

Figure 6.8

Meiosis with cytokinesis creates haploid daughter cells

The homologous chromosomes of a diploid cell after interphase occurs are paired and then separated into two haploid cells in meiosis I. Sister chromatids of these haploid cells are separated during meiosis II, each into two haploid cells. Cytokinesis occurs at the end of both meiosis I and meiosis II. ▶️

Q1: Is a daughter cell haploid or diploid after the first meiotic division? How about after the second meiotic division?

Q2: What is the difference between homologous chromosomes and sister chromatids?

Q3: If the skin cells of house cats contain 38 homologous pairs of chromosomes, how many chromosomes are present in the egg cells they produce?

a wide variety of consumer products and that we are probably all exposed to—can cause an increased risk of miscarriage and babies with birth defects."

Shuffling the DNA

As a glance at a pair of parents and their biological children will tell you, the offspring resulting from sexual reproduction are similar to their parents, but—unlike the clones resulting from asexual reproduction—they are *not* identical. Because each half of a sexually reproducing organism's DNA comes from a different parent, meiosis and fertilization maintain the constant chromosome number of a species while allowing for genetic diversity within the population.

Meiosis generates genetic diversity in two ways: *crossing-over* between the paternal and maternal members of each homologous pair of chromosomes, as well as *independent assortment* of these paired-up chromosomes during meiosis I. **Crossing-over** is the physical exchange of identical chromosomal segments during meiosis I between the nonsister chromatids in each duplicated homologous pair. These nonsister chromatids make physical contact at random sites along their length, and each exchange the exact same segments of DNA (**Figure 6.9**). The chromatids are said to be *recombined*, and the exchange of DNA segments is known as **genetic recombination**. Without crossing-over, every chromosome inherited by a gamete would be just the way it was in the parent cell.

The **independent assortment** of chromosomes—the random distribution of the homologous chromosomes into daughter cells during meiosis I—also contributes to the genetic variety of the gametes produced. Each homologous chromosome pair in a given meiotic cell orients itself independently when it lines up at the imaginary equatorial plane known as the metaphase plate during meiosis I, leading to many possible combinations of maternal and paternal chromosomes in the daughter cells (**Figure 6.10**).

As with crossing-over, the independent assortment of chromosomes creates gametes that are likely to be genetically different from the parent, and also from each other. Then, during fertilization, the fusion of two gametes adds a tremendous amount of genetic variation because it combines a one-in-a-million egg with a one-in-a-million sperm. These three processes together—crossing-over, independent assortment, and fertilization—give each of us our genetic uniqueness.

What Can You Do?

There are things you can do to reduce your own risk of exposure to BPA. Today, the U.S. Food and Drug Administration recommends that individuals not put hot or boiling liquid intended for consumption in plastic containers made with BPA. (Some, but not all, plastics that are marked with the recycling code 3 or 7 may be made with BPA.) The organization also recommends discarding all bottles with scratches, which may harbor bacteria and, if the plastic contains BPA, may lead to greater release of the chemical.

"Get educated about your world," says Heather Patisaul, a BPA researcher at North Carolina State University. "You can either become completely reliant on the information you get from media and government, or you can educate yourself, which is vastly more useful." If you are concerned about BPA exposure, it is possible to cut it down by making lifestyle changes like not eating canned food, not drinking bottled water, and not putting plastic in the microwave. "You can be empowered," says Patisaul. "Those types of things can effect great change."

Still, BPA and similar chemicals are ubiquitous in modern life, says Ana Soto. Ultimately, the best way to avoid them will be for government regulators to take a stand and outlaw the use of these chemicals in consumer products, she notes. She encourages individuals to contact their representatives and ask them to push legislation limiting the use of BPA in manufacturing.

Ten Years Later

Hunt and her team published their results in 2003. "We got a firestorm," she recalls with a grimace. The press reported the findings extensively, and many people and companies were upset over the allegations that BPA was toxic. Members of the plastics industry who did not agree with the paper's conclusions criticized Hunt's work. But there was more supporting research to come. Soto and Sonnenschein had also turned their attention to BPA because in everyday products it is a far more common chemical than nonylphenol and is therefore of greater concern.

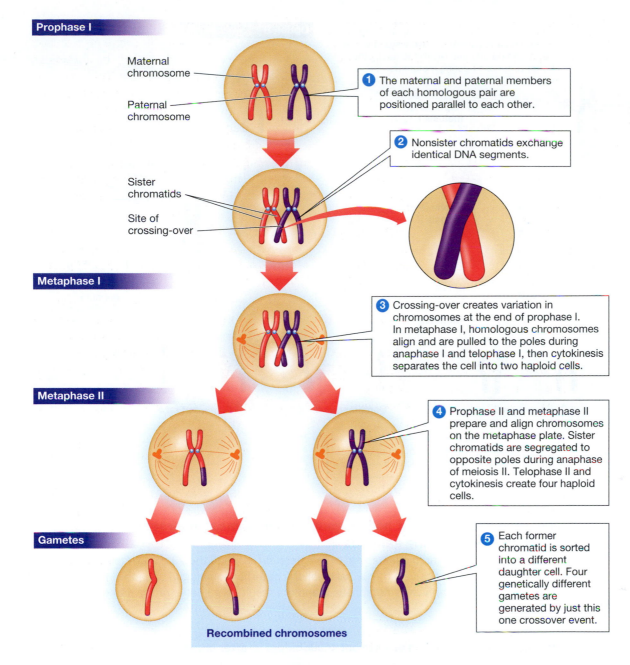

Prophase I

Maternal chromosome

Paternal chromosome

1 The maternal and paternal members of each homologous pair are positioned parallel to each other.

2 Nonsister chromatids exchange identical DNA segments.

Sister chromatids

Site of crossing-over

Metaphase I

3 Crossing-over creates variation in chromosomes at the end of prophase I. In metaphase I, homologous chromosomes align and are pulled to the poles during anaphase I and telophase I, then cytokinesis separates the cell into two haploid cells.

Metaphase II

4 Prophase II and metaphase II prepare and align chromosomes on the metaphase plate. Sister chromatids are segregated to opposite poles during anaphase of meiosis II. Telophase II and cytokinesis create four haploid cells.

Gametes

5 Each former chromatid is sorted into a different daughter cell. Four genetically different gametes are generated by just this one crossover event.

Recombined chromosomes

Figure 6.9

Crossing-over produces chromosomes with new combinations of DNA

Only one maternal and one paternal chromosome are depicted here, rather than the 23 pairs of homologous chromosomes found in humans. Only the resulting cells of prophase I, metaphase I, metaphase II, and the resulting gametes of the process of meiosis are depicted here.

Q1: Why is the term "crossing-over" appropriate for the exchange of DNA segments between homologous chromosomes?

Q2: At what stage of meiosis (I or II) does crossing-over occur?

Q3: What would be the effect of crossing-over between two sister chromatids?

Option A

Option B

Maternal homologue

Paternal homologue

If we consider just two pairs of homologous chromosomes, there are two ways in which they can be oriented relative to each other after prophase I during metaphase I.

Metaphase of meiosis I

The maternal and paternal homologues of each pair are sorted into two daughter cells through anaphase I, telophase I, and cytokinesis.

Metaphase of meiosis II

At the end of meiosis II, sister chromatids are segregated into separate daughter cells.

Gametes

Combination 1

Combination 2

Combination 3

Combination 4

Four possible combinations of chromosomes in gametes generated by meiosis II.

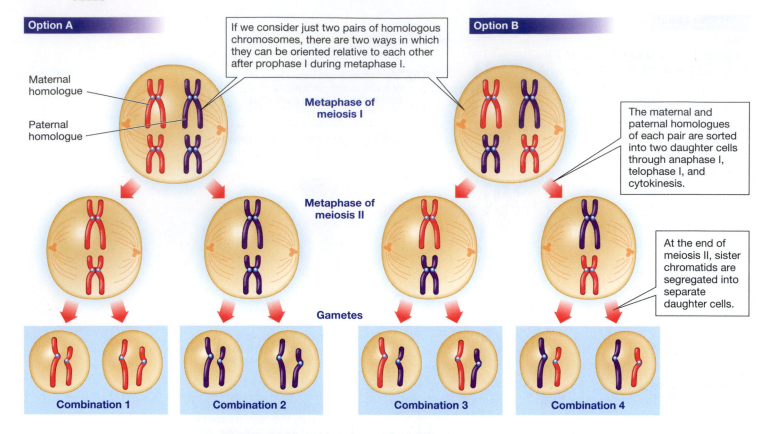

Figure 6.10

The independent assortment of homologous chromosomes generates chromosomal diversity among gametes

Only two pairs of homologous chromosomes are shown here, rather than the 23 homologous pairs in human cells. Each gamete will receive either a maternal or a paternal homologue of each chromosome. Only the resulting cells of metaphase I, metaphase II, and the resulting gametes of the process of meiosis are depicted here.

Q1: During meiosis, does random assortment occur before or after crossing-over?

Q2: What would be the effect on genetic diversity if homologous chromosomes did not randomly separate into the daughter cells during meiosis?

Q3: With two pairs of homologous chromosomes, four kinds of gametes can be produced. How many kinds of gametes can be produced with three pairs of homologous chromosomes? What does this suggest for the 23 homologous pairs of chromosomes in human cells?

At the same time that Soto, Sonnenschein, and Hunt were doing their work, Frederick vom Saal at the University of Missouri found that male mice that had been exposed to BPA in utero—even at very low doses—had dramatically enlarged prostates in adulthood that were hypersensitive to hormones. This study suggests that men are also at risk of health effects from BPA.

In 2007, Hunt followed up her original work with a study that she says made the first paper look like "child's play." Her team exposed pregnant mice to BPA just as their female fetuses were producing a supply of eggs in their ovaries. When that second generation of females became adults, their eggs were also damaged, Hunt found, demonstrating that BPA exposure affects not just adult females, but two generations of their offspring.

Exposure aside, not everyone agrees that BPA is toxic. Numerous companies that manufacture plastics have conducted studies whose results do not match Hunt's and vom Saal's results. To reach a scientific consensus, on November 28, 2006, Soto, Sonnenschein, Hunt, vom Saal, and 34 other researchers from across the United States gathered at the University of North Carolina in Chapel Hill to summarize the research on BPA. The result of their two-day meeting was the "Chapel Hill Bisphenol A Consensus Statement," summarizing hundreds of studies done in vitro and in vivo over the previous 10 years. Completing this analysis, the group concluded firmly that BPA exposure at current levels in our environment presents a risk to human health (see "What Can You Do?" on page 108). "It was quite clear that there is a serious problem," says Soto.

Over time, many baby-bottle manufacturers took BPA out of their bottles, even as government regulators were slower to respond. "As scientists, our role is to call attention to what is wrong, but it is the role of the politicians to act on it and try to straighten it out," says Sonnenschein. Then, in July 2012, the FDA banned BPA from baby bottles and children's drinking cups, though the prohibition does not apply to the use of BPA in other types of containers (**Figure 6.11**). There is still concern, however, from many scientists about the chemicals that have replaced BPA, some of which are also estrogen mimics.

In the last several years, scientists all over the world have shown that BPA disrupts meiosis and mitosis and causes a plethora of health problems in mice and rats, including breast and prostate cancer, miscarriage and birth defects, diabetes and obesity, and even behavioral problems such as attention deficit hyperactivity disorder. Whether BPA is causing similar diseases in humans remains unknown; such hypotheses are hard to test because most people already have BPA in their bodies, making experimental control groups difficult to set up. BPA has been found in human blood, urine, breast milk, and amniotic fluid. In 2016, Hunt and her colleagues found "near-universal exposure" to BPA in a group of pregnant women in the United States. A major source of that exposure, they found, was cash register receipts, which can be hard to avoid touching.

Hunt, Soto, and Sonnenschein continue to explore the effects of BPA; they are studying how exposure to low doses of BPA affects monkeys, a model animal that more accurately represents the human system. "We're slowly raising awareness," says Hunt, "and slowly changing things."

Figure 6.11

BPA-free bottles and cans are now widely available
If you are concerned about being exposed to BPA, check labels before you buy.

Cancer's Big 10

When the cell cycle spirals out of control, cancer emerges: abnormal cells divide in a frenzy and can invade other tissues. There are more than 100 types of cancer, but some are more prevalent than others. And some are more deadly than others, because of their location in the body or how quickly the cells divide. New treatments, screening procedures, and vaccines can reduce these rates.

Assessment available in smartwork5

Top 10 cancer sites by rate of incidence

Incidence rates per 100,000

Male

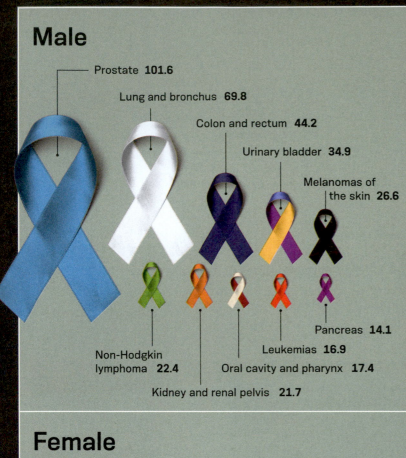

Prostate **101.6**
Lung and bronchus **69.8**
Colon and rectum **44.2**
Urinary bladder **34.9**
Melanomas of the skin **26.6**
Non-Hodgkin lymphoma **22.4**
Kidney and renal pelvis **21.7**
Oral cavity and pharynx **17.4**
Leukemias **16.9**
Pancreas **14.1**

Female

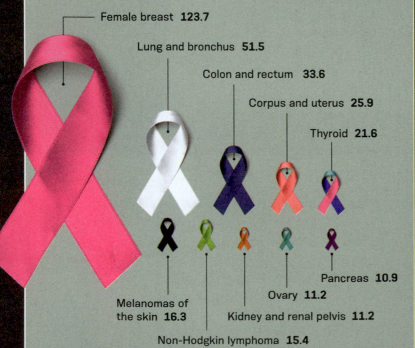

Female breast **123.7**
Lung and bronchus **51.5**
Colon and rectum **33.6**
Corpus and uterus **25.9**
Thyroid **21.6**
Melanomas of the skin **16.3**
Non-Hodgkin lymphoma **15.4**
Kidney and renal pelvis **11.2**
Ovary **11.2**
Pancreas **10.9**

Top 10 cancer deaths by rate of incidence

Incidence rates per 100,000

Male

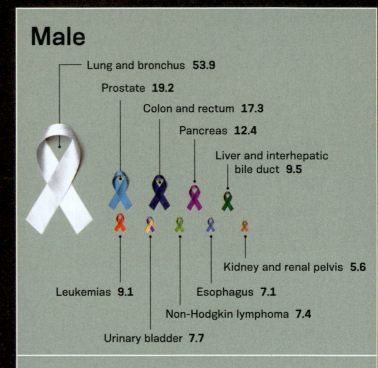

Lung and bronchus **53.9**
Prostate **19.2**
Colon and rectum **17.3**
Pancreas **12.4**
Liver and interhepatic bile duct **9.5**
Leukemias **9.1**
Urinary bladder **7.7**
Non-Hodgkin lymphoma **7.4**
Esophagus **7.1**
Kidney and renal pelvis **5.6**

Female

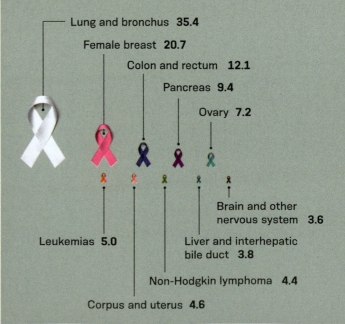

Lung and bronchus **35.4**
Female breast **20.7**
Colon and rectum **12.1**
Pancreas **9.4**
Ovary **7.2**
Leukemias **5.0**
Corpus and uterus **4.6**
Non-Hodgkin lymphoma **4.4**
Liver and interhepatic bile duct **3.8**
Brain and other nervous system **3.6**

REVIEWING THE SCIENCE

- The **cell cycle** is the set sequence of events over the life span of a eukaryotic cell that will divide. **Interphase** and **cell division** are the two main stages of the cell cycle. Interphase is longer, and consists of the G_1, **S**, and G_2 **phases**. DNA is replicated in the S phase. Cells that will not divide exit the cell cycle and enter a G_0 **phase**.

- Cell division is necessary for growth and repair in multicellular organisms, and for **asexual** and **sexual reproduction** in all types of organisms. Many prokaryotes divide through **binary fission**, a form of asexual reproduction.

- Each **chromosome** in a cell contains a single DNA molecule compacted by packaging proteins. The **somatic cells** of eukaryotes have two of each type of chromosome matched together in **homologous pairs**. One chromosome in each pair is inherited from the mother; the other, from the father. Chromosomal replication produces two identical **sister chromatids** that are held together firmly at the **centromere**.

- Eukaryotes perform cell division through **mitosis** followed by **cytokinesis**, producing daughter cells that are genetically identical to each other and to the parent cell. The four main phases of mitosis are prophase, metaphase, anaphase, and telophase. Through these phases, the chromosomes of a parent cell are condensed and positioned appropriately, and the sister chromatids are separated to opposite ends of the cell. During cytokinesis, the cytoplasm of the parent cell is physically divided to create two daughter cells.

- The cell cycle is carefully regulated. Checkpoints ensure that the cycle does not proceed if conditions are not right. Unregulated cell growth and cancer occur when checkpoints fail.

- **Meiosis** is critical for sexual reproduction. In animals, the products of meiosis are sex cells, called **gametes**, that fuse during **fertilization** to give rise to a **zygote**. Meiosis—consisting of two rounds of nuclear and cytoplasmic divisions—produces **haploid** gametes containing only one chromosome from each homologous pair. When two gametes fuse during fertilization, a **diploid** zygote is formed, having one of each homologous chromosome from both parents.

- During **meiosis I**, the maternal and paternal members of each homologous pair are sorted into two daughter cells. **Meiosis II** is similar to mitosis in that sister chromatids are segregated into separate daughter cells at the end of cytokinesis.

- Meiosis produces genetically diverse gametes through **crossing-over** of homologous chromosomes, leading to **genetic recombination** and then the **independent assortment** of homologous chromosomes. Meiosis and fertilization together introduce genetic variation into populations.

THE QUESTIONS

Challenge Yourself

1 Suppose a scientist has recently identified a new protein that prevents a cell from entering mitosis if there are any signs of DNA damage. This protein would be classified as a type of _____ protein.

(a) chromatid

(b) cell cycle checkpoint

(c) benign

(d) malignant

(e) angiogenesis

2 You've been reduced in size by a misfire from a "shrink-inator" gun! You realize that you are inside a cell during prophase I of meiosis. You see two linear molecules of DNA compacted and attached to one another by a centromere. What exactly are you looking at?

(a) a homologous pair of chromosomes

(b) sister chromatids

(c) the metaphase plate

(d) cell cycle checkpoints

(e) gametes

3 Scientists are able to isolate cells in various phases of the cell cycle. During an experiment in which Dr. Patrick McGroyn is causing cultured tumor cells to go through mitotic cell division, he isolates a group of cells that contain one and a half times more DNA than cells isolated in the G_1 phase have. What phase must these cells be in? How do you know?

Try Something New

4 Domesticated sheep cells have a total of 54 chromosomes (versus 46 chromosomes in human cells).

(a) How many separate DNA molecules are present in a sheep liver cell at the end of the G_1 phase?

(b) How many separate DNA molecules are present in the daughter cells after meiosis I in the ovary of the sheep?

(c) How many separate DNA molecules are present in the daughter cells after meiosis II in the ovary of the sheep?

5 Biopsies from aggressive cancers often have cells that contain several nuclei per cell when viewed through a microscope. Which scenario could explain how such a multinucleated cell might have come to be?

(a) The cell underwent repeated mitosis with simultaneous cytokinesis.

(b) The cell had multiple S phases before it entered mitosis.

(c) The cell underwent repeated mitosis, but cytokinesis did not occur.

(d) The cell underwent repeated cytokinesis but no mitosis.

(e) The cell actually went through meiosis and not mitosis.

6 Describe the likely consequences of bypassing the G_1 and G_2 checkpoints in the cell cycle. Why do compounds like nonylphenol lead to the multiplication of abnormal cells?

Leveling Up

7 **What do *you* think?** Cancer begins with a single cell that breaks loose of normal restraints on cell division and starts dividing rapidly to establish a colony of rogue cells. As cancer cells spread through the body, they disrupt the normal functions of tissues and organs; unchecked, cancer can cause death through failure of multiple organ systems. Many cancers could be prevented by not smoking or chewing tobacco, eating less meat and processed foods, eating more fruits and vegetables, drinking alcohol only moderately if at all, exercising regularly, and maintaining a healthy weight. Only 5–10 percent of cancers are directly attributable to genetic causes.

A "sin tax" is a tax on a product or activity that has negative effects on others, as a way to offset some of those effects. Common targets of sin taxes are tobacco and alcohol because of their public health costs. Proponents of a sin tax on tobacco argue that such a tax would decrease the amount that people smoke (because of the increased cost) and could also partially fund the costs of medical care necessitated by increased rates of cancer and other diseases caused by smoking. Critics point out that sin taxes have historically triggered smuggling and black markets, and have a disproportionate effect on poor people because the wealthy can more easily afford to pay the higher prices.

What do you think? Should we institute taxes on tobacco? Would fewer people smoke, or would they smoke less, if tobacco were more expensive? What about higher health care premiums for smokers, based on their higher risk for cancer? Would increasing the premiums cause more people to stop smoking? Would you support such a policy?

8 *Write Now* **biology: BPA effects** The studies of BPA described in this chapter used an inbred strain of mice that is known to be especially susceptible to estrogen and estrogen-like chemicals, such as BPA. The plastics industry maintains that the susceptibility of this strain of mice to estrogen renders these studies invalid as a basis for estimating the effects of BPA on humans. BPA researchers respond that the current situation, exposing millions of people to unknown levels of BPA, constitutes a massive uncontrolled experiment. They maintain that even a small risk of harm is too great to be allowed when so many people are exposed. Bills banning the use of BPA in food and beverage containers were introduced in Congress in 2009, 2011, 2013, 2014, and 2015. All failed to pass. Another bill was introduced to the House of Representatives on September 28, 2016.

a. Go to the Embryo Project Encyclopedia (http://embryo.asu.edu), an NSF-funded online repository of information about embryo research. Search for the term "BPA" to find the most up-to-date information on BPA research.

b. Go to the website of the American Chemistry Council (http://www.americanchemistry.com), a trade organization that advocates for the chemical industry. Search for the term "BPA" to find its most up-to-date policy statement on the safety of BPA.

c. Draft a letter urging your congressional representative to support or oppose the most recent BPA bill, explaining why you believe the weight of scientific evidence makes your position a prudent response to the situation.

Thinking Multidimensionally

9 The superhero of the amphibians might be the axolotl, also called the Mexican salamander. Unlike other amphibians, it doesn't undergo metamorphosis, which means it stays aquatic, it is resistant to getting cancer, and it has a remarkable healing ability—it can regenerate amputated limbs!

a. Draw and label how the somatic cells of the axolotl are unique, enabling it to regrow body parts.

b. What other organisms also have the ability to regenerate body parts?

c. How can science use the knowledge gained from this model organism to benefit humanity?

d. Research other organisms that also are immune to cancer.

10 A new meat is gracing the tables of Britain, the eastern gray squirrel. Originally native to North America, their introduction to England has devastated the native red squirrel population. In an effort to control their exponential growth, a British campaign was launched with the motto "Save a red, eat a grey!"

a. How is this invasive species' destruction analogous to how cancer spreads and impacts healthy cells?

b. Cancerous cells lower cellular pH and disrupt homeostasis. Based on this information, design a potential cancer treatment.

For more, visit digital.wwnorton.com/bionow2 for access to:

Dog Days of Science

Two canine-loving researchers unravel the genetic secrets of man's best friend.

After reading this chapter you should be able to:

- Distinguish between the genotype and phenotype of a given genetic trait.
- Describe the importance of Gregor Mendel's experiments to our understanding of inheritance.
- Illustrate Mendel's laws of segregation and independent assortment.
- Create a Punnett square to predict the phenotype of offspring from parents with a known genotype—both for single genes and for two independent genes.
- Give examples of Mendelian traits and of traits with complex inheritance.
- Explain how an individual's phenotype may be determined by multiple genes that interact with one another and with the environment.

CHAPTER
07

PATTERNS OF INHERITANCE

DRIVING QUESTION

What can variation in dogs teach us about heredity (or inheritance of traits)?

- HS-LS3.A: Inheritance of Traits
- HS-LS3.B: Variation of Traits

117

Gordon Lark's best friend was dying. Soft and shaggy, with tousled black hair, Georgie hadn't left Lark's side in 10 years, since his daughter had first purchased Georgie as a puppy from two kids by the side of the road. But as she aged, Georgie had become ill with Addison's disease, a disorder in which her body's immune system began to attack and destroy her own tissues. Georgie passed away in 1996.

Lark, a scientist at the University of Utah in Salt Lake City, was heartbroken. To help heal the wound, he decided to adopt another dog of the same breed—a Portuguese water dog (PWD), named for its history of helping Portuguese fishermen with their work (**Figure 7.1**). He contacted Karen Miller, a PWD breeder on a farm in rural New York. As part of the owner screening process, Miller asked Lark about his profession. "I said I was a soybean geneticist," Lark recalls, "but all she heard was 'blah, blah, genetics, blah, blah.' And she got really excited."

As a breeder, Miller was keenly interested in how dogs inherit characteristics from their parents, so Miller and Lark began talking by phone each week about genetics. When it came time to pick up his new puppy, Mopsa, Lark requested the bill. But Miller didn't want Lark's money. She had something else in mind: she gave

him Mopsa free of charge, in the hope that Lark would start researching dog genetics. "That's silly," Lark told her. He wasn't a dog researcher. Lark had spent a career studying the genetic traits of bacteria and soybeans.

A **genetic trait** is any inherited characteristic of an organism that can be observed or detected in some manner. Some genetic traits are **invariant**, meaning they are the same in all individuals of the species. All soybeans, for example, have pods that contain seeds. Other genetic traits are **variable**; for example, soybean seeds occur in various sizes and colors, including black, brown, and green.

Apart from his love for canines, Lark wasn't intimately familiar with dogs' physical and biochemical traits. **Physical traits**, such as the shape of a dog's face, are easy to observe. **Biochemical traits**, on the other hand, such as a dog's susceptibility to Addison's disease, are often more difficult to observe. It is easy to collect physical and biochemical information from a field of soybeans but far more difficult to collect it from domesticated animals in homes all over the country. And to study dogs, Lark would also need data on **behavioral traits**, such as shyness and extroversion—factors he didn't have to take into account with soybeans. All of these traits—physical, biochemical, and behavioral—are influenced by genes.

Figure 7.1

A Portuguese Water Dog

Getting to the Genes

A **gene** is the basic unit of information affecting a genetic trait. At the molecular level, a gene consists of a stretch of DNA on a **chromosome**—a threadlike molecule, made of DNA and proteins, that is found in the nucleus of a eukaryotic cell (**Figure 7.2**). A gene contains the information, or "code," for a specific protein; the protein then causes or contributes to a particular genetic trait. To study PWD traits, Lark would need dog DNA, which can be obtained from blood or saliva. Once he had that DNA, he could begin to search for **alleles**—different versions of a given gene—and link them to genetic traits. Alleles of a gene arise by **mutation**, which is any change in the DNA that makes up a gene (see Chapter 9 for more on mutation). Genetic diversity in a species, whether it is soybeans or dogs or humans, comes about because the species as a whole contains many different alleles of its genes.

Dogs are the champions of genetic variation. All dogs are the same species, yet a Pekingese weighs only a couple of pounds, while a Saint Bernard can weigh over 180 pounds. Dogs, in fact, are reported to have more variation in the size and shape of their species than any other living land mammal on Earth, with the possible exception of humans.

On a whim, Lark agreed to dabble in dog genetics, but to do so, he would need to compare the **genotypes** of individual dogs (their genetic makeup, controlled by their combinations of alleles) with their **phenotypes** (the physical expression of their genetic makeup). The genotype of a given trait is the pair of alleles that codes for its phenotype. To identify genes responsible for dog traits, Lark would need both types of information for each of many dogs.

Miller was already on the case. Three months after Lark got Mopsa, Miller sent him 5,000 PWD pedigrees—detailed health and breeding records for individual dogs. Lark was astonished. It was the first of many times that the enthusiasm and generosity of dog owners would contribute to his research.

"That was literally how this started," says Lark. Today, Lark and Miller's unlikely partnership has blossomed into a national research project producing valuable knowledge about the genetic basis of health and disease in both man and man's best friend. What's more, their effort has demonstrated how tiny genetic changes can create huge variation in a single species.

Pet Project

The Georgie Project, as Lark fondly named it, officially began in 1996. Lark's first task was to collect genotypes and phenotypes from PWDs. To his pleasant surprise, PWD owners were enthusiastic and began flooding him with pedigrees, blood samples, and X-rays taken by their veterinarians. In short order, Lark had DNA from more than 1,000 dogs and detailed body measurements for over 500. Then the hard work began.

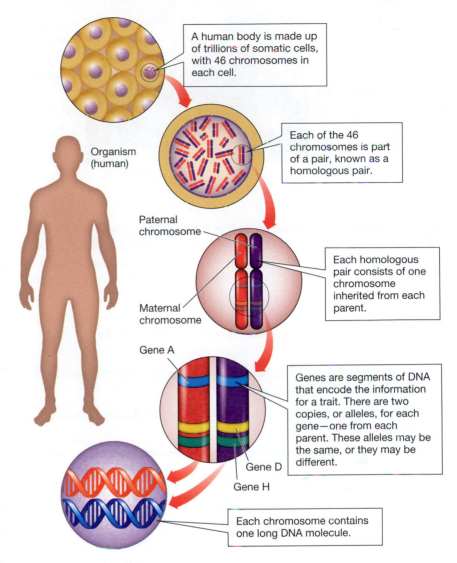

Organism (human)

A human body is made up of trillions of somatic cells, with 46 chromosomes in each cell.

Each of the 46 chromosomes is part of a pair, known as a homologous pair.

Paternal chromosome

Maternal chromosome

Each homologous pair consists of one chromosome inherited from each parent.

Gene A

Genes are segments of DNA that encode the information for a trait. There are two copies, or alleles, for each gene—one from each parent. These alleles may be the same, or they may be different.

Gene D

Gene H

Each chromosome contains one long DNA molecule.

Figure 7.2

Genes are segments of DNA that confer a genetic trait
Somatic cells (cells of the body) have two copies of most genes.

Q1: What is the physical structure of a gene?

Q2: How many copies of each gene are found in the diploid cells in a woman's body?

Q3: With 46 chromosomes in a human diploid cell, how many chromosomes come from the person's mother and how many from the father?

Using the dogs' genotypes and phenotypes, Lark set out to pinpoint the alleles for particular traits. Some genes have alleles that are **dominant** when paired with another allele; that is, one allele prevents a second allele from affecting the phenotype when the two alleles are paired together. The

Phenotype:

Genotype: bb   BB or Bb

Figure 7.3

Poodles illustrate variation in the coat color gene

These poodles, close cousins of the Portuguese water dog, may have a black coat (dominant allele B) or a brown coat (recessive allele b). Other coat colors, with different inheritance patterns, are found in poodles and other dog breeds.

Q1: Which might you observe directly: the genotype or the phenotype?

Q2: Which poodle could be heterozygous: the one with the black coat or the one with the brown coat?

Q3: Can you identify with certainty the genotype of a black poodle? A brown poodle?

black-fur allele (*B*), for example, is dominant in dogs. An allele that has no effect on the phenotype when paired with a dominant allele is said to be **recessive**. In dogs, the brown-fur allele (*b*) is recessive. (When a gene has one dominant and one recessive allele, we generally use an uppercase letter for the dominant allele and a lowercase letter for the recessive allele.)

An individual that carries two copies of the same allele (such as *BB* or *bb*) is **homozygous** for that gene. An individual whose genotype consists of two different alleles for a given phenotype (*Bb*) is **heterozygous** for that gene. Having one dominant allele and one recessive allele, a heterozygous individual will show the dominant phenotype;

GORDON LARK

A geneticist at the University of Utah in Salt Lake City, Gordon Lark initiated the Georgie Project in 1996 to study the genetics of Portuguese water dogs. The national research project has led to valuable knowledge about the genetic basis of health and disease in humans and dogs.

a dog that is heterozygous for fur color (*Bb*), for example, will be black (**Figure 7.3**).

The first dog trait that Lark decided to investigate was size. What makes a Great Dane large and a Chihuahua small? To find out, Lark asked for help from the "mother of all dog projects," as Lark calls her—a researcher named Elaine Ostrander, whose entry into dog research was almost as strange as Lark's.

Crisscrossing Plants

In 1990, Ostrander was a young, enthusiastic researcher who had just completed her postdoctoral studies in molecular biology at Harvard University and was ready to start her own laboratory in California. But first she had to decide which organism to study. Typical choices included fruit flies, worms, or plants—organisms that are easy to grow and manipulate. Ostrander picked plants, just as Gregor Johann Mendel, an Austrian monk who later became known as the "father of modern genetics," had done in the mid-1800s.

Mendel famously bred pea plants in a garden at his monastery. Through his work with pea plants, Mendel discovered patterns of inheritance that today form the foundation of genetics for scientists like Ostrander. "Mendel's laws," as they are now called, describe how genes are passed from parents to offspring. These laws allow us to use parental genotypes to predict offspring genotypes and phenotypes.

Each time Mendel bred two pea plants together, he was performing a **genetic cross**, or just "cross" for short. A genetic cross is a controlled mating experiment performed to examine how a particular trait is inherited. In a series of genetic crosses, the organisms involved in the first cross are called the **P generation** ("P" for "parental").

For example, Mendel investigated the inheritance of flower color by crossing pea plants that had different flower colors (**Figure 7.4**). He had noticed that some plants always "bred true" for flower color; that is, the offspring always produced flowers that had the same color as the parents and were therefore homozygous. He performed a genetic cross with a P generation

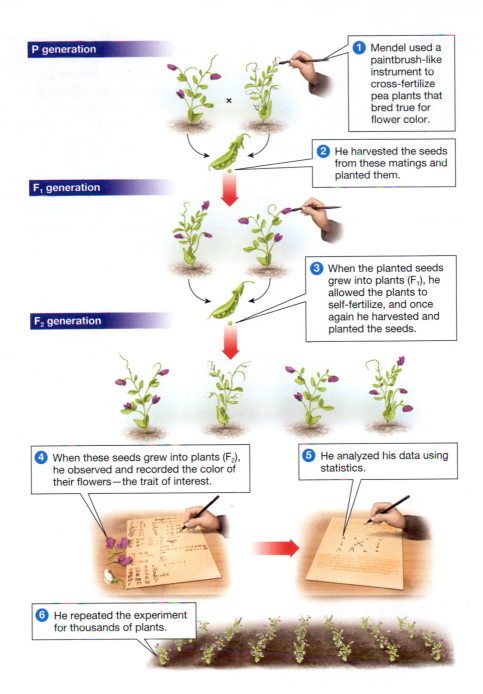

P generation

1. Mendel used a paintbrush-like instrument to cross-fertilize pea plants that bred true for flower color.

2. He harvested the seeds from these matings and planted them.

F₁ generation

3. When the planted seeds grew into plants (F₁), he allowed the plants to self-fertilize, and once again he harvested and planted the seeds.

F₂ generation

4. When these seeds grew into plants (F₂), he observed and recorded the color of their flowers—the trait of interest.

5. He analyzed his data using statistics.

6. He repeated the experiment for thousands of plants.

Figure 7.4

Mendel's careful experiments

Mendel was meticulous in conducting his research and making observations, following a very careful protocol. Controlled breeding is possible in flowering plants because they have both male and female reproductive structures. Mendel manipulated his P-generation plants by removing all the female structures from one plant and all the male structures from the other, thus preventing self-fertilization. He could then perform the initial cross by transferring pollen from the "male" plant to the flower of the "female" plant. ▶

Q1: What would you predict about the color of the F₁ plants' flowers?

Q2: Why was it important that Mendel begin with pea plants that he knew bred true for flower color? Why couldn't he simply cross a purple-flowered plant and a white-flowered plant?

Q3: Over the years, Mendel experimented with more than 30,000 pea plants. Why did Mendel collect data on so many plants? Why didn't he study just one cross? *Hint:* Read "What Are the Odds?" on page 124 before answering.

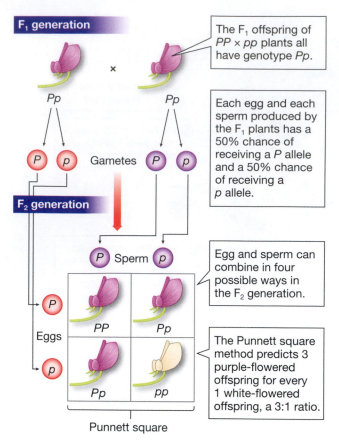

The F₁ offspring of *PP* × *pp* plants all have genotype *Pp*.

Each egg and each sperm produced by the F₁ plants has a 50% chance of receiving a *P* allele and a 50% chance of receiving a *p* allele.

Egg and sperm can combine in four possible ways in the F₂ generation.

The Punnett square method predicts 3 purple-flowered offspring for every 1 white-flowered offspring, a 3:1 ratio.

F₁ generation

Pp × *Pp*

Gametes

F₂ generation

P Sperm *p*

Eggs

	PP	Pp
P	PP	Pp
p	Pp	pp

Punnett square

Figure 7.5

Punnett squares predict the offspring of genetic crosses

Q1: Why did Mendel's entire F₁ generation look the same?

Q2: The phenotype ratio in the F₂ generation is 3:1 purple-to-white flowers. What is the genotype ratio?

Q3: Draw a Punnett square for a genetic cross of two heterozygous, black-coated dogs. What are the phenotype and genotype ratios of their offspring?

shows all possible ways that two alleles can be brought together through fertilization. To create a Punnett square showing how a trait is inherited, list the alleles of the male genotype across the top of the grid, writing each unique allele just once. List the alleles of the female genotype along the left edge of the grid, again writing each unique allele only once. In the case of Mendel's cross of the F₁ generation, the male genotype (*Pp*) is mated with the female genotype (also *Pp*).

Next, fill in each box (or "cell") in the grid by combining the male allele at the top of each column with the female allele listed at the beginning of each row. The Punnett square shows all four ways in which the two alleles in the sperm can combine with the two alleles found in the eggs. The four genotypes shown within the Punnett square are all equally likely outcomes of this cross.

Using the Punnett square method, we can predict that ¼ of the F₂ generation is likely to have genotype *PP*, ½ to have genotype *Pp*, and ¼ to have genotype *pp*. Because the allele for purple flowers (*P*) is dominant, plants with *PP* or *Pp* genotypes have purple flowers, while plants with *pp* genotypes have white flowers. Therefore, we predict that ¾ (75 percent) of the F₂ generation will have purple flowers and ¼ (25 percent) will have white flowers—a 3:1 (¾:¼) ratio of phenotypes. This prediction is very close to the actual results that Mendel obtained. Of a total 929 F₂ plants that Mendel raised, 705 (76 percent) had purple flowers and 224 (24 percent) had white flowers.

Results like these supported Mendel's first law, the **law of segregation**, which, in modern terms (Mendel did not know about DNA), states that the two alleles of a gene are separated during meiosis, the specialized type of cell division during sexual reproduction that was discussed in Chapter 6, and end up in different gametes—egg or sperm cells. One of the two alleles is found on one of the chromosomes in a homologous pair, and the other allele is found on the other chromosome in the pair (see **Figure 7.2**). Remember, homologous chromosome pairs are partitioned into separate daughter cells during meiosis I. Mendel's law of segregation can be used to predict how a single trait will be inherited.

You can try this on your own by making a Punnett square to predict the ratio of black

in which one parent bred true for purple flowers (*PP*) and the other bred true for white flowers (*pp*). The first generation of offspring of a genetic cross is called the **F₁ generation** ("F" is for "filial," a word that refers to a son or daughter). When the individuals of the F₁ generation are crossed with each other, the resulting offspring are said to belong to the **F₂ generation**. Mendel allowed the F₁-generation pea plants to self-fertilize to produce the F₂ generation.

We can predict the results of an experimental cross by using a grid-like diagram called a **Punnett square** (**Figure 7.5**). A Punnett square and brown offspring that would result if two heterozygous (*Bb*) black-coated dogs were mated. It is important to understand that the predicted ratios simply give the *probability* that a particular offspring will have a certain phenotype or genotype; the actual ratio will vary (see "What Are the Odds?" on page 124).

Peas in a Pod

Mendel's research on pea seeds led to his second law, the **law of independent assortment**. This law states that when gametes form, the two alleles of any given gene segregate during meiosis independently of any two alleles of other genes. For example, pea seeds can have a round or wrinkled shape, and they can be yellow or green. Two different genes control the two different traits—the *R* gene, with alleles *R* (round) and *r* (wrinkled), controls seed shape, while the *Y* gene, with alleles *Y* (yellow) and *y* (green), controls the color of the seed—but neither gene affects the inheritance of the other.

Mendel tested the idea of independent assortment in the set of experiments illustrated in **Figure 7.6**. He tracked seed shape, the trait controlled by the *R* and *r* alleles, and seed color, controlled by the *Y* and *y* alleles. The test of his hypothesis came when he examined the phenotypes of the offspring produced by crossing the heterozygous F$_1$ plants (*RrYy*). As predicted by the hypothesis, two new phenotypic combinations were found among the F$_2$ offspring: plants with round, green seeds (*RRyy* or *Rryy*) and plants with wrinkled, yellow seeds (*rrYY* or *rrYy*). Figure 7.6 summarizes the ratios of the two parental phenotypes and the two novel, nonparental phenotypes.

Traits controlled by a single gene and unaffected by environmental conditions are called **Mendelian traits**. But when Mendel described his laws of inheritance, he had no idea what genes were made of, where they were located within a cell, or how they segregated and independently assorted. Today we know that genes are located on chromosomes and that these chromosomes are the basis for all inheritance. We call these assertions the **chromosome theory of inheritance**, which explains the mechanism underlying

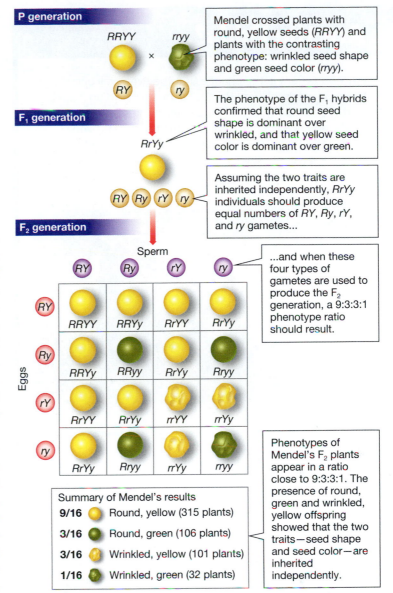

Figure 7.6

Independent assortment of pea color and shape

Mendel used two-trait breeding experiments, called **dihybrid crosses**, to test the hypothesis that the alleles of *two different genes* are inherited independently from each other. ▶

Q1: List all the possible offspring genotypes and phenotypes.

Q2: What is the offspring phenotype ratio?

Q3: Complete a Punnett square for a genetic cross of two true-breeding Portuguese water dogs—one with a black, wavy coat (homozygous dominant, *BBWW*) and one with a brown, curly coat (homozygous recessive, *bbww*). What is the phenotype ratio of their offspring (F$_1$)? Now fill out another Punnett square, crossing two of the offspring. What is the phenotype ratio of the F$_2$ generation?

What Are the Odds?

The *probability* of an event is the chance that the event will occur. For example, there is a probability of 0.5 that a coin will turn up "heads" when it is tossed. A probability of 0.5 is the same thing as a 50 percent chance, or ½ odds, or a ratio of one heads to one tails (1:1). If you toss the coin only a few times, the observed percentage of heads may differ greatly from 50 percent. But if you toss it many, many times, that observed percentage will be very close to 50 percent. Each toss of a coin is an independent event, in the sense that the outcome of one toss does not affect the outcome of the next toss. The probability of getting two heads in a row is a product of the separate probabilities of each individual toss: 0.5 × 0.5, which is 0.25. In our cross, the probability of getting a brown puppy is ¼, or 0.25. To go

back to a Punnett square to predict the ratio of puppies from a genetic cross of two heterozygous (*Bb*) black-coated dogs, the probability of getting a black puppy is ¾, or 0.75.

We cannot know with certainty what the actual phenotype or genotype of a particular offspring is going to be, except when true-breeding individuals are crossed. For example, two brown dogs, both of whom have a *bb* genotype, will have only *bb*-genotype, brown-phenotype offspring. Moreover, the probability that a particular offspring will display a specific phenotype is completely unaffected by how many offspring there are. The likelihood that we will see the 3:1 black-to-brown outcome, however, increases when we analyze a larger number of offspring, just as Mendel analyzed thousands of pea plants.

Mendel's laws by identifying chromosomes as the paired factors (where each homologous chromosome in a pair has one allele for a gene) that are shuffled and recombined, and then separated randomly into sperm and egg cells during meiosis (see Figures 6.7 and 6.8 in Chapter 6). Then, during fertilization, a one-in-a-million sperm fuses with a one-in-a-million egg to create a unique individual. That is how offspring can have genotypes and phenotypes that were not present in either parent, such as a brown puppy born to two black dogs.

Going to the Dogs

Like Mendel, Elaine Ostrander planned to study plants to unravel the secrets of genetics and inheritance. But when she arrived at UC Berkeley to open her lab, the space was not yet available. So she wandered down the hall and into the office of Jasper Rine, a geneticist who normally studied yeast but was looking for someone to start a mammalian genome research project. Ostrander volunteered.

But which mammal to study? "I was allergic to cats, and I didn't know enough about cows or pigs or horses," she recalls, so she picked dogs. Not only was Ostrander a dog lover, but the American Kennel Club had just begun offering funding to researchers trying to identify genes associated with dog diseases.

In 1993, Ostrander began identifying all the genes unique to dogs—that is, making a map of the dog **genome**. Some colleagues said she was nuts, that no one would give her money to support the research. But Ostrander is nothing if not persistent, and she knew the potential value of the research: dogs have more than 350 inherited diseases, and up to 300 of those are similar to conditions in people, including cancer, epilepsy, heart disease, and Addison's disease, the illness that killed Lark's dog Georgie. The genetics of bladder cancer is difficult to study in humans, for example, but the disease is quite common in Scottish terriers and would be easier to study in a dog species. By cracking the genetic code of dogs, Ostrander hoped to uncover causes and potential treatments for human diseases.

In 2005, she published the first full dog genome sequence, for a female boxer named Tasha. The achievement gained her scientific fame and raised awareness among scientists of the importance of dog genetics to human health. "Of the more than 5,500 mammals living today, dogs are arguably the most remarkable," Ostrander's coauthor, Eric Lander, a professor of biology at the Massachusetts Institute of Technology, said when the first dog genome sequence was published. "The incredible physical and behavioral diversity of dogs—from Chihuahuas to Great Danes—is encoded in their genomes. It can uniquely help us understand embryonic development, neurobiology, human disease, and the basis of evolution" (**Figure 7.7**).

Shadow, a standard poodle, was the first dog to have its genome partially (about 80%) sequenced.

Tasha, a boxer, was the first dog to have its complete genome sequenced. Boxers are vulnerable to hip, thyroid, and heart problems. Scientists identified a gene for cardiomyopathy in boxers, a heart disorder also found in humans.

Pembroke Welsh corgis may develop a fatal neurodegenerative condition similar to amyotrophic lateral sclerosis (ALS) in humans. The human gene mutation associated with ALS was also found in corgis with the condition.

Psychiatric disorders often have a genetic component. Doberman pinschers are susceptible to canine compulsive disorder, similar to obsessive-compulsive disorder in humans. The responsible gene in Dobermans has been linked to autism disorders in humans.

Golden retrievers are prone to cancers of the bone marrow. Ostrander's research group is analyzing the genomes of hundreds of goldens with and without cancer, hoping to identify the genes responsible.

Figure 7.7

Man's best friend

The Dog Genome Project has identified the genetic basis of several diseases and conditions in dogs, and in some cases it has been able to link the gene to a similar gene in humans.

Q1: Boxers are far more inbred than poodles. Why does that inbreeding make the former a better target for genetic studies of disease than the latter?

Q2: Explain why a geneticist interested in finding a gene linked to cancer would want to look at the DNA of senior golden retrievers with *and* without cancer?

Q3: Obsessive-compulsive disorder (OCD) in humans is characterized by obsessive thoughts and compulsive behavior, such as pacing. Canine compulsive disorder (CCD) is characterized by compulsive behavior such as "flank sucking," sometimes seen in Doberman pinschers. Would you predict that the medications given to humans with OCD would decrease compulsive behaviors in CCD dogs? Why or why not?

But years before Ostrander completed the dog genome, she had begun a different pet project. In 2001, Ostrander received a call from a scientist in Utah who wanted to talk about dogs. It was Gordon Lark, who told her he was collecting genetic trait information about PWDs. "The day I met Gordon was the best day of my life," says Ostrander. "I knew it was golden."

In 2002, the duo published a paper pinpointing genes that control dog body shape, from the tall, lanky look of a greyhound to the short, stocky frame of a pit bull. In the acknowledgments of the paper, they thanked Karen Miller and all the PWD owners who had contributed pedigree information.

In the spring of 2006, Lark and Ostrander began their second collaboration, this time to identify the genetic basis of dog size. Lark collected skeletal measurements of 92 PWDs and DNA samples from each dog. Ostrander used that genotype and phenotype information to identify a key gene for body size—*IGF1*, which controls the activity of a growth factor and is known to influence body size in mice and humans. This gene's two alleles are called *I* and *B*. Lark and Ostrander discovered that PWDs homozygous for allele *I* (*II*) were usually large dogs, and those homozygous for allele *B* (*BB*) were always small dogs. That single gene accounted for whether a PWD was large or small.

Interestingly, neither *IGF1* allele is dominant or recessive. Instead, heterozygous dogs, with an *IB* genotype, are medium-sized dogs. This is an example of a trait inherited by **incomplete dominance**—in which neither allele is able to exert its full effect, so a heterozygote displays an intermediate phenotype. Dogs with an *IB* genotype aren't large or small, but rather medium-sized (**Figure 7.8**).

Early in the twentieth century, geneticists identified yet another type of interaction among alleles—codominance—that Mendel had not observed among his pea plants. A pair of alleles shows **codominance** when the effect of the two alleles is equally visible in the phenotype of the heterozygote. In dogs, gum color is codominant. A dog's gums can be pink, black, or pink with black spots; in the latter case, both alleles are fully on display, and neither is diminished or diluted by the presence of the other allele (as in incomplete dominance) or suppressed by a dominant allele (as in the case of dominant and recessive alleles). In humans, the blood type AB is a codominant trait.

Genotype: *II*
Phenotype: Large

Genotype: *BB*
Phenotype: Small

Genotype: *IB*
Phenotype: Medium

Figure 7.8

Incomplete dominance of body size alleles

Great Danes and Chihuahuas illustrate the extreme size variation found in domestic dogs (see the chapter-opening photo). Unlike the case with Mendelian traits, dogs heterozygous for the main body size gene show an intermediate size like the Cocker Spaniel pictured here on the right.

Q1: What are the genotypes of a large and a small dog?

Q2: Is it possible to have a heterozygous large dog? Explain why or why not.

Q3: Crossing a Great Dane and a Chihuahua is likely to be unsuccessful, even though they are members of the same species (and thus have compatible sperm and egg). Why is that? What are some potential risks of such a cross?

It's Complicated

Many of the traits people tend to be curious about—body weight, intelligence, athleticism, and musical talent, to name a few—are yet more complicated. A **complex trait** is a genetic trait whose pattern of inheritance cannot be predicted by Mendel's laws of inheritance. Complex traits do not fit the straightforward single-gene, single-phenotype pattern discussed so far.

Sometimes a *single gene* influences a number of *different traits*. Such cases are examples of

Most Chronic Diseases Are Complex Traits

A disease is a condition that impairs health. It may be caused by external factors, such as infection by viruses, bacteria, or parasites, or injury from harmful chemicals or high-energy radiation. Nutrient deficiency can also lead to disease. Inadequate vitamin C consumption, for example, produces scurvy, once common among sailors and pirates. Disease may also be caused by the malfunction of one or more genes. Diseases caused exclusively by gene malfunction are described as genetic disorders, distinguishing them from infections and other types of diseases.

But many of the diseases that are most common in industrialized countries—heart disease, cancer, stroke, diabetes, asthma, and arthritis, for example—are caused by multiple genes interacting in complex ways with each other and with external factors. They are complex traits: malfunctions in key genes make a person susceptible to developing these diseases, but environmental factors affect whether the disease will actually appear and how severe the symptoms will be. A large percentage of the estimated risk of developing chronic diseases is preventable by lifestyle choices such as maintaining good nutrition, exercising regularly (see graph), and avoiding tobacco. (The word "chronic" means "unceasing"—a reference to the fact a person who develops one of these diseases will have it for the rest of their lives.)

A major goal of modern genetics is to identify genes that contribute to human disease. Researchers have identified alleles associated with increased risk of a number of common ailments, including high blood pressure, heart disease, diabetes, Alzheimer disease, several types of cancer, and schizophrenia. The hope is that one day soon, genetic tests will tell us whether we are predisposed to a disease before we become ill with it. Then, a person carrying a risky allele might take preventive measures to reduce the chance of developing the condition, and treatment could be customized to fit the particular allele involved. This tailored approach to treatment, called "personalized medicine," is already being used to treat breast cancer and other chronic diseases.

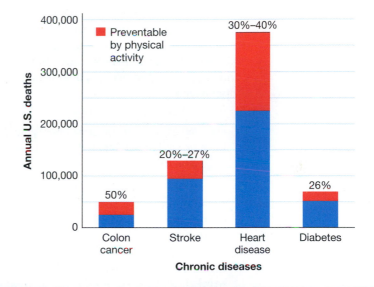

---

pleiotropy (*pleio*, "many"; *tropy*, "change"). In PWDs, Lark found that single genes can control multiple related skeletal traits. The shape of a dog's head and the shape of its limb bones are controlled by a single gene. That connection makes sense, says Lark, since a small head and long legs are advantageous for a fast dog, while a strong dog, like a pit bull, uses both its massive jaw and its short, thick legs for power.

Another good example of pleiotropy comes from a long-term breeding experiment to tame Russian silver foxes. Researchers found that as foxes became tamer and tamer, they also developed floppy ears instead of straight ones and had shorter legs and curlier tails than ordinary foxes have (see "The New Family Pet?," page 129).

Patterns of inheritance can get even more complicated. *Single traits* governed by the action of *more than one gene* are called **polygenic traits**. In humans, polygenic traits include eye and skin color, running speed, blood pressure, body size, and more. Of the thousands of human genetic traits, governed by an estimated 24,000 genes, fewer than 4,000 are known or suspected to be controlled by a single gene with a dominant and a recessive allele. The rest are polygenic traits.

Another twist on inheritance is **epistasis**, which occurs when the phenotypic effect of a gene's alleles depends on the presence of alleles for another, independently inherited gene. Labrador coat color, for example, is affected by epistasis (**Figure 7.9**). Dog fur, as mentioned earlier, has a dominant allele (*B*) that leads to black fur and a recessive allele (*b*) that produces brown fur. But the effects of these alleles (*B* and *b*) can be eliminated completely,

Genotype:	B-, E-	bb, E-	--, ee
Phenotype:	Black	Brown	Yellow

Figure 7.9

Epistasis in coat color

These Labrador retrievers show complex inheritance of coat color. The yellow dog carries two alleles that interfere with the deposition of melanin in hair. Both the brown and the black dogs must carry at least one allele that allows melanin deposition. A dash indicates that the allele is unknown, based on phenotype.

Q1: What are the possible genotypes (at both genes) of the black dog? The yellow dog? The brown dog?

Q2: Draw a Punnett square showing possible matings between the black dog (assuming it is heterozygous at both genes) and the yellow dog (assuming it is heterozygous at the *B* gene). List all the possible phenotypes of their offspring. (See Figure 7.6 for an example of a Punnett square made with two traits.)

Q3: If you wanted the most variable litter possible, what colors of Labrador retrievers would you cross?

ELAINE OSTRANDER

The "mother of all dog projects," Elaine Ostrander is chief and NIH distinguished investigator of the Cancer Genetics and Comparative Genomics Branch, as well as head of the Comparative Genetics Section at the National Institutes of Health. She studies genes important to growth, size variation, and cancer in dogs.

depending on which allele of the expression gene (*E* or *e*) is present. Dogs with a dominant *E* allele deposit a pigment called melanin in their fur and are therefore able to express whatever fur color genotype is present. But a recessive *ee* genotype blocks the deposition of melanin in fur, so the dog is yellow, regardless of the genotype at the *B/b* gene (*BB*, *Bb*, or *bb*).

If the environment affects the phenotype, it becomes nearly impossible to predict the phenotype when given only the genotype of an individual or its parents. The effects of many genes depend on internal and external environmental conditions, such as body temperature, carbon dioxide levels in the blood, external temperature, and amount of sunlight.

For example, cats have a gene that codes for an enzyme called tyrosinase, which is involved in melanin production. Siamese cats have a special C^t allele of the gene. The C^t allele codes for a tyrosinase that works well at colder temperatures ($\leq$35°C) but does not function at warmer temperatures ($\geq$37°C), so the production of melanin depends on the temperature of the surroundings (**Figure 7.10**). Because a cat's extremities tend to be colder than the rest of its body, melanin is produced there, and hence the paws, nose, ears, and tail of a Siamese cat tend to be dark. If a patch of light hair is shaved from the body of a Siamese cat and the skin is covered with an ice pack, the hair that grows back will be dark. Similarly, if dark hair is shaved from the tail and allowed to grow back under warm conditions, it will be light-colored.

Man's Best Friend

After describing the inheritance of size in Portuguese water dogs, Lark and Ostrander looked at the *IGF1* gene in over 350 dogs representing 14 small breeds and 9 giant breeds. The genotype *BB* was common in small dogs and virtually nonexistent in large dogs. "All small dog breeds had them. It didn't matter when they were bred or how; they all had the exact same pattern," says Ostrander.

"It was amazing," adds Lark. Breeders have, over time, been selecting for these alleles to create smaller and smaller dogs. "What mankind can do, without any genetic tools but just knowledge of heritability, is just extraordinary," he says.

Colder extremities (<37°C)

Body temperature (37°C)

Colder extremities (<37°C)

Figure 7.10

The environment can alter the effects of genes

Coat color in Siamese cats is controlled by a temperature-sensitive allele.

Q1: The gene that brings about the pale Siamese body fur is also responsible in part for the typical blue eyes of the species. What is the term for this type of inheritance?

Q2: Siamese kittens that weigh more tend to have darker fur on their bodies. Why might this be?

Q3: The Siamese cat pictured is called a "seal point" because it has seal-colored (dark brown) extremities. Some Siamese cats show the same color pattern, but the dark areas are of a lighter color or even a different shade—for example, lilac point, red point, blue point. What results would you predict if the experiments described in the text (shaving the cat and then increasing or decreasing temperature) were conducted on cats with these color patterns?

After their success with dog size, Lark and Ostrander identified genes responsible for other traits—fur color, leg length, skull shape, and more. They also identified genes related to cancer and other complex traits that might tell us something about human disease (see "Most Chronic Diseases Are Complex Traits," page 127). In border collies, Ostrander's team identified a gene

The New Family Pet?

The silver fox is the same species as the more familiar red fox. Because of its soft, silver coat, it has been bred in captivity for over 100 years to provide fur coats, stoles, and hats for the wealthy.

In 1959, a Russian geneticist, Dmitry Belyaev, began to conduct breeding experiments on silver foxes he had purchased from a fur breeder, pairing only the tamest individuals of each generation. He determined how tame a fox was by observing its response when approached and offered food. As the foxes became tamer in each generation, they did not show a "fear response" until they were older—9 weeks instead of 6 weeks. (Domestic dogs develop a fear response at about 8–12 weeks.) In addition, in the tame foxes, the hormones associated with a fear response did not increase until later. These traits were all clearly influenced by the same gene or genes— an example of pleiotropy.

Another surprising result was that the foxes' appearance began to change along with the changes in behavior. They began to develop shorter tails, wider faces, and floppier ears. All of these made the adult

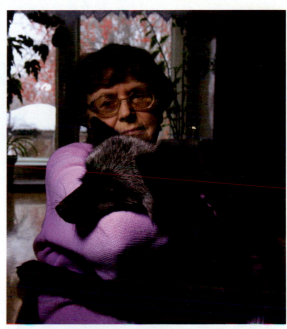

Only 50 years and 35 generations separate this tame silver fox from its wild relative

foxes look more puppy-like, and they are similar to the differences that can be observed when comparing the domestic dog to its ancestor, the wolf. Scientists conjecture that in both cases, tameness and associated changes in development, physiology, and anatomy were brought about by breeding for juvenile features.

GEORGIE

Georgie was Gordon Lark's first Portuguese water dog, a fiercely loyal and playful friend.

Does Bigger Mean Better?

Genome size is the total amount of DNA in one copy of an organism's genome, typically measured in millions of base pairs (Mb). There is a huge range in genome sizes across plants and animals, from living organisms of 150,000 Mb to the *E. coli* genome of 4.6 Mb. Dogs and humans have smaller genomes than many species, making them easier to study genetically than, say, a lungfish.

Assessment available in smartwork5

Lungfish
130,000 Mb

Red-Spotted Newt
39,100 Mb

White Spruce
23,600 Mb

Human
3,200 Mb

Dog
2,400 Mb

Chicken
1,200 Mb

Legend

▬ = 100 megabases

A megabase (Mb) is a unit of length for DNA equal to 1 million base pairs of nucleotides.

involved in an eye disease that causes blindness in both humans and dogs. Her lab also identified a gene involved in kidney cancer in dogs that causes a similar syndrome in humans.

Today, Lark and Ostrander both continue their canine research, and the Georgie Project lives on. In one investigation, the team is collecting data from tissues of deceased PWDs to determine the dogs' state of health at time of death. From data gleaned, they hope to learn how the health of aging dogs (and maybe people) is associated with their genetic makeup. "I'm really happy with the growth of the field," says Ostrander, "and there's a lot of research coming down the pike."

Mopsa, the puppy that Karen Miller gave Lark in 1996 in return for studying PWDs, died in April 2012, just a week short of her sixteenth birthday (**Figure 7.11**). But Lark has a new best friend, a PWD puppy he named Chou (pronounced "shoo"), for the French *petit chou*, meaning "little cabbage."

"We often make a mistake and call Chou 'Mopsa,' because Chou looks so similar," says Lark. After all, he adds, PWDs share similar genotypes and thus similar phenotypes. And these genetic traits make them the cuddly, devoted pets that they are.

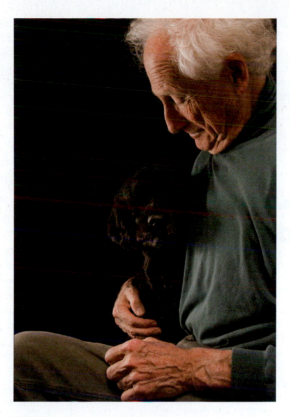

Figure 7.11

Gordon Lark and Mopsa

REVIEWING THE SCIENCE

- A **genetic trait** is any characteristic that is inherited and may be **physical**, **biochemical**, or **behavioral**. All of these types of traits are either **invariant** (the same in all members of a species) or **variable** (different in different members of a species).

- A **gene** is a stretch of DNA that affects one or more genetic traits. Genes are formed on **chromosomes**, threadlike molecules made of DNA and proteins.

- The **genotype** is an individual's genetic makeup or, more specifically, the pair of different versions of a given gene (**alleles**) that determine a given trait. The **phenotype** is the physical expression of an individual's genetic makeup or, more specifically, the expression of a version of the given trait.

- An allele is **dominant** when it prevents a second allele from affecting the phenotype. This second allele is said to be **recessive** because it has no effect on the phenotype when paired with a dominant allele.

- When a genotype consists of two copies of the same allele, it is **homozygous** for that gene. A genotype that consists of two different alleles is **heterozygous** for that gene.

- A grid-like diagram called a **Punnett square** helps predict the probability of genotypes and phenotypes resulting from a **genetic cross**.

- Mendel's experiments enabled him to deduce two laws of inheritance: The **law of segregation** states that two alleles of a gene are separated during meiosis and end up in different gametes. The **law of independent assortment** states that during meiosis, the two alleles of any given gene segregate independently of any two alleles of any other gene.

- **Mendelian traits** are genetic traits controlled by a single gene and unaffected by environmental conditions.

- The **chromosome theory of inheritance** explains how Mendel's laws arise: genes occupy specific locations on chromosomes, and those chromosomes are randomly shuffled and recombined during meiosis.

- When a heterozygote displays an intermediate phenotype, neither allele is dominant—a condition known as **incomplete dominance**, in which neither allele is able to exert its full effect. When the effect of the two alleles is equally visible in the phenotype of the heterozygote, the pair of alleles shows **codominance**.

- **Complex traits** have patterns of inheritance that Mendel's laws cannot predict. In **pleiotropy**, for example,

a single gene influences a number of different traits. **Polygenic traits**, by contrast, are governed by the action of more than one gene. And in **epistasis**, the phenotypic effect of the alleles of one gene depends on the presence of alleles for another, independently inherited gene.

THE QUESTIONS

Challenge Yourself

1 A riddle: In my type of inheritance, the F_1 offspring of a true-breeding black parent and a true-breeding white parent are all gray. What type of inheritance am I?

2 Some time ago, you noticed that the sunflowers in your garden were either tall or short—nothing in between. You bred the tall sunflowers for many generations, until you felt confident that they "bred true," and you did the same for the short sunflowers. You then set up a parental cross (P), and all of the resulting F_1 offspring were short sunflowers. From this experiment, you conclude that the short phenotype is

(a) pleiotropic.

(b) recessive.

(c) true-breeding in the F_1 generation.

(d) dominant.

(e) incompletely dominant.

3 In chickens, a mutant gene called *frizzle* causes not only feathers that curl outward like a Labradoodle's fur, but also an abnormal body temperature, an increased metabolism, and fewer eggs laid than by a normal chicken. From this information, you can conclude that the *frizzle* gene is _____.

Try Something New

4 The silver fox (see "The New Family Pet?" on page 129) belongs to the same species as the red fox: *Vulpes vulpes*. Two silver foxes

always breed true for silver offspring. A silver fox bred to a red fox will produce either all red offspring or, occasionally, half red and half silver offspring. Red foxes bred together usually produce all red offspring, but they occasionally produce silver offspring in the ratio of 3 red to 1 silver. (*Hint:* Draw Punnett squares showing these predicted results.) Which of the following statements is/are consistent with the information provided here about inheritance of coat color in *Vulpes vulpes*? (Select all that apply.)

(a) Red foxes are all homozygous.

(b) Silver foxes are all homozygous.

(c) Red is dominant to silver.

(d) Some silver foxes are homozygous and some are heterozygous.

(e) Some red foxes are homozygous and some are heterozygous.

5 In your garden you grow Big Boy (round) and Roma (oval) tomatoes. You love the taste of Big Boys, but you think it's easier to slice Roma tomatoes. You decide to cross-pollinate a Big Boy and a Roma to see whether you can create a new strain of "Long Boys." In the first generation, all of the tomatoes are round. How would you explain this result? What would your next cross be? Write out the cross in a Punnett square, using parental genotypes. What proportion of the next generation, if any, would be oval?

6 For several hundred years, goldfish have been selectively bred in China and Japan for body color and shape, tail shape, bulging eyes, and even fleshy head growths.

Wild goldfish   Pet-shop goldfish   Black moor goldfish

Imagine that you have a tank of pet-shop goldfish and have just added a couple of black moor goldfish, hoping that they will breed. When the eggs laid by the black moor female (P generation) hatch and the young fish (F_1 generation) begin to develop, you are shocked to see that they are orange. How would you explain this result in terms of the inheritance of body color in goldfish? What breeding experiment could you conduct to test your hypothesis?

7 In 2009, a large team of researchers including Elaine Ostrander and Gordon Lark published the results of its research on coat inheritance in dogs. The study began by focusing on dachshunds and Portuguese water dogs, but then widened to more than 80 breeds. The scientists were able to explain 95 percent of the variation in dog coat types with just two alleles at each of three genes, each inherited independently of the other. These genes coded for hair length (*L/l*), wave or curl in the coat (*W/w*), and the presence of "furnishings" (*F/f*), which are the moustache and eyebrows often seen in wire-haired dogs (see photo). Long-haired dogs carry two copies of the long-hair allele, which is recessive to the short-hair allele. Dogs with furnishings can be either homozygous or heterozygous for the furnishings allele; dogs without furnishings are homozygous for the no-furnishings allele.

a. At the hair length and furnishings genes, what is the genotype of a long-haired dog without furnishings?

b. At the hair length and furnishings genes, what are all the possible genotypes of a short-haired dog with furnishings?

c. Create a Punnett square of two dogs heterozygous for hair length and furnishings. What is the offspring phenotype ratio for those two traits?

Leveling Up

8 **Doing science** Do you want to get involved in dog research? If you have a purebred as a pet, you can. Find out whether the Dog Genome Project at the National Institutes of Health is doing research on your pet's breed. If they are, you can send in a swab of your dog's saliva and contribute to science. Visit the NIH website (http://research.nhgri.nih.gov/dog_genome) for more information.

9 **Is it science?** The November 18, 2003, issue of *Weekly World News* printed a story about a woman who, after repeatedly watching the movie *Shrek* while taking fertility drugs, gave birth to a baby who looked like the main character, an ogre named Shrek. Like Shrek, the newborn had dull green skin, a large flat nose, and ears protruding from stems. From what you know about genetics, do you think it's possible for a developing fetus to change so drastically (from a normal-looking baby to a "Shrek" baby) because its mother was obsessed with a movie? Why or why not? How would you explain your answer to someone who believed this news report?

10 **What do *you* think?** Many people are critical of those who breed or purchase purebred dogs, arguing that there are many mixed-breed dogs waiting to be adopted from shelters. They also point out that mixed-breed dogs are less likely than purebred dogs to suffer from genetic diseases. Those who prefer a particular breed argue that there is a strong genetic influence on dog personality and behavior, and that they don't want any surprises when they add a new member to their family. What do you think?

Thinking Multidimensionally

11 Meet the Hawaiian happy face spider, *Theridion grallator*, named for the happy face pattern on its abdomen. The pattern is due to the presence of a dominant allele, while plain yellow abdomens result from homozygous recessive alleles. This inheritance is controlled by simple Mendelian traits.

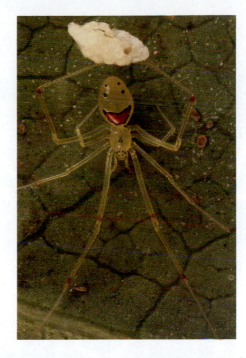

a. Using a Punnett square, cross two heterozygous Hawaiian happy face spiders and record the genotype and phenotype ratios.

b. If a yellow happy face spider is mated with a patterned happy face spider and the offspring are 50 percent patterned to 50 percent yellow, what were the genotypes of the parent spiders?

12 The ancestor of maize (corn) is a plant called teosinte (top). When teosinte is cross-bred with modern maize, a hybrid cob is produced (middle).

a. What type of inheritance (Mendelian trait, codominance, or incomplete dominance) is displayed by this hybrid?

b. Roughly five genes are responsible for the notable differences between teosinte and maize. Describe the phenotypic differences between teosinte and maize.

For more, visit digital.wwnorton.com/bionow2 for access to:

A Deadly Inheritance

How researchers identified a mysterious genetic disorder, and their risky effort to develop a cure.

After reading this chapter you should be able to:

- Interpret a human pedigree to determine whether a given condition is recessive, dominant, or sex-linked.

- Review a human karyotype to identify the sex chromosomes and any abnormalities in chromosome number.

- Diagram a chromosome, identifying genes, alleles, and loci.

- Explain how sex is genetically determined in humans, and how sex determination relates to the inheritance of sex-linked traits.

- Describe the genotype and phenotype of a "genetic carrier" and use this term appropriately for sex-linked traits only.

- Compare and contrast the inheritance of recessive, dominant, and sex-linked genetic disorders.

- Calculate the probability of inheriting a particular genetic disorder by using a Punnett square.

CHAPTER 08

CHROMOSOMES AND HUMAN GENETICS

DRIVING QUESTION

How can gene therapy potentially cure genetically inherited diseases such as Wiskott-Aldrich syndrome and cystic fibrosis?

- HS-LS3.B: Variation of Traits
- HS-ETS1.B: Developing Possible Solutions

Felix clutches his mother's side. Her arm is wrapped tightly around him. Lying in a rumpled white hospital bed, Felix looks away from the two red tubes protruding from his body. Blood pumps out one side of his body into a humming machine next to the bed and is pumped back into the other side. Dr. Christoph Klein steps into the room, clothed head to toe in blue scrubs, and offers a smile and a reassuring word (**Figure 8.1**).

Shortly after his birth in 2005, Felix began to bleed. His parents rushed him to an intensive care unit. The bleeding eventually stopped, but the hospital visits did not. Three years later, Felix was diagnosed with a rare and deadly disease: Wiskott-Aldrich syndrome (WAS). "Wiskott-Aldrich was the very last diagnosis I would ever want to receive," Felix's mother later said. "Every day I prayed, 'Please, Lord, let this chalice pass from us.'"

Felix's pediatrician feared the worst. WAS patients suffer from recurring infections, pneumonia, bleeding, and rashes; they often develop leukemia or lymphoma and die of complications due to infections. Some patients can be treated with a bone marrow transplant. However, if the donor is not a matched sibling or a close unrelated match, survival rates are low.

So, in 2009, Felix's family appeared in Klein's office at the Hannover Medical School in Germany, holding tightly to the hope that he might be able to save Felix's life. Klein, a pediatrician who has made a career of studying rare diseases, was running a clinical trial testing a new treatment for WAS. The treatment was risky and the results unknown, but it was Felix's only hope to be cured of his deadly disease—one that had perplexed scientists for decades.

A Mysterious Malady

More than 70 years earlier, in 1937, three young brothers had come to see Dr. Alfred Wiskott, also a German pediatrician. At first, Wiskott had no idea what was causing their symptoms. The boys bled abnormally: their blood was unable to clot, and they had bloody diarrhea. They also had recurring ear infections and blistering, weeping rashes on their skin. Wiskott recorded their symptoms, but he could not help the boys. All three died at a young age.

What had killed the brothers? Their parents also had four daughters, all of whom were healthy, so it was unlikely that an infection, a toxin, or an environmental factor had caused the illness. Instead, Wiskott suspected, the boys might have inherited a disease from their parents.

Thanks to the chromosome theory of inheritance, Wiskott knew that the boys had inherited hereditary material, in the form of chromosomes, from their parents. Recall from Chapter 7 that offspring inherit one chromosome from the mother and one from the father. Wiskott suspected that a gene on one of the inherited chromosomes was causing the mysterious illness.

A disease caused by an inherited mutation, passed down from a parent to a child, is a **genetic disorder**. Wiskott recognized the importance of researching genetic disorders, since such studies could lead to the prevention of or cure for a disease. But daunting problems have long plagued the study of human genetic disorders. From a biological point of view, humans have a long generation time, select their own mates, and decide whether and when to have children. In addition, human families tend to be much smaller than would be ideal for a scientific study. From an ethical point of view, geneticists and physicians cannot intervene and perform experiments directly on humans to determine how genetic disorders are inherited.

Figure 8.1

Felix and Dr. Klein

Dr. Christoph Klein examines Felix, who suffers from a rare genetic disease.

Painful Pedigree

Twenty years after Wiskott described the first cases of the mysterious disease and determined that it was inherited, an American pediatrician, Dr. Robert Anderson Aldrich, solved the next piece of the puzzle. Aldrich met a six-month-old boy with anemia, bloody diarrhea, and general weakness. After several emergency room visits, the baby died.

Aldrich sat down with the boy's mother to review possible causes. After an hour of asking questions, he still had no idea what might have brought on the illness. Finally, he asked about other relatives who might have had a similar illness. The boy's grandmother, who had tagged along to the meeting, exclaimed sadly, "Just like all the rest of them." Other male infants in the family, it turned out, had died under similar circumstances.

Given that information, Aldrich worked with the mother and grandmother to trace the family's history back six generations by drawing a **pedigree**, a chart similar to a family tree that shows genetic relationships among family

CHRISTOPH KLEIN

A medical doctor and cancer researcher, Christoph Klein is now chair of pediatrics at the University of Munich. In 2010, Klein began testing a new gene therapy to treat young children with Wiskott-Aldrich syndrome, a rare and life-threatening disease. The therapy, though still experimental, has been very successful.

members over two or more generations of a family's medical history (**Figure 8.2**). Pedigrees provide scientists with a way to analyze information in order to learn about the inheritance of a particular genetic trait or disorder. Aldrich found that 16 male infants in the family, but no females, had died of the syndrome (**Figure 8.3**).

Because of the "remarkable family history," as he described it, Aldrich concluded, like Wiskott, that the illness was a genetic disorder, caused by a mutation passed down from parent to child. Genetic disorders can be caused by mutations in

Generation

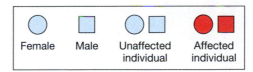

All geneticists who study human inheritance use a standard set of symbols in pedigrees:

- Roman numerals identify different generations
- Circles represent females; squares represent males
- Open symbols (blue here) represent unaffected individuals; filled symbols (red here) represent affected individuals
- Numbers listed below the symbols identify individuals of a given generation

Female	Male	Unaffected individual	Affected individual

Figure 8.2

Patterns of inheritance can be analyzed in family pedigrees

This cystic fibrosis pedigree shows six children (generation III), two of whom are affected with the disease.

Q1: Which two children in this pedigree have cystic fibrosis? How do you know?

Q2: Does either parent of these two children have cystic fibrosis? If so, which one(s)? How do you know?

Q3: Do any of the grandparents of these two children have cystic fibrosis? If so, which one(s)? How do you know?

Generation

Figure 8.3

Pedigree of a family with a history of Wiskott-Aldrich syndrome

As in Figure 8.2, circles represent females and squares represent males in this pedigree. Individuals affected by WAS are shaded red.

Q1: How many male and how many female descendants (individuals that did not join the family by marriage) does generation IV of Aldrich's pedigree contain?

Q2: What proportions of the male and female descendants in generation IV were affected by the disorder?

Q3: Why did Aldrich hypothesize that the disease was X-linked? (You will need to read ahead to answer this question.)

Figure 8.4

A human karyotype

To research chromosomal abnormalities, a scientist will take a photograph of a cell's chromosomes during mitosis and then pair up each set of homologous chromosomes to create a **karyotype**. In humans, the autosomes are numbered 1 through 22, and the sex chromosomes are designated X or Y.

Q1: Is this the karyotype of a male or a female?

Q2: How would the karyotype of a person with Down syndrome differ from this karyotype?

Q3: The size of a chromosome correlates roughly with the number of genes residing on it. Why are an extra copy of chromosome 21 and a missing Y chromosome two of the least damaging chromosomal abnormalities?

individual genes or by abnormalities in chromosome number or structure.

Every species has a characteristic number of chromosomes: humans have 23 pairs of homologous chromosomes, for a total of 46, while mosquitoes, for example, have only 3 pairs, or 6 in total. One of the 23 pairs of human chromosomes consists of the **sex chromosomes**; they determine whether a person

is male or female. All other chromosomes are called **autosomes**.

Autosomes are homologous chromosomes exactly alike in terms of length, shape, and the genes they carry (**Figure 8.4**). Human autosomes are labeled with the numbers 1 through 22 (for example, chromosome 4). Sex chromosomes are assigned letter names; in humans, males have one X chromosome and one Y

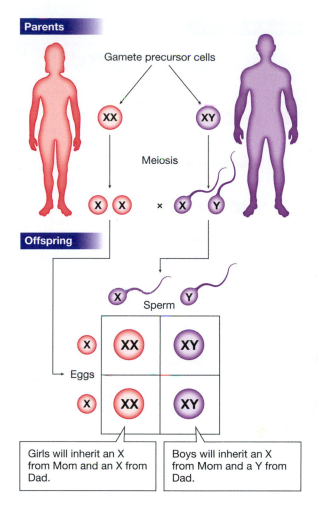

Parents

Gamete precursor cells

XX × XY

Meiosis

X X × X Y

Offspring

X Sperm Y

Eggs

	X (sperm)	Y (sperm)
X (egg)	XX	XY
X (egg)	XX	XY

Girls will inherit an X from Mom and an X from Dad.

Boys will inherit an X from Mom and a Y from Dad.

Figure 8.5

Dad's chromosomes determine baby's sex

Q1: What are the odds that a given egg cell will contain an X chromosome? A Y chromosome? What are those odds for a sperm cell?

Q2: If a couple has two daughters, does that mean their next two children are more likely to be sons? Explain your reasoning. (*Hint:* Refer back to "What Are the Odds?" on page 124.)

Q3: Sisters share the same X chromosome inherited from their father, but they may inherit different X chromosomes from their mother. What is the probability that brothers share the same Y chromosome? What is the probability that brothers share the same X chromosome?

ALFRED WISKOTT

A German pediatrician who lived from 1898 to 1978, Alfred Wiskott described the cases of three young brothers with a serious bleeding disease—what would later be called Wiskott-Aldrich syndrome. Wiskott deduced that the boys had inherited a gene from their parents that was causing the illness.

chromosome, whereas females have two X chromosomes. The Y chromosome in humans is much smaller than the X chromosome.

Because human females have two copies of the X chromosome, all the gametes (eggs) they produce contain one X chromosome, passed on to their offspring. Males, however, have one X chromosome and one Y chromosome, so half of their gametes (sperm) will contain an X chromosome and half will contain a Y chromosome (**Figure 8.5**).

Each chromosome has a particular structure, with genes arranged on it in a precise sequence. Any change in the chromosome number or structure, compared to what is typical for a species, is considered a **chromosomal abnormality**. The two most common types of chromosomal abnormalities in humans are changes in the overall number of chromosomes and changes in chromosome structure, such as a change in the length of an individual chromosome (**Figure 8.6**).

Changes in the number of chromosomes in humans are usually lethal, although Turner

syndrome, in which females receive only one X chromosome, is an exception to this rule. Turner syndrome individuals tend to live long, healthy lives with mild to moderate reproductive issues. Similarly, Down syndrome individuals receive three copies of chromosome 21 and can live relatively long lives, but they still suffer from mild to moderate intellectual and developmental disabilities. Changes in chromosome structure can also have dramatic effects. For example, cri du chat syndrome, caused by a deletion on chromosome 5, results in slowed growth, a small head, and developmental delays.

Looking for Loci

In his 1954 paper describing the disease, Aldrich suggested that since primarily males inherit the syndrome, it is caused by a mutation on a sex chromosome. Aldrich recognized that the family's disease was tightly linked to gender and concluded that the disease was caused by

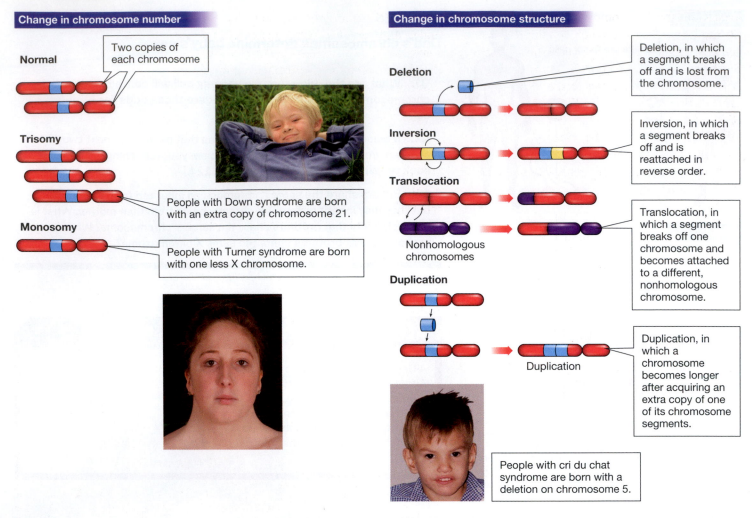

Change in chromosome number

Normal

Two copies of each chromosome

Trisomy

People with Down syndrome are born with an extra copy of chromosome 21.

Monosomy

People with Turner syndrome are born with one less X chromosome.

Change in chromosome structure

Deletion

Deletion, in which a segment breaks off and is lost from the chromosome.

Inversion

Inversion, in which a segment breaks off and is reattached in reverse order.

Translocation

Nonhomologous chromosomes

Translocation, in which a segment breaks off one chromosome and becomes attached to a different, nonhomologous chromosome.

Duplication

Duplication

Duplication, in which a chromosome becomes longer after acquiring an extra copy of one of its chromosome segments.

People with cri du chat syndrome are born with a deletion on chromosome 5.

Figure 8.6

Chromosomal abnormalities can cause serious genetic disorders

Any increase or decrease in the number of chromosomes almost invariably results in spontaneous abortion of the fetus, which is estimated to occur in up to 20 percent of all pregnancies. Down syndrome and a missing or additional sex chromosome are exceptions. Changes in chromosome structure may have relatively minor or more severe effects, depending on the size and location of the change.

Q1: Why are changes in chromosome *number* almost always more severe than changes in chromosome structure?

Q2: In which part of meiosis would you predict that chromosomal abnormalities are produced? (Refer back to Chapter 6 if necessary.)

Q3: Create a mnemonic to help remember the four kinds of structural changes (for example, Doctors Improve Treatment Daily).

sex-linked inheritance—an inherited mutation on the X or Y chromosome. But where exactly was the mutation located?

The physical location of a gene on a chromosome is called its **locus** (plural "loci"). Because a gene can occur in different versions, or alleles, a diploid cell can have two different alleles at a given locus on a pair of homologous chromosomes. If the two alleles at a locus are different, the cell is heterozygous for the gene. If the two alleles at a locus are identical, the cell is homozygous for the gene (**Figure 8.7**).

The sex chromosomes, however, are different. Roughly 1,240 of the estimated 20,000 human genes are found on the X and Y chromosomes. Approximately 1,180 of those 1,240 genes are located on the X chromosome, while only about 60 are located on the much smaller Y chromosome. These 1,240 genes are said to be **sex-linked**. Sex-linked genes on the X chromosome are **X-linked**. Sex-linked genes on the Y chromosome are **Y-linked**. One of these Y-linked genes is the *SRY* gene (short for "sex-determining region of Y"). *SRY* functions as the "master sex switch," committing the sex of the developing embryo to male. In the absence of this gene, a human embryo develops as a female.

The *SRY* gene does not act alone; in both males and females, other genes on the autosomes and sex chromosomes directly influence the development of the sexual characteristics that distinguish men and women. Still, the *SRY* gene plays a crucial role because when present, it causes other genes to produce male sexual characteristics, but when absent, those genes produce female sexual characteristics. Individuals that have XY chromosomal makeup but a nonfunctional SRY gene are considered intersex; they are not reproductively functional males or females, although they appear phenotypically female (**Figure 8.8**).

Interestingly, there are no well-documented cases of disease-causing Y-linked genes. X chromosomes, however, contain genes known to be involved in many human genetic traits and

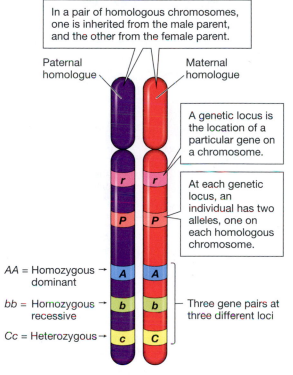

Figure 8.7

Genetic loci on homologous chromosomes

The genes shown here take up a larger portion of the chromosome than they would if they were drawn to scale. The average human chromosome has more than a thousand different genes interspersed with large stretches of noncoding DNA.

Q1: How do we know whether two chromosomes are homologous?

Q2: In one sentence, explain how the terms "gene," "locus," and "chromosome" are related.

Q3: If hair color were determined by a single gene, what would be an example of the gene's alleles?

Figure 8.8

Advocates for intersex youth: interACT

Founded by attorney Anne Tamar-Mattis in 2006 under the name "Advocates for Informed Choice," interACT was created with a focused mission of ending harmful medical interventions on intersex children. Its mission statement reads "interACT uses innovative legal and other strategies, to advocate for the human rights of children born with intersex traits."

disorders. Aldrich correctly concluded that the gene leading to the family's bleeding disease was located on the X chromosome, so it was X-linked.

X Marks the Spot

Thanks to advances in molecular biology tools, in 1994, researchers determined that the gene causing WAS is on the X chromosome and called it, unsurprisingly, *WAS* (gene names are typically italicized, while disorder names, in this case "WAS," are not). *WAS*, they discovered, is the genetic code for a protein crucial for the formation and function of blood cells and immune system cells. Without a healthy copy of this gene, individuals acquire blood and immune system disorders, are susceptible to infections, and have increased risk of lymphoma, cancer of the lymph nodes.

Felix inherited a mutated version of *WAS* from his mother and no allele from his father. We can use a Punnett square to illustrate how the X-linked recessive mutation for WAS is inherited. We label the recessive mutated *WAS* allele *a*, and in the Punnett square we write this allele as X^a to emphasize the fact that it is on the X chromosome. Then we label the dominant, healthy allele *A* and write this allele as X^A in the Punnett square (**Figure 8.9**). Individuals like Felix's mother, who have only one copy of a recessive allele, are said to be **genetic carriers** of the disorder. Carriers can pass on the disorder allele, but they do not have the disease.

If a carrier female like Felix's mother, with genotype $X^A X^a$, has children with a normal male (with genotype $X^A Y$), each of their sons will have a 50 percent chance of getting the disorder. Felix had a 50 percent chance of getting WAS, and he did.

Males of genotype $X^a Y$ suffer from the condition because the Y chromosome does not have a

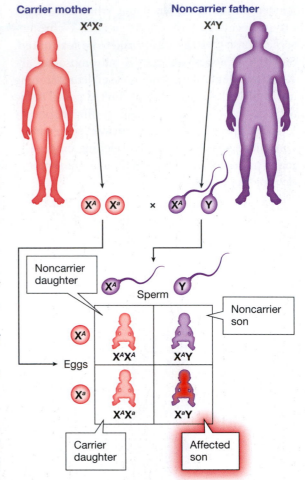

Figure 8.9

X-linked recessive conditions are more common in males

The recessive disorder allele (*a*) is located on the X chromosome and is denoted by X^a. The dominant normal allele (*A*) on the X chromosome is denoted by X^A. ▶️

Q1: Which of the children specified in this Punnett square represents Felix? What is his genotype?

Q2: Explain why Felix is neither homozygous nor heterozygous for the *WAS* gene.

Q3: Create a Punnett square to illustrate the offspring that could result if Felix had children with a noncarrier woman. What is the probability that a son would have WAS? What is the probability that a daughter would be a carrier of WAS?

ROBERT ALDRICH

Robert Aldrich was an American pediatrician who lived from 1917 to 1998. By creating a family pedigree, Aldrich demonstrated that the mysterious bleeding syndrome afflicting infant boys was a sex-linked, recessive disorder.

Genetic Diseases Affecting Americans

Wiskott-Aldrich syndrome is only one of many genetic conditions seen in humans. Here are some of the most common genetic diseases in the United States, most of which can be identified in newborns through genetic testing.

U.S. births per year: 4,000,000
• = 1 birth

6,037 births
Down Syndrome

1,140 births
Cystic Fibrosis

800 births
Marfan Syndrome

Male U.S. births per year: 2,050,000
• = 1 male birth

590 births
Duchenne Muscular Dystrophy

400 births
Hemophilia

400 births
Fragile X Syndrome

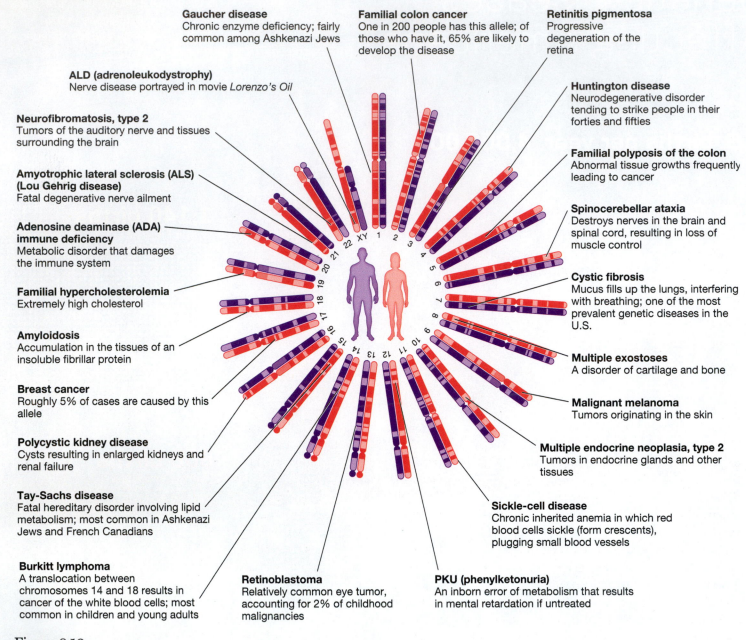

Gaucher disease
Chronic enzyme deficiency; fairly common among Ashkenazi Jews

Familial colon cancer
One in 200 people has this allele; of those who have it, 65% are likely to develop the disease

Retinitis pigmentosa
Progressive degeneration of the retina

ALD (adrenoleukodystrophy)
Nerve disease portrayed in movie *Lorenzo's Oil*

Huntington disease
Neurodegenerative disorder tending to strike people in their forties and fifties

Neurofibromatosis, type 2
Tumors of the auditory nerve and tissues surrounding the brain

Familial polyposis of the colon
Abnormal tissue growths frequently leading to cancer

Amyotrophic lateral sclerosis (ALS) (Lou Gehrig disease)
Fatal degenerative nerve ailment

Spinocerebellar ataxia
Destroys nerves in the brain and spinal cord, resulting in loss of muscle control

Adenosine deaminase (ADA) immune deficiency
Metabolic disorder that damages the immune system

Cystic fibrosis
Mucus fills up the lungs, interfering with breathing; one of the most prevalent genetic diseases in the U.S.

Familial hypercholesterolemia
Extremely high cholesterol

Amyloidosis
Accumulation in the tissues of an insoluble fibrillar protein

Multiple exostoses
A disorder of cartilage and bone

Breast cancer
Roughly 5% of cases are caused by this allele

Malignant melanoma
Tumors originating in the skin

Polycystic kidney disease
Cysts resulting in enlarged kidneys and renal failure

Multiple endocrine neoplasia, type 2
Tumors in endocrine glands and other tissues

Tay-Sachs disease
Fatal hereditary disorder involving lipid metabolism; most common in Ashkenazi Jews and French Canadians

Sickle-cell disease
Chronic inherited anemia in which red blood cells sickle (form crescents), plugging small blood vessels

Burkitt lymphoma
A translocation between chromosomes 14 and 18 results in cancer of the white blood cells; most common in children and young adults

Retinoblastoma
Relatively common eye tumor, accounting for 2% of childhood malignancies

PKU (phenylketonuria)
An inborn error of metabolism that results in mental retardation if untreated

Figure 8.10

Single-gene disorders

Mutations of single genes that lead to genetic disorders are found on the X chromosome and on each of the 22 autosomes in humans. In each of these mutations, the healthy allele at that locus codes for an important function; for example, the sickle-cell allele is a mutation in the gene that codes for the hemoglobin protein, critical for carrying oxygen in the blood. For clarity, only one such genetic disorder per chromosome is shown.

Q1: Which chromosome contains the gene for cystic fibrosis? For Tay-Sachs disease? For sickle-cell disease?

Q2: No known genetic disorders are encoded on the Y chromosome. Why do you think this is?

Q3: In your own words, explain why most single-gene disorders are recessive rather than dominant.

copy of that gene. In other words, because males cannot be heterozygous for any X-linked genes, the effects of an *a* allele cannot be masked. In general, males are more likely than females to get recessive X-linked disorders, because they need to inherit only a single copy of the disorder allele to exhibit the disorder. Females, on the other hand, must inherit two copies to be affected. X-linked recessive inheritance explains why boys are more likely to get WAS than girls are.

Other X-linked genetic disorders in humans include red-green color blindness, hemophilia, and Duchenne muscular dystrophy—a lethal disorder that causes muscles to waste away, often leading to death at a young age. All of these X-linked disorders are caused by recessive alleles.

More Common, but No Less Deadly: Zoe's Story

X-linked disorders like WAS are rare compared to autosomal disorders. Both sexes are equally likely to be affected by **recessive genetic disorders** on autosomes, since both males and females have two copies of autosomal chromosomes and therefore identical odds of being homozygous or heterozygous for a disorder allele. Several thousand human genetic disorders are inherited as recessive traits on autosomes. These include sickle-cell disease, Tay-Sachs disease, and the most common fatal genetic disease in the United States, cystic fibrosis (**Figure 8.10**).

Scott and Jada first began to suspect something was wrong when their newborn daughter, Zoe, didn't put on any weight. Every time Zoe ate, her belly became hard and bloated. She screamed in pain. "I was beside myself," recalls Jada. "I knew something wasn't right."

Like Felix, Zoe spent the first year of her life in and out of the pediatrician's office, until the eve of her first birthday, April 6, 2005. On that day, Scott and Jada sat with Zoe in a children's hospital in Florida, waiting for a second opinion. The doctor came in and asked them to sit down. The diagnosis was cystic fibrosis (**Figure 8.11**).

Cystic fibrosis (CF) is a lethal recessive genetic disorder caused by one or more mutations in the

Figure 8.11

Cystic fibrosis damages lung tissue

The panel on the left shows a chest x-ray from a cystic fibrosis patient and the panel on the right shows a normal chest x-ray. Notice the damaged fibrotic lung tissue in the cystic fibrosis x-ray.

cystic fibrosis transmembrane regulator gene (*CFTR*). A *CFTR* mutation causes the body to produce abnormally thick, sticky mucus, which clogs the airways and leads to lung infections. The thick mucus also obstructs the pancreas, preventing enzymes from reaching the intestines, where they are needed to break down and digest food. The average life span of people with CF who live to adulthood is approximately 35 years. There is no cure.

Recessive genetic disorders vary in severity; some, like cystic fibrosis, are lethal, whereas others have relatively mild effects. Adult-onset lactose intolerance, for example, is caused by a single recessive allele that leads to a shutdown in the production of lactase, the enzyme that digests milk sugar.

The only individuals who get a disorder caused by an autosomal recessive allele (*a*) are those who have two copies of that allele (*aa*). Usually, when a child inherits a recessive genetic disorder, both parents are heterozygous; that is, they both have the genotype *Aa* (**Figure 8.12**). It is also possible for one or both parents to have the genotype *aa* and thus the disease. Because the *A* allele is dominant and does not cause the disorder, heterozygous individuals (*Aa*) like Scott and Jada, Zoe's

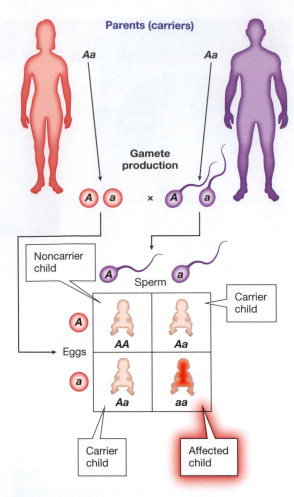

Parents (carriers)

Aa × Aa

Gamete production

A a × A a

Noncarrier child

Sperm

A a

Carrier child

Eggs

	A	a
A	AA	Aa
a	Aa	aa

Carrier child

Affected child

Figure 8.12

Inheritance of cystic fibrosis, an autosomal recessive disorder

The patterns of inheritance for a human autosomal recessive genetic disorder are the same as for any recessive trait (compare this figure with the pattern shown by Mendel's pea plants in Figure 7.5). Recessive disorder alleles are denoted *a*. Dominant, normal alleles are denoted *A*. Here, the parents are a carrier female (genotype *Aa*) and a carrier male (genotype *Aa*).

Q1: Which of the children in this Punnett square represents Zoe? What is her genotype?

Q2: If Zoe's parents had another child, what is the probability that the child would have cystic fibrosis? That the child would be a CF carrier?

Q3: If Zoe is able to have a child of her own someday, and the other parent is not a carrier of cystic fibrosis (he would likely be tested before they chose to have children), what is the probability that the child would have cystic fibrosis? That the child would be a carrier?

parents, are genetic carriers of the disorder; they carry the disorder allele (*a*) but do not have the disease.

If two carriers of a recessive genetic disorder have children, the patterns of inheritance are the same as for any other recessive trait. Each child, male or female, has a 25 percent chance of not carrying the disorder allele (genotype *AA*), a 50 percent chance of being a carrier (genotype *Aa*), and a 25 percent chance of inheriting the disorder (genotype *aa*). Zoe did not beat the odds.

These percentages reveal one way in which lethal recessive disorders such as cystic fibrosis can persist in the human population. Although homozygous recessive individuals (with genotype *aa*) often die before they are old enough to have children, carriers (with genotype *Aa*) are not harmed by the disorder. In a sense, the *a* alleles can hide in heterozygous carriers, and those carriers are likely to pass the disorder allele to half of their children. An estimated one in 29 European Americans has a mutated *CFTR* gene. Recessive genetic disorders can also arise in the human population because new mutations can produce new copies of the recessive alleles.

Deadly with One Allele

Cystic fibrosis is an example of a recessive genetic disorder, in which a child, like Zoe, inherits two recessive copies of a disorder allele. A more rare type of inherited disease is a **dominant genetic disorder**, caused by an autosomal dominant allele (*A*). In this case, the allele that causes a disorder cannot "hide" in the same way that a recessive allele can: *AA* and *Aa* individuals get the disorder; only *aa* individuals are symptom-free (**Figure 8.13**). Dominant genetic disorders are more rare than recessive disorders because a dominant disorder often produces serious negative effects immediately upon birth, and individuals with the *A* allele may not live long enough to reproduce. Hence, few people with a dominant genetic disorder pass the allele on to their children.

For this reason, most cases of a dominant genetic disorder are produced by a new mutation in a generation. For example, achondroplasia, a form of dwarfism, is caused by a mutation

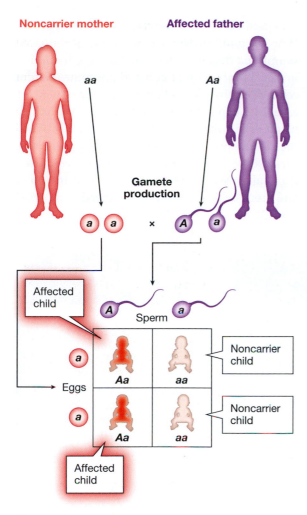

Noncarrier mother   **Affected father**

aa   Aa

Gamete production

a   a   ×   A   a

Sperm

A   a

Affected child

Eggs

a   **Aa**   **aa**   Noncarrier child

a   **Aa**   **aa**   Noncarrier child

Affected child

Figure 8.13

Inheritance of an autosomal dominant disorder

The pattern of inheritance for a human autosomal dominant genetic disorder is the same as for any other dominant trait. This Punnett square shows the possible children of a normal female (genotype *aa*) and an affected male (genotype *Aa*).

Q1: What is the probability that a child with one parent who has an autosomal dominant disorder will inherit the disease?

Q2: Why are there no carriers with a dominant genetic disorder?

Q3: Because dominant genetic disorders are rare, it is extremely rare for both parents to have the condition (genotype *Aa*). Draw a Punnett square with two *Aa* parents. What proportion of the offspring would have the disorder? What proportion would be normal?

in a gene involved in bone growth. People with achondroplasia have a decreased life span, so few of them live long enough to pass the mutation on to their offspring. Instead, infants with achondroplasia are born to unaffected parents at a rate of between one in 10,000 and one in 100,000. Almost all of these births are to fathers older than 35 years, who produce this mutation during sperm production.

Huntington disease, a dominant genetic disorder, is an exception to that rule because symptoms of the disease—uncontrolled movements and loss of intellectual faculties caused by dying brain cells—arise later in life, in one's forties, after the person carrying that allele has had the opportunity to reproduce. In this way the allele is readily passed from one generation to the next. This is why some couples whose families have a history of Huntington's may choose to screen their developing fetus for the gene that causes the disorder.

Replacing Deadly Genes: A Work in Progress

Most inherited genetic disorders, including cystic fibrosis and Huntington disease, have no cure. Patients and their families do everything they can simply to manage the disease. At the age of 7, Zoe was taking 25 pills each day just to digest food. At the age of 10, she underwent four extended hospitalizations, and almost her entire summer was spent on drip IVs. Zoe continues to take antibiotics and mucus thinners, uses nasal sprays, and spends 2 hours a day doing breathing therapy. "We don't talk about the future," says Jada, Zoe's mother. Her voice cracks. "But she is aware that she's not like everybody else."

Scientists have not given up the race to find effective treatments, even cures, for genetic disorders. Thanks to Christoph Klein, WAS is one of the few such disorders for which an effective treatment has been identified.

It was in 2003 that Klein, then at Boston Children's Hospital, saw the first glimmer of hope that there might be a therapy for boys afflicted with WAS. He and colleagues believed

GENETICS

that they might be able to treat WAS using **gene therapy**, a technique for correcting defective genes responsible for disease development. Gene therapy is a type of **genetic engineering**, the permanent introduction of one or more genes into a cell, tissue, or organism. Klein hoped that by correcting the defective *WAS* gene in young affected boys, he could offer a short-term treatment, and potentially even a permanent cure, for the disorder.

First, Klein and his team tested their plan in mice. They bred mice that lacked the entire *WAS*

gene and thus could not produce the important WAS protein. These mice had some of the same symptoms that Felix had, including a reduction in the number of blood and immune system cells. Then the researchers used a virus to insert a healthy copy of the *WAS* gene into the mice's blood cells, where they hoped it would produce the WAS protein. Months later, Klein was thrilled to find that the mouse cells were expressing healthy WAS protein, and that mature blood and stem cells were being produced in normal quantities.

Prenatal Genetic Screening

How is the baby? This is one of the first questions we ask after a child is born. Usually everything is fine, but sometimes, as with Felix and Zoe, the answer can be devastating. Today, some parents choose to have prenatal genetic screening tests performed to check their baby's health before birth.

In **amniocentesis**, a needle is inserted through the abdomen into the uterus to extract a small amount of amniotic fluid from the pregnancy sac that surrounds the fetus. This fluid contains fetal cells (often sloughed-off skin cells) that can be tested for genetic disorders. Another method is **chorionic** (kohr-ee-AH-nik) **villus sampling** (**CVS**), in which a physician uses ultrasound to guide a narrow, flexible tube through a woman's vagina and into her uterus, where the tip of the tube is placed next to the villi, a cluster of cells that attaches the pregnancy sac to the wall of the uterus. Cells are removed from the villi by gentle suction and then tested for genetic disorders.

Risks associated with amniocentesis and CVS—including vaginal cramping, miscarriage, and premature birth—have declined dramatically in recent years because of advances in technology and more extensive training. Recent studies suggest that the risk of miscarriage after CVS and amniocentesis is essentially the same: about 0.06 percent. The tests are widely used by parents who know they face an increased chance of giving birth to a baby with a genetic disorder. Older parents, for example, might want to test for Down syndrome, since the risk of that condition increases with the age of the mother and perhaps the father too. A couple in which one parent carries an allele for a dominant genetic disorder (such as Huntington disease), or both parents are carriers for a recessive genetic disorder (such as cystic fibrosis), might also choose prenatal genetic screening.

Couples who elect to have such tests performed have only two choices if their fears are confirmed: they can abort

the fetus, or they can give birth to a child with a genetic disorder. Prior to conception, however, couples at risk of having a child with a genetic disorder have options to minimize that risk.

- If they are willing and can afford the procedure, a couple can choose to have a child by **in vitro fertilization** (**IVF**), in which an egg is fertilized by a sperm in a petri dish, after which one or more embryos are implanted into the mother's uterus.

- During **preimplantation genetic diagnosis** (**PGD**), one or two cells are removed from the developing embryo in the dish, usually 3 days after fertilization. The cell or cells removed from the embryo are tested for genetic disorders. Finally, one or more embryos that are free of disorders are implanted into the mother's uterus, and the rest of the embryos, including those with genetic disorders, are frozen. Typically, parents who opt for PGD either have a serious genetic disorder or carry alleles for one.

Like all other genetic screening methods, the use of PGD raises ethical issues. People who support PGD feel that amniocentesis and CVS provide parents with a bleak set of moral choices: if the fetus has a serious genetic disorder, the parents can either abort the fetus or allow the birth of a child who will live a life that may be short and full of suffering. In their view, discarding an embryo at the 4- to 12-cell stage is morally preferable to aborting a well-developed fetus, or to giving birth to a child that will suffer the devastating effects of a serious genetic disorder. Those opposed to PGD agree that the moral choices are bleak, but they argue that once fertilization has occurred, a new life has formed and it is immoral to end that life, even at the 4- to 12-cell stage. What do you think?

Bolstered by these results, the scientists tested the technique in human cells and finally, in 2005, began a human clinical trial in Germany. The first two boys, both 3 years old, were admitted in 2006. Between 2006 and 2009, 10 boys were admitted to the clinical trial, with Felix coming on board in 2009.

During the first phase of the trial, Felix had to lie still for 9 hours as a machine pumped blood out of his body in order to extract cells. These cells were then taken to a laboratory where scientists used a genetically modified virus to insert a healthy copy of *WAS* into them (**Figure 8.14**). The doctors then pumped the genetically engineered cells back into Felix's body. And the waiting began.

Klein published the early results in 2010. After gene therapy, nine of the boys improved with regard to bleeding and infections. They were able to play soccer like their healthy peers, they responded to vaccinations, and they did not develop severe infectious diseases. Klein and his colleagues showed that gene therapy for the fatal WAS diagnosis is feasible, and that WAS can be corrected.

A new therapy does not come without risks, though. A few years after receiving the gene therapy, seven of the patients developed leukemia

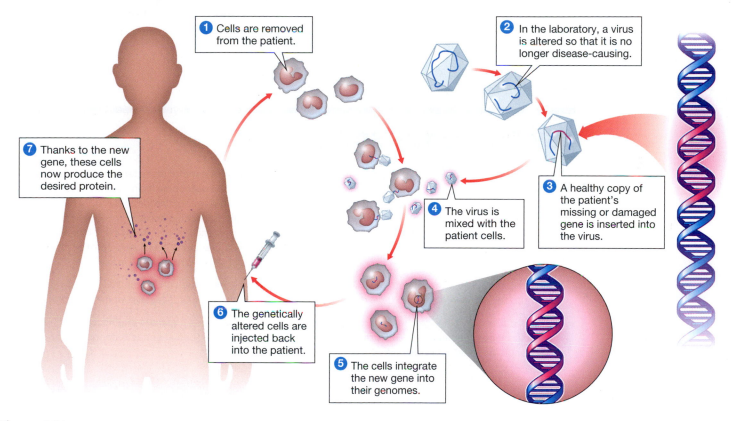

Figure 8.14

Gene therapy

In gene therapy, genetic information is transferred into cells to achieve a desired effect. Gene therapy may be used to compensate for a genetic mutation in a cell that causes the cell to malfunction. ▶▤

Q1: Which gene was missing or damaged in Felix's case? From what chromosome would a healthy copy be taken?

Q2: Why did Dr. Klein's group first conduct gene therapy on mice rather than on humans? What are the advantages and limitations of this approach?

Q3: If Felix has children of his own someday, will they run the risk of inheriting his disorder, or has gene therapy removed that possibility? Explain your reasoning.

from the type of virus used to insert the healthy gene into their cells. Three of the boys died, either from leukemia or from complications after bone marrow transplants. For this reason the trial was halted. Together with other physicians and researchers, Klein is investigating the causes of leukemia after gene therapy, and he is looking for safer ways to insert the healthy gene into cells.

A Happy Ending for Felix

Numerous gene therapy trials continue around the world for other genetic diseases, using other methods to insert new genes as a way to avoid risks like leukemia. In 2012, European health officials green-lit gene therapy for a recessive genetic disorder called lipoprotein lipase deficiency (LPLD), the first commercially approved gene therapy treatment in either Europe or the United States. Many others nip at its heels, including treatments for hemophilia and ADA-SCID, also known as the "bubble boy disease."

As for Felix, a year after the gene therapy, he returned to the hospital for tests. The treatment had worked. Today, his body is producing healthy, functional blood cells, and most of his symptoms are completely resolved. Felix now enjoys a normal life. His inherited genetic disorder was cured by gene therapy. "The most beautiful moment is when my son smiles, hugs me, and says, 'Mum, I love you so much,'" said Felix's mother after the treatment. "We are grateful that there was a treatment that helped our son."

REVIEWING THE SCIENCE

- A **genetic disorder** is a disease caused by an inherited mutation in a gene, passed down from a parent to a child. Genetic disorders can be caused by mutations in individual genes or by **chromosomal abnormalities** (changes in chromosome number or structure).

- Every person has two **sex chromosomes**: males have one X and one Y chromosome, and females have two X chromosomes. The *SRY* gene on the Y chromosome is required for human embryos to develop as males.

- The physical location of a gene on a chromosome is called its **locus**. Genes found solely on the X or Y chromosome are said to be **sex-linked**.

- A **dominant genetic disorder** is caused by a dominant allele on an **autosome** (non–sex chromosome). Dominant genetic disorders are more rare than **recessive genetic disorders**, which are caused by two autosomal recessive alleles. Several thousand human genetic disorders are inherited as recessive traits on autosomes.

- A family **pedigree** can be used to determine whether a given condition is recessive, dominant, or sex-linked.

- **Gene therapy** is a **genetic engineering** technique for correcting defective genes responsible for disease development.

THE QUESTIONS

Challenge Yourself

1 Sickle-cell disease is inherited as a recessive genetic disorder in humans; the normal hemoglobin allele (*H*) is dominant to the sickle-cell allele (*h*). For two parents of genotype *Hh* (carriers), construct a Punnett square to show the possible genotypes of their children.

2 Which of the following correctly predicts the possible genotypes of the children from question 5?

(a) $\frac{1}{4}$ *HH*, $\frac{1}{4}$ *Hh*, $\frac{1}{2}$ *hh*

(b) $\frac{1}{2}$ *HH*, $\frac{1}{2}$ *Hh*

(c) $\frac{1}{2}$ *HH*, $\frac{1}{2}$ *hh*

(d) ¼ *HH*, ½ *Hh*, ¼ *hh*

(e) ¼ *HH*, ¾ *Hh*

3 Which of the following correctly predicts the possible phenotypes of the children from question 5?

(a) ¼ normal, ¾ sickle-cell disease

(b) all normal

(c) all sickle-cell disease

(d) ½ normal, ½ sickle-cell disease

(e) ¾ normal, ¼ sickle-cell disease

4 Given that the normal hemoglobin allele (*H*) is dominant to the sickle-cell allele (*h*), each time two *Hh* individuals have a child together, what is the chance that the child will have sickle-cell disease?

(a) 0%

(b) 75%

(c) 25%

(d) 50%

(e) 100%

5 Which of the following karyotypic sex chromosome abnormalities result(s) in a male phenotype? (Select all that apply.)

(a) XO

(b) XXY

(c) XXX

(d) XYY

(e) XXXY

(f) XY with a complete deletion of the *SRY* gene

Try Something New

6 On the Indonesian island of Bali, about ¾ of the feral (stray) cats have a stumpy tail, while only ¼ of the cats have a long tail. Experiments in which a bunch of random stumpy-tailed cats were mated with each other yielded some stumpy-tailed kittens and some long-tailed kittens. The same experiments with long-tailed cats produced only long-tailed kittens. What do these results show about the genotype of the long-tailed cats? (*Hint:* Draw Punnett squares using the letter "t" for the tail trait to help you determine which allele is dominant and which is recessive.)

(a) They must all be *tt*.

(b) They are both *Tt* and *tt*.

(c) They are both *Tt* and *TT*.

(d) They must all be *Tt*.

7 Recall that human females have two X chromosomes, while human males have one X chromosome and one Y chromosome.

a. Do males inherit their X chromosome from their mother or from their father?

b. If a female has one copy of an X-linked recessive allele for a genetic disorder, does she have the disorder?

c. If a male has one copy of an X-linked recessive allele for a genetic disorder, does he have the disorder?

d. Assume that a female is a carrier of an X-linked recessive disorder. With respect to the disorder allele, how many types of gametes can she produce?

e. Assume that a male with an X-linked recessive genetic disorder has children with a female who does not carry the disorder allele. Could any of their sons have the genetic disorder? How about their daughters? Could any of their children be carriers for the disorder? If so, which sex(es) could they be?

8 For the pedigree shown here, the disorder is caused by a recessive (*g*) allele on the X chromosome. Label each of the following individuals with the correct genotype ($X^G Y$, $X^g Y$, $X^G X^G$, $X^G X^g$, $X^g X^g$).

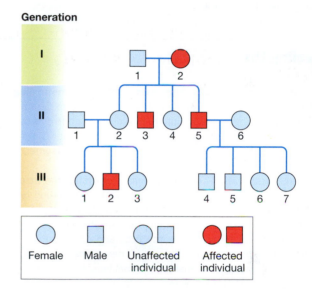

a. Generation I, number 2

b. Generation II, number 2

c. Generation II, number 5

d. Generation III, number 2

e. Generation III, number 6

9 For the pedigree shown here, the disorder is caused by a recessive (*g*) autosomal allele. Label each of the following individuals with the correct genotype (*GG*, *Gg*, *gg*).

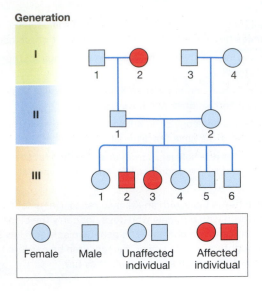

Generation

I

II

III

○ Female	□ Male	○□ Unaffected individual
		● ■ Affected individual

a. Generation I, number 2

b. Generation II, number 1

c. Generation II, number 2

d. Generation III, number 3

Leveling Up

10 *Write Now* **biology: debating preimplantation genetic diagnosis (PGD)** This assignment is designed to give you the experience of applying your knowledge of biology to a current controversy or topic of interest. You will apply the same sort of reasoning that you should be able to use as an informed citizen and consumer when making decisions that involve biology.

The scenario: The U.S. Senate Committee on Commerce, Science and Transportation (CCST) is considering proposing legislation on preimplantation genetic diagnosis (PGD; see "Prenatal Genetic Screening" on page 148). The chair of the committee has invited special-interest groups to present testimony on the pros and cons of PGD. You will contribute a position paper defining PGD, describing your group's position on the technology, and making a recommendation for legislation.

Choose one of the following special-interest groups to represent (or your instructor will assign one to you):

(a) **Reproductive Specialist Group (medical doctors).** You will describe how PGD and in vitro fertilization (IVF) can be used to screen for genetic disorders or for sex selection and "family balancing," and to increase fertility in women of advanced maternal age. You will argue that legislation is not necessary, since medical association guidelines are already in place, and that it is not the place of government to judge individuals' reasons for undergoing PGD.

(b) **Genetics and Public Policy Group.** You will present data on the beliefs of the American public in this matter. You will also present the status of current legislation in the United States and how it compares to PGD legislation internationally. You will propose limited legislation based on these findings.

(c) **Parents of Down Syndrome Children.** You will argue that people with disabilities already suffer from prejudice, and that widespread use of such testing will cause even more prejudice. You will propose legislation prohibiting PGD for

almost all conditions, with the exception of deadly, infant-onset disease.

(d) **Americans with Cystic Fibrosis.** You will present the case of a couple, both of whom are carriers for CF, and discuss their options. You will propose limited legislation that disallows PGD for sex selection and nondisease conditions.

11 **What do *you* think?** *Gattaca*, a science fiction movie from 1997, is based in a future in which preimplantation genetic diagnosis (PGD) is used to create genetically superior children, and social status is determined mainly by one's DNA. Watch the movie and take careful notes as you observe how PGD has been misused by society. Do you think these outcomes are likely, if we continue to develop PGD? Are the benefits worth this risk?

12 **Life choices** While you cannot control the genes that you inherited from your parents, you can control your exposure to external factors that increase your chances of developing a disease that is only partially genetically determined. For this assignment, you will estimate your life expectancy, as determined mainly from your lifestyle choices. Go to the "Living to 100" website (http://www.livingto100.com) and complete the Life Expectancy Calculator. Answer all questions honestly (no one will see your answers), and go to the "feedback" page at the end of the questionnaire. Take notes so that you can answer the following questions.

a. What is your calculated life expectancy, given the answers you gave?

b. List the recommendations to increase your life expectancy, and how much they could increase your life span.

c. What change would increase your life span the most?

d. Identify one way that you could feasibly increase your life span. Will you do that? Why or why not?

Thinking Multidimensionally

13 Ellis–van Creveld syndrome is a disease that results in very short stature (dwarfism) and, in some cases, extra fingers or toes (polydactyly) because of mutations in two genes on chromosome 4 called *EVC* and *EVC2*. This is an autosomal recessive disorder that is common among the Old Order Amish of Pennsylvania.

a. If two individuals who are carriers of this disease conceive a child, what is the percent chance that their child will inherit this disease?

b. While Ellis–van Creveld syndrome occurs within different ethnic groups worldwide, why is it more common among the Amish population in Pennsylvania?

 From 1932 to 1972, the Public Health Service, working with the Tuskegee Institute, began a study on the effects of untreated syphilis in black men ages 25 and older in Macon County, Alabama. Researchers told the men they were being treated for "bad blood," but in reality they were not being given penicillin to cure their illness, despite its being readily available by 1947.

a. What questions do you have about this experiment?

b. Why were these men not given proper treatment by the researchers?

c. How has this study changed the way we conduct research on human subjects?

For more, visit digital.wwnorton.com/bionow2 for access to:

Pigs to the Rescue

Using CRISPR, a hot new genome-editing tool, scientists hope to create a steady stream of transplant organs—from pigs.

After reading this chapter you should be able to:

- Describe the structure of DNA, using appropriate terminology.
- Use the base-pairing rules to determine a complementary strand of DNA based on a given template strand.
- Describe how the genome editing tool CRISPR-Cas9 works.
- Label a diagram of DNA replication, identifying the location of each step in the process.
- Identify when PCR and gene sequencing technology should be used in an experiment.
- Explain the cause of DNA replication errors, and describe how they are repaired.
- Give an example of a mutation and its potential effects on an organism.

CHAPTER
09

WHAT GENES
ARE

DRIVING QUESTION

How does CRISPR work, and why are pigs suitable candidates for the human organ shortage crisis?

- HS-LS1.A: Structure and Function
- HS-LS3.A: Inheritance of Traits
- HS-LS3.B: Variation of Traits
- HS-ETS1.A: Defining and Delimiting Engineering Problems
- HS-ETS1.B: Developing Possible Solutions

"Pigs," thought Marc Güell. "What if we could modify pigs?"

Güell, a young biochemist at Harvard Medical School in Massachusetts, had been working with colleagues on techniques to manipulate **DNA** (**deoxyribonucleic acid**), the genetic code of life. In 2012, Güell and his colleague Luhan Yang, along with their boss, the geneticist George Church, had begun adding and deleting genes from organisms, a process known as genome editing, using a new technological breakthrough called CRISPR-Cas9—CRISPR for short.

Before the discovery of CRISPR (pronounced "crisper"), editing DNA was difficult and expensive, and it was typically done only in model organisms such as mice and fruit flies—species for which biologists had developed a solid tool kit for genetic manipulation. CRISPR-Cas9, a simple but creative combination of a specific protein and two single-stranded **RNA** (**ribonucleic acid**) molecules, made genome editing inexpensive and available to all, not just to well-funded genetics labs. It also enabled researchers to quickly and efficiently change the DNA of nearly any organism—like a "molecular scalpel," as one of the CRISPR inventors called it—making it possible to edit genes in fungi, plants, humans, you name it. "It was the biggest change in my career," says Güell. "With CRISPR, everything is easier, faster, and cheaper."

CRISPR also provided the new opportunity to edit more than one gene at a time, and the team picked a unique initial specimen: *Sus scrofa domesticus*, the domestic pig. "We started to think about what would be a good application to use this technology," says Güell. "One of the things that seemed like a very interesting, very difficult problem was the lack of organs for transplantation. We thought, what if we could modify pigs to make them compatible enough to be an unlimited source of organs?" (**Figure 9.1**).

Figure 9.1

Experimental pig-to-baboon lung transplant

A lung from a pig engineered by CRISPR to prevent rejection is tested for safety and efficacy in primates by being transplanted into a baboon.

Deep in the DNA

Each day in the United States, an average of 22 people die waiting for an organ transplant, and 119,000 men, women, and children are currently on the national transplant waiting list, each hoping their name is called before it's too late.

Researchers have explored many ways to grow and store organs for transplantation—from freezing them to building them from scratch—but one of the most promising, if you can look past the mud and flies, is pigs. Our porcine friends have long been considered an excellent potential source of organs because their organs are relatively close in size to human organs—including the heart, liver, and kidneys—and because pigs and humans have similar organ anatomies (**Figure 9.2**). In addition, pigs are an easier sell to the public: in

MARC GÜELL AND GEORGE CHURCH

Marc Güell (right) is a research fellow at the Wyss Institute at Harvard Medical School, where he has been working to engineer the pig genome for human compatibility. He works in the lab of geneticist George Church (left), a pioneer in genome engineering and synthetic biology.

general, people tend to prefer the idea of transplants from pigs over transplants from mammals more closely related to us, such as baboons. If we were able to transplant organs from animals into humans, a process called *xenotransplantation*, healthy organs could be available on demand and in essentially limitless supply.

Yet there has been a barrier to harvesting pig organs for humans: the pig genome is dotted with DNA from a family of viruses called porcine endogenous retroviruses, or PERVs. DNA is built from two parallel strands of repeating units called **nucleotides**. Each nucleotide is composed of the sugar deoxyribose, a phosphate group, and one of four **bases**: adenine, cytosine, guanine, or thymine. We identify nucleotides by their bases, using "adenine nucleotide" as shorthand for "nucleotide with an adenine base."

The nucleotides of a single strand are connected by covalent bonds between the phosphate group of one nucleotide and the sugar of the next nucleotide. The two DNA strands are connected by hydrogen bonds linking the bases on one strand to the bases on the other, like the rungs that connect the two sides of a ladder (**Figure 9.3**). The term **base pair** (or nucleotide pair) refers to two nucleotides held together by one of these bonds between their bases; that is, a base pair corresponds to one rung of the DNA ladder. The ladder twists into a spiral called a **double helix** (**Figure 9.4**). Within the long, winding double helix of the pig genome, short sections of DNA from PERVs are scattered about—sections made up of the same four nucleotides but encoding information for viral proteins rather than pig proteins.

Nucleotides do not form base pairs willy-nilly. The adenine (A) nucleotide on one strand can pair only with thymine (T) on the other strand (see **Figure 9.3**); similarly, cytosine (C) on one strand can pair only with guanine (G) on the other strand. These **base-pairing rules**, which provide **complementary base-pairing** between two nucleic acid strands, have an important consequence: when the sequence of nucleotides on one strand of the DNA molecule is known, the sequence of nucleotides on the other, complementary strand of the molecule is automatically known as well. The fact that A can pair only with T and that C can pair only with G allows the original strands

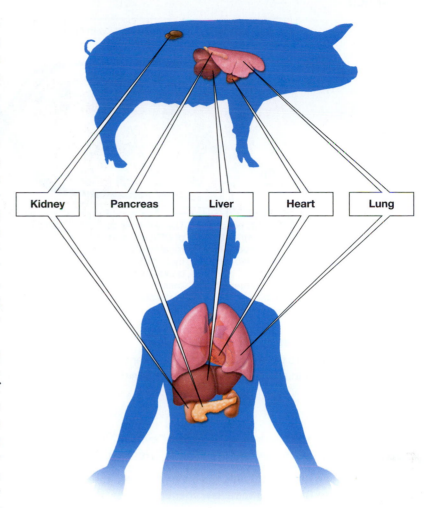

Figure 9.2

Pig and human organs are remarkably similar in size

Q1: Name one reason why, for a potential transplant, matching the size of organs shown would be important.

Q2: Name a tissue transplant for which size matching would not be as important.

Q3: Name a tissue transplant for which size matching would not be important at all.

to serve as "template strands" on which new strands can be built through complementary base-pairing. (We will see more in Chapter 10 about building new DNA strands, including how RNA can pair with DNA, which CRISPR takes advantage of.)

Still, the four nucleotides can be arranged in any order along a single strand of DNA, and each DNA strand is composed of millions of these nucleotides, so a tremendous amount of

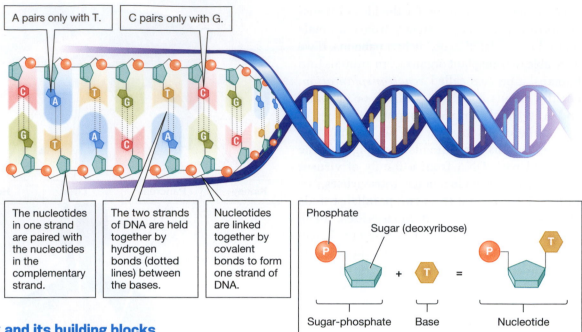

A pairs only with T.

C pairs only with G.

The nucleotides in one strand are paired with the nucleotides in the complementary strand.

The two strands of DNA are held together by hydrogen bonds (dotted lines) between the bases.

Nucleotides are linked together by covalent bonds to form one strand of DNA.

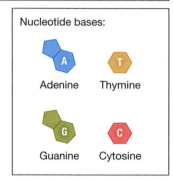

Phosphate

Sugar (deoxyribose)

P

T

=

P

T

Sugar-phosphate   Base   Nucleotide

Figure 9.3

The DNA double helix and its building blocks

A molecule of DNA consists of two complementary strands of nucleotides that are twisted into a spiral around an imaginary axis, rather like the winding of a spiral staircase. ▶️

Nucleotide bases:

A — Adenine   T — Thymine

G — Guanine   C — Cytosine

Q1: Name two base pairs.

Q2: Why is the DNA structure referred to as a "ladder"? What part of the DNA represents the rungs of the ladder? What part represents the sides?

Q3: Is the hydrogen bond that holds the base pairs together a strong or weak chemical bond? (*Hint:* Refer to Chapter 3 to review chemical bonds, if necessary.)

Figure 9.4

What DNA actually looks like

In November 2012, Italian researchers used an electron microscope to directly visualize DNA for the first time. This is the single thread of double-stranded DNA that they saw.

information can be stored in a DNA sequence and in a genome. The genome of the domestic pig, for example, has about 3 billion base pairs, the human genome has about 3.2 billion base pairs, a tomato has only about 900 *million* base pairs, and the bacterium *Escherichia coli* has a measly 4.6 million. The sequence of nucleotides in DNA differs among species and among individuals within a species, and these differences in genotype can result in different phenotypes (**Figure 9.5**).

In the mid-1990s, scientists became very excited about the idea of using pig organs in humans, but testing stalled because of the fear that humans would become infected with PERVs. Just breeding pigs in sterile conditions can't get rid of the virus; it's integrated right there in the double helix. The Harvard team believed CRISPR might be able to solve that

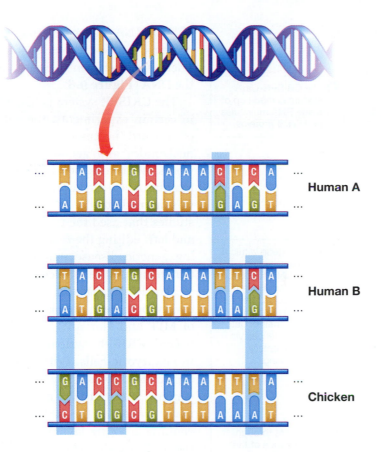

Figure 9.5

The sequence of bases in DNA differs among species and among individuals within a species

The sequence of bases in a hypothetical gene is compared for two humans (A and B) and a chicken. Base pairs highlighted in blue are variant; that is, they differ between the genes of persons A and B, and between the same genes in humans and chickens.

Q1: If all genes are composed of just four nucleotides, how can different genes carry different types of information?

Q2: Would you expect to see more variation in the sequence of DNA bases between two members of the same species (such as humans) or between two individuals of different species (for example, humans and chickens)? Explain your reasoning.

Q3: Do different alleles of a gene have the same DNA sequence or different DNA sequences?

problem by destroying the PERV DNA in pig cells once and for all.

Precise Cuts

The DNA sequences called **CRISPR,** a blessedly easy acronym for *clustered regularly interspaced short palindromic repeats*, are actually part of a defense system used by bacteria. Bacteria are constantly under bombardment from viruses that try to sneak into and take over their genomes, so bacteria evolved a set of defensive measures, including a tool to recognize and cut foreign, interloping DNA. Pioneered by microbiologists Jennifer Doudna of UC Berkeley and Emmanuelle Charpentier at Umeå University in Sweden, the CRISPR-Cas9 system is composed

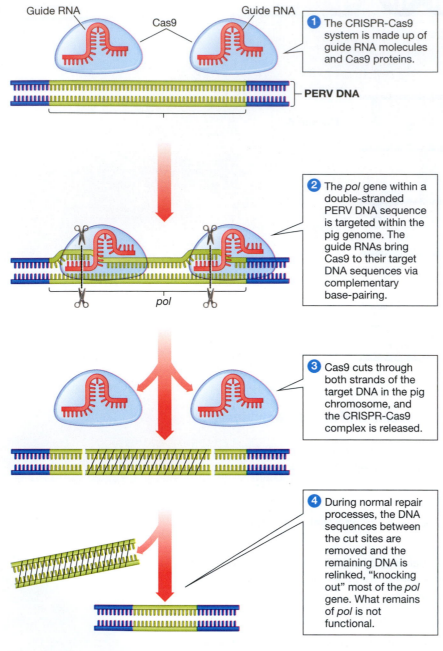

Guide RNA

Cas9

Guide RNA

1 The CRISPR-Cas9 system is made up of guide RNA molecules and Cas9 proteins.

PERV DNA

2 The *pol* gene within a double-stranded PERV DNA sequence is targeted within the pig genome. The guide RNAs bring Cas9 to their target DNA sequences via complementary base-pairing.

pol

3 Cas9 cuts through both strands of the target DNA in the pig chromosome, and the CRISPR-Cas9 complex is released.

4 During normal repair processes, the DNA sequences between the cut sites are removed and the remaining DNA is relinked, "knocking out" most of the *pol* gene. What remains of *pol* is not functional.

Figure 9.6

Genome editing with CRISPR-Cas9, an efficient and cost-effective tool

Q1: What common mechanism is employed by the guide RNA to find its target DNA sequence?

Q2: How many strands of DNA must Cas9 cut to be effective?

Q3: Does Cas9 also cause the deletion of DNA from the genome?

of two RNA molecules that guide two Cas9 proteins to chosen, precise sites in a genome. There, Cas9 efficiently cleaves both strands of the DNA (**Figure 9.6**).

The CRISPR system is like a "Delete" key or, in certain experiments, the cut-and-paste tool of a word processing program—except with nucleotides and genes instead of letters and words. While most new laboratory tools can take months or years to become widely used, CRISPR was immediately a lab favorite, with genetic studies that used the tool being published right and left: editing the zebra fish genome, correcting genetic disease mutations in adult mice, enhancing pest resistance in wheat, and more.

Shortly after Doudna and Charpentier's discovery, both Feng Zhang (at the Broad Institute of MIT and Harvard) and Church's lab engineered CRISPR systems for genome editing in eukaryotic cells. With that innovation, the Church team was ready to start editing the pig genome, but first they needed to identify how many copies of the PERV DNA the genome contained. Using laboratory tools to sequence the pig DNA, they identified 62 copies of PERVs scattered throughout the genome.

Sixty-two may sound like a lot, but it's a drop in the bucket compared to the number of genes in a mammalian genome. Pigs have over 1,600 genes for their sense of smell alone. An estimated 19,000 protein-coding genes are packed into the human genome, surrounded by even more DNA that does not code for proteins. Cells are adept at stuffing an enormous amount of DNA into a small space; they use a variety of packaging proteins to wind, fold, and compress the DNA double helix, going through several levels of packing to create the DNA-protein complex that we call a chromosome (**Figure 9.7**).

Short lengths of double-stranded DNA are wound around "spools" of proteins, known as **histone proteins**, to create a beads-on-a-string structure consisting of many histone beads, called **nucleosomes**, connected by strings of DNA. This beads-on-a-string structure is compressed and coiled into a more compact form, known as the **chromatin fiber**, by yet other types of packaging proteins. The chromatin fiber is then looped back and forth to further condense and coil again to form the resulting chromosome in the interior of the nucleus.

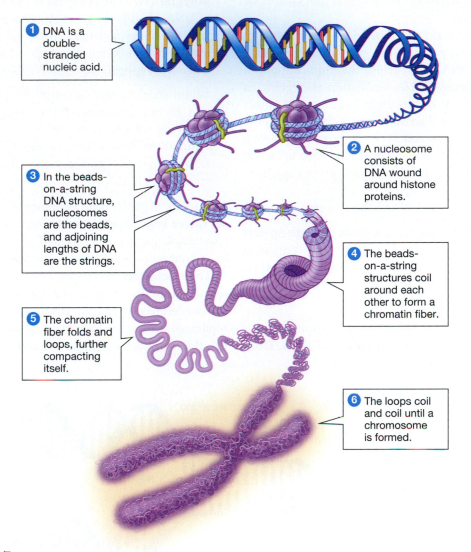

1 DNA is a double-stranded nucleic acid.

2 A nucleosome consists of DNA wound around histone proteins.

3 In the beads-on-a-string DNA structure, nucleosomes are the beads, and adjoining lengths of DNA are the strings.

4 The beads-on-a-string structures coil around each other to form a chromatin fiber.

5 The chromatin fiber folds and loops, further compacting itself.

6 The loops coil and coil until a chromosome is formed.

Figure 9.7

Chromosomes are meticulously organized DNA-protein complexes

The DNA double helix is continuously coiled and packaged around proteins until it is compacted into a chromosome.

Q1: What are the structures that result from the first level of coiling around proteins called?

Q2: What makes up a "bead" and what makes up the "string" in the beads-on-a-string structure of DNA?

Q3: What is the name for the structure that is more compact than the beads-on-a-string structure but less compact than an actual chromosome?

Double or Nothing

DNA from viruses is actually present in all mammalian genomes that we know of, including the human genome, yet not all these viral instructions behave the same. In humans, viral DNA was integrated into the genome a long, long time ago, and although it remains in the human genome, it is no longer active. In pigs, viruses entered and wove their DNA into the pig genome much more recently, and that viral DNA is still active.

All this viral DNA remains in these genomes because it is passed from generation to generation via **DNA replication**, the duplication of a DNA molecule. DNA replication is ongoing in our bodies: It occurs right before a cell enters mitosis, so that there is a copy of the DNA to pass along to the new cell. It occurs feverishly when a new embryo is being formed. And it occurs when viruses hijack our cell machinery and copy their own DNA.

Cells replicate DNA in three steps:

1. The DNA molecule unwinds through special proteins that bind the DNA at sequences known as **origins of replication**, which then break the hydrogen bonds connecting the two strands of DNA.

2. Each strand is then used as a template for the construction of a new strand of DNA. **DNA**

polymerase—a key enzyme in the replication of DNA—builds the two new strands of DNA, starting from **primers** that pair complementary nucleotides at specific sites near the origins of replication.

3. When construction is completed, there are two identical copies of the original DNA molecule. Each copy is composed of a template strand of DNA (from the original DNA molecule) and a newly synthesized strand of DNA.

This mode of replication is known as **semiconservative replication** because one "old" strand (the template strand) is retained, or "conserved," in each new double helix (**Figure 9.8**).

The mechanics of copying DNA are far from simple. More than a dozen enzymes and proteins are needed to unwind the DNA, to stabilize the separated strands, to start the replication process, to attach nucleotides that are complementary

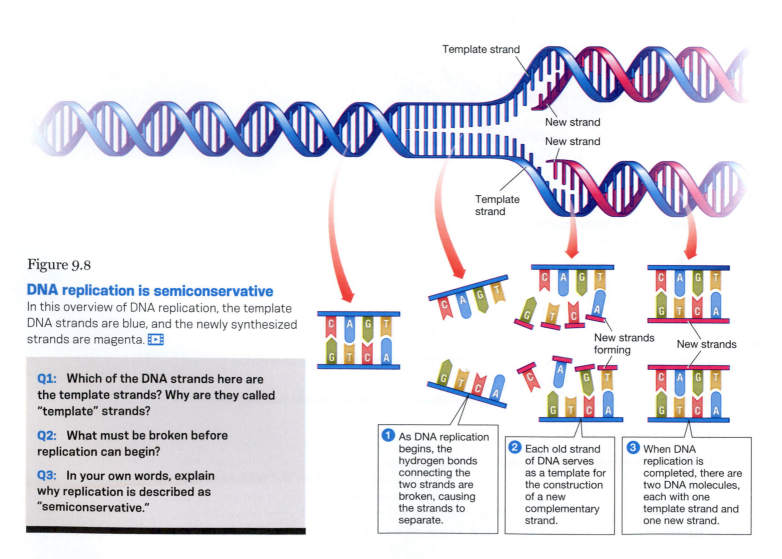

Figure 9.8

DNA replication is semiconservative

In this overview of DNA replication, the template DNA strands are blue, and the newly synthesized strands are magenta. ▶️

Q1: Which of the DNA strands here are the template strands? Why are they called "template" strands?

Q2: What must be broken before replication can begin?

Q3: In your own words, explain why replication is described as "semiconservative."

Template strand

New strand

New strand

Template strand

New strands forming

New strands

1 As DNA replication begins, the hydrogen bonds connecting the two strands are broken, causing the strands to separate.

2 Each old strand of DNA serves as a template for the construction of a new complementary strand.

3 When DNA replication is completed, there are two DNA molecules, each with one template strand and one new strand.

to the template strand, to "proofread" the results, and to join partly replicated fragments of DNA to one another, ultimately forming a completely replicated chromosome.

Despite the complexity of this task, cells can copy DNA molecules containing billions of nucleotides in a matter of hours—about 8 hours in humans (over 100,000 nucleotides per second). This speed is achieved in part by starting the replication of the DNA molecule at thousands of different origins of replication at once.

Before CRISPR was developed, the greatest genetic technology was a tool for replicating DNA quickly in a test tube. In the 1980s, scientists developed a way to mimic natural DNA replication in a laboratory—fast. The **polymerase chain reaction** (**PCR**) is a technique that makes it possible to produce millions of copies of a targeted DNA sequence—to "amplify" the DNA—in just a few hours, even with an extremely small initial amount of DNA (**Figure 9.9**).

PCR relies on heat, not special proteins, to cause the DNA to unwind and its strands to separate. PCR also uses targeted primers to start the replication process at points determined by the scientist, rather than at DNA's natural origins of replication. The researcher

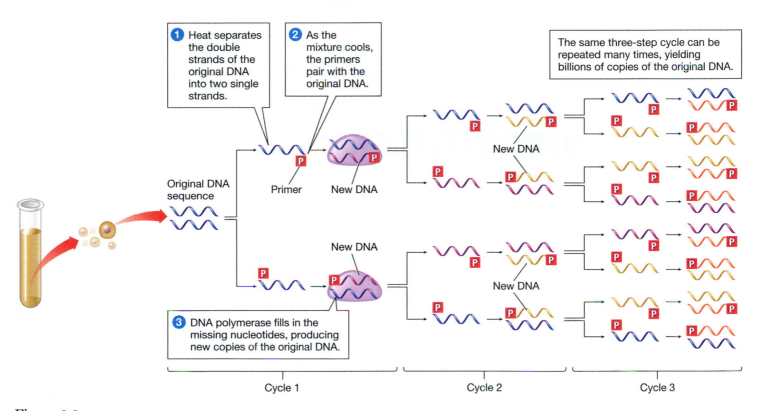

Figure 9.9

PCR can amplify small amounts of DNA more than a millionfold

Short primers consisting of synthetic DNA segments are mixed in a test tube with a sample of the target DNA, the enzyme DNA polymerase, and all four nucleotides (A, C, G, and T). The primers form base pairs with the two ends of a gene of interest. A machine then processes the mixture and doubles the number of double-stranded versions of the template sequence. The doubling process can be repeated many times (only three cycles are shown here).

Q1: PCR replicates DNA many times to increase the amount available for analysis. Why is this process called "amplification"?

Q2: During the PCR cycle, what causes the DNA strands to separate?

Q3: Identify a difference between how PCR and DNA replication are accomplished.

New strand

Template strand

... T A C T G C A A A A C T C A T G ...
... A T G A C G T T T C G A G T A C ...

DNA damage

1 Repair proteins detect and tag the damaged DNA strand.

DNA polymerase

... T A C T G C A A A A C T C A T G ...
... A T G A C G T T T C G A G T A C ...

Cut   Cut

2 Repair enzymes cut the DNA on both sides of the damage.

... T A C T G C A A A A C T C A T G ...
... A T   C ...

G A C G T T T C G A G T A

3 The damaged segment of DNA is removed and degraded.

... T A C T G C A A A A C T C A T G ...
... A T G A C G T T T G A G T A C ...

4 A repair DNA polymerase fills the gap in the DNA with the correct sequence of bases.

creates a solution containing DNA polymerase, primers, and loose nucleotides, and the solution goes into a PCR machine, a device developed for this specific process. The machine cycles between heating, which separates the strands, and cooling, which enables the added enzymes to build two new strands of DNA through complementary base-pairing. These heating and cooling steps are repeated 25–40 times to replicate millions of exact copies of the targeted region of DNA.

PCR remains a central tool in most biology labs. The Harvard team, for example, used PCR to replicate pig DNA for their sequencing reactions, which require many copies of the same sequence in order to succeed. DNA sequencing machines, which produce a data file listing the complete sequence of nucleotides in a given strand of DNA, were used to identify all the places in the pig genome where PERV had been integrated.

Making Mutations

To inactivate all 62 copies of PERV in the pig genome at once, the Church team decided to try to prevent DNA polymerase from successfully replicating the PERV DNA. To that end, the team planned to use CRISPR to *mutate* a small region of the PERV DNA: a gene named *pol* (see **Figure 9.6**). A change to the sequence of nucleotides in an organism's DNA is called a **mutation**. The extent of a mutation can range from the change of one nucleotide in a single base

Figure 9.10

Repair proteins fix DNA damage

Large complexes of DNA repair proteins work together to fix damaged DNA.

Q1: Summarize how DNA repair works and why the repair mechanisms are essential for the normal function of cells and whole organisms.

Q2: Is DNA repair 100 percent effective?

Q3: What would happen to an organism if its DNA repair became less effective?

pair (known as a **point mutation**) to the addition or deletion of one or more whole chromosomes (a *chromosomal abnormality*, described in Chapter 8).

Güell was developing techniques to artificially mutate genes, yet point mutations occur naturally and randomly often, especially during DNA replication. When DNA is copied right before mitosis occurs in a cell, there are many opportunities for mistakes to be made. The enzymes that copy DNA sometimes insert an incorrect nucleotide in the newly synthesized strand. In addition, DNA in cells is constantly being damaged by chemical, physical, and biological agents, including energy from radiation or heat, collisions with other molecules in the cell, attacks by viruses (like PERVs), and random chemical accidents (some of which are caused by environmental pollutants, but most of which result from normal metabolic processes).

Replication errors and damage to DNA—especially to essential genes—disrupt normal cell functions. If not repaired, DNA damage leads to malfunctioning proteins, such as Felix's WAS protein in Chapter 8. DNA damage can also cause the death of cells and, ultimately, the death of an organism. Thankfully, cells have a way to recover: DNA polymerase immediately corrects almost all mistakes during DNA replication, "proofreading" complementary base pairs as they form.

DNA polymerase is not infallible. When an incorrect nucleotide is added but escapes proofreading by DNA polymerase, a mismatch error has occurred. This happens about once in every 10 million nucleotides. But cells have another backup safety program: repair proteins that correct 99 percent of mismatch errors, reducing the overall chance of an error to one mistake in every *billion* nucleotides (**Figure 9.10**).

On the rare occasions when a mismatch error is not corrected by repair proteins, the DNA sequence is changed, and the new sequence is reproduced the next time the DNA is replicated. If the mutation occurs within a gene, it will result in the formation of a new allele. Most new alleles are either neutral or harmful, but occasionally a mutation may be beneficial.

Three types of point mutations can alter a gene's DNA sequence: substitutions, insertions, and deletions. In a **substitution** point mutation, one nucleotide is substituted for another in the DNA sequence of the gene. An **insertion** or **deletion** point mutation occurs when a nucleotide is, respectively, inserted into or deleted from a DNA sequence. Sickle-cell disease, a human genetic blood disorder, is caused by a substitution point mutation (**Figure 9.11**). Sometimes,

Normal hemoglobin DNA

Sickle-cell hemoglobin DNA

···G A C T C C T···

···G A C A C C T···

Normal hemoglobin

Sickle-cell hemoglobin

Normal red blood cells

A sickled red blood cell

Figure 9.11

A point mutation in the hemoglobin gene leads to sickle-cell disease

In people with the genetic disorder sickle-cell disease, a single base in the gene that makes hemoglobin, an important protein involved in oxygen transport in red blood cells, is altered. The red blood cells of people with sickle-cell disease become curved and distorted under low-oxygen conditions and can clog blood vessels, leading to serious effects, including heart and kidney failure.

Q1: What are the three types of point mutations?

Q2: Sickle-cell disease is an autosomal recessive genetic disorder. How many mutated hemoglobin alleles do people with sickle-cell disease have?

Q3: Because of improved treatments, individuals with sickle-cell disease are now living into their forties, fifties, or longer. How might this extension of life span affect the prevalence of sickle-cell disease in the population?

however, changing a few nucleotides in a gene's DNA sequence has little or no effect. In such cases, a mutation is said to be "silent" because it produces no change in the function of the protein, and therefore no change in the phenotype of the organism. (We further explore mutations and their effect on the resulting proteins in Chapter 10.)

Insertions and deletions can be point mutations, but they can also involve more than one nucleotide; sometimes thousands may be added or deleted. Large insertions and deletions almost always result in the synthesis of a protein that cannot function properly. Güell's goal was to use CRISPR in the pig genome to delete bits of DNA in the *pol* gene so that the PERVs could no longer function. The *pol* gene is essential for PERVs to replicate, so a large mutation should halt the formation of viral particles. In addition to being essential to the virus, *pol* is present in all 62 copies of PERVs but not elsewhere in the pig genome, so targeting CRISPR to mutate *pol* would spare damage to other parts of the pig DNA. "My biggest concern is always not to destroy the genome," says Güell. So, he and the team designed two CRISPR RNA guides to target the complex right to *pol*, and to mutate the heck out of it.

After much trial and error, the team inserted its CRISPR-Cas9 construct into pig embryos, and the system began its work. The CRISPR system made 455 different cuts in the PERV *pol* genes throughout the genome, resulting in deletions ranging from 1 to 148 base pairs. About 80 percent of the engineered mutations were small deletions of fewer than nine base pairs—tiny, but enough to disrupt the gene.

Pigs Are People Too?

In October 2015, Güell, Church, and the team presented the results of their initial work: Using CRISPR, they had inactivated 62 PERVs in pig embryos. Recall that in previous non-CRISPR-edited experiments, PERVs from the pig cells infected human cells when put in close proximity. This time, when the team placed the CRISPR-edited pig cells next to human cells, there was zero transmission of the virus. "The gene editing completely disrupted the ability of the virus DNA to replicate and form viral particles," says Güell.

Although the team has successfully destroyed the PERV genes in the pig genome, there are still a few more steps before pig organs will be ready and safe for human transplantation. Other genes need to be manipulated, says Güell, such as immune system proteins so that the human immune system doesn't reject the organs as foreign. However, the next immediate step, says Güell, is to edit an embryo, implant it back into a female pig, and raise a live, genetically engineered animal. "We're trying hard," says Güell. "It could be in the next year."

Other researchers are pursuing a different path: growing actual human organs in pigs. Again, CRISPR is the workhorse for the experiment. In 2016, a team at UC Davis used CRISPR to remove the genes that encode for the pig pancreas in a pig embryo. The procedure created a void in the embryo, which they then filled with human stem cells. Stem cells have the potential to develop into almost all human cells and organs—including, in this case, a fully functional human pancreas.

Some research groups are trying to use this method to grow other organs. **Figure 9.12** shows the process by which scientists could use CRISPR and human stem cells to grow a pig with human kidneys, which would then be used to donate a kidney to a patient in need of a transplant. No researchers have yet to complete this process, and the work remains controversial, especially because of concerns that the human cells might migrate to the developing pig's brain and make the pig, even in some small way, more human.

Overall, CRISPR has revitalized the idea of a safe, clean, limitless source of healthy organs. "It opens up the possibility of not just transplantation from pigs to humans but the whole idea that a pig organ is perfectible," Church told the BBC in 2016. "Gene editing could ensure the organs are very clean, available on demand, and healthy, so they could be superior to human donor organs."

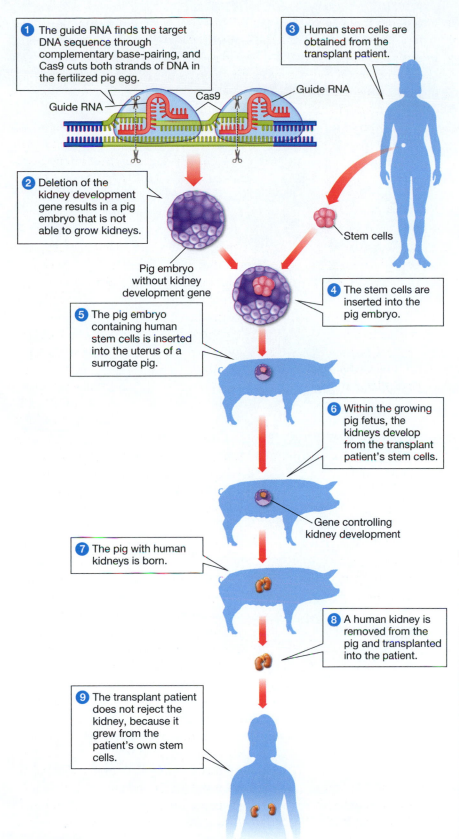

1 The guide RNA finds the target DNA sequence through complementary base-pairing, and Cas9 cuts both strands of DNA in the fertilized pig egg.

Guide RNA   Cas9   Guide RNA

2 Deletion of the kidney development gene results in a pig embryo that is not able to grow kidneys.

3 Human stem cells are obtained from the transplant patient.

Stem cells

Pig embryo without kidney development gene

4 The stem cells are inserted into the pig embryo.

5 The pig embryo containing human stem cells is inserted into the uterus of a surrogate pig.

6 Within the growing pig fetus, the kidneys develop from the transplant patient's stem cells.

Gene controlling kidney development

7 The pig with human kidneys is born.

8 A human kidney is removed from the pig and transplanted into the patient.

9 The transplant patient does not reject the kidney, because it grew from the patient's own stem cells.

Figure 9.12

Human organs could grow up in pigs

Growing a human organ (here, a kidney) in pigs modified by CRISPR to lack that organ could help meet transplant needs.

Q1: Name a step in this process that is similar to the original CRISPR method that removed the PERVs from the pig genome. (*Hint*: Review Figure 9.6.)

Q2: Name a step in this process that is not included in the original CRISPR method that removed the PERVs from the pig genome.

Q3: What parts of this process would scientists need to change in order to develop several different human organs in a single pig?

The Meteoric Rise of CRISPR

The genome editing tool CRISPR, short for "clustered regularly interspaced short palindromic repeats," has taken molecular biology laboratories by storm over the past 5 years. It has been used to edit the genomes of crops and livestock to improve breeding and production, to control populations of disease-carrying insects, to silence genetic disorders in animal models, and more. Here are a few highlights from the short but shining history of CRISPR.

Assessment available in smartwork5

PubMed search results for "CRISPR" by year

CRISPR

Search results article count

Year	Count
2002	1
2003	1
2004	0
2005	5
2006	6
2007	12
2008	21
2009	32
2010	45
2011	79
2012	126
2013	282
2014	607
2015	1,258
2016	2,143

1987 — CRISPR repeats are first observed in bacterial genomes. Their significance is not yet known.

2002 — The term "CRISPR" is coined by researchers in Spain and the Netherlands.

2006 — Researchers propose that CRISPR functions in nature as part of a bacterial adaptive immune system.

2011 — The final necessary piece for the genome editing system is identified: a second small RNA needed to guide Cas9 to its targets.

2013 — The CRISPR-Cas9 system is used to edit targeted genes in both human and mouse cells, and later plant cells.

2015 — In China, scientists use CRISPR-Cas9 to edit preimplantation human embryos, repairing a mutated gene that would cause a blood disorder. Subsequently, an international ban prohibits the use of genome editing to make changes to the human genome.

2016 — The first human trial to use CRISPR genome editing gets approval from the National Institutes of Health, in a cancer therapy trial to edit a patient's own immune system cells.

REVIEWING THE SCIENCE

- Genes are composed of **DNA**, which consists of two parallel strands of repeating units called **nucleotides** twisted into a **double helix**.

- The four nucleotides of DNA contain the **bases** adenine (A), cytosine (C), guanine (G), and thymine (T). The nucleotides exhibit **complementary base-pairing** according to **base-pairing rules**: A can pair only with T, and C can pair only with G.

- DNA is wrapped around **histone proteins**, forming **nucleosomes**. The nucleosome structures can further compact the DNA by coiling around themselves to form a **chromatin fiber**. Chromatin fibers further coil around themselves to form chromosomes.

- The **CRISPR**-Cas9 editing system is composed of two pieces of **RNA** designed to form base pairs at precise locations in a gene. This DNA-RNA interaction guides the Cas9 proteins to the sites where they efficiently cut the DNA, resulting in a gene deletion after normal repair processes take place. Additional genetic manipulations are required to generate a gene insertion.

- **DNA replication** occurs in all living organisms prior to mitosis. The double helix unwinds, and the two strands break apart. Each strand of DNA serves as a template from which a new strand is copied. **DNA polymerase** builds each new strand of DNA using **primers** located near the **origins of replication**.

- The **polymerase chain reaction**, or **PCR**, is a laboratory technique to amplify the DNA from a small initial amount to millions of copies. Amplified DNA can then be sequenced to examine specific genes or mutations.

- DNA is subject to damage by physical, chemical, and biological agents, and errors in DNA replication are common. DNA polymerase "proofreads" the DNA during replication and corrects most mistakes. Repair proteins are a backup repair mechanism and correct any errors that DNA polymerase misses.

- A change to the sequence of bases in an organism's DNA is called a **mutation**. Three types of mismatch mutations can alter a gene's DNA sequence: **substitutions**, **insertions**, and **deletions**. If only a single base is altered, it is a **point mutation**.

THE QUESTIONS

Challenge Yourself

1 If a strand of DNA has the sequence CGGTATATC, then the complementary strand of DNA has the sequence

(a) ATTCGCGCA.

(b) GCCCGCGCT.

(c) GCCATATAG.

(d) TAACGCGCT.

2 Place the following steps of DNA replication and repair in the correct order by numbering them from 1 to 5.

_____ a. A template strand begins to be replicated.

_____ b. If the incorrect base is not identified and replaced, it remains as a point mutation in the DNA.

_____ c. DNA polymerase identifies and replaces most incorrect bases with the correct base, complementary to the base on the template strand.

_____ d. An incorrect base is added to the growing strand of DNA.

_____ e. Proteins identify and replace any incorrect bases missed by DNA polymerase.

3 Given the template DNA sequence GCAGCATGTT, identify each of the following mutations to the complementary strand as an insertion (I), a deletion (D), or a substitution (S).

_____ a. CGTCGTACA

_____ b. CGTGGTACAA

_____ c. CGTCGTACTAA

4 If the target sequence for PCR on one strand of DNA is ATGCAAATCCTGG, what is the sequence of each of the two strands of the double-stranded DNA produced?

Try Something New

5 Using base-pairing rules to guide you, for a DNA double helix that contains 20 percent adenine (A), specify the percentage of (a) thymine, T; (b) guanine, G; and (c) cytosine, C.

6 Why do scientists need to use CRISPR to remove PERVs from pig embryos, rather than just editing the genes in the organs of adult pigs?

7 Scientists estimate that genes encoding proteins make up less than 1.5 percent of the human genome. Other genes in our cells encode different types of nonprotein RNA molecules. The rest of our genome consists of various types of **noncoding DNA**, defined as DNA that does not code for any kind of functional RNA. Some of the remaining DNA has regulatory functions—for example, controlling gene expression. Some of it has architectural functions, such as giving structure to chromosomes or positioning them at precise locations within the nucleus. Given that these sections of DNA are noncoding, does it matter if they contain replication errors? Explain your reasoning.

Leveling Up

8 **What do *you* think?** In another line of research directed at solving the shortage of organs available for transplantation, scientists have been working feverishly on generating human tissues and organs using human induced pluripotent stem cells (iPSCs). These cells, which have the ability to transform into many different cell types, start as a patient's own skin cells and are treated to attain an undifferentiated, or embryonic-like, state. Scientists then direct them with growth factors to become kidney cells, liver cells, or any other cell type needed. For more information, see https://www.cirm.ca.gov/patients/creating-new-types-stem-cells.

Read the article "Eight Tiny Organs Grown by Scientists," about the potential of iPSCs to fill the transplantable organ void (http://www.insight.mrc.ac.uk/2015/07/20/eight-tiny-organs-grown-by-scientists). Using the information in this article, make a list of pros and cons for both pig organs and iPSC organs. Reflect on which option you think is the better course of action.

9 ***Write Now* biology: next-generation GMO** Read the article "CRISPR Brings an Early Harvest" (http://www.genengnews.com/gen-news-highlights/crispr-brings-an-early-harvest/81253507), which describes a new way to produce GMO tomatoes without using standard gene transfer techniques. Write a letter to your state or federal legislator expressing your opinion either for or against this new technology for creating GMO foods. Cover the following points in your letter.

a. Explain how CRISPR-Cas9 technology works, comparing and contrasting it to traditional GMO technology.

b. Describe how the CRISPR system was applied to the plants in the article.

c. In your final paragraph, try to persuade your government representative to either endorse or condemn this technology for improving crop yield and quality, using specific examples.

Thinking Multidimensionally

10 In 1922, King Tutankhamen's tomb was discovered in Egypt. Rumors of "King Tut's curse" soon spread when Lord Carnarvon, the British sponsor of the archaeological dig, died shortly after attending the opening of the pharaoh's tomb.

Curses aside, imagine you're a scientist and your job is to investigate if ancient pathogenic microbes were uncovered when the door to the tomb was unsealed. You know that bacteria are capable of creating durable structures to protect themselves and that these endospores can remain dormant for centuries.

a. If 3,000-year-old bacterial endospores were uncovered, what portion of the bacterium would be most important to protect to ensure it could be revived?

b. Do you think these bacteria would be the same or different from bacteria today? Why?

c. Describe the process you would use to amplify the DNA of the bacteria to compare their nucleotide sequences to modern-day bacteria.

11 In Greek mythology, a chimera was part lion, part goat, and part snake. Chimerism occurs when an organism has more than one set of DNA.

Human-animal chimeras have been created in U.S. research farms, where scientists are growing human tissue inside livestock. The process involves adding human cells to animal embryos and harvesting the organs for transplantation.

a. Why are pigs a good candidate for this technology?

b. How can humanity benefit from creating human-animal chimeras?

c. What ethical issues does this create?

d. If you could design a chimera, which organisms would you combine? Explain your rationale.

For more, visit digital.wwnorton.com/bionow2 for access to:

Tobacco's New Leaf

One company's quest to produce tomorrow's drugs in today's plants.

After reading this chapter you should be able to:

- Explain how an organism uses gene expression to create a phenotype from a genotype.
- Caption a diagram of transcription, identifying each step in the process and the relevant molecules.
- Caption a diagram of translation, identifying each step in the process and the role of each type of RNA.
- Determine the correct amino acid from a given codon.
- Justify how the genetic code fits the definitions of "redundant" and "universal."
- Describe how a cell can increase or decrease its expression of particular genes, and why this ability is important for an organism.

CHAPTER 10

HOW GENES WORK

DRIVING QUESTION

How can tobacco plants be genetically engineered to produce vaccines?

- HS-LS1.A: Structure and Function
- HS-LS3.A: Inheritance of Traits
- HS-LS3.B: Variation of Traits
- HS-ETS1.A: Defining and Delimiting Engineering Problems

The greenhouse is vast, the size of half a football field. A loud, steady thrum reverberates in the room as massive metal fans push air through the hot, humid space. Mike Wanner, a tall, serious man with gray hair and piercing brown eyes, walks through rows of leafy plants, each a foot and a half tall. He stoops to rub a leaf between his fingers, then raises his hand to his nose and sniffs. "That's what tobacco smells like," Wanner shouts, straining to be heard over the noise of the fans.

Wanner is the North Carolina site manager for operations at Medicago, a Canadian biotechnology company growing tobacco at a facility outside Durham, once home to Lucky Strike cigarettes (**Figure 10.1**). But these plants are not being grown to smoke, chew, or dip. Instead, they serve a dramatically different purpose: this tobacco will be used to make flu vaccines.

Influenza vaccines are normally grown in chicken eggs—a process that takes months. As an alternative, companies and researchers have begun experimenting with "biopharming"— manufacturing vaccine proteins in plants. Since

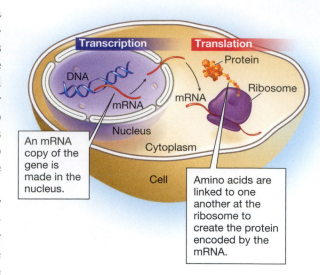

Figure 10.2

An overview of gene expression

Genetic information flows from DNA to RNA to protein during gene expression, which occurs in two steps: first *transcription* and then *translation*. The transcription of a protein-coding gene produces an mRNA molecule, which is then transported to the cytoplasm, where translation occurs and the protein is made with the help of ribosomes.

most genes contain instructions for building proteins, scientists insert a gene that codes for the protein that interests them. This production of a protein from a gene occurs via **gene expression**—the process by which genes are transcribed into RNA and then translated to make proteins (**Figure 10.2**).

Proteins participate in virtually every process inside and between cells, so gene expression is the fundamental way by which genes influence the structure and function of a cell or organism. It is the process through which an organism's genotype gives rise to its phenotype (see Chapter 7). All prokaryotes, eukaryotes, and viruses utilize gene expression.

In addition to being involved in every process in a cell, proteins are the key components of some vaccines. Injected pieces of viral proteins activate the human immune system to defend against that virus in the future (see Figure 2.1). To make these vaccines, scientists must produce large quantities of viral proteins. The traditional method is to inject chicken eggs with a virus, let the virus multiply in the chicken cells, and then

Figure 10.1

Growing tobacco to treat the flu

Tobacco plants grow in a Medicago greenhouse in Quebec City, Canada.

Figure 10.3

Tobacco or egg?

Using tobacco plants to produce proteins for influenza vaccines has several advantages over the traditional approach using chicken eggs.

	Tobacco	Egg
Speed from outbreak to vaccine production:	1 month	6 months
Cost to produce 50 million flu vaccines:	$36 million	$400 million
Allergy risks:	Minimal	Individuals with egg allergies cannot receive the vaccine
Availability:	Not yet approved by the Food and Drug Administration	Approved and currently the most widely used technology to produce flu vaccines

Q1: Which is the faster way to produce vaccines: biopharming with plants or creating vaccines in eggs? Why is this important?

Q2: How much cheaper is biopharming with plants than creating vaccines in eggs? Why is this important?

Q3: Why must tobacco-derived vaccines, or any new medications for that matter, be approved by the FDA?

extract the virus, remove its genetic material, and prepare a vaccine from the leftover viral proteins. It's a long, cumbersome process.

Unlike chicken eggs, however, plants can be grown in vast quantities, and they grow rapidly—often in just days or weeks. "The big advantage of plant systems is that they can produce massive amounts of proteins very inexpensively," says James Roth, director of the Center for Food Security and Public Health at Iowa State University, who studies biopharming in plants and animals (and is not associated with Medicago). In the event of a pandemic flu outbreak, Medicago could produce vaccines 6 times faster and 12 times cheaper than traditional egg manufacturing could, the company claims (**Figure 10.3**).

In April 2012, Medicago put its tobacco plants to the test, running the North Carolina manufacturing facility at full tilt for 30 days. The U.S. Department of Defense had given the company millions of dollars to test whether it could quickly produce enough pandemic flu vaccine from the tobacco to stem an outbreak. The pressure is on, said Wanner, standing inside the facility. He looked out over his crop. "We'll see what happens."

Fighting the Flu with Tobacco

In 1997, Louis-Philippe Vézina, then a research scientist at Agriculture and Agri-Food Canada and an associate professor at Université Laval in Quebec City, Canada, decided to start a company. A plant biotechnologist, Vézina wanted to explore the possibility of manufacturing proteins in alfalfa plants, and thus he called his company Medicago, after the genus to which alfalfa belongs. Later, as the company grew, Vézina and his team discovered that tobacco produces higher yields of proteins in a shorter time frame than alfalfa does, so the company switched plants.

MICHAEL WANNER

Michael Wanner is chief financial officer and U.S. site manager for Medicago, an innovative biotechnology company that develops vaccines in the leaves of tobacco plants.

Figure 10.4

Swine flu

In 2009, people around the world donned masks as a precaution against influenza virus H1N1, nicknamed "swine flu" because it contained DNA from bird, swine, and human flu viruses. The virus caused a pandemic, killing an estimated 284,500 people.

By 2005, Medicago had begun to receive calls asking about its product and when it would begin clinical trials, the first step toward getting a drug approved by the U.S. Food and Drug Administration (FDA). "I was surprised. Even the FDA was asking, 'When are we going to see your vaccine?'" recalls Vézina. "They were keen on technologies like this because they know how much vaccines cost, and they're interested in anything that can decrease that cost."

One of the first vaccines that Medicago produced was a vaccine for influenza virus H1N1, or swine flu, the most common cause of the flu in 2009 (**Figure 10.4**). In the previous H1N1 pandemic, it had taken months for vaccines grown in chicken eggs to reach the market. By

contrast, Medicago produced its vaccine, ready for testing, in just 19 days.

To make a flu vaccine in plants, Vézina and his team use genetic engineering in much the same way that it was used to create a gene therapy for Wiskott-Aldrich syndrome, as described in Chapter 8. They identify and synthesize a single viral gene that codes for hemagglutinin, a protein found on the surface of flu viruses. To make large quantities of that viral protein, the scientists at Medicago first insert the hemagglutinin gene into small, rod-shaped bacteria called "agrobacteria," which infect plants. Loaded with the hemagglutinin gene, these agrobacteria are then exposed to the tobacco plants: a robotic arm lifts a tray containing 5-week-old tobacco plants secured to the surface, flips the tray upside down, and dips the plants into a liquid solution swimming with the bacteria. Once the plants are immersed in the liquid, the technicians turn on a vacuum, sucking air out of the leaves and pulling bacteria into them—like dipping a sponge into water, squeezing it, and then releasing it to soak up the liquid.

When the plants are returned to the greenhouse, their leaves are floppy and almost translucent, like wet tissue paper. Now things are cooking. Once the bacteria are inside the leaves, the bacterial cells release the viral hemagglutinin gene into the plant cells, where it is transported into plant cell nuclei to begin the process of gene expression: making a protein from DNA via the two steps of transcription and translation.

"When you get the *Agrobacterium* solution into the leaf, it will invade the cells and there will be a burst of viral gene expression in the plant cell," said Vézina. "The plant takes over and uses its machinery to produce the protein" (**Figure 10.5**).

Two-Step Dance, Transcription: DNA to RNA

The first step in gene expression—in this case, for a flu gene in a plant cell—is **transcription**, the synthesis of RNA based on a DNA template. In the nucleus, an enzyme called

LOUIS-PHILIPPE VÉZINA

Plant researcher Louis-Philippe Vézina cofounded Medicago in 1997 and served as the company's chief scientific officer until 2014, overseeing all scientific research and development.

RNA polymerase binds to a segment of DNA near the beginning of the gene that is called a **promoter**. The promoter contains a specific sequence of DNA bases that the RNA polymerase recognizes and binds. At Medicago, scientists attach specific promoters to the hemagglutinin gene so that the plant cell's RNA polymerase can identify the gene and actively transcribe it, maximizing the rate of transcription, says Vézina.

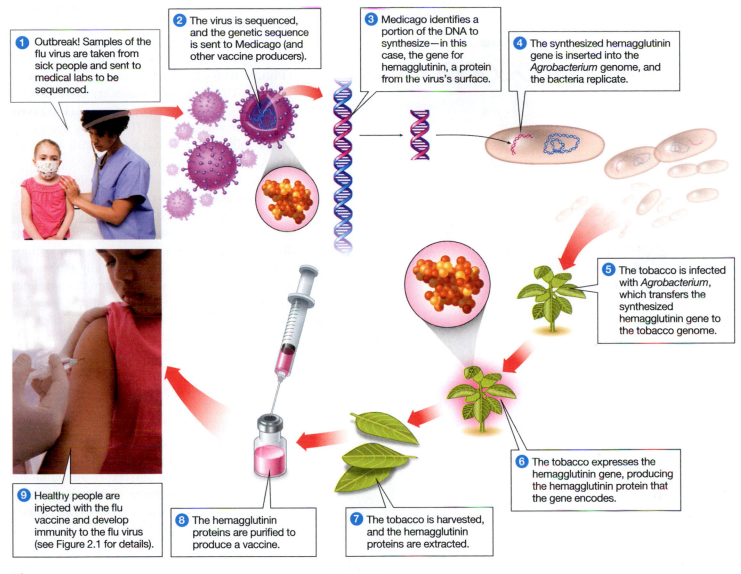

1. Outbreak! Samples of the flu virus are taken from sick people and sent to medical labs to be sequenced.

2. The virus is sequenced, and the genetic sequence is sent to Medicago (and other vaccine producers).

3. Medicago identifies a portion of the DNA to synthesize—in this case, the gene for hemagglutinin, a protein from the virus's surface.

4. The synthesized hemagglutinin gene is inserted into the *Agrobacterium* genome, and the bacteria replicate.

5. The tobacco is infected with *Agrobacterium*, which transfers the synthesized hemagglutinin gene to the tobacco genome.

6. The tobacco expresses the hemagglutinin gene, producing the hemagglutinin protein that the gene encodes.

7. The tobacco is harvested, and the hemagglutinin proteins are extracted.

8. The hemagglutinin proteins are purified to produce a vaccine.

9. Healthy people are injected with the flu vaccine and develop immunity to the flu virus (see Figure 2.1 for details).

Figure 10.5

From outbreak to vaccine

During flu season or any other viral outbreak, a large network of medical professionals and scientists is activated to create a vaccine quickly and effectively.

Q1: In which of the step(s) illustrated here does DNA replication occur? In which step(s) does gene expression occur?

Q2: Why do vaccine producers not simply replicate the entire viral genome, instead isolating the gene for one protein and replicating only that gene?

Q3: What role do the bacteria play in this process? Why are they needed?

1 Transcription begins when RNA polymerase binds to the promoter.

RNA polymerase

Hemagglutinin gene

Promoter (in pink)

Terminator (in pink)

Direction of transcripton

2 An mRNA molecule is produced as RNA polymerase moves down the template strand of DNA.

RNA polymerase

Template strand of DNA

RNA nucleotides

New RNA strand

Direction of transcripton

3 Transcription ends when RNA polymerase reads through the terminator.

Tobacco cell

Figure 10.6

Plants making proteins, I: Transcription

RNA polymerase transcribes the hemagglutinin gene into a molecule of RNA.

Q1: Why is only one strand of DNA used as a template?

Q2: If a mutation occurred within the promoter or terminator region, do you think it would affect the mRNA transcribed? Why or why not?

Q3: The template strand of part of a gene has the base sequence TGAGAAGACCAGGGTTGT. What is the sequence of RNA transcribed from this DNA, assuming that RNA polymerase travels from left to right on this strand?

Once bound to the promoter, the RNA polymerase unzips the DNA double helix at the beginning of the gene, separating a short portion of the two strands. Only one of the two DNA strands is used as a template, and thus it is called the **template strand**. The RNA polymerase begins to move down the DNA template strand, constructing a **messenger RNA (mRNA)** molecule, a strand of nucleotides complementary to the DNA, from free nucleotides floating around in the nucleus (**Figure 10.6**). RNA does not have all the same bases as DNA: its four bases are adenine (A), cytosine (C), guanine (G), and uracil (U). Those bases pair with the four DNA bases according to the following rules: RNA's A with DNA's T, C with G, G with C, and U with A.

Part of the reason that tobacco cells produce so much hemagglutinin so quickly is that the hemagglutinin gene inserted into the cells contains a special DNA sequence that triggers multiple RNA polymerases to transcribe a hemagglutinin gene at a single time. As for any gene, as an RNA polymerase moves away from the promoter and travels down the template strand, another RNA polymerase can bind at the promoter and start synthesizing a second mRNA on the heels of the first. At any given time, therefore, many RNA polymerases can be traveling down a DNA template simultaneously, each synthesizing an mRNA.

Transcription stops when the RNA polymerase reads through a special sequence of bases called a **terminator**. In eukaryotic cells, the mRNA then undergoes an elaborate sequence of modifications that prepare it to leave the nucleus. These steps include chemical modification of both ends of the mRNA, as well as a process called RNA splicing.

Most eukaryotic genes (and many viral genes) are embedded with stretches of sequences that don't code for anything, called **introns**. The stretches of DNA in a gene that carry instructions for building the protein are called **exons**. Because of this patchwork construction, with genes made of introns and exons interspersed, newly transcribed mRNA (pre-mRNA) is also a patchwork of coding sequences intermixed within noncoding sequences. During **RNA splicing**, the introns are snipped out of a pre-mRNA and the remaining pieces of mRNA—the exons—are joined to generate the

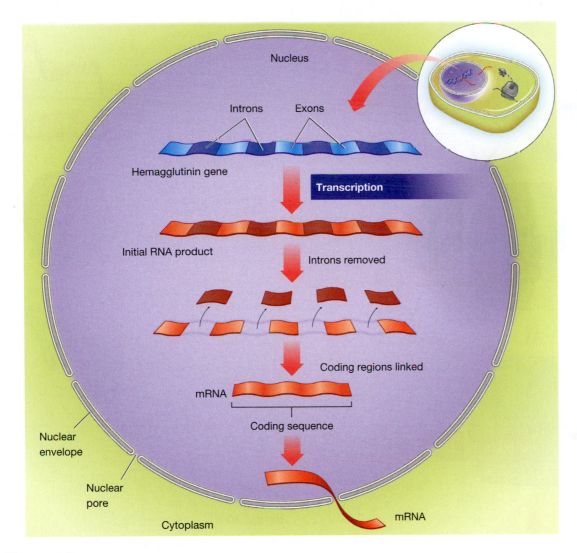

Figure 10.7

Processing mRNA for export to the cytoplasm

In eukaryotes, introns must be removed before an mRNA leaves the nucleus.

Q1: In your own words, define RNA splicing. When during gene expression does it occur?

Q2: What do you predict would happen if the introns were not removed from RNA before translation? Why would it be a problem if the introns were not removed?

Q3: Where is the mRNA destined to go once it has been transported out of the nucleus?

mature mRNA (**Figure 10.7**). This mRNA is then ready to leave the nucleus.

To review, transcription occurs when RNA polymerase binds to a promoter, unzips the DNA helix, and constructs a strand of mRNA based on the DNA template strand. Transcription ends at the terminator sequence, and the mRNA is then processed, at which time noncoding introns are spliced out of the sequence. Voilà! mRNA is created from a gene.

Two-Step Dance, Translation: RNA to Protein

The microscopic molecular dance inside the tobacco cells continues with translation. Once the hemagglutinin gene has been transcribed into mRNA in the nucleus, it is time to make

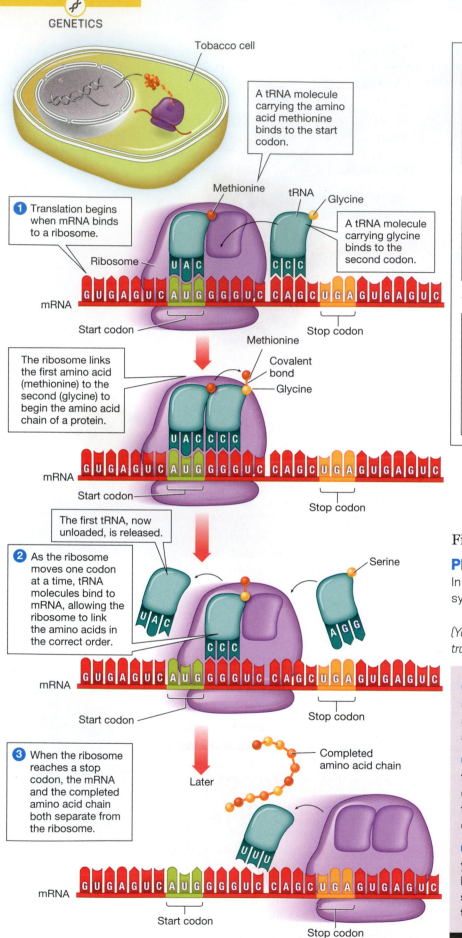

Tobacco cell

A tRNA molecule carrying the amino acid methionine binds to the start codon.

Methionine   tRNA   Glycine

1 Translation begins when mRNA binds to a ribosome.

A tRNA molecule carrying glycine binds to the second codon.

Ribosome

UAC   CCC

mRNA   GUGAGUCAUGGGGUCCAGCUGAGUGAGUC

Start codon   Stop codon

The ribosome links the first amino acid (methionine) to the second (glycine) to begin the amino acid chain of a protein.

Methionine

Covalent bond

Glycine

UACCCC

mRNA   GUGAGUCAUGGGGUCCAGCUGAGUGAGUC

Start codon   Stop codon

The first tRNA, now unloaded, is released.

2 As the ribosome moves one codon at a time, tRNA molecules bind to mRNA, allowing the ribosome to link the amino acids in the correct order.

Serine

UAC   AGG

CCC

mRNA   GUGAGUCAUGGGGUCCAGCUGAGUGAGUC

Start codon   Stop codon

3 When the ribosome reaches a stop codon, the mRNA and the completed amino acid chain both separate from the ribosome.

Completed amino acid chain

Later

UUU

mRNA   GUGAGUCAUGGGGUCCAGCUGAGUGAGUC

Start codon   Stop codon

Messenger RNA (mRNA)

Specifies the order of amino acids in a protein using a series of three-base codons, where different amino acids are specified by particular codons.

GUGAGUC

Ribosomal RNA (rRNA)

As a major component of ribosomes, assists in making the covalent bonds that link amino acids together to make a protein.

Transfer RNA (tRNA)

Transports the correct amino acid to the ribosome, using the information encoded in the mRNA; contains a three-base anticodon that pairs with a complementary codon revealed in the mRNA.

AGG

UAC

Figure 10.8

Plants making proteins, II: Translation

In translation, the hemagglutinin mRNA directs the synthesis of the hemagglutinin protein. ▶️

(You will need to finish reading the section on translation to answer the following three questions.)

Q1: Which amino acid always begins an amino acid chain? Which codon and anticodon are associated with that amino acid?

Q2: Each of the codons for stopping translation binds to a tRNA molecule that does not carry an amino acid. How would the binding of a stop codon cause the completed amino acid chain to be released?

Q3: Given the partial mRNA sequence that you specified in Figure 10.6's question 3 as being transcribed from the DNA template strand, what is the amino acid sequence that would be translated?

the protein, the actual product that will be extracted from the tobacco leaves. First the mRNA is transported from the nucleus, where it was made, to the sites of protein synthesis: the cell structures called **ribosomes** in the cytoplasm. To escape the nucleus, the long strand of mRNA passes through a nuclear pore, like a noodle slipping through the hole of a colander. Once the mRNA molecule arrives in the cytoplasm, the information it contains must be translated, with the help of ribosomes, from the language of mRNA (nitrogenous bases) to the language of proteins (amino acids). **Translation** is the process by which ribosomes convert the information in mRNA into proteins.

During translation, ribosomes "read" the mRNA code like a grocery list (bread, milk, etc.), and collect the corresponding amino acids, linking them in the precise sequence dictated by mRNA (**Figure 10.8**). Ribosomes read the mRNA information in sets of three bases at a time, and each unique sequence of three mRNA bases is called a **codon**. The hemagglutinin gene has about 1,770 bases, of which 1,695 code for the protein. That makes 565 codons (1,695 divided by 3), and therefore the hemagglutinin protein is composed of 565 amino acids.

The four bases of mRNA (A, C, G, U) can be arranged to create a three-base sequence in 64 different ways (because $4^3 = 64$). Therefore, there are 64 possible codons (**Figure 10.9**). Most of the 64 codons specify a particular amino acid. A couple of amino acids are specified by only one codon, while other amino acids are specified by anywhere from two to six different codons. Some codons do not code for any amino acid and instead act as signposts that communicate to the ribosomes where they should start or stop reading the mRNA. The **start codon** (AUG) is the ribosome's starting point on the mRNA strand, and there are three possible **stop codons** (UAA, UAG, and UGA). By beginning and ending at fixed points, the cell ensures that the mRNA message is read in precisely the same way every time.

The information specified by all 64 possible codons is the **genetic code** (see **Figure 10.9**). The genetic code has several significant characteristics. First, it is *unambiguous*: each codon specifies only one amino acid. It is also *redundant*: since there are a total of 64 codons but only 20 amino acids, several different codons call for the same amino acid, as already mentioned. Finally, the genetic code is virtually *universal*: nearly every organism on Earth uses the same code, from agrobacteria to tobacco cells to human cells—a feature that illustrates the common descent of all organisms.

Making a protein from an mRNA strand requires two additional types of RNA: The first is **ribosomal RNA (rRNA)**, an important component of ribosomes. The second is **transfer RNA (tRNA)**, which is the caddy for the process, delivering specific amino acids to the ribosomes as the codons are read off the mRNA "list."

Each tRNA specializes in binding to a specific amino acid and recognizes and pairs with a specific codon in the mRNA, like a puzzle piece that fits one amino acid on one end and one codon on the other. At one end of a tRNA molecule, a special sequence of three nitrogenous bases, called an **anticodon**, binds the correct codon on the mRNA. At the other end, the specific amino acid attaches (see **Figure 10.8**).

Let's recap. For translation to occur, an mRNA molecule must first bind to a ribosome. The ribosomal machinery then "scans" through the mRNA until it finds a start codon (AUG). Next the ribosome recruits the appropriate tRNAs one by one, as determined by the codons read in the mRNA sequence. A special site on the ribosome facilitates the linking of one amino acid to another, like beads on a string. Finally, the ribosome reaches a stop codon. The amino acid chain cannot be extended further, because none of the tRNAs will recognize and pair with any of the three stop codons. At this point the mRNA molecule and the completed amino acid chain separate from the ribosome. The new protein then folds into its compact, specific three-dimensional shape and is ready to go to work in the cell.

But this process does not always go as planned. In Chapter 9 we learned that a mutation is a change in the sequence of DNA bases. Mutations affect an organism by disrupting or preventing the healthy formation of a protein. For example, a mutation can cause a DNA sequence not to be translated or transcribed, prompt the amino acid chain to end prematurely, or make the final protein fold incorrectly,

Figure 10.9

The genetic code

(Top) The genetic code is composed of the 64 possible codons found in the mRNA. Each codon specifies an amino acid or is a signal that starts or stops translation. *(Bottom)* The genetic code is used during the translation of mRNA to protein.

Q1: How many codons code for isoleucine? For tryptophan? For leucine?

Q2: What codons are associated with asparagine? With serine?

Q3: From the partial mRNA sequence that you specified in Figure 10.6's question 3 as being transcribed from the DNA template strand, remove only the first A. What amino acid sequence would be translated as a result of this change? How does that sequence compare to the amino acid sequence you translated from the original mRNA sequence? *Bonus:* What kind of mutation is this? (*Hint:* See Chapter 9.)

among other possibilities. While single-base substitutions are not always a problem, single-base insertions and deletions cause a genetic "frameshift," shifting all subsequent codons "downstream" by one base (**Figure 10.10**). This shift scrambles the entire downstream DNA message, in turn scrambling the entire RNA message and causing the ribosomes to assemble a very different sequence of amino acids from the mutation point onward—as if every letter in this phrase was shifted to the left one space while the word length and spaces between words were retained (a si fever ylette ri nthi sphras ewa sshifte dt oth erigh ton espac ewhil eth ewor dlengt han dspace sbetwee nword swer eretaine d).

Tweaking Gene Expression

Inside a tobacco cell, as in most living cells, the expression of many genes can be turned on or off, slowed down (**down-regulated**), or sped up (**up-regulated**). This **gene regulation** enables organisms to change which genes they express in response to internal signals (from inside the body) or external cues in the environment. In this way, by producing different proteins as needed, organisms can adapt to their surroundings.

All cells in a multicellular individual have essentially the same DNA, yet different cells express different sets of genes, and within a given cell the pattern of gene expression can change over time. Single-celled organisms, such as bacteria, face a more difficult challenge: they are directly exposed to their environment, and they have no specialized cells to help them deal with changes in that environment. One way they meet this challenge is to express different genes at different times.

The expression of most genes in prokaryotes and eukaryotes is regulated by both internal and external signals. Many genes are also developmentally regulated, meaning that their expression can change, sometimes dramatically, as an organism grows and develops. Gene expression is regulated at many different points in the cell, including DNA packing (the way DNA is compressed or unwound in the genome), transcription, mRNA processing, and several

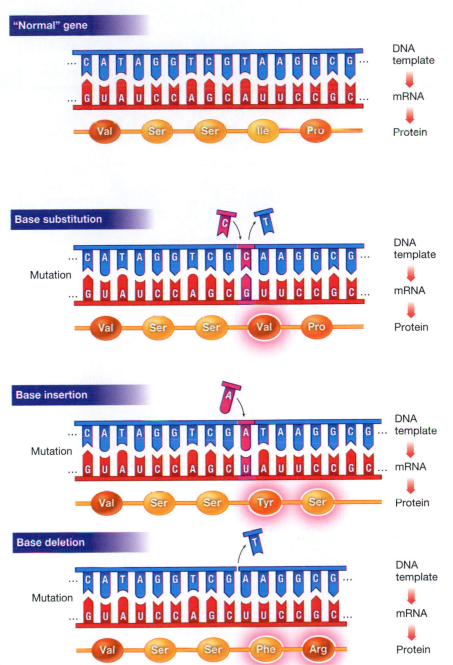

Figure 10.10

Effects of point mutations

Q1: Why is an insertion or a deletion in a gene more likely to alter the protein product than a substitution, such as A for C, would?

Q2: Which would you expect to have more impact on an organism: a point mutation as shown here, or the insertion or deletion of a whole chromosome (discussed in Chapter 8)?

Q3: Which mechanisms in a cell prevent mutations? (*Hint:* Refer back to Chapter 6 if needed.)

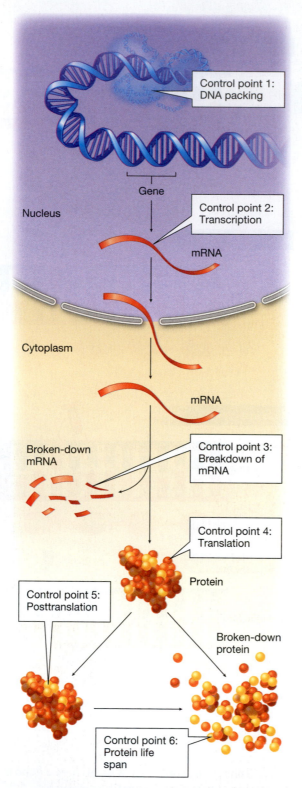

Control point 1:
DNA packing

Gene

Nucleus

Control point 2:
Transcription

mRNA

Cytoplasm

mRNA

Broken-down
mRNA

Control point 3:
Breakdown of
mRNA

Control point 4:
Translation

Protein

Control point 5:
Posttranslation

Broken-down
protein

Control point 6:
Protein life
span

points during translation (**Figure 10.11**). But for all living cells, a few genes are always expressed at a low level; their transcription is not regulated, because these genes are needed at all times.

At Medicago, the company takes advantage of gene regulation in tobacco cells to produce as much hemagglutinin protein as possible. After the agrobacteria are vacuum-sucked into the leaves and transcription and translation begin to actively occur in the tobacco cells, the plants are sent to grow in an incubation room, where technicians can alter the humidity, temperature, and level of light to maximize the amount of protein expressed by the plants.

"We've been able to tweak environmental conditions of the plants to boost gene expression," says Vézina. "You name it, we've tried it." It's an important step in the process, adds Wanner. "We determined what the best conditions for protein expression are."

To the Market

It's the end of April, and the final steps are being taken to isolate and purify the hemagglutinin protein from the tobacco plants, to see how much vaccine can be made in a month. After the plants have incubated for several days, the leaves are stripped and diced into green confetti, then digested with enzymes to

Figure 10.11

How gene expression is regulated

In eukaryotes, gene expression can be controlled at several points along the pathway from gene to protein to phenotype: before transcription, at transcription, during RNA processing, or at translation. Gene expression can also be regulated after translation, by control of the activity or life span of the protein.

Q1: As illustrated here, at what control point is transcription regulated?

Q2: What is a possible advantage of regulating gene expression before transcription, versus after?

Q3: If you wanted to up-regulate production of the hemagglutinin protein in a tobacco plant carrying the hemagglutinin gene, at which control point(s) would that be possible? Justify your reasoning.

The Deadly Price of a Pandemic

The Spanish flu of 1918 devastated populations across Europe, killing not only the young and old, but also healthy adults. Since then, new strains of flu have reared their ugly heads, and doctors and vaccine manufacturers struggle to anticipate the next outbreak.

1918
H1N1

50M
dead worldwide

650K
dead in U.S.

1957
H2N2

1.5M
dead worldwide

70K
dead in U.S.

1968
H3N2

750K
dead worldwide

34K
dead in U.S.

2009
H1N1

285K
dead worldwide

12K
dead in U.S.

break up the leaf material so that the desired proteins are released into solution. The resulting solution, which resembles green-pea soup, is filtered several times to isolate clusters of hemagglutinin, which will then be processed into a vaccine product that is safe to inject into people.

Medicago is not the first company to produce a human drug using a plant. The first human-like enzyme was produced from tobacco back in 1992 at Virginia Polytechnic Institute. Numerous other plant biopharming companies have sprung up since then, experimenting with various plant species, including corn, soybean, duckweed, and more. But the field is not without risks and controversy. "The main concern and risk is spread—that the gene will get out into nature and spread," says Roth. "But there are techniques to make sure that doesn't happen, and it is closely regulated."

One of those techniques is to grow the plants in contained environments, where there is no risk of contaminating food crops. Israel-based biotech company Protalix Biotherapeutics, for example, grows carrot cells inside in large hanging bags of fluid and cells. In May 2012, the FDA approved the first biopharmed drug for humans (produced by Protalix)—a therapy for Gaucher disease, a rare genetic disorder—grown in the company's carrot cells. "This approval demonstrates a proof of concept for the power of this technology," the CEO of Protalix told the journal *Nature*.

Today, Medicago has completed safety trials for its pandemic flu vaccine and has had positive results in the first of two large clinical trials in humans that tested its seasonal flu vaccine's efficacy. When the April manufacturing test was completed, the company had produced an astounding 10 million doses of flu vaccine in a single month. It would have taken 5–6 months to produce the same amount using the traditional method of growing vaccines in chicken eggs. With that success under its belt, in 2015 the company began building a second production complex. Based in Quebec City, the manufacturing facility will have the capacity to deliver up to 50 million doses of seasonal flu vaccine.

Medicago is also developing novel vaccines against rotavirus and rabies virus, and the company recently received a U.S. government contract to manufacture antibodies to treat Ebola virus infection. "It might have taken a bit longer than we thought for biopharming to be accepted, but it deserves the visibility and attraction it has now," says Vézina. "Biopharming is here to stay. I'm convinced of this."

REVIEWING THE SCIENCE

- Most genes contain instructions for building proteins. **Gene expression** is the process by which genes are transcribed into RNA and then translated into a protein.

- During **transcription**, which occurs in the nucleus, RNA polymerase binds the **promoter** of a gene and produces a **messenger RNA (mRNA)** version of the gene sequence from free nucleotides.

- Next, during **RNA splicing**, **introns** are snipped out of the pre-mRNA sequence, and the remaining **exons** are joined. The mRNA is transported out of the nucleus.

- During **translation**, which occurs in the cytoplasm, ribosomes convert the sequences of bases in an mRNA molecule to the sequence of amino acids in a protein, with the help

of **ribosomes** composed of **ribosomal RNA (rRNA)** and proteins, and associated with **transfer RNA (tRNA)**.

- Ribosomes read the mRNA information in sets of three bases at a time, called **codons**. There are 64 possible codons, including a **start codon** (AUG) and three possible **stop codons** (UAA, UAG, and UGA).

- The **genetic code** is redundant because there are 64 codons and only 20 amino acids, and universal because nearly all organisms on Earth use the same code.

- Gene expression is regulated at many different points in the pathway from gene to protein. Organisms rely on **gene regulation** to respond to signals inside the body and to external cues in the environment.

THE QUESTIONS

6 The following nucleic acid is an entire primary transcript (pre-mRNA not yet processed): ACGCAUGCGaugaugccccucag GUCUguuuccgugaUGCCGUUGACCUGA. The nucleotides in capitals are exons; and the nucleotides in lowercase type are introns. Appropriately splice this primary transcript.

(a) ACGCAUGCGGUCUUGCCGUUGACCUGA

(b) augccuuucagguuuccguga

(c) augccuuucagGUCUguuuccguga

(d) ACGCAUGCGaugGUCUUGCCGUUGACCUGA

(e) ACGCAUGCGaugccuagGUCUguuuccgugaUGCCGUUGACC UGA

7 How is gene expression similar to DNA replication, and how is it different? Give at least one similarity and one difference.

Leveling Up

8 *Write Now* **biology: genotype to phenotype** Your friend, who is also taking a biology class, has become a little confused. He informs you that the genetic code is known to be ambiguous because a given genotype may give rise to a variety of phenotypes during gene expression (for example, his twin brother is an inch taller and more tan than he is). You like your friend and would like him to pass his next biology exam, so you decide to help him. Write him a brief note explaining (a) why the genetic code is not, in fact, ambiguous and (b) how gene expression derives a phenotype from a genotype.

9 **What do *you* think?** Most people carry two copies of a normal gene that codes for an enzyme, glucosylceramidase, that is involved in breaking down lipids no longer needed in cells. (Enzymes are proteins that cause specific chemical changes; they are biological catalysts.) One in 100 people in the United States carries a recessive mutation that codes for a defective glucosylceramidase enzyme. And about one in 40,000 people carries two copies of the mutation and displays the symptoms of Gaucher disease. These symptoms, caused by the accumulation of lipids in cells, include anemia, enlarged organs, swollen glands and joints, and, in severe cases, neurological problems and early death.

Enzyme replacement therapy is effective but very expensive—about $200,000 annually—and must be continued, every 2 weeks, for life. The Israeli biotech company Protalix Biotherapeutics, working with the U.S.-based Pfizer Pharmaceuticals, has developed a process to genetically modify carrots to produce a replacement enzyme. The biopharmed enzyme will cost about 25 percent less than the standard enzyme therapies, which are grown in mammalian cell lines. Protalix is now working on treatments for other enzyme deficiency diseases.

The FDA's May 2012 approval of the drug developed by Protalix alarmed some environmental activists and health advocates, who fear that the company's genetically modified carrot is just the beginning of a wedge that will lead to an underregulated and potentially dangerous industry. There is some legitimacy to their concerns: the U.S. Department of Agriculture (USDA) does not require an environmental impact assessment for biopharmed crops; nor does it require biotech companies to share the location of their test fields or the identity of the biopharmed molecules being produced. Furthermore, the USDA is not sufficiently staffed to effectively monitor companies involved in biopharming.

What do you think? Should biopharming be allowed in the United States? If so, under what conditions and with what limits? For example, should it be allowed to produce drugs for only life-threatening illnesses, or only under highly controlled conditions? Be prepared to discuss your observations and reflections in class.

10 **Life choices** Go to the Centers for Disease Control and Prevention (CDC) influenza website (http://www.cdc.gov/flu) and read the pages "Key Facts About Influenza (Flu)" (under "Flu Basics") and "Key Facts About Seasonal Flu Vaccine" (under "Prevention—Flu Vaccine"). You can also go to the Mayo Clinic's influenza website (http://www.mayoclinic.org/diseases-conditions/flu/home/ovc-20248057). Then answer the following questions.

a. What is the flu? How is it passed on?

b. What are the possible symptoms and complications of the flu?

c. How can you decrease your chance of getting the flu, and what treatments are available if you become infected?

d. What are the benefits and risks of the flu vaccine?

e. Why is there a new flu vaccine every year?

f. Why is the flu vaccine more effective in some years than in others?

g. Who would you recommend should get the flu vaccine? Explain your reasoning.

h. Do you get a flu vaccine every year? Why or why not?

Thinking Multidimensionally

Standard Whippet

Bully Whippet

11 Bully whippets might just be the Incredible Hulk of the dog world because of a naturally occurring mutation to the myostatin (*MSTN*) gene and selective breeding.

AATTACTGCTCTGGAGAGTGTGAATTTGTG   Normal Myostatin gene

AATTACTGCTCTGGAGAGTGAATTTGTGTT   Mutated Myostatin gene

Look at the DNA sequences of the normal and mutant forms of the *MSTN* gene in whippets above.

a. What type of mutation creates the bully whippet?

b. Determine the amino acids that are created from each of these segments of the *MSTN* gene, and describe how the changes would impact the three-dimensional structure of the myostatin protein and its function.

c. Bully whippets are the result of a homozygous recessive mutation to the *MSTN* gene. Use a Punnett square to cross two heterozygous whippets to determine the percent chance of getting a bully whippet puppy.

12 In 1918, a son of the American architect Frank Lloyd Wright invented Lincoln Logs after being inspired by his famous dad's design of the Imperial Hotel in Tokyo. Each small piece in a Lincoln Log set can be put together to create different buildings. How is this analogy similar to DNA's structure and to DNA's being universal among organisms?

For more, visit digital.wwnorton.com/bionow2 for access to:

Whale Hunting

Fossil hunters discover Moby Dick's earliest ancestor—a furry, four-legged land lover.

- -

After reading this chapter you should be able to:

◆ Define evolution and list the six types of evidence for evolution.

◆ Compare and contrast artificial selection and natural selection.

◆ Summarize the argument that the fossil record provides evidence in support of evolution.

◆ Give an example of a homologous or vestigial trait, and explain how such traits support the theory of common descent.

◆ Explain why even distantly related species have similar DNA.

◆ Use your knowledge of evolution and continental drift to make a prediction about the geographic location of a given set of fossils.

◆ Relate similarities in embryonic development among species to their shared evolutionary past.

DRIVING QUESTION

What evidence is there that modern-day whales evolved from terrestrial ancestors?

- HS-LS4.A: Evidence of Common Ancestry and Diversity
- HS-LS4.B: Natural Selection

Fossils break all the time. This time, the 50-million-year-old ear bone of a small, deerlike mammal called *Indohyus* snapped clean off the skull. Sheepishly, the young laboratory technician cleaning the fossil handed the broken piece to his boss, paleontologist and embryologist J. G. M. "Hans" Thewissen at Northeast Ohio Medical University. Thewissen tenderly turned the preserved animal remains over in his hand. Then, as the tech reached for the fossil to glue it back onto the animal's skull, Thewissen went rigid.

"Wow, that is weird," said Thewissen. The *Indohyus* ear bone, which should have looked like the ear bone of every other land-living mammal—like half a hollow walnut shell, but smaller—was instead razor thin on one side and very thick on the other (**Figure 11.1**). "Wow," repeated Thewissen. This wasn't the ear of a deer, or any other land mammal. Thewissen squinted closer. "It looks just like a whale," he said.

Although they live in the ocean like fish, whales are mammals like us. So are dolphins and porpoises (**Figure 11.2**). Like all mammals, whales are warm-blooded, have backbones, breathe air, and nurse their young from mammary glands. Numerous fossils have been found documenting whales' unique transition from land-living mammals to the mammoths of the sea, during which whale populations developed longer tails and shorter and shorter legs. But one crucial link in the fossil record was missing: the closest land-living relatives of whales. What did the ancestors of whales look like before they entered the water? Staring at the strange fossil in his hand, Thewissen realized he could be holding the ear of that missing link.

Whales are but one of the many organisms that share our planet. Every species is exquisitely fit for life in its particular environment: whales in the open ocean, hawks streaking through the sky, tree frogs camouflaged in the green leaves of a rainforest. There is a great diversity of life on Earth—animals, plants, fungi, and more—with each species well matched to its surroundings. This diversity of life is due to evolution.

"Evolution," in everyday language, means "change over time." In science, biological **evolution** is a change in the inherited characteristics of a

J. G. M. "HANS" THEWISSEN

Paleontologist and embryologist J. G. M. "Hans" Thewissen is a professor and whale expert at Northeast Ohio Medical University in the Department of Anatomy and Neurobiology. He and his lab study ancestral whale fossils and modern whale species.

Indohyus

When the *Indohyus* skull broke, Thewissen saw its very thick medial tympanic wall, like those found in all whales but no other living mammals.

Tympanic wall

Sediment filling middle ear

Whale

Thick medial tympanic wall

White-tailed deer

Thin medial tympanic wall

Figure 11.1

The mysterious ear bone

The *Indohyus* fossil ear bone (top) looks more like the ear bone of whales (middle) than that of any modern land mammal (bottom). (Source: *Indohyus* and whale photos courtesy of Thewissen Lab, NEOMED.)

Figure 11.2

A montage of mammals
Earth boasts an amazing diversity of life, as evidenced here. These are only a few of the almost 5,500 species of mammals currently living on the planet.

group of organisms over generations. Whales, for example, evolved from four-legged land-living animals into sleek emperors of the ocean, slowly becoming suited for the water over tens of millions of years. Whales changed as a population. Populations evolve; individuals do not.

You may wonder how we can be sure of evolutionary change, especially for a transformation as extreme as that of a furry, four-legged beast into a whale. There is strong evidence for evolution not only from fossils like those that Thewissen studies, but also from features of existing organisms, common patterns of how embryos develop, DNA evidence, geographic evidence, and even direct observation of organisms evolving today—including man's best friend.

Artificial to Natural

As we saw in Chapter 7, all dogs are a single species. That species is *Canis lupus familiaris*, a subspecies of the gray wolf. Scientists estimate that domestication of the gray wolf began about 16,000 years ago as humans habituated the animals to civilization. From there, people bred the wolf for desired qualities, such as decreased agression and the ability to follow commands. Millennia later, dogs were selectively bred for specific traits like the long legs of a greyhound or the short snout of a bulldog. This selective process has led to incredible variation in the size and shape of dog breeds, from 6-pound Chihuahuas to 200-pound Great Danes (**Figure 11.3**). Dogs are an example of evolution that we can directly observe, and they evolved via artificial selection. **Artificial selection** is brought about by **selective breeding**, in which humans allow only individuals with certain inherited characteristics to mate.

Through selective breeding, humans have crafted enormous evolutionary changes within not only dogs, but many other domesticated organisms, including ornamental flowers, pet birds, and food crops. We can observe evolution happening through artificial selection via selective breeding.

Artificial selection happens when humans choose which individuals of a particular species are allowed to breed. Without the intervention of humans, can the environment itself

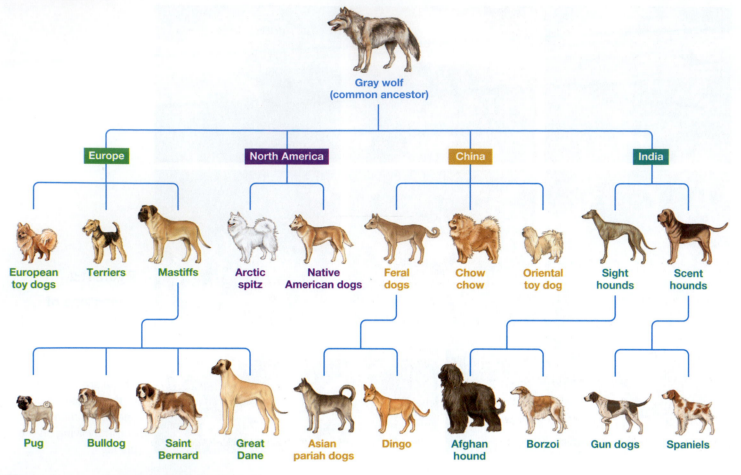

Figure 11.3

Selective breeding of dogs produces myriad traits

Dogs were domesticated only a few times, and always from gray wolves. Thus, the remarkable diversity of dogs represents the effects of selective breeding on a small number of lineages of domesticated wolves.

Q1: What is selective breeding, and how does it work?

Q2: Explain how selective breeding leads to artificial selection.

Q3: Name as many organisms as you can whose current characteristics are due to artificial selection.

"choose" who survives and breeds? It does. In nature, evolution occurs mainly via *natural* selection, as well as through other mechanisms we will discuss later. **Natural selection** is the process by which individuals with advantageous inherited characteristics for a particular environment survive and reproduce at a higher rate than do individuals with other, less useful characteristics. In other words, whoever has the most kids wins!

After the environment "chooses" the winners—those that successfully breed the most—the characteristics of those individuals become more common in successive generations because they have produced more offspring. For example, in 1977 a terrible drought struck the Galápagos Islands off the coast of Ecuador. One species of small ground finches—petite birds with sharp, pointy beaks—starved as the small, tender seeds they ate became scarce. But some heat-loving, drought-resistant plants still produced large, hard seeds. The finches with larger beaks could eat those seeds. They survived and reproduced, and by 1978, in just one generation, the average beak size in the population had increased (**Figure 11.4**).

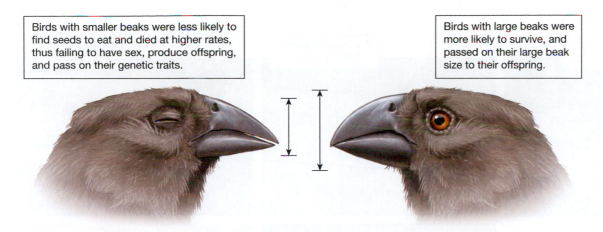

Birds with smaller beaks were less likely to find seeds to eat and died at higher rates, thus failing to have sex, produce offspring, and pass on their genetic traits.

Birds with large beaks were more likely to survive, and passed on their large beak size to their offspring.

Figure 11.4

Natural selection results in larger beak size in finches

After a drought, only birds with larger beaks were able to eat the available food: large, hard seeds. In the span of just one generation, the average size of the species' beak was visibly larger.

Q1: What is natural selection?

Q2: If heavy rains caused an abundance of small, tender seeds and fewer large seeds, what do you predict would happen to the average beak size of the finches?

Q3: Compare and contrast artificial selection and natural selection. Name two ways in which they are similar. How are they different?

This is one of many examples of how a population can evolve via natural selection so that more and more individuals have beneficial traits, and fewer and fewer have disadvantageous traits. This is called **adaptation**—an evolutionary process by which a population becomes better matched to its environment over time. The finch population quickly adapted to its new, drier environment. Over time, the small-beaked finches died off and the large-beaked birds survived and reproduced; the finch population adapted to its environment in just a few years. Other adaptations take millions of years, such as whale ancestors adapting to aquatic life.

It is important to realize that biological evolution includes human evolution. Surveys taken since 2013 reveal that about 40 percent of adults in the United States do not believe that humans evolved from earlier species of animals. This statistic is startling because evolution has been a settled issue in science for nearly 150 years. Scientists like Thewissen go to work every day and see evolution in action. In fact, the vast majority of scientists of all nations agree that the evidence for evolution is overwhelming.

Six lines of evidence provide compelling support for biological evolution:

1. Direct observation of evolution through artificial selection

2. Fossil evidence

3. Shared characteristics among living organisms

4. Similarities and differences in DNA

5. Biogeographic evidence

6. Common patterns of embryo development

Nowhere is all this evidence more present and intriguing than in one of the most dramatic transitions to occur on Earth: the evolution of small, land-living mammals into dolphins, porpoises, and mighty whales.

Fossil Secrets

Fossils are the mineralized remains of formerly living organisms or the impressions of formerly living organisms (**Figure 11.5**). For Thewissen, getting his hands on the

Soft-bodied animals such as this one dominated life on Earth 600 million years ago (mya).

A fossil of a trilobite that lived between 410 and 355 mya.

Fossilized leaf of a 300-million-year-old seed fern.

This 20-million-year-old termite is preserved in amber, the fossilized resin of a tree.

A fossil of a *Velociraptor* entangled with a *Protoceratops*, which bit down on the predator's claw, locking both in a death grip.

Once solid wood has fossilized into solid rock, it is known as petrified wood.

Figure 11.5

Fossils through the ages

Myriad fossils exist, ranging from imprints of organisms, to preserved organisms, to completely mineralized bone and wood. Each fossil can be dated, and when the results are compiled, they can tell the life history of Earth.

Indohyus fossils was no easy task. Beginning in 2003, he made an annual pilgrimage to Dehradun, India, a city nestled in the foothills of the Himalayas. There he visited the widow of A. Ranga Rao, an Indian geologist who had hoarded piles of fossils excavated from Kashmir, a disputed border area between India and Pakistan. Most early whale fossils have come from the India-Pakistan region, where whales first evolved. But because of political tensions, in recent decades it has been too dangerous to travel to Kashmir, much less dig for fossils. Thewissen was frustrated by his inability to collect fossils in Kashmir.

The fossil record enables biologists to reconstruct the history of life on Earth, and it provides some of the strongest evidence that species have evolved over time. The relative depth or distance from the surface of Earth at which a fossil is found indicates its *order* in the fossil record. The ages of fossils correspond to their order: older fossils are found in deeper, older rock layers.

The fossil record contains excellent examples of how major new groups of organisms arose from previously existing organisms. The record includes numerous **transitional fossils**, evidence of species with some similarities to the ancestral group (land-living mammals) and some similarities to the descendant species (whales). Thewissen spent decades studying these intermediates—from the first known whale, the wolflike *Pakicetus* that waded in shallow freshwater; to the larger crocodile-like *Ambulocetus* that stalked its prey underwater; to the fully aquatic *Dorudon*, with its blowhole, flippers, and tail (**Figure 11.6**). Yet Thewissen and others had long been searching for the

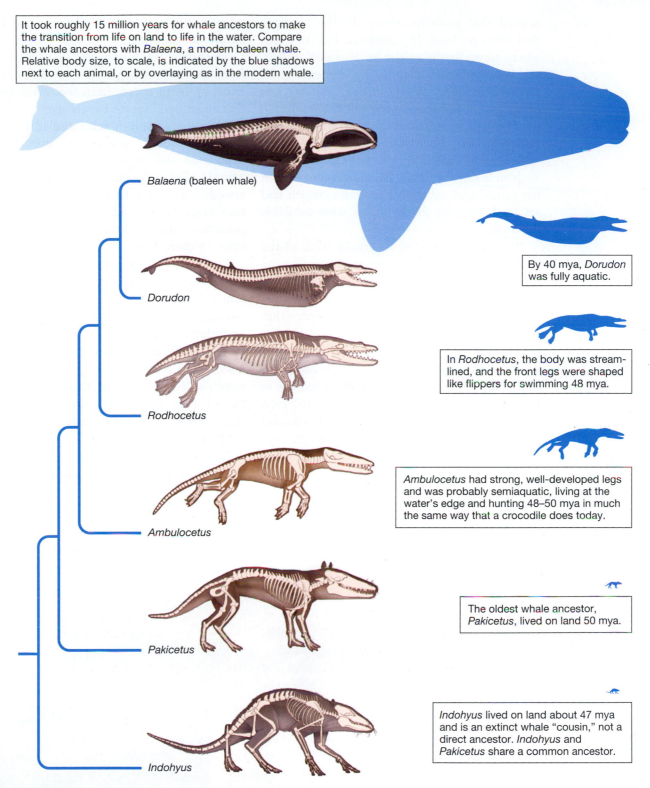

It took roughly 15 million years for whale ancestors to make the transition from life on land to life in the water. Compare the whale ancestors with *Balaena*, a modern baleen whale. Relative body size, to scale, is indicated by the blue shadows next to each animal, or by overlaying as in the modern whale.

Balaena (baleen whale)

Dorudon

By 40 mya, *Dorudon* was fully aquatic.

Rodhocetus

In *Rodhocetus*, the body was streamlined, and the front legs were shaped like flippers for swimming 48 mya.

Ambulocetus

Ambulocetus had strong, well-developed legs and was probably semiaquatic, living at the water's edge and hunting 48–50 mya in much the same way that a crocodile does today.

Pakicetus

The oldest whale ancestor, *Pakicetus*, lived on land 50 mya.

Indohyus

Indohyus lived on land about 47 mya and is an extinct whale "cousin," not a direct ancestor. *Indohyus* and *Pakicetus* share a common ancestor.

Figure 11.6

Skeletons and body sizes of modern whales and fossil ancestors

These reconstructed skeletons from modern whales (top) and various ancestors are in chronological order.

Q1: What is the general definition of a fossil?

Q2: How are the ancestors of modern whales different from their present form?

Q3: What is meant by the term "intermediate fossil" when referring to the fossil record?

animal that preceded them all—the ancestor that lived on land. If Thewissen had to guess where those fossils might be, it was Kashmir, and potentially in Ranga Rao's basement.

Unfortunately, Ranga Rao's widow was protective of the fossils, worried that someone might steal her husband's property and legacy. But each year, Thewissen visited her, chatted with her, and gained her trust. When she passed away in 2007, she made Thewissen cotrustee of her estate, and suddenly the fossils, which had sat in dusty piles for 30 years, were available for study.

"I focused on taking the rocks back to the U.S., and having my fossil preparers remove the fossils from the rocks, which is very difficult," says Thewissen. From Ranga Rao's collection, Thewissen identified more than 400 bones that belonged to *Indohyus*. By collecting a thighbone here and a jawbone there, his team compiled a Frankenstein-like skeleton of a single *Indohyus* individual (**Figure 11.7**). After the discovery of the whalelike ear, the researchers looked even more carefully at the features of the fossils and found additional evidence that *Indohyus* was a relative of whales. This unassuming little animal, with a pointy snout and slender legs

tipped with hooves, lived close to and had an affinity for the water.

The first clues about *Indohyus*'s lifestyle came from its teeth. Oxygen in the molecules that make up teeth comes from the water and food that an animal ingests. Levels of oxygen isotopes in *Indohyus*'s teeth match those of water-going mammals today, suggesting that *Indohyus* lived near and potentially spent a significant amount of time in the water. It also had large, crushing molars with levels of carbon isotopes that suggest it grazed on plants, as do hippopotamuses and muskrats that graze near and in water (**Figure 11.8**).

Lisa Cooper, a graduate student in Thewissen's lab at the time, identified another adaptation to the water: *Indohyus*'s leg bones. From the outside, the limbs of *Indohyus* look like those of any other mammal walking around on land. But on the inside, it's another story. Cooper cut out a section of bone from a limb, ground it down until she could see light through it, and then looked at the bone under a microscope. She saw that a thick layer of bone was wrapped around the bone marrow.

"Hans already had lots of bones of the earliest whales, and they all had extraordinarily

Figure 11.7

Fossilized skeleton of *Indohyus*, oldest cousin of the whales
A reconstructed fossilized skeleton of *Indohyus* was compiled from multiple sources and locations. The illustration is an artist's depiction of the living animal about 47 million years ago. (Source: Photo courtesy of Thewissen Lab, NEOMED.)

Crushing basins

Crushing basins

Crushing basins

The molars of *Indohyus* (top left) are similar to the shape of molars in contemporary aquatic plant-eating animals like hippos (top right and bottom left). These molars have crushing basins for grinding up tough plant fibers.

Figure 11.8

Comparing the skulls and jaws of fossilized *Indohyus* and a modern hippopotamus
These organisms' teeth indicate their ability to eat plant material.

thick bones," says Cooper, now an assistant professor at Northeast Ohio Medical University. Modern animals that live in shallow water, such as manatees and hippos, also have thick bones, which help prevent them from floating and enable them to dive quickly (**Figure 11.9**). "It isn't just isolated to whales," says Cooper. "Bones have thickened again and again as different groups of vertebrates entered the water. When you trace back through the fossil record, there is a pretty good correlation between thickness of bone and whether something was living in the water."

Indohyus's thick bones are an example of an **adaptive trait**, a feature that gives an individual improved function in a competitive environment. By being able to easily wade and dive in water, *Indohyus* had an advantage over other organisms in escaping predators and accessing plants to eat on the river floor. Adaptive traits take many forms, from an anatomical feature

LISA COOPER

Lisa Cooper is an assistant professor at Northeast Ohio Medical University in the Department of Anatomy and Neurobiology. She earned her PhD in Thewissen's lab.

Figure 11.9

Cross sections of the femurs of *Indohyus*, water-dwelling animals, and land-dwelling animals

Aquatic organisms have thick, dense bone around a narrow marrow space, while terrestrial organisms have thin bone and a large marrow space. *Indohyus* bone structure is a trait that is shared with other water-dwelling animals.

Q1: Why do water-dwelling animals have thicker bones than land-dwelling animals?

Q2: Why does this thick-bone adaptation suggest a water-dwelling lifestyle?

Q3: How did this adaptation likely increase survival or reproduction in *Indohyus*?

Pakicetus, an extinct water-dwelling whale ancestor

Manatee, a water-dwelling mammal

Indohyus, an extinct water-dwelling whale cousin

Hippopotamus, a land- and water-dwelling animal

Polar bear, a land-dwelling mammal

Rat, a land-dwelling mammal

like *Indohyus*'s bones, to behaviors, to the functions of individual proteins. Echolocation in bats, for example, is an adaptation for catching insects in the dark. Stick insects physically and behaviorally mimic the plants they live on—an adaptation that helps them avoid detection by predators (**Figure 11.10**).

Figure 11.10

Stick insects avoid detection by predators

Stick insects are well adapted to their environment; they move slowly and look just like the branches they live on.

The Ultimate Family Tree

Thick bones are not restricted to just the water-loving ancestors of whales, as we noted, but can be seen in many other animals, such as hippos. This similarity across organisms is another type of evidence for evolution—shared characteristics among species. Many shared characteristics—such as thick bones for animals that take to shallow water, or sexual reproduction via egg and sperm, or eukaryotic cells—result from organisms sharing traits that evolved from a **common ancestor**, an organism from which many species have evolved. A group of organisms have **common descent** if they share a common ancestor.

When one species splits into two, the two resulting species share similar features, or **homologous traits**, because they have common descent—though these features may begin to look different from one another over time

(**Figure 11.11**). For example, whales are so different from humans that it can be difficult to identify similarities, but we evolved from a common mammalian ancestor and do share homologous traits: humans and whales both nurse their young and have a single lower jawbone because our common ancestor had those traits.

Vestigial traits are another type of trait that many organisms have because of a common ancestor. These features are a piece of the evolutionary past, inherited from a common ancestor but no longer used. Vestigial traits may appear as reduced or degenerated parts whose function is hard to discern (**Figure 11.12**). For example, many modern whales have vestiges of thighbones, also called femurs, embedded in the skin next to the pelvis. In land mammals, birds, and other tetrapod vertebrates, these bones are critical for walking, running, and jumping. Aquatic whales have no need of this bone, yet its traces remain.

Whales also have small muscles devoted to nonexistent external ears, likely from a time when they were able to move their ears as land animals such as dogs do for directional hearing. Vestigial traits are not adaptations. In fact, they can be detrimental. Most humans no longer need wisdom teeth to replace lost teeth during adolescence, yet most people still have them. They tend to erupt around the twentieth year of life, often causing severe pain and displacing other teeth, and they usually require removal.

Clues in the Code

Within the cells of every organism is one of the strongest pieces of evidence for evolution: DNA. Living things universally use DNA as hereditary, or genetic, material (see Chapter 8 for review). The fact that all organisms on Earth—even those as different as bacteria, redwood trees, and humans—use the same genetic code is further evidence that the great diversity of living things evolved from a common ancestor.

Researchers have analyzed the DNA sequences of whales and other animals and shown that, of all animals, whales are most

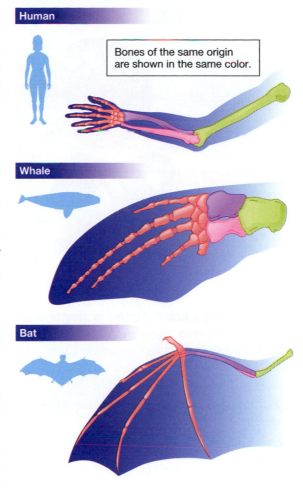

Bones of the same origin are shown in the same color.

Human

Whale

Bat

Figure 11.11

Homologous traits are shared characteristics inherited from a common ancestor

The human arm, the whale flipper, and the bat wing are homologous structures due to common descent. All three structures have a matching set of five digits and a matching set of arm bones that have been altered by evolution for different functions.

Q1: What is meant by the term "common ancestor"? Give an example.

Q2: Why are homologous structures among organisms evidence for evolution?

Q3: Aside from skeletal structural similarities, what other commonalities among organisms might be considered homologous?

External surface

Skeletal structure

Reduced
hind legs

Figure 11.12

Vestigial traits are reduced or degenerated remnants with no apparent function

Snakes are limbless reptiles with no apparent use for the degenerated remnants of hind legs that they still have. The python shown here has extremely reduced hind legs that are only barely visible externally. ▶

Q1: Why are vestigial structures among organisms evidence for evolution? Give an example of another vestigial structure.

Q2: Are vestigial structures also homologous structures? Explain.

Q3: Why do vestigial structures still exist if they are no longer useful?

closely related to even-toed ungulates—a group of hoofed mammals that includes modern deer, giraffes, camels, pigs, and hippos. *Indohyus* is an example of an extinct even-toed ungulate. Molecular studies therefore confirm the prediction that whales and *Indohyus* share a common ancestry.

According to DNA sequence similarity, hippos are whales' closest living relatives. Whale DNA is more similar to hippo DNA than to the DNA of other marine mammals, such as seals and sea lions. **DNA sequence similarity** is a measure of how closely related two DNA molecules are to each other. For example, in the DNA sequences of the same insulin gene in humans and mice, 83 percent of the nucleotides are identical at corresponding positions (**Figure 11.13**). We share a more recent common ancestor with mice than we do

with chickens, for whom the DNA sequence of the same gene is only 72 percent identical to ours. In fact, the insulin gene of our closest living relative on Earth, the chimpanzee, has 98 percent similarity to the human insulin gene. That comparability implies that humans and chimpanzees share a very recent common ancestor.

The fact that these separate lines of evidence—anatomical features and DNA—yield the same result over and over again for diverse groups of organisms is strong evidence for evolution.

Whale evolution is "one of the best case studies documenting how a vertebrate can go from a terrestrial to an aquatic environment," says Lisa Cooper. Another type of evidence supporting whale evolution comes from the locations where whale fossils have been found.

Figure 11.13

DNA sequence similarities of the insulin gene

The complete coding sequence of the human insulin gene contains 333 nucleotides. Only the first 50 nucleotides are shown here. Unshaded paired sequences are identical nucleotides at that position; those shaded in yellow are different.

> **Q1:** If a sequence from another species showed a 96 percent sequence similarity to humans, would that species be more or less closely related to humans than chimpanzees are?
>
> **Q2:** Should similarities in the DNA sequences of genes be considered evolutionary homology? Explain.
>
> **Q3:** How is the increased similarity in the DNA sequences of genes between more closely related organisms—and the decreased similarity between less closely related organisms—evidence for evolution? Use the example in this figure to support your answer.

Birthplace of Whales

Earth's continents are on massive tectonic plates, which slowly move over time in a process called continental drift or plate tectonics. About 250 million years ago, South America, Africa, and all of the other landmasses of Earth had drifted together to form one giant continent called Pangaea. About 200 million years ago, Pangaea slowly began to split up, separating to ultimately form the continents as we know them. That separation continues today; for example, each year South America and Africa drift farther apart by about an inch.

We can use knowledge of evolution and continental drift to make predictions about the **biogeography** of a species—the geographic locations where its fossils will be found. For example, today the lungfish *Neoceratodus fosteri* is found only in northeastern Australia, but its ancestors lived during the time of Pangaea. As predicted, therefore, fossils of those ancestors are found on all continents except Antarctica (**Figure 11.14**).

The biogeography of whale fossils matches the pattern predicted by evolution: all early

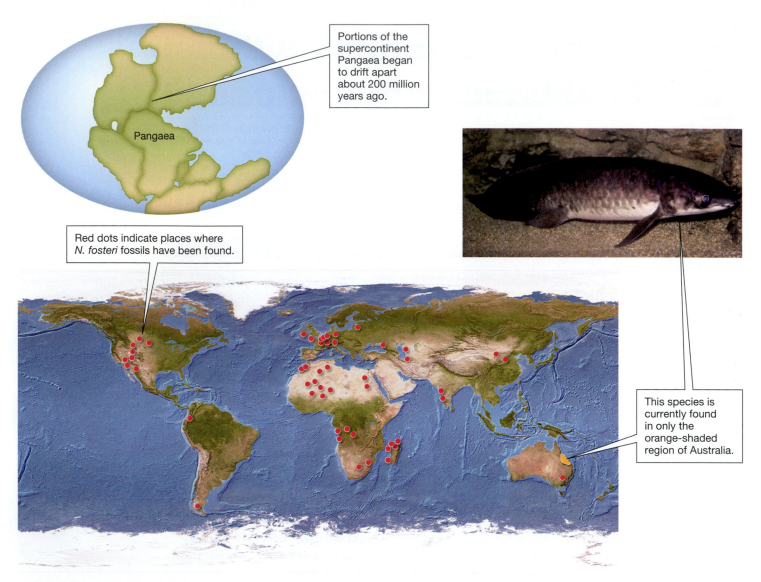

Portions of the supercontinent Pangaea began to drift apart about 200 million years ago.

Pangaea

Red dots indicate places where *N. fosteri* fossils have been found.

This species is currently found in only the orange-shaded region of Australia.

Figure 11.14

Biogeography can reveal a species' evolutionary past

Ancestors of the freshwater lungfish *Neoceratodus fosteri* lived during the time of Pangaea. *N. fosteri* fossils have been found on all continents except Antarctica.

Q1: Why should we expect to find *N. fosteri* fossils all over the world, given that it first evolved in Pangaea?

Q2: Can we use biogeographic evidence to support evolution without using fossil evidence? Give examples.

Q3: How might we use DNA sequence similarities together with biogeography as evidence for evolution?

species of whales, such as the crocodile-like *Pakicetus*, that lived in rivers and lakes but did not swim in the ocean are found near India and Pakistan. "It makes sense," says Thewissen. "You don't have crocodiles crossing the Atlantic." But fossils of fully aquatic protocetids, which emerged about 40 million years ago, are geographically much more widespread; they have been found as far away from Pakistan as Canada. "Protocetids are good swimmers, so we find their fossils all around the world," says Thewissen.

Growing Together

Though Thewissen has built a career on finding and describing whale fossils, he has recently become enamored with another vein of evolutionary evidence: embryology. A major prediction of evolution is that organisms should carry within themselves evidence of their evolutionary past, and they do. Evidence of evolution can be observed in shared patterns of **embryonic development**.

Once again, these common patterns are caused by descent from a common ancestor. Rather than evolving new organs "from scratch," new species inherit structures that may have been modified in form and sometimes even in function.

Upon fusion of sperm and egg, an animal embryo begins to grow and develop. The manner in which an embryo develops, especially at the early stages, may mirror early developmental stages of ancestral forms. For example, anteaters and some baleen whales do not have teeth as adults, but as fetuses they do. And the embryos of fishes, amphibians, reptiles, birds, and mammals (including humans) all develop pharyngeal pouches or gill slits (**Figure 11.15**). In fish, the pouches develop into gills that adult fish use to absorb oxygen underwater. In human embryos, these same features become parts of the ear and throat.

"I was interested to get embryos to look at some of these processes that we see happen in evolution, to see if they happen in development," says Thewissen. The first trait he examined was

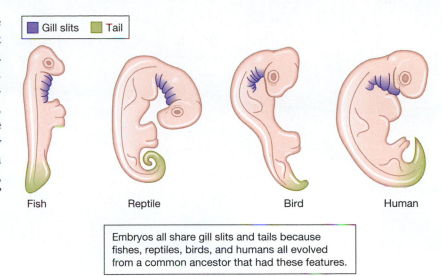

Gill slits ■ **Tail**

Fish   Reptile   Bird   Human

Embryos all share gill slits and tails because fishes, reptiles, birds, and humans all evolved from a common ancestor that had these features.

Figure 11.15

Evolutionary history can be extrapolated from similarities in embryo development

Complex structures in descendant species are generally elaborations of structures that existed in their common ancestor.

Q1: Why are the similarities among organisms during early development evidence for evolution? Give an example.

Q2: Are the similar structures among vertebrate species during embryogenesis homologous structures? Explain.

Q3: Why do embryonic structures still exist at points during embryogenesis if they are not used after birth?

hind limbs. "We know, from fossil evidence, that early whales lose their hind limbs," says Thewissen. So he wondered if hind limbs exist in whale or dolphin embryos (dolphins are also mammals and are closely related to whales). And if so, what makes them subsequently disappear before the animal is born.

Examining spotted dolphin embryos, Thewissen saw that when the embryos are the size of a pea, they do develop hind limb buds, but by the time they grow into the size of a bean, the limb buds are gone. In 2006, he and researchers at several other universities studied the genes that are active in whale and dolphin embryos and concluded that whales' hind limbs regressed over millions of years through small changes in

Watching Evolution Happen

In the laboratory, scientists can manipulate populations of organisms to watch evolution in real time via artificial selection. In 2012, researchers at UC Irvine manipulated the growing environment of *Escherichia coli* bacteria, exposing them to far hotter temperatures than normal, to see if they would adapt. Most of the bacteria adapted via one of just two primary pathways.

Assessment available in smartwork5

1 | Scientists grow **115** separate, genetically identical populations of *E. coli* at a comfortable temperature of **37**°C, or **98.6**°F.

2 | The populations reproduce through **2,000** generations while being subjected to an increased heat of **42.2**°C, or **107.96**°F.

3 | One sample is taken from each of the **115** populations, and the genome is sequenced.

4 | **1,258** molecular changes, averaging **11** genetic mutations per clone, are detected. To help survive the heat, the *E. coli* tend to mutate along two different trajectories:

- Mutations in the RNA polymerase complex, an enzyme that transcribes RNA.

- Mutations in the *rho* gene, which encodes a protein that stops RNA transcription.

5 | The research continues. The next step is to figure out how the mutations in the RNA polymerase complex and in *rho* helped the *E. coli* survive the heat.

a number of genes relatively late in embryonic development (**Figure 11.16**). The loss of hind limbs in the embryo corresponds to the disappearance of hind limbs in the fossil record. "It was an awesome way to combine embryology with fossils," says Thewissen. In 2015, researchers at the Smithsonian Institution likewise found that the development of fetal ear bones in the womb paralleled changes observed in the whale fossil record.

With their discovery of *Indohyus*, Thewissen, Cooper, and their team bridged a 10-million-year gap in the fossil record, identifying an important transition species to whales. It is another rock in the mountain of evidence surrounding whale evolution, which was further fortified in April 2015 with the discovery of fossils from a new species of extinct pygmy sperm whale found in Panama.

Evolution is supported by mutually reinforcing, independent lines of evidence: direct observation, fossils, shared characteristics among living organisms, similarities and divergences in DNA, biogeographic evidence, and common patterns of embryo development. Just as the theory of gravity forms the foundation of physics, so evolution is the central tenet of biology.

There is no question that evolution happens. The intriguing, fascinating question is *how* does it happen? Through scientific research, we know that whales descended from a group

Forelimb   Hind limb

Dolphins maintain their front limb buds, which become flippers.

Dolphins' hind limb buds do not develop into limbs or flippers.

Figure 11.16

Dolphin embryonic development

The embryos of dolphins from weeks 4–9 of development show the formation and then subsequent loss of the hind limb buds. (Source: Based on a photo by Thewissen Lab, NEOMED, permission granted.)

of land-living mammals including the petite *Indohyus*, munching on freshwater plants and splashing through the shallows. But how did a population of small, furry animals become the mammoths of the sea? Next we investigate the mechanisms of evolution—how and why it works.

REVIEWING THE SCIENCE

- Biological **evolution** is the change in characteristics of a population of organisms over inherited generations.

- **Artificial selection** results in biological evolution. Humans choose which organisms survive and reproduce—a process known as **selective breeding**.

- **Natural selection** is the process by which individuals with advantageous genetic characteristics for a particular environment survive and reproduce at a higher rate than competing individuals with other, less useful characteristics.

- **Adaptation** is an evolutionary process by which a population becomes better matched to its environment over time. An adaptation is also a trait— an **adaptive trait**—that has evolved as a result of this process.

- **Fossils** are the preserved remains (or their impressions) of formerly living organisms. The fossil record enables biologists to reconstruct the history of life on Earth, and it provides some of the strongest evidence that species have evolved over time.

- Many similarities among organisms are due to the fact that the organisms evolved via **common descent** from a **common ancestor**. When one species splits into two, the two resulting species share similar features, called **homologous traits**. If a homologous trait is no longer useful, it is called a **vestigial trait**.

- The fact that organisms as different as bacteria, redwood trees, and humans show **DNA sequence similarity** is evidence that the great diversity of living things evolved from a common ancestor.

- **Biogeography** uses knowledge about evolution and plate tectonics to make predictions about the geographic locations where fossils will be found.

- Similarities in **embryonic development** of different organisms show that the characteristics of modern organisms arose through evolutionary modifications of traits inherited from common ancestors.

THE QUESTIONS

Challenge Yourself

1 All mammals have tailbones and muscles for moving a tail. Even humans have a reduced tailbone and remnant tail-twitching muscles, though these features have no apparent usefulness. These traits in humans would best be described as

(a) convergent structures.

(b) fossil evidence.

(c) evidence from biogeography.

(d) vestigial traits.

2 Reduced tailbones and the associated remnant muscles in humans are an example of what type of evidence for common descent?

(a) artificial selection

(b) homologous traits

(c) biogeography

(d) fossil evidence

3 When a population evolves to be better fitted to its environment, this is an example of _____, which is brought about by the process of _____.

4 In one sentence each, identify the similarities and the differences between artificial selection and natural selection.

Try Something New

5 Cat DNA is much more similar to dog DNA than to tortoise DNA. Why?

(a) Cats and dogs are both carnivores and take in similar nutrients.

(b) Cats and dogs have lived together with humans for a long period of time, so they have grown more similar.

(c) Cats and dogs have more offspring during their lifetime than tortoises have, so their DNA changes less rapidly.

(d) Cats and dogs have a common ancestor that is more recent than the common ancestor of cats and tortoises.

6 DNA sequences were analyzed from humans and three other mammals: species X, Y, and Z. Which of these mammals is most closely related to humans? (*Note:* Regions identical to human DNA are shown in bold type.)

Human:
AATGCTTTGGGGGATCGCGAGCGCAGCGC

Species X:
GGGTT**TTT**ATCGCTATATATATATA

Species Y:
AATGCTTTGGGGGATCGCGAGCGCATATA

Species Z:
AATGCGGGTTTTT**ATC**TATATATATATA

7 Which of the following is *not* an example of artificial selection?

(a) Your younger sibling got a hamster as a birthday present, and it turned out to be pregnant. Several of the offspring had long, silky hair, and your sibling put them together in an enclosure to try to produce more baby hamsters with long, silky hair. Your sibling continued to breed long- and silky-haired hamsters to each other and now, several years later, your sibling's bedroom is full of cages containing long- and silky-haired hamsters.

(b) Your mother has been saving the best seeds from her lima bean plants every summer and replanting them the next year. She likes seeds that are plump and bright green and saves only these each year. Within 10 years, almost all of her lima beans have become plump and bright green.

(c) Farmer Brown has a duck that can type. He sets up an online dating profile for his duck to find a female duck that can also type. Seven years after the arranged wedding of the two typing ducks, Farmer Brown has an entire flock of ducks, most of which can type.

(d) Female fish in a natural pond have variable skill at depositing their eggs near the murky shore, where the eggs are better hidden from predators. Eggs that are not deposited near the murky shore are quickly devoured by fish. Female hatchlings from the eggs deposited near the murky shore grow up to be good at depositing eggs near the shore and therefore have a survival advantage.

8 The fossil record shows that the first mammals evolved 220 million years ago. The supercontinent Pangaea began to break apart 200 million years ago. On which continents would you predict that fossils of the first mammals will be found?

Leveling Up

9 **What do *you* think?** The prerequisites for medical school application always include courses in cell biology, genetics, and biochemistry but rarely include a formal course in evolution. Do you think medical schools should require a formal course in evolution as a prerequisite for admission? Why or why not? Research the issue and support your case using information you find from credible sources.

10 *Write Now* biology: evidence for evolution This assignment is designed to expand your knowledge of the evidence for evolution. View the following videos, found at http://www.pbs.org/wgbh/

evolution/educators/teachstuds/svideos.html, and answer the questions accompanying each one here.

Video 1: "Isn't Evolution Just a Theory?"
Why is evolution not just a theory? Use specifics from the video to defend your answer.

Video 2: "Who Was Charles Darwin?"
Why do you think Charles Darwin's ideas and book *On the Origin of Species* were so groundbreaking and "revolutionary"? Use specifics from the video to defend your answer.

Video 3: "How Do We Know Evolution Happens?"
Describe how the video portrays whale evolution. Include specific examples of transitional fossils described in the video, and explain why the scientists at the time the video was made considered the fossils to be whale ancestors.

Video 5: "Did Humans Evolve?"
DNA sequences of different species can be used to provide evidence of common descent. Using examples from the video, explain why DNA sequence similarity is the best evidence of evolution on Earth.

Video 6: "Why Does Evolution Matter Now?"
Describe why the theory of evolution matters to the field of medicine and to individual doctors. Use the example of tuberculosis from the video to support your answer. How does this video affect your answer to question 13?

Thinking Multidimensionally

11 Turkeys destined for dinner plates have changed drastically over the last 100 years because of selective breeding, muscle-building diets, and vaccinations aimed to keep them free from disease. Research how mankind has changed domesticated turkeys and the ramifications these alterations have had on the animal.

12 Write a science journal entry by examining the modern "tree of life" image.

Consider the following as you write your journal:

a. What are the three domains, and how do they differ?

b. Identify and describe the characteristics of the first animals.

c. Regarding plant evolution, how do gymnosperms and angiosperms differ?

d. If extinct organisms were included on this tree, where would dinosaurs be positioned? Why?

e. Place the following organisms in the order that they evolved: yeast, whale, fern, frog, *E. coli*, turtle, amoeba, *Homo sapiens*, salmon, dragonfly, robin.

13 In visiting a natural history museum, you come across a display on horse evolution and are surprised to see so many intermediate forms and that the ancestor of modern horses was no bigger than a typical house cat! Determined to uncover what prompted these changes, you analyze the fossils of horse teeth and hooves below and research horse evolution further on your own.

a. What patterns emerge when scientists analyze horse teeth and hoof fossils?

b. Research the environmental pressures that influenced larger body sizes and increased speed in horses.

c. Critics argue that some museum displays do not accurately reflect current views on horse evolution since the horses are often depicted in a linear arrangement. Design and draw an updated museum display to communicate to visitors at a natural history museum a better representation of horse evolution.

		Fore Foot	Hind Foot	Teeth
Quaternary or Age of Man	Recent			
	Pleistocene	One Toe — Splints of 2nd and 4th digits	One Toe — Splints of 2nd and 4th digits	Long-Crowned, Cement-covered
	Pliocene			
Tertiary or Age of Mammals	Miocene	Three Toes — Side toes not touching the ground	Three Toes — Side toes not touching the ground	
	Oligocene	Three Toes — Side toes touching the ground; splint of 5th digit	Three Toes — Side toes touching the ground	Short-Crowned, without Cement
		Four Toes		
	Eocene	Four Toes — Splint of 1st digit	Three Toes — Splint of 5th digit.	
Age of Reptiles { Cretaceous Jurassic Triassic		Hypothetical Ancestors with Five Toes on Each Foot and Teeth like those of Monkeys etc.		

For more, visit digital.wwnorton.com/bionow2 for access to:

Battling Resistance

An antibiotic-resistant superbug is evolving to overcome our drugs of last defense. How do we beat a champion pathogen?

After reading this chapter you should be able to:

- Clarify why evolution occurs only in populations, not in individuals.
- Distinguish among directional, stabilizing, and disruptive selection.
- Explain how natural selection brings about increased reproductive success of a population in its environment.
- Compare and contrast convergent evolution and evolution by common descent, and give an example of each.
- Describe how DNA mutations can create new alleles at random.
- Illustrate how gene flow works and how it may inhibit evolution.
- Relate the process of genetic drift to genetic bottlenecks and the founder effect, using examples.

CHAPTER

12

MECHANISMS
OF EVOLUTION

DRIVING QUESTION

*How do staph infections develop
resistance to antibiotics?*

- HS-LS4.B: Natural Selection

Dawn Sievert vividly remembers June 14, 2002—the day disaster struck. Sievert is an epidemiologist, or "disease detective"—someone who studies the patterns and causes of human disease. At the time, she was working at the Michigan Department of Community Health, monitoring reported cases of antibiotic-resistant bacteria, a major health care concern. Sievert spent a significant amount of her time investigating outbreaks of an increasingly common and worrisome microbe called MRSA, short for "methicillin-resistant *Staphylococcus aureus*."

S. aureus, commonly known as "staph," is a small, round bacterium that usually lives benignly in our nostrils and on our skin. But on rare occasions, staph slips beneath the surface of a burn or cut and causes an infection, which can be especially dangerous for individuals with suppressed immune systems, such as the elderly and patients on chemotherapy.

Staph is one of the most common causes of hospital infections today, and it is treated with antibiotics, drugs that kill bacteria but not human cells. Penicillin (**Figure 12.1**) was the first antibiotic used against staph, but the wily microbe evolved resistance to penicillin even before the drug became commercially available to the public in the 1940s.

When penicillin stopped working against staph, doctors switched to an antibiotic called methicillin. Methicillin worked for about 20 years, until populations of staph evolved widespread resistance to that antibiotic as well. To the chagrin of doctors and patients everywhere, bacteria are able to adapt rapidly to new antibiotics, thanks to their short generation time and ability to share resistance genes among themselves. (See Chapter 15 for more on bacteria.) These tiny microbes are the Navy SEALs of evolution—the best of the best at evolving.

As we saw in Chapter 11, biological *evolution* is a change in the frequencies of inherited traits in a population over generations. Staph adapted to the presence of methicillin: Only bacteria with genetic traits protecting the microbe from the antibiotic survived. These traits were passed among populations and down from one generation to the next until methicillin resistance was frequent across staph populations (**Figure 12.2**).

Today, MRSA is rampant in hospitals, so doctors have been forced to turn to one of medicine's last lines of defense against the superbug. Vancomycin, a strong, blunt antibiotic that was first isolated from the mud of the Borneo jungle, is considered one of the "drugs of last resort" for fighting these serious infections—which brings us back to June 14, 2002, and Dawn Sievert.

On that day, a lab technician at a dialysis center in Detroit, Michigan, took two swabs of an infected foot ulcer belonging to a 40-year-old diabetic woman (**Figure 12.3**). The patient had previously suffered from numerous foot infections, including MRSA, which had been treated with vancomycin for 6½ weeks. The swabs of this latest infection were sent to a local laboratory, where technicians grew the bacteria in a dish to test its susceptibility to various antibiotics.

When the results of the first test came in, the laboratory staff immediately picked up the phone and called Sievert's office. The bacteria, they told Sievert and the health department, appeared to be resistant to vancomycin. "First, we needed laboratory confirmation and had to

DAWN SIEVERT

Dawn Sievert is an infectious disease epidemiologist who worked for the Centers for Disease Control and Prevention until 2016. While at the Michigan Department of Community Health previously, Sievert investigated the first-ever VRSA infection.

Figure 12.1

Penicillin, produced by mold, kills bacteria
The mold *Penicillium*, the fuzzy white growth with blue spores growing on this petri dish, secretes the antibiotic penicillin into the agar medium, killing the red bacteria surrounding it.

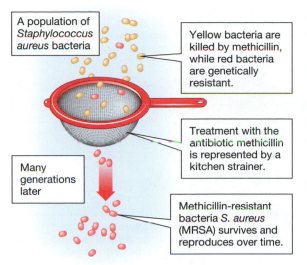

A population of *Staphylococcus aureus* bacteria

Yellow bacteria are killed by methicillin, while red bacteria are genetically resistant.

Treatment with the antibiotic methicillin is represented by a kitchen strainer.

Many generations later

Methicillin-resistant bacteria *S. aureus* (MRSA) survives and reproduces over time.

Figure 12.2

Natural selection results in resistance to antibiotics

The use of antibiotics allows any bacteria (in this case, *S. aureus*) that are resistant to the antibiotic to survive and reproduce. Over time, the frequency of resistant bacteria increases in the surviving populations.

Q1: What is natural selection selecting for here?

Q2: Why do bacteria that are not genetically resistant to antibiotics die out when exposed to antibiotics?

Q3: Why is the antibiotic represented by a kitchen strainer in this figure?

Figure 12.3

MRSA infection in the foot of a patient with diabetes

control any potential for panic," says Sievert. Her team asked the local lab to run its own test again and to send the health department a sample to independently test. Both teams waited.

The tests came back from each lab. Both were positive; the woman's foot was infected with the first reported vancomycin-resistant strain of *S. aureus*. VRSA had arrived.

Birth of a Superbug

"At that point, it was the first-ever VRSA in the world," says Sievert. Unfortunately, it was not the last.

Through evolution, staph first survived penicillin, then methicillin, then vancomycin (**Figure 12.4**). The evolution of *Staphylococcus*

MRSA (methicillin-resistant *Staphylococcus aureus*) bacteria growing on an agar plate

Paper tab infused with vancomycin. The surrounding area it has diffused into, killing off all bacteria in the area, is called the **zone of inhibition**.

VRSA (vancomycin-resistant *Staphylococcus aureus*) bacteria growing on an agar plate

Paper tab infused with vancomycin. Note that the zone of inhibition is much smaller, demonstrating that the bacteria can grow in the presence of the antibiotic.

Figure 12.4

MRSA versus VRSA in a vancomycin-resistance test

In the presence of vancomycin, MRSA (top) does not grow, while VRSA (bottom) grows relatively well.

Q1: What is the difference between MRSA and VRSA?

Q2: Why is there a clear zone (the "zone of inhibition") around the paper disk in the top dish but not in the bottom dish?

Q3: Why is the lack of a clear zone around the paper disk in the bottom dish so alarming?

aureus is a profound example of a species changing over time.

In the mid-1800s, two English biologists, Charles Darwin and Alfred Russel Wallace, studied the diversity of life and concluded that species were not, as was generally thought at that time, the unchanging result of separate acts of creation. Instead, both men came to the bold new conclusion that species "descend with modification" from ancestor species; that is, new species arise from previous species.

Descent with modification, we know today, occurs not only in populations of large organisms like whales and finches, but also in the tiniest single-celled bacteria and viruses. We generally think of evolution happening over millions of years, but some evolutionary changes, such as adaptations for antibiotic resistance, take place over very short time spans as particular alleles spread rapidly through a population. Recall from Chapter 7 that *alleles* are different versions of the same gene (sequences of DNA) produced by random mutation, and therefore *allele frequencies* are the proportions of specific alleles in a population (**Figure 12.5**). Evolution corresponds to changes in the relative proportions or frequencies of alleles in a population over time.

When allele frequencies in a population change, becoming more or less common, the attributes or phenotypes of the population change as well; that is, the population evolves. As more and more staph containing the allele for methicillin resistance survived and reproduced, the whole population of staph evolved, becoming new, more powerful bacteria. But how exactly does this happen? Where do new alleles come from, and how do the frequencies of alleles in a population change?

Sievert and a host of other researchers and doctors experienced the emergence of a new allele firsthand. When the results of the foot ulcer test came in positive for VRSA, Sievert's team immediately called the Centers for Disease Control and Prevention (CDC), a government agency that investigates disease outbreaks and makes public health recommendations. The Michigan Health Department and the CDC converged on the dialysis center where VRSA had been found. They pored over the medical history of the patient, examined her wound, took swabs from the nostrils and wounds of anyone

> Because there are 15 mice, the gene pool has 30 allele possibilities (15 mice × 2 alleles per mouse).

> Of the 30 total alleles in this population, 13 are white-fur-pigment alleles, so the white-allele frequency is 13/30 = 0.43, or 43%.

●●●●●●●●●●●●●●●●● 17/30 = 57%

○○○○○○○○○○○○○ 13/30 = 43%

Figure 12.5

Allele frequencies are calculated as percentages in a population

These mice have two white-fur-pigment alleles and appear white, have two black-fur-pigment alleles and appear gray, or have one black and one white allele and appear gray. To calculate the white-fur-pigment allele frequency in the population, the number of white alleles is counted and divided by the total number of alleles.

Q1: What would the white-fur-pigment allele frequency be if three of the homozygous black allele mice (having two black alleles) were heterozygous (having one white and one black allele) instead?

Q2: What would the white-fur-pigment allele frequency be if all of the white mice died and were therefore removed from the population? Would the black-fur-pigment allele frequency be affected? If so, how?

Q3: What would the white-fur-pigment allele frequency be if all of the gray mice died and were therefore removed from the population?

who had come in contact with her, and then waited anxiously to see whether the dangerous microbe had spread.

Thankfully, they found that the vancomycin-resistant bug had not yet spread. "We're lucky.

If that highly resistant organism had the ability to spread rapidly, we'd have some very sick people at risk in hospitals and other health care settings," says Sievert.

Yet it was not the end of VRSA. Since 2002, there have been 13 additional cases of vancomycin-resistant staph infections in the United States: in the urine of a woman in New York with multiple sclerosis, in the toe wound of a diabetic man in Michigan, in the triceps wound of a woman in Michigan, and more. So far, each infection has been isolated; the microbe has never been transmitted from person to person, as MRSA has. VRSA, while dangerous, doesn't appear to spread through the human population.

But that's not to say it won't evolve to do so. When MRSA first emerged, it seemed restricted to hospital settings. Today, however, clinicians have been horrified to see cases of MRSA pop up from simple scrapes on a playground, suggesting that the microbe is out in the community, where it can do widespread damage. In theory, the same is possible for VRSA, raising the bone-chilling specter of the superbug evolving into an "apocalyptic bug," as one reporter called it.

How did staph acquire vancomycin resistance 14 separate times? How likely is vancomycin resistance to become more widespread? We can find answers to these questions by understanding four mechanisms of evolution:

1. Natural selection

2. Mutation

3. Gene flow

4. Genetic drift

Bacteria are ideal organisms for examining these evolutionary mechanisms, for the same reason that they are so dangerous: because they evolve incredibly fast.

to emerge, says Gilmore. He knew that once vancomycin was widely used to kill staph, the microbe population would evolve resistance to the poison. The process by which a population gains one or more alleles that enable it to survive better than other populations is called *natural selection*, as we saw in Chapter 11. Darwin and Wallace were the first to propose natural selection, and today we know that it is the central driver of evolution.

During natural selection, individuals with particular inherited characteristics survive and reproduce at a higher rate than other individuals in a population. Natural selection acts by favoring some phenotypes over others (**Figure 12.6**). For example, in an environment where bacteria are exposed to vancomycin, the bacteria that can resist the antibiotic will continue to live and reproduce, while those that cannot will perish. Although natural selection acts directly on the phenotype, not on the genotype, of a population, the alleles that code for a phenotype favored by natural selection tend to become increasingly common in future generations. Bacteria that survive an antibiotic attack, for example, pass on alleles that confer that resistance to their offspring.

Natural selection acts on adaptive traits, and therefore a population can become better suited to survive and reproduce in its environment. And unfortunately for us, natural selection is the mechanism by which staph is adapting to our use of vancomycin.

Because staph's vancomycin-resistance adaptation is very recent, scientists are studying the patterns of how natural selection gives rise to VRSA. In nature we observe three common patterns of natural selection: *directional selection*, *stabilizing selection*, and *disruptive selection*. All types of natural selection operate by the same principle: individuals with certain forms of an inherited trait have better survival rates

Rising Resistance

Harvard Medical School microbiologist Michael Gilmore has long tracked the ways that bacteria evolve antibiotic resistance. After staph evolved widespread resistance to methicillin in the 1980s and vancomycin began to be used in hospitals, "we waited and waited and waited" for vancomycin resistance

MICHAEL GILMORE

Michael Gilmore is a microbiologist at Harvard Medical School. He and his laboratory uncovered the genetic basis for the recent emergence of VRSA.

A population of *Staphylococcus aureus* bacteria.

After treatment with methicillin, shown as a red kitchen strainer, the resistant bacteria (MRSA) survive and reproduce.

Additional treatments with methicillin do not reduce the MRSA population.

After treatment with vancomycin, shown as a purple kitchen strainer, the resistant bacteria (VRSA) survive.

The frequency of the VRSA bacteria containing the resistance allele increases dramatically. This is evolution.

Figure 12.6

Evolution happens

Imagine there is a population of *Staphylococcus aureus* bacteria living on your skin. Most of them are susceptible to the antibiotic methicillin (red strainer). A few, however, are randomly resistant to methicillin (red), like the ones in Figure 12.2. A few of these bacteria, in turn, could also be resistant to vancomycin (purple).

Q1: Why does the population of *S. aureus* bacteria *not* pose a life-or-death health threat outright?

Q2: Why do vancomycin-resistant bacteria have a higher frequency in the population after treatment with vancomycin?

Q3: If this figure used the mouse example of allele frequency from Figure 12.5, and the white mice increased in numbers like the vancomycin-resistant bacteria here did, what would happen to the frequencies of the white-fur-pigment and black-fur-pigment alleles?

and produce more offspring than do individuals with other forms of that trait.

Directional selection is the most common pattern of natural selection, in which individuals at *one extreme* of an inherited phenotypic trait have an advantage over other individuals in the population. The peppered moth provides a vivid example: Before 1959, the number of dark-colored moths in both England and the United States increased after industrial pollution blackened the bark of trees, causing dark-colored moths to be harder for bird predators to find than light-colored moths. A reduction in air pollution following clean-air legislation, enacted in 1956 in England and in 1963 in the United States, caused the bark of trees to become lighter, and suddenly the reverse occurred: light-colored moths became harder for predators to find than dark-colored moths. As a result, the proportion of dark-colored moths plummeted because they were easily seen and eaten by predators (**Figure 12.7**). Similarly, when methicillin became widely used to fight staph, MRSA evolved via directional selection: the bacteria in hospital settings that were resistant to the antibiotic survived, while those that were not perished.

In cases of **stabilizing selection**, individuals with *intermediate values* of an inherited phenotypic trait have an advantage over other individuals in the population. Birth weight in humans provides a classic example of this pattern of natural selection (**Figure 12.8**). Historically, light or heavy babies did not survive as well as babies of average weight, and as a result there was stabilizing selection for intermediate birth weights. Today, however, this stabilizing trend is not as strong, because advances in the care of low-birthweight babies and an increase in the use of cesarean deliveries for large babies have allowed babies of all weights to thrive.

Finally, **disruptive selection** occurs when individuals with *either extreme of an inherited trait* have an advantage over individuals with an intermediate phenotype. This type of selection is the least commonly observed in nature, but one example of a trait affected by disruptive selection is the beak size within a population of birds called African seed crackers (**Figure 12.9**). During one dry season, birds with large beaks survived on hard seeds and

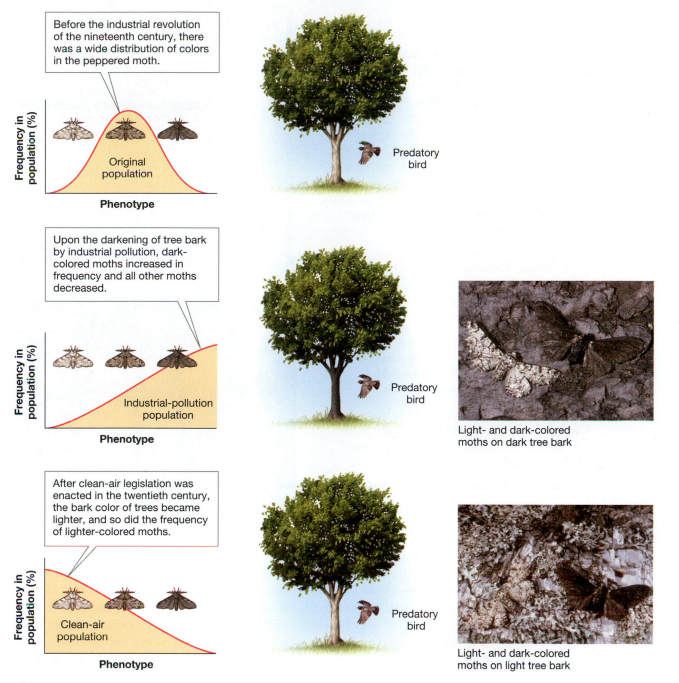

Frequency in population (%)

Before the industrial revolution of the nineteenth century, there was a wide distribution of colors in the peppered moth.

Original population

Phenotype

Predatory bird

Upon the darkening of tree bark by industrial pollution, dark-colored moths increased in frequency and all other moths decreased.

Industrial-pollution population

Phenotype

Predatory bird

Light- and dark-colored moths on dark tree bark

After clean-air legislation was enacted in the twentieth century, the bark color of trees became lighter, and so did the frequency of lighter-colored moths.

Clean-air population

Phenotype

Predatory bird

Light- and dark-colored moths on light tree bark

Figure 12.7

The peppered moth has undergone directional selection two different times in the last 200 years

The industrial revolution of the nineteenth century caused extreme air pollution in both England and the United States, due mostly to soot from the mass burning of coal.

Q1: If one extreme phenotype makes up most of a population after directional selection, what happened to the individuals with the other phenotypes?

Q2: What do you think would happen to the phenotypes of the peppered moth if the tree bark was significantly darkened again by disease or pollution?

Q3: What do you think would happen to the phenotypes of the peppered moth if the tree bark became a medium color, neither light nor dark? (You will need to read the next paragraph to answer this question.)

EVOLUTION

Figure 12.8

Stabilizing selection and human birth weight from 1935 to 1946

This graph is based on data for 13,700 babies born between 1935 and 1946 in a hospital in London. In countries that can afford intensive medical care for newborns, the strength of stabilizing selection has been greatly reduced in recent years. However, even with improved care of low-birthweight babies and cesarean deliveries of large babies, babies at the extremes of newborn weight still survive at a lower rate than those closer to the median weight.

Q1: Think of another example of stabilizing selection in human biology. Has modern technology or medicine changed its impact on the resulting phenotypes?

Q2: How do you think a graph of birth weight versus survival for a developing country with little health care would compare to the graph shown here?

Q3: How do you think a graph of birth weight versus survival for an affluent area of the United States today would compare to the graph shown here?

Babies that weighed 8 pounds at birth survived at a higher rate than did lighter or heavier babies.

birds with small beaks survived on soft seeds, but birds with intermediate-sized beaks fed inefficiently on both types of seeds. As a result, the birds with intermediate-sized beaks died, and the birds with large and small beaks lived. Therefore, natural selection favored both large-beaked and small-beaked birds over birds with intermediate beak sizes.

Any of these patterns of natural selection can cause distantly related organisms to evolve similar structures because they survive and reproduce under similar environmental pressures. This type of evolution, called **convergent evolution**, results in organisms that appear very much alike despite vastly dissimilar genetics. Cacti found in North American deserts and distantly related plants (euphorbias) found in African and Asian deserts offer an excellent example of convergent evolution. These two types of desert plants have very different genetics, but they look similar and function in a similar manner. In

another example, sharks and dolphins are only distantly related (sharks are fish, and dolphins are marine mammals), yet they both evolved for success as predators in the ocean and share common characteristics, such as a streamlined body (**Figure 12.10**). When species share characteristics due to convergent evolution and not because of modification by descent from a recent common ancestor, those characteristics are called **analogous traits** (instead of homologous traits).

Enter *Enterococcus*

Intent on determining how VRSA was evolving after the first Michigan infection, Gilmore began to track the appearance of the bug. He and others wanted to know where the allele for vancomycin resistance came from, and how staph had acquired it 14 different times.

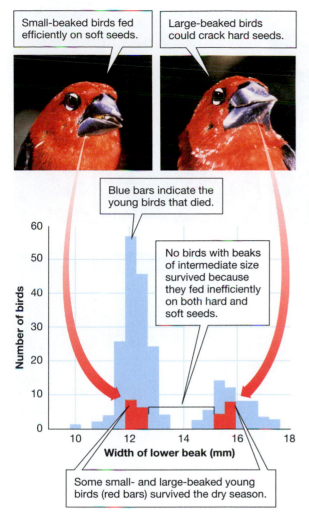

Small-beaked birds fed efficiently on soft seeds.

Large-beaked birds could crack hard seeds.

Blue bars indicate the young birds that died.

No birds with beaks of intermediate size survived because they fed inefficiently on both hard and soft seeds.

Some small- and large-beaked young birds (red bars) survived the dry season.

Figure 12.9

Disruptive selection for beak size

Among a group of young African seed crackers hatched in one year, only those with small or large beaks survived the dry season, when seeds were scarce. Although many of the small- and large-beaked birds did not survive, none of the intermediate-beaked birds survived.

Q1: Almost all birds starved during the dry season depicted here. What type of selection would have been present if only the intermediate-beaked birds had survived (instead of the small- and large-beaked birds)?

Q2: Describe a scenario in which African seed crackers would experience directional selection for either smaller- or larger-beaked birds. What kind of environmental conditions might bring about such a situation?

Q3: Of the three patterns of natural selection presented in this discussion, which one always results in two different phenotypes in the following generations?

As Gilmore searched for patterns among the cases of VRSA infection, he noticed some peculiar things. First, most of the infections were turning up in people with diabetes, typically in bad foot wounds. Second, in most cases, when scientists looked closely at the samples, they saw not only staph but also a small spherical bacterium called *Enterococcus*, which had evolved resistance to vancomycin years earlier (**Figure 12.11**). Upon close observation, Gilmore and others noted that the enterococci, growing cozily side by side with staph, contained a vancomycin-resistance gene identical to the one in the staph. The presence of this gene suggested that staph had acquired vancomycin resistance directly from *Enterococcus*, rather than via random mutations in its own genome.

New alleles in a species emerge via mutation. A **mutation** is a change in the sequence of any segment of DNA in an organism, and it is the only means by which new alleles are generated. DNA mutations create new alleles at random, thereby providing the raw material for evolution. In this sense, all evolutionary change depends ultimately on mutation. Mutations can stimulate the rapid evolution of populations by providing new genetic variation—differences in genotypes between individuals within a population. Then, natural selection and other mechanisms of evolution act on the resulting phenotypes.

In sexually reproducing species, genetic mutations that occur in an organism's germ line cells—the cell lineage that produces gametes such as eggs and sperm—can contribute to evolution. Mutations in other cells of the body, such as skin or blood cells, can affect the individual by causing cancer or other problems, but those mutations are not passed to that individual's offspring. If mutations are not passed to offspring, they cannot contribute to evolution.

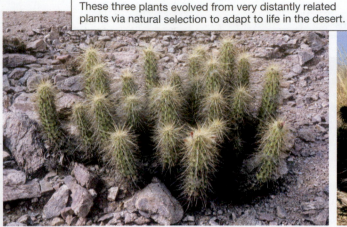

These three plants evolved from very distantly related plants via natural selection to adapt to life in the desert.

Natural selection has caused two distantly related animals—sharks (a type of fish, left) and dolphins (a type of mammal, right)—to look very similar.

Figure 12.10

Natural selection can result in convergent evolution

Q1: How is convergent evolution different from evolution by common descent?

Q2: What is the main difference between a homologous trait (see Figure 11.11) and an analogous trait?

Q3: Why are convergent traits considered evidence for evolution (see Chapter 11)?

Figure 12.11

***Staphylococcus aureus* (left) and *Enterococcus* (right)**

The same does not hold true for bacteria, which are single cells and reproduce via asexual reproduction. All genetic mutations in a bacterial cell are passed on to the offspring. (See Chapter 6 for a review of cellular replication.) If a mutation passed to an offspring increases the individual's ability to reproduce, that mutation is favored by natural selection. These favored mutated genes are passed from parent to offspring, spreading through future generations in a way that alters the population as a whole.

But bacteria don't always wait for the right random mutations to pop up in their genomes. Sometimes they simply borrow new alleles from other organisms. That appeared to be the case with VRSA, researchers found. "In most cases, VRSA has developed in a perfect storm," says Sievert. It emerges within "a very bad wound that's not healing and is a soup mix of organisms coming together and sharing genes."

Horizontal gene transfer is the process by which bacteria pass genes to one another (**Figure 12.12**). Bacteria store these genes on small, circular pieces of DNA called *plasmids*. Some bacteria send these plasmids to each other through small tunnels, like dropping a package down a trash chute.

Horizontal gene transfer is one example of another mechanism by which evolution occurs: gene flow. **Gene flow** is the exchange of alleles between populations. Gene flow can occur between two different species—in this case, between a population of staph and a population of *Enterococcus*—or between two populations of the same species, such as strains of staph passing methicillin resistance among themselves in the community. An individual that migrates between two otherwise isolated populations of a species may facilitate gene flow as well (**Figure 12.13**). Gene flow can also occur when only gametes move from one population to another, as happens when wind or pollinators transport pollen from one population of plants to another.

The introduction of new alleles via gene flow can have dramatic effects. Two-way gene flow consists of an exchange of alleles between one population and another, so it tends to make the genetic composition of different populations more similar. If one strain of staph shares the methicillin-resistance allele with another strain of staph, for example, the two populations

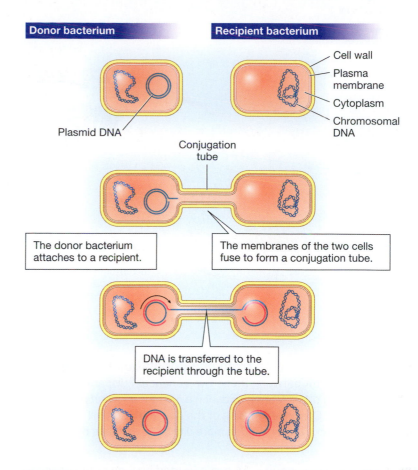

The donor bacterium attaches to a recipient.

The membranes of the two cells fuse to form a conjugation tube.

DNA is transferred to the recipient through the tube.

These bacteria are connected by a conjugation tube and are in the process of passing plasmid DNA from one to another.

Figure 12.12

Horizontal gene transfer accelerates the rate of evolution in prokaryotes

The diagram depicts the horizontal gene transfer of plasmid DNA through **conjugation**, the physical process of transferring genetic material through direct contact. In the case of VRSA, *Staphylococcus aureus* acquired the vancomycin-resistance gene through horizontal gene transfer from *Enterococcus*.

Figure 12.13

Migrants can move alleles from one population to another

As a result of gene flow between these two genetically disparate populations of geese, the populations will eventually become more similar to each other. Mating between a migrant *aa* genotype goose and a population 2 goose of *AA* genotype will result in *Aa* genotype offspring. Continued mating of the *Aa* genotypes will result in all three possible genotypes (*AA*, *Aa*, and *aa*). The resulting population 2 gene pool will look much more similar to the gene pool of population 1.

Q1: If a goose with genotype *AA* had migrated instead of the goose with genotype *aa*, would the scenario described here still be considered gene flow? Why or why not?

Q2: If a goose with genotype *Aa* had migrated instead of the goose with genotype *aa*, would the scenario still be considered gene flow? Why or why not?

Q3: If the goose with genotype *aa* had migrated to population 2 as shown but had failed to mate with any of the *AA* individuals, would the scenario still be considered gene flow? Why or why not?

become more alike. A mutual exchange of alleles through gene flow can *counteract* the effects of the other mechanisms, such as mutation, that tend to make populations more different from one another.

Gene flow appeared responsible for the emergence of VRSA; staph picked up an allele for vancomycin resistance from *Enterococcus*. But Gilmore made another observation that suggested why this horizontal gene transfer was happening now, and not back in the 1980s when vancomycin had first begun to be widely used.

Gilmore noted that all the VRSA samples taken from patients were clonal cluster 5 (CC5) strains of *S. aureus*. "There were implications that there was something special about clonal cluster 5 strains that was leading to this vancomycin resistance bubbling up in these strains," says Gilmore. Compared to other strains of staph, CC5 strains appeared to have evolved the ability to readily take up and use vancomycin-resistance alleles from *Enterococcus*.

Primed for Pickup

In 2012, Gilmore and his team at Harvard analyzed the DNA of 11 of the 12 known cases of VRSA at the time (samples from one of the 12 weren't available, and the thirteenth and fourteenth cases had yet to occur). In the genome of every sample of VRSA, they found three traits that demonstrate how the CC5 strains of staph evolved to effectively pick up the allele for vancomycin resistance.

First, all the vancomycin-resistant CC5 staph bacteria have the same mutation in a gene called *DprA*. *DprA* appears to be involved in preventing horizontal gene transfer. A mutation in this gene might make it easier for the staph to take up DNA from other bacteria, such as *Enterococcus*.

Second, the CC5 strains lack a set of genes that encode an antibiotic that kills other bacteria. Perhaps this antibiotic normally kills *Enterococcus* near the staph, which would explain why horizontal transfer between the two species does not typically occur.

Finally, Gilmore and his team found that in place of that missing set of antibiotic genes, the vancomycin-resistant CC5 staph have a unique cluster of genes encoding proteins that confuse the human immune system. These proteins could make it easier for staph and other bacteria to grow in a wound because the host immune system would be less able to fight them off.

The lack of the antibiotic genes and the presence of new genes create the ideal conditions for a mixed infection, in which different species of pathogens mingle in a festering soup of contamination. Mixed infections are breeding grounds for antibiotic resistance because they are sites of gene flow among different organisms. In this explanation, the CC5 staph evolved via the three usual mechanisms—mutation, natural selection, and gene flow (via horizontal transfer)—to be more susceptible to take up the vancomycin-resistance allele from *Enterococcus*.

In addition to natural selection (and its intriguing permutation called *sexual selection*; see "Sex and Selection"), mutation, and gene flow,

Sex and Selection

Staphylococcus and other bacteria replicate asexually by copying their DNA and dividing in two. But adding sex to the equation complicates reproduction. Another mechanism by which species evolve is called sexual selection. In **sexual selection**, nature selects a trait that increases an individual's chance of mating—even if that trait decreases the individual's chance of survival.

Sexual selection favors individuals that are good at getting mates, and it often helps explain differences between males and females in size, courtship behavior, and other traits. Species whose males and females are distinctly different in appearance, such as peacocks, lions, and ducks, are said to exhibit **sexual dimorphism**. In many species, the members of one sex—often females—are choosy about whether to mate. In birds, for example, brightly colored males may perform elaborate displays in their attempts to woo a mate. In other species, males may attract attention by other means, such as calling vigorously. Females then select as their mates the males with the loudest calls.

Yet, some characteristics that increase an individual's chance of mating can *decrease* its chance of survival. For example, male túngara frogs perform a complex mating call that may or may not end in one or more "chucks." Females prefer to mate with males that emit chucks, but frog-eating bats use that same sound to help them locate their prey. As a result, a frog's attempt to locate a mate can end in disaster.

A male peacock performs an elaborate mating display to a potential mate.

Male túngara frogs face an ecological trade-off: the same type of call that is most successful at attracting females for mating also attracts predatory bats to dinner.

In this small population of upland burrowing crayfish, similar numbers of blue, orange, and green crayfish survive under normal environmental conditions.

In this small population of upland burrowing crayfish, several more blue crayfish than orange or green crayfish survive after a random catastrophe kills off most of the population.

Time

After generations of reproduction, the population contains frequencies of blue, orange, and green crayfish similar to those of the original population.

After generations of reproduction, the rebounded population contains a higher frequency of blue crayfish than orange or green crayfish.

Figure 12.14

A genetic bottleneck is a type of genetic drift

Two small populations of upland burrowing crayfish are compared. The one on the right experiences a genetic bottleneck event. This is genetic drift—a change in the frequency of a trait that is not associated with natural selection. In fact, the blue crayfish could be less well adapted to the environment than the other crayfish are.

Q1: Why do you think a genetic bottleneck is more likely to occur in a small population than in a large population?

Q2: Genetic drift is often described as a "chance event." Give other examples of chance events that could cause a genetic bottleneck.

Q3: Which resulting population has more genetic diversity?

there is one other mechanism by which organisms evolve. Although alleles that code for beneficial traits like antibiotic resistance for bacteria are usually selected for and maintained in a population by the nonrandom action of natural selection, in some cases, chance events can cause allele frequencies to change from a parent generation to the next generation.

Genetic drift is a change in allele frequencies caused by random differences in survival and reproduction among the individuals in a population. Just by chance, some individuals in a generation may leave behind more descendants than other individuals do. In this case, the genes of the next generation will be the genes of the "lucky" individuals, not necessarily the "better" individuals. Genetic drift occurs in all populations, including mammals and bacteria, but it is more likely to cause evolution in a small population than in a large one.

Two ways in which strong genetic drift may occur include genetic bottlenecks and the founder effect. A **genetic bottleneck** is a drop in the size of a population, for at least one generation, that causes a loss of genetic variation (**Figure 12.14**). A genetic bottleneck can threaten the survival of a population.

In the 1970s, for example, the population of the endangered Florida panther plummeted because of hunting and habitat destruction. The species barely escaped extinction. At one point, experts believed that only six wild individuals in the whole species were still alive. This rapid population reduction created a genetic bottleneck in which an estimated half of the genetic variation within the species was lost, and severe inbreeding among the individuals that were left resulted in maladies including low sperm counts and abnormally shaped sperm in male panthers (**Figure 12.15**).

Thankfully, panther numbers have increased to about 80–100 individuals in recent years, in part because of captive breeding programs. Most

Figure 12.16

Early Dutch colonists in South Africa

Figure 12.15

Abnormally shaped sperm in the rare Florida panther

Florida panthers have more abnormal sperm than do panthers from other populations—a possible effect of a genetic bottleneck promoting the increase of harmful alleles in the Florida panther population.

Abnormal panther sperm

Normal panther sperm

of the members of this rebounded population are incredibly similar genetically, because of the limited variety of alleles present in the original six individuals. In an effort to restore gene flow to the population, in 1995 researchers released eight female Texas cougars, close genetic cousins to the Florida panthers, into the panthers' habitat. Five of those females produced eight litters of kittens—more genetically diverse, stronger, and longer-lived than their fathers.

The **founder effect** occurs when a small group of individuals establishes a new population isolated from its original, larger population. For example, in South Africa a population of people called Afrikaners descended primarily from a few Dutch colonists (**Figure 12.16**). Today, the Afrikaner population has an unusually high frequency of the allele that causes Huntington disease because those original colonists by chance carried the allele when they settled in the area.

After Vancomycin

When the initial case of VRSA popped up back in 2002, the patient's foot was carefully treated with an old topical drug that is rarely used today. The only other options were one or two

EVOLUTION

new drugs that hadn't yet been released on the market. "There are still some antibiotics we can use for treatment, but we recommend those be used very responsibly in order to maintain their effectiveness," says Sievert. "You don't want to give them out when not necessary because you don't want to see resistance to them. You save them as the end of the line, so something is available if the commonly used drugs end up failing because of resistance." After 9 months of careful treatment, the patient healed and was able to keep her foot. Other cases of VRSA have similarly been resolved by the use of the newer antibiotics and the delicate care of doctors.

But what happens when even these drugs no longer work against staph? Pharmaceutical companies are developing few new antibiotics, says Frank DeLeo, acting chief of the

Laboratory of Human Bacterial Pathogenesis at the National Institute of Allergy and Infectious Diseases, and bacteria evolve more quickly than we can develop new antibiotics. It's a vicious cycle, he says. "As we continue to use antibiotics, populations of microbes will develop resistance to antibiotics." Recently, microbes appear to be winning the struggle. In July 2016, researchers reported the second U.S. appearance of a superbug resistant to colistin, an antibiotic of last resort. "The golden age of antibiotics appears to be coming to an end," the *Los Angeles Times* wrote.

VRSA has not yet become as widespread as MRSA. Scientists believe the reason may be that staph does not handle the resistance allele for vancomycin very well; the gene from *Enterococcus* is bulky and difficult to manage, so the staph individuals that receive the gene are often outcompeted by other staph individuals in the environment. "Though it allows them to survive, they don't do well with it. There's a [survival] cost to having this resistance," says Gilmore. In the absence of vancomycin, staph is better off without the resistance allele, so perhaps staph individuals lose it or don't pick it up when not in the presence of the drug. VRSA therefore remains "a rare, unstable organism in the environment that isn't very hardy, so it doesn't last and spread," says Sievert. "That's the good news."

But there's still trepidation that staph may adapt and get comfortable with this resistance gene, just as staph adapted to methicillin resistance, says Gilmore. If staph survives and reproduces more easily with the resistance gene, the gene will likely spread to staph throughout the host/human population. "We don't know if that's a possibility," says Gilmore. "My suspicion is that it is." (See "How Can *You* Make a Difference? Help Prevent Antibiotic Resistance!")

In fact, the thirteenth VRSA case, from a man in Delaware in July 2012, was not a CC5 strain, says Gilmore—suggesting that vancomycin resistance is spreading to other strains of staph. "It's disconcerting," he adds. It seems that other populations of staph may now also be evolving vancomycin resistance. "The main worry is that this would move into a strain that was highly transmissible in the community," adds DeLeo. "The potential is there."

How Can *You* Make a Difference? Help Prevent Antibiotic Resistance!

The U.S. Centers for Disease Control and Prevention (CDC) recommends the following guidelines:

- Take antibiotics exactly as the doctor prescribes. Do not skip doses. Complete the prescribed course of treatment, even when you start feeling better.

- Only take antibiotics prescribed for you; do not share or use leftover antibiotics. Antibiotics treat specific types of infections. Taking the wrong medicine may delay treatment and allow bacteria to multiply.

- Do not save antibiotics for the next illness. Discard any leftover medication once the prescribed course of treatment is completed.

- Do not ask for antibiotics when your doctor thinks you do not need them. Remember that antibiotics have side effects. When your doctor says you don't need an antibiotic, taking one might do more harm than good.

- Prevent infections by practicing good hand hygiene and getting recommended vaccines.

SOURCE: http://www.cdc.gov/Features/AntibioticResistance.

Race against Resistance

"Each year in the United States, more than 2 million people become infected with antibiotic-resistant bacteria, including methicillin-resistant *Staphylococcus aureus* (MRSA), resulting in at least 23,000 deaths. To combat antibiotic resistance, doctors need novel antibiotics, yet fewer and fewer new antibiotics come to pharmacies each year. In 2011, Congress passed the Generating Antibiotic Incentives Now (GAIN) Act to stimulate the development of new antibiotics, and it's working. So far, six new antibiotics have been approved through that program.

Assessment available in smartwork**5**

Antibacterial drugs approved by the FDA

19	11	11	11	4	3	8
1980–84	1985–89	1990–94	1995–99	2000–04	2005–09	2010–15

Time line of antibiotic resistance

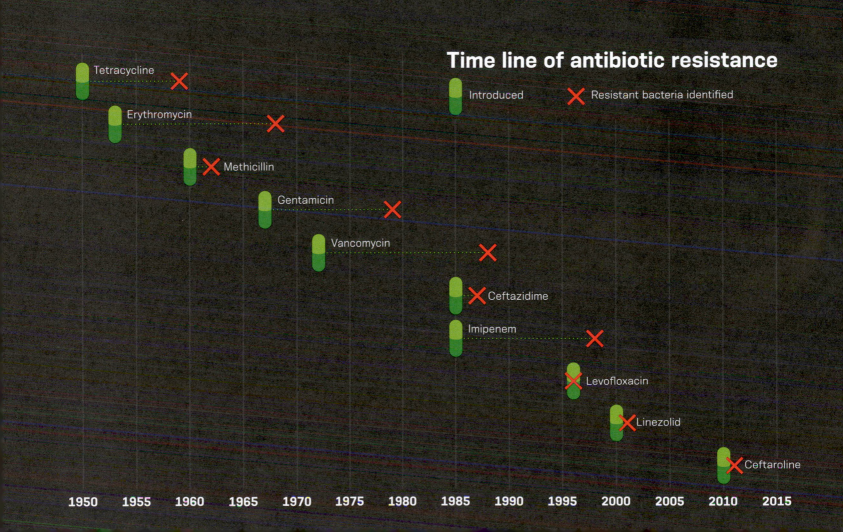

Introduced   ✕ Resistant bacteria identified

Tetracycline
Erythromycin
Methicillin
Gentamicin
Vancomycin
Ceftazidime
Imipenem
Levofloxacin
Linezolid
Ceftaroline

1950 1955 1960 1965 1970 1975 1980 1985 1990 1995 2000 2005 2010 2015

REVIEWING THE SCIENCE

- Natural selection for inherited traits occurs in three common patterns: directional, stabilizing, and disruptive.

- In **directional selection**, individuals at one phenotypic extreme of a given genetic trait have an advantage over all others in the population.

- In **stabilizing selection**, individuals with intermediate phenotypes have an advantage over all others in the population.

- During **disruptive selection**, individuals with either extreme phenotype have an advantage over those with an intermediate phenotype.

- In **convergent evolution**, distantly related organisms (those without a recent common ancestor) evolve similar structures in response to similar environmental challenges.

- All mechanisms of evolution depend on the genetic variation provided by new alleles created by **mutation**.

- **Sexual selection** occurs when a trait increases an individual's chance of mating even if it *decreases* that individual's chance of survival.

- **Gene flow** is the exchange of alleles between separate populations.

- **Genetic drift** is a change in allele frequencies produced by *random* differences in survival and reproduction in a small population, and most dramatically occurs through one of two processes: a genetic bottleneck or the founder effect.

- A **genetic bottleneck** occurs when a drop in the size of a population leads to a loss of genetic variation in the new, rebounded population.

- The **founder effect** occurs when a few individuals from a large population establish a new population, leading to a loss of genetic variation in the new, isolated population.

THE QUESTIONS

Challenge Yourself

1 In a population, which individuals are most likely to survive and reproduce?

(a) The individuals that are the most different from the others in the population.

(b) The individuals that are best adapted to the environment.

(c) The largest individuals in the group.

(d) The individuals that can catch the most prey.

2 A study of a population of the goldenrod wildflower finds that large individuals consistently survive and reproduce at a higher rate than small or medium-sized individuals. Assuming size is an inherited trait, the most likely evolutionary mechanism at work here is

(a) disruptive selection.

(b) directional selection.

(c) stabilizing selection.

(d) natural selection, but it is not possible to tell whether it is disruptive, directional, or stabilizing.

3 Explain how, because of sexual selection, an individual might be very successful at surviving (natural selection), but not pass on genes to the next generation.

Try Something New

4 Two large populations of the same species found in neighboring locations that have very different environments are observed to become genetically more similar over time. Which of the four main evolutionary mechanisms is the most likely cause of this trend? Justify your answer.

5 The Tasmanian devil, a marsupial indigenous to the island of Tasmania (and formerly mainland Australia as well), experienced a population bottleneck in the late 1800s when farmers did their best to eradicate it. After it became a protected species, the population rebounded, but it is now experiencing a health crisis putting it at risk for disappearing again. Many current Tasmanian devil populations are plagued by a type of cancer called devil facial tumor disease, which occurs inside individual animals' mouths. Afflicted Tasmanian devils can actually pass their cancer cells from one animal to another during mating rituals that include vicious biting around the mouth.

Unlike the immune systems of other species, including humans, the Tasmanian devil's immune system does not reject the passed cells as foreign or nonself (as we reject a liver transplant from an unmatched donor), but accepts them as if they were their own cells. Why would a population bottleneck result in the inability of one devil's immune system to recognize another devil's cells as foreign?

6 Global warming is causing more and more ice to melt each year at far-northern latitudes, exposing more bare ground than ever before. These vast areas of brown ground coloration make polar bears (which are white) much more conspicuous to their prey. Recently, an infant polar bear was born with brown fur. This polar bear survived to adulthood and has sired several offspring with brown fur. Which of the following is a plausible explanation of how the brown fur trait appeared in these polar bears?

(a) A polar bear realized it would be better to be brown in order to hide more effectively. It induced mutations to occur in its fur pigment gene, which resulted in a change in pigment from white to brown fur.

(b) One or more random mutations occurred in the fur pigment gene in an individual polar bear embryo, which resulted in a change in pigment from white to brown fur.

(c) Increased temperatures due to global warming caused targeted mutations in the fur pigment gene in an individual polar bear embryo, which resulted in a change in pigment from white to brown fur.

(d) A female polar bear realized it would be better for her offspring to be brown and therefore mated with a grizzly bear to achieve this result.

7 In the garden shed belonging to one of this text's authors, stabilizing selection has occurred over the past 10 years in the house mouse, *Mus musculus*. Which of the following scenarios is an example of stabilizing selection?

(a) Small and medium-sized mice cannot reach the seed shelf in the shed and therefore are at a disadvantage for finding food, so they do not survive and reproduce as well as large mice do.

(b) Small mice cannot reach the seed shelf, and large mice are easily seen by hawks circling above. Medium-sized mice therefore survive and reproduce better than both small and large mice.

(c) Small mice can easily cross the yard to the vegetable garden, and large mice can easily reach the seed shelf. Medium-sized mice have trouble with the seed shelf and are seen by hawks in the yard. Small and large mice survive and reproduce much better than medium-sized mice.

(d) All of these are examples of stabilizing selection.

(e) None of these are examples of stabilizing selection.

Leveling Up

8 **What do *you* think?** One way to prevent a small population of a plant or animal species from going extinct is to deliberately introduce some individuals from a large population of the same species into the smaller population. In terms of the evolutionary mechanisms discussed in this chapter, what are the potential benefits and drawbacks of transferring individuals from one population to another? Do you think biologists and concerned citizens should take such actions?

9 *Write Now* biology: **mechanisms of evolution** This assignment explores the mechanisms of evolution through five selected short stories from *Welcome to the Monkey House* by Kurt Vonnegut Jr. Answer the questions associated with each story.

"Harrison Bergeron"
What message is this story trying to send? Cite examples from the story and relate them to the mechanisms of evolution from this chapter.

"Welcome to the Monkey House"
Is this story an example of sexual selection? Why or why not? Cite examples from the story and from this chapter to support your thinking.

"The Euphio Question"
If technology could produce such an instrument, how would it affect the evolution of humans? What about the evolution of other species on Earth?

"Unready to Wear"
Relate this story to as many of the mechanisms of evolution from this chapter as you can. Cite examples from the story and the chapter to support your thinking.

"Tomorrow and Tomorrow and Tomorrow"
Do you think these types of drugs are a good or bad thing? Where would you draw the line on technology's ability to extend life? How would drugs like these affect the natural selection and evolution of humans? What about the evolution of other species?

Thinking Multidimensionally

10 In 1801, Jean-Baptiste Lamarck proposed a mechanism for evolution, the theory of inheritance of acquired characteristics, which postulates that if an organism changes during its lifetime, those changes will be passed on to the progeny. For example, Lamarck thought that giraffes evolved long necks by stretching to reach leaves on trees and passed this trait on to their offspring.

a. How would Darwin explain how giraffes got long necks?

b. We also know that young male giraffes spar with their long necks. Based on your understanding of sexual selection, what is another reason giraffes may have evolved long necks?

11 One of the most cited examples of natural selection in biology textbooks is the changes Bernard Kettlewell reported on moth coloration against sooty and lichen-covered tree trunks. In the 1950s, Kettlewell pinned dead moths on the different tree trunks and observed birds' feeding preferences.

a. Figure 12.7 describes this classic account of evolution via directional selection. Some critics argue that this example of "evolution in action" is fraudulent. What flaws are there to this experimental design?

b. Compare when moths are active to when birds are predominantly so.

c. What are the main predators of moths, and how do they find their prey?

For more, visit digital.wwnorton.com/bionow2 for access to:

Fast Lizards, Slow Corals

The rapid evolution of lizards and the slow growth of corals offer clues about how species are born.

After reading this chapter you should be able to:

- Diagram the relationships among adaptive traits, reproductive fitness, and evolution by natural selection.

- Defend the utility of the biological species concept, and discuss its shortcomings.

- Define speciation and the role of genetic divergence among populations in creating new species.

- Differentiate between allopatric and sympatric speciation.

- Describe the two kinds of reproductive isolating mechanisms and explain their role in speciation.

- Define and give an example of coevolution.

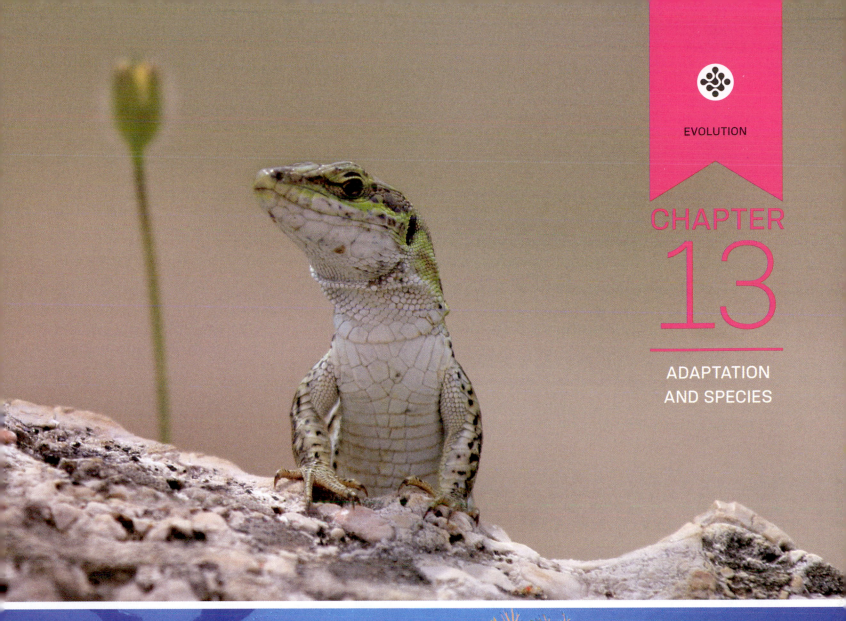

CHAPTER 13

ADAPTATION AND SPECIES

DRIVING QUESTION

What events created new species of island-dwelling lizards and sea fan corals?

- HS-LS4.B: Natural Selection
- HS-LS4.C: Adaptation

I n 1971, the United States was at war in Vietnam, a gallon of gas cost 40 cents, a computer engineer sent the first e-mail—and an Israeli biologist named Eviatar Nevo captured 10 lizards on Pod Kopište, a small, rocky island off the coast of Croatia. Each lizard was about the length of a pinky finger and as heavy as a nickel. There was nothing remarkable about them.

Nevo and his team released the captured lizards, a common Mediterranean species called *Podarcis sicula*, on a nearby island just 3 miles away, within sight of the original island but separated by a deep ocean gulf (**Figure 13.1**). Pod Mrčaru is a smaller, plant-covered island that was already inhabited by two other lizard species. Nevo was curious to find out what would happen when the three species began to compete for resources on Pod Mrčaru. But Nevo never got the chance to return to the island. Shortly after he and his team departed, unrest broke out across Croatia, and war held the region in a stranglehold throughout the 1980s and '90s.

Scientists did not return to Pod Mrčaru until 2004. While visiting the University of Antwerp in Belgium, biologist Duncan Irschick from the University of Massachusetts, Amherst, and his Belgian colleagues decided to investigate the mysterious island. One of the researchers, familiar with an obscure paper that Nevo had published in 1972, thought it would be interesting to see what, if anything, had happened to the lizards. Further, Irschick would be able to complete some research on lizard behavior that he had planned to do in Europe anyway. "We didn't know what we'd find," recalls Irschick. "We decided, 'Let's just go check.'" This time, the lizards on the island were quite remarkable.

Figure 13.1

View of Pod Kopište from Pod Mrčaru
The island Pod Kopište (top), the original home of the *Podarcis sicula* lizards (bottom), is only 3 miles from their new home on the island Pod Mrčaru.

Leaping Lizards

Back in the 1970s, Nevo's ten *Podarcis sicula* lizards had been well adapted to the rocky, sparsely vegetated environment of Pod Kopište. They had specific **adaptive traits**, inherited characteristics that enabled them to survive and reproduce successfully on the island. Adaptive traits can be structural features, biochemical traits, or behaviors. In this case, the lizards were fast, with long legs that may have helped them catch the insects that made up most of their diet on the island. They were also territorial, fighting with other lizards over space and mating partners. These adaptive traits, among others, enabled the lizards to survive and reproduce better than competitors lacking those traits on Pod Kopište.

When transplanted to Pod Mrčaru, however, the lizards faced a new environment. Pod Kopište is large and rocky with few plants, and the lizards ate primarily insects; Pod Mrčaru is smaller yet has an abundant supply of plants, including leaves and stems of local shrubs and grasses (**Figure 13.2**). Either the lizards would adapt to their new surroundings, or they would die. The term **adaptation** is commonly applied to adaptive traits *or* the process of evolution through natural selection that brings about adaptive traits, as discussed in Chapter 12. Therefore, an adaptation can be a trait that is advantageous to an individual or a population or, in broader terms, it can be the evolutionary

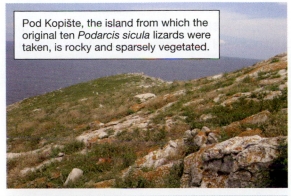

Pod Kopište, the island from which the original ten *Podarcis sicula* lizards were taken, is rocky and sparsely vegetated.

Pod Mrčaru is a smaller island covered with lush vegetation.

Figure 13.2

Pod Kopište's sparse vegetation and Pod Mrčaru's lush vegetation

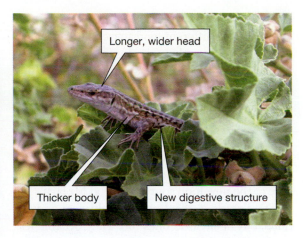

Longer, wider head

Thicker body

New digestive structure

Figure 13.3

***Podarcis sicula* lizard on Pod Mrčaru, 33 years after introduction of the species**

process of natural selection that enables a good match between a population of organisms and their environment. If the population of lizards was going to survive, it would need to adapt to its new environment.

When Irschick and his team landed on Pod Mrčaru in 2004, they set to work catching lizards. It wasn't hard; the island was swarming with them. Irschick could sit on a rock and simply pick up lizards as they ran by. There were several thousand lizards on the 7-acre island. When Irschick and his team looked closely, they realized that *all* the lizards were *P. sicula*. The ten transplants had wiped out the other two species of lizards. What's more, the adapted lizards looked strange compared to *P. sicula* lizards on other islands nearby (**Figure 13.3**). "They were these really big, chunky lizards," recalls Anthony Herrel, then a postdoctoral fellow in Irschick's lab and now a researcher at the French National Centre for Scientific Research in Paris. "They were unlike lizards on any of the other islands."

Twice a year for 3 years, the group returned to Pod Mrčaru to collect, weigh, and measure the lizards. They also took small pieces of the tails (which grow back) to test the lizards' DNA and compare it to the original population of *P. sicula* on Pod Kopište. Those DNA tests confirmed that the lizards on Pod Mrčaru were indeed descended from the original ten that Nevo had transported to the island. In addition to taking size and weight measurements, the team tested the strength of the lizards' bite. They also dissected two samples of dead lizards.

What the researchers found was unprecedented. In about 33 years, from the time Nevo stepped off the island to the time Irschick stepped on, the species had evolved dramatically. The descendant lizards' heads were larger and shaped differently, making the lizard's bite much stronger than that of the original lizards brought to the island. The descendant lizards also had a unique digestive-tract structure called a cecal valve, a set of muscles between the large and small intestine that slow down food digestion, enabling the lizards to better process the cellulose of plants. It's rare for lizards to have these structures, says Herrel. Only a few plant-eating lizard species, like iguanas, have cecal valves.

DUNCAN IRSCHICK

Duncan Irschick is a biologist at the University of Massachusetts, Amherst, who studies animal function and evolution, specializing in research on animal movement and gecko adhesion.

The new traits were the result of the lizards adapting to a different food source, says Irschick: on Pod Kopište, they ate insects, but on Pod Mrčaru, they began to feast on plants. Flushing the stomachs of a few lizards revealed that the lizards' diet was made up of two-thirds plant material. Over time, the lizards that had evolved a new head shape and digestive structures survived and reproduced better than those that had not, because these traits enabled them to take advantage of the new food sources on Pod Mrčaru.

The lizards evolved not only physical adaptations, but behavioral ones too. They became less territorial and less aggressive and were mating more often, probably because more space and food were available. Adaptations, whether behavioral, physical, or biochemical, have three important characteristics. First, they closely match an organism with its environment; in this case, the lizards evolved to match the ecosystem on Pod Mrčaru. Second, adaptations are often complex, such as the lizards' new gut structure. Finally, adaptations help the organism accomplish important functions, such as feeding and mating.

The lizards on Pod Mrčaru show that evolution by natural selection can improve the adaptive fit of organisms to their environment not only over long periods of time, as with the evolution of whales described in Chapter 11, but also over surprisingly short periods of time. In just 33 years, the *P. sicula* lizard population evolved both physical and behavioral traits that helped it flourish on the new island. "It was really, truly rapid evolution across multiple facets," says Irschick.

What Makes a Species?

Despite how impressive the lizards' quick adaptation may be, natural selection does not always result in a perfect match between an organism and its environment. In many cases, animals fail to adapt successfully. In fact, scientists estimate that 99 percent of all species that have ever lived are now extinct. Every extinct species is a silent testament to a failure to adapt in the face of adversity.

It's tempting to label the descendant lizards as a new species, but Irschick is hesitant to do so without more testing. Most commonly, the term **species** is used to refer to members of a group that can mate with one another to produce fertile offspring. According to the **biological species concept**, a species is a group of natural populations that can interbreed to produce fertile offspring and cannot breed with other such groups; that is, they are **reproductively isolated** from other populations (**Figure 13.4**). Irschick and Herrel have yet to test whether the Pod Mrčaru lizards can still mate with their cousins back on Pod Kopište. According to the

Figure 13.4

A male and a female South Pacific rattlesnake confirm that they are the same species
After performing this mating ritual, these snakes successfully mated, producing viable offspring.

Q1: How do we know that these rattlesnakes are members of the same species?

Q2: How would you design an experiment to determine whether two populations of snakes are distinct species according to the biological species concept?

Q3: For which types of populations does the biological species concept *not* work as a way of determining how they are related?

ANTHONY HERREL

Anthony Herrel worked as a research fellow in Duncan Irschick's lab before starting his own lab at the French National Centre for Scientific Research in Paris, where he studies the evolution of feeding and locomotion in vertebrates.

Figure 13.5

One species or two?

Though they look quite different, these tree frogs are genetically similar enough to be considered distinctly colored variations of the same species. Breeding these frogs with each other would determine their classification under the biological species concept.

Q1: List three kinds of information that scientists use to distinguish between species.

Q2: What differences can you observe between the individuals in the photos? Why are these differences not enough to confirm that they are from two different species?

Q3: How is genetic divergence among populations determined?

biological species concept, if individuals from the two populations of lizards can still breed and produce fertile offspring, they are not different species. This same idea holds true for all sexually reproducing organisms. (Incidentally, it's not obvious why sexual reproduction is so common; see "Why Sex?")

The definition of "species" is not a black-and-white issue; it is a frequently discussed, multi-faceted topic in evolution, and the biological species concept doesn't always apply. For example, not all species can be defined by their ability to interbreed, such as prokaryotes like bacteria, which reproduce asexually. To handle these cases and others, scientists may use biogeographic information, DNA sequence similarity, and **morphology**—the organisms' physical characteristics—to identify and distinguish species (**Figure 13.5**).

Why Sex?

Sex is ubiquitous in the animal and plant kingdoms. An estimated 99 percent of multicellular eukaryotes are capable of sexual reproduction, which involves the joining of two haploid gametes produced through meiosis. Yet sex is very costly for individuals, so scientists have struggled to explain why it is so prevalent compared to asexual reproduction. And no, it's not because sex feels good; the first eukaryotes to engage in sex were single-celled protists some 2 billion years ago, long before animals developed neurons capable of giving an individual a sense of pleasure.

Costs of Sex

1. Time and energy must be invested to find or attract a mate.

2. Parents pass on only 50 percent of their genetic material to offspring, as opposed to the 100 percent that is passed on through asexual reproduction.

3. Gene combinations that have benefited the parents may be shuffled and broken apart during meiosis and recombination.

Possible Benefits of Sex

1. The genetic diversity created by sexual reproduction is critical for adaptation to new environments.

2. Sexual reproduction can help a population eliminate detrimental alleles and generate new beneficial alleles.

3. Rapid genetic change that occurs through sexual recombination can help a population evolve resistance to parasitic infections.

The lizards that adapted to the island of Pod Mrčaru are physically different enough from their cousins on Pod Kopište to be considered a different species, says Herrel, yet their DNA sequences are close to identical, so he believes they might still be able to interbreed. **Speciation** is the process by which one species splits to form two species or more. Fundamentally, speciation occurs because of **genetic divergence**, the accumulation of differences in the DNA sequences of genes in two or more populations of organisms over time; as a result, the populations become more and more genetically dissimilar.

The two populations of lizards may continue to diverge, becoming more and more different because of their **geographic isolation** from one another, and may eventually become two unarguably distinct species. This is one of the most common ways that new species form: individuals of a single population become geographically separated from one another. This process can begin when a newly formed geographic barrier, such as a river, a canyon, or a mountain chain, isolates two populations of a single species (**Figure 13.6**). Such geographic isolation can also occur when a few members of a species colonize a region that is difficult to reach, such as an island located far outside the usual geographic range of the species.

Geographically isolated populations are disconnected genetically; there is little or no gene flow between them. Without gene flow, the other mechanisms of evolution we saw in Chapter 12—mutation, genetic drift, and natural selection—can more easily cause populations to diverge from one another. If populations remain isolated long enough, they can evolve into new species. The formation of new species from geographically isolated populations is called **allopatric speciation** (*allo*, "other"; *patric*, "country"), as shown in **Figure 13.7**.

But what happens when there isn't a physical barrier between two populations? In the ocean, for example, plants and animals can drift

The Kaibab squirrel is confined to the North Rim of the Grand Canyon.

Abert's squirrel lives on the South Rim and other southern locations, all the way into Mexico

Figure 13.6

The Grand Canyon is a geographic barrier for squirrels

The Kaibab squirrel population became isolated from the Abert's squirrel population when the Colorado River cut the Grand Canyon—which is as deep as 6,000 feet in some places. With gene flow between them blocked, probably beginning about 5 million years ago, the two populations eventually accumulated enough genetic differences to become two distinct species.

Q1: What is the definition of gene flow? How was gene flow blocked between these species?

Q2: Name as many types of geographic barriers as you can. Which do you think would be the best at blocking gene flow?

Q3: Are geographic barriers universal for all species? If not, name a geographic barrier that might block gene flow for one species but not another.

around or swim extremely long distances. There are few physical barriers yet many separate, unique species. How do new species form when populations are free to mix and mingle?

Caribbean Corals

Carlos Prada did not spend much of his time at the University of Puerto Rico in a classroom. Instead, he spent it in the ocean, studying corals, a type of marine invertebrate. On regular dives onto the Caribbean reef, the graduate student documented the morphology of a species of coral named *Eunicea flexuosa*, commonly called "sea fans." At 12 different locations along the Puerto Rican coast, Prada observed different morphologies of the coral at different depths. In shallow areas, less than 15 feet from the surface, the sea fans were wide, with a broad network of thick branches like a bush. In deeper areas, between 15 and 55 feet from the surface, the sea fans were much taller and spindly, resembling trees with networks of thin branches.

"They're supposed to be the same species, but they really looked different at different depths," says Prada. Curious about whether the morphologies were due to genetics or a result of the surrounding environment shaping the animals as they grew, Prada began

Figure 13.7

Physical barriers can produce allopatric speciation by blocking gene flow

Allopatric speciation can occur when populations are separated by a geographic barrier, such as a rising sea.

Q1: What factors must be present for allopatric speciation to occur?

Q2: If a geographic barrier is removed and the two reunited populations intermingle and breed, what attributes must the offspring have in order for the two populations, according to the biological species concept, to be considered still the same species?

Q3: If the two populations in question 2 are determined to still be the same species, did allopatric speciation occur?

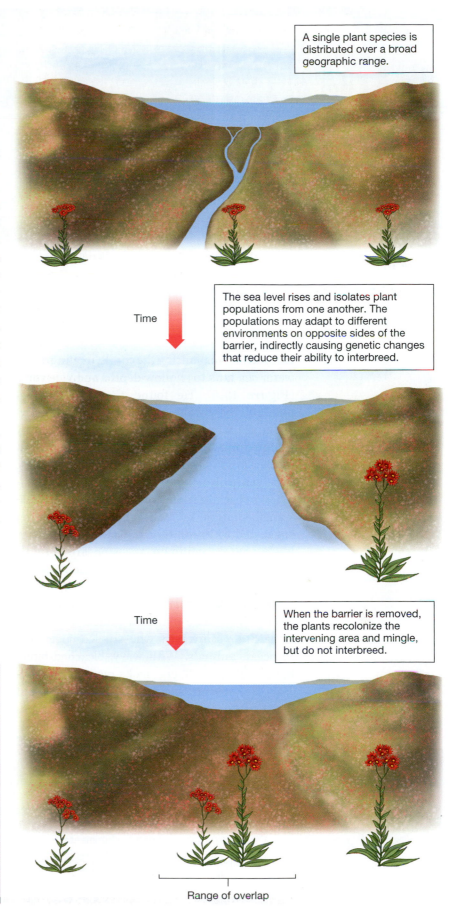

A single plant species is distributed over a broad geographic range.

Time

The sea level rises and isolates plant populations from one another. The populations may adapt to different environments on opposite sides of the barrier, indirectly causing genetic changes that reduce their ability to interbreed.

Time

When the barrier is removed, the plants recolonize the intervening area and mingle, but do not interbreed.

Range of overlap

Shallow-water sea fan

Deep-water sea fan

Carlos Prada transplants coral to waters of different depths.

Figure 13.8

Different corals at different depths

These two corals were once considered the same species, *Eunicea flexuosa*, commonly called a "sea fan." (Source: Photos courtesy of Carlos Prada.)

carefully transplanting the corals, moving deep-water sea fans to shallow depths and vice versa (**Figure 13.8**). He found that when transplanted, the corals did change. The shallow-water sea fans became taller and more spindly when planted in deep water, and the deep-water sea fans became wider in shallow waters, but—critically—neither made a complete transition to the alternate shape. The lack of a total transformation by either form to the other suggested to Prada that the corals, while they likely share a common ancestor, are actually two species that have adapted to their respective water depths.

When Prada finished his graduate work in Puerto Rico, he e-mailed a professor at Louisiana State University who studied speciation in ocean animals. With wavy, bleached-blond hair, Michael Hellberg looks more like a California surfer than a professor, but this evolutionary biologist has long been fascinated with how one species splits to become two, especially in the ocean. "Say you have a new lake forming, and a species becomes isolated in the lake. Then it's pretty obvious there's not going to be a lot of interbreeding to fight against, and the species just adapts. To me, there's no mystery in that," says Hellberg. This would be an example of allopatric speciation. "I've always tried to target groups where species look closely related and where ranges of the species overlap. That makes things a lot harder."

Hellberg welcomed Prada into his crew, and the two set out to extend Prada's work to find out whether his idea—that coral evolve different adaptive traits at different depths in the ocean, leading to the formation of new species—was unique to coral reefs in Puerto Rico or could be observed in other areas around the Caribbean. With Hellberg's support, Prada traveled to the Bahamas, Panama, and Curaçao to observe and take samples from sea fan colonies.

As he waited for Prada to return home with the data, Hellberg remained skeptical of the idea of **ecological isolation**, that two closely related species in the same territory could be reproductively isolated by slight differences in habitat. But

MICHAEL HELLBERG AND CARLOS PRADA

Michael Hellberg (right) is an evolutionary biologist at Louisiana State University who studies how species evolve in marine environments. Carlos Prada (left) was a graduate student in Hellberg's lab, and is now a postdoctoral researcher at Penn State studying how organisms cope with changes in the environment.

as Prada sampled more and more locations, the evidence was convincing. Indeed, everywhere he looked, Prada saw the same thing he had observed in Puerto Rico: broad, leaflike coral at shallow depths, and tall, sticklike coral in deeper waters. Distance between the populations did not matter; some shallow-water and deep-water fans were close enough together that Prada could reach out and bend them to touch each other. But depth did matter. The corals were physically close enough to easily interbreed in the water, yet the species had somehow become specialized to two different depths (**Figure 13.9**).

Prada brought his coral samples back to the lab and performed tests to see how genetically similar the two groups of corals were. Prada and Hellberg found that all the shallow-water sea fans across the Caribbean were more closely related to each other than they were to any of the deep-water sea fans. The same held true for the deep-water sea fans: they were more closely related to each other than to any of the shallow-water fans. In the DNA data, Hellberg had seen some genetic exchange between the two populations in the past, yet that exchange of genetic materials had been limited across the two groups, and each species had bred almost exclusively among its own populations.

Different Depths, Different Habitats

How did a few yards of depth produce habitats so different that the corals evolved different adaptations to each depth? The scientists are still investigating the differences between the shallow and deep habitats, but they have some hypotheses. One is that the coral have adapted their morphologies, and possibly their biochemistry, to suit different symbiotic algae that grow on them at different depths.

Symbiotic algae live on coral and, through photosynthesis, use sunlight to produce energy and organic compounds that coral use to maintain and grow calcium carbonate skeletons. In turn, the coral provide the algae with a sheltered place to live and produce carbon dioxide that the algae use during photosynthesis. (This cooperative relationship between species is known as mutualism; see Chapter 20.) But different species of algae have

In ecological isolation, the two species are usually physically close enough to breed, but somehow do not.

Shallow

Deep

Figure 13.9

Depth of water provides ecological isolation for sea fans
The two depths shown here differ slightly in their quantity and quality of light, the force of the waves or current, the amount of sediment deposited on them, the number and type of predators, and the availability and type of food.

different light and nutrient requirements, and organisms 15 feet below the surface receive a lot of sunlight, while those at 30 feet receive less light, and light of different wavelengths. Prada found different types of algae living on the two coral populations (**Figure 13.10**), so it's possible that the sea fans evolved to be better hosts for the most successful algae at their depth.

In many cases in the ocean, as with coral and algae, the interaction between two species so strongly influences their survival that they have evolved in tandem—a phenomenon known as **coevolution**. The term "coevolution" encompasses a wide variety of ways in which an adaptation in one species evolves alongside a complementary adaptation in another species.

Figure 13.10

Different species of corals and their resident algae
Although all these algae (single-celled eukaryotic organisms) look the same under a microscope, they are actually different species with different light requirements and photosynthetic capacities.

Flowers that coevolved with hummingbirds produce nectar, are colored to attract the birds, and are shaped specifically to fit the hummingbirds' bills.

Figure 13.11

Coevolution at its finest

This hummingbird's bill fits perfectly into this flower for easy access to the hummingbird's favorite food, nectar. The hummingbird feasts and then distributes the flower's pollen by carrying the pollen along to its next meal. It's a win-win situation!

Q1: Describe how coevolution, as with the hummingbird bill and hummingbird-pollinated flowers, is different from the kind of evolution described in Chapters 11 and 12.

Q2: Is coevolution the same thing as convergent evolution, described in Chapter 12? Why or why not?

Q3: Do you think one species' adapting over time to feed specifically and extremely successfully on another species is an example of coevolution? Why or why not?

So Many Chromosomes

New plant species can form in a single generation as a result of **polyploidy**, a condition in which an individual gains an extra full set or two (or three) of chromosomes. Humans and most other eukaryotes are *diploid* (having two sets of chromosomes), but some organisms are *triploid* (three sets) or *tetraploid* (four sets), or have an even higher number of chromosomes. Polyploidy is invariably fatal in people, but in many plant species it is not lethal.

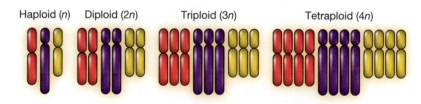

Haploid (*n*)   Diploid (2*n*)   Triploid (3*n*)   Tetraploid (4*n*)

Polyploidy can occur when two different species hybridize and produce an offspring with an odd number of chromosomes, because of improper alignment of the chromosomes during mitosis (see Chapter 6). This increase in chromosome sets can lead to reproductive isolation because the chromosome number in the gametes of the new polyploid individual no longer matches the number in the gametes of either of its parents.

Polyploidy has had a large effect on life on Earth: more than half of all plant species alive today are descended from species that originated by polyploidy. A few animal species also appear to have originated by polyploidy, including several species of lizards, salamanders, and fish.

Another example of coevolution is the relationship between hummingbirds and certain species of flowers (**Figure 13.11**).

Alternatively, or in addition to hosting specific algae, Hellberg hypothesizes that the shallow-water and deep-water sea fans may have adapted to the type and density of sedimentation found at different ocean depths. Or perhaps predators are more abundant at certain depths. Whatever the major habitat factors driving the corals' adaptation, natural selection led to the formation of two different coral populations tailored to the ocean depths at which they grow.

Today, Hellberg considers the two populations different species. "What makes this interesting is that you have populations that could potentially interbreed, yet they maintain their differences in the face of each other," says Hellberg. "For just about any biological question you want to ask, they are independent entities." The formation of new species in the *absence* of geographic isolation is called **sympatric speciation** (from *sym*, "together"; **Figure 13.12**). Sympatric speciation is a particularly important process in plants (see "So Many Chromosomes").

Still, a question lingers: Why don't the two coral species interbreed? They appear to be capable of it, because of evidence of some genetic

Haplochromis chilotes feeds on insects.

Haplochromis macrognathus feeds on other fishes.

Macropleurodus bicolor feeds on snails and other mollusks.

Astatotilapia elegans is a generalized bottom feeder.

Figure 13.12

Sympatric speciation drives diversity among Lake Victoria cichlid species

Scientists have described some 500 species of cichlid fishes in Lake Victoria. Genetic analyses indicate that they all descended from just two ancestor species over the past 100,000 years. These four species show some of the differences in feeding behavior and morphology.

Q1: What is the main difference between allopatric and sympatric speciation?

Q2: Name two events that must happen for both allopatric speciation and sympatric speciation to occur.

Q3: Do you think all of the 500 species in Lake Victoria arose through sympatric speciation? Why or why not?

mixing in the past, yet there is no evidence that interbreeding is common; only a few rare hybrids grow on the ocean floor between them. When two species are reproductively isolated from each other, we say that **reproductive barriers** exist between those species. Reproductive barriers are often divided into two categories: *prezygotic* and *postzygotic*.

Barriers that prevent a male gamete (such as a human sperm) and a female gamete (such as a human egg) from fusing to form a zygote are **prezygotic barriers**. Prezygotic barriers act *before* the zygote exists (**Figure 13.13**). Barriers that prevent zygotes from developing into

Blue-footed boobies point their beaks, wings, and tails upward in a mating dance called "sky pointing."

Figure 13.13

The blue-footed booby courtship dance is a prezygotic, behavioral reproductive barrier

This species of booby has a unique ritual dance that must be accurately completed before mating. Other booby species do not perform exactly the same dance and therefore do not mate with the blue-footed booby.

Q1: What does "prezygotic" mean?

Q2: How is the booby's ritual dance a prezygotic reproductive barrier?

Q3: What are some other prezygotic reproductive barriers besides a mating dance?

Table 13.1

Reproductive Barriers That Isolate Two Species in the Same Geographic Region

Type of Barrier	Description	Effect
Prezygotic		
Ecological isolation	The two species breed in different portions of their habitat, in different seasons, or at different times of day.	Mating is prevented.
Behavioral isolation	The two species respond poorly to each other's courtship displays or other mating behaviors.	Mating is prevented.
Mechanical isolation	The two species are physically unable to mate.	Mating is prevented.
Gametic isolation	The gametes of the two species cannot fuse, or they survive poorly in the reproductive tract of the other species.	Fertilization is prevented.
Postzygotic		
Zygote death	Zygotes fail to develop properly, and they die before birth.	No offspring are produced.
Hybrid sterility	Hybrids survive but are unable to produce viable offspring.	No offspring are produced.
Hybrid performance	Hybrids survive poorly or reproduce poorly.	Hybrids are not successful.

healthy and fertile offspring are called **postzygotic barriers**, which act *after* the zygote is formed.

A wide variety of cellular, anatomical, physiological, and behavioral mechanisms generate pre- and postzygotic reproductive barriers, but they all have the same overall effect: little or no mating occurs, and therefore few or no alleles are exchanged between species (**Table 13.1**). Hellberg suspects that some prezygotic barriers exist between the two sea fan species—perhaps the gametes of the two species don't fuse successfully, as in some sea urchins (**Figure 13.14**)—but there may also be a postzygotic barrier such as generation time.

It seems likely that the reason hybrids between the two species do not succeed is the uniquely long generation time for these corals, says Prada. Sea fans don't reach reproductive age until they are 15–20 years old, and they

continue reproducing until they are 60 years old or more. So, while coral gametes and larvae can and do disperse far from their parents and may interbreed or take root at incorrect depths, natural selection then has 15–20 years to winnow out the less successful offspring. If there are small differences in survival rates between the two species at a particular depth— if a deep-water coral fares slightly worse in a shallow area than a shallow-water coral does— those differences will become amplified over 15 years, and by the time the corals reach reproductive age it is likely that only the shallow-water species will still be alive in the shallow area.

New species form and adapt to their environments in many different ways, from island-bound lizards to depth-dependent corals. And from those varied beginnings comes the vast diversity of life that exists on our planet today, the result of billions of years of evolution. In

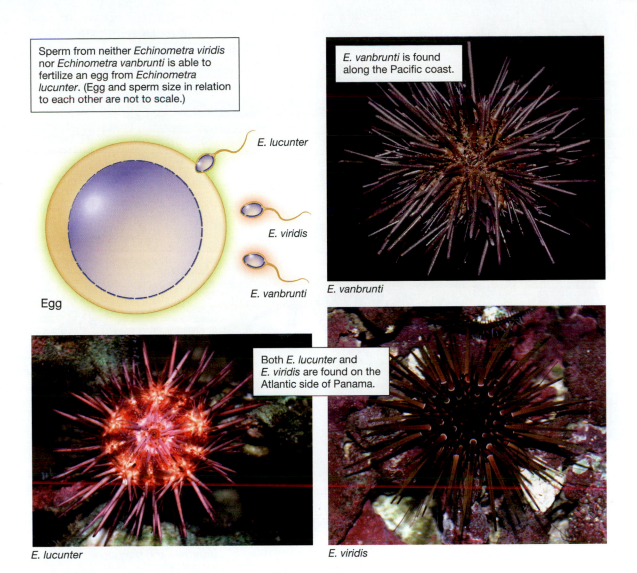

Sperm from neither *Echinometra viridis* nor *Echinometra vanbrunti* is able to fertilize an egg from *Echinometra lucunter*. (Egg and sperm size in relation to each other are not to scale.)

E. lucunter

E. viridis

E. vanbrunti

Egg

E. vanbrunti is found along the Pacific coast.

E. vanbrunti

Both *E. lucunter* and *E. viridis* are found on the Atlantic side of Panama.

E. lucunter

E. viridis

Figure 13.14

Gametic isolation among sea urchin species

The sea urchin species found along both coasts of Panama illustrate gametic isolation as a reproductive barrier. At one point long ago, the Panama landmass was not continuous and did not separate the two oceans. These three species likely evolved through both allopatric and sympatric speciation from a recent common ancestor after the oceans were separated.

Q1: Which species is/are sympatric with *Echinometra lucunter*?

Q2: Which species is/are allopatric with *E. lucunter*?

Q3: If two individuals have incompatible gametes, what will be the result of a mating event between them?

the next section of this book, we explore the dramatic transitions in life-forms that have walked, crawled, and swum around our planet, from small dinosaurs with wings to the most successful, most cunning species ever to traverse the planet: humans.

On the Diversity of Species

There are an estimated 8.74 million eukaryotic species on Earth, yet we are familiar with only a fraction of those. Scientists estimate that 86% of land species and 91% of aquatic species have not been discovered. Of the known species, insects top the chart with an estimated 1 million different species.

Numbers of known species

Fish
31,200

Crustaceans
47,000

Mollusks
85,000

Fungi
99,000

Arachnids
102,200

Mammals
5,500

Sponges
6,000

Amphibians
6,500

Reptiles
8,700

Jellyfish and polyps
9,800

Birds
10,000

Millipedes and centipedes
16,100

Segmented worms
16,800

Flatworms
20,000

Insects
1,000,000

Plants
310,100

Roundworms
25,000

REVIEWING THE SCIENCE

- **Adaptive traits,** or adaptations, are inherited characteristics that improve an individual's chances of surviving and reproducing in a specific environment.

- Natural selection leads to **adaptation,** genetic changes that improve the reproductive success of organisms in their environment over time. Adaptation does not produce "perfection."

- According to the **biological species concept,** a **species** is a group of populations that interbreed and can produce live and fertile offspring.

- The process of **speciation,** one species splitting into two or more species, is the by-product of **reproductive isolation** and **genetic divergence** between populations.

- **Allopatric speciation** occurs when populations of a species become **geographically isolated,** limiting gene flow and making genetic divergence more likely.

- Speciation that occurs between populations lacking geographic isolation is called **sympatric speciation**.

- In **ecological isolation,** two closely related species in the same area are reproductively isolated by slight differences in habitat.

- When adaptive traits of one species evolve in concert with adaptive traits in another species, **coevolution** has occurred.

- Reproductive isolating mechanisms can be **prezygotic barriers** (before zygote formation) or **postzygotic barriers** (after zygote formation).

THE QUESTIONS

Challenge Yourself

1 Natural selection

(a) leads to species being better adapted to their environment.

(b) may lead to speciation if there is no gene flow between populations.

(c) may lead to genetic divergence among populations in differing environments.

(d) all of the above

(e) none of the above

2 Which of the following examples describe an adaptive trait? (Select all that apply.)

(a) A rainforest tree that is vulnerable to storm damage.

(b) A male bird that is more successful than others at attracting a female mate.

(c) A rabbit that is better camouflaged in its environment.

(d) A desert plant that is able to survive drought.

(e) A frog that is more noticeable to predators.

3 Place the following events in the order in which they are most likely to occur by numbering them from 1 to 5.

_____ a. Speciation occurs.

_____ b. Genetic divergence takes place.

_____ c. Geographic barrier arises.

_____ d. Gene flow is discontinued.

_____ e. Species adapts to a new environment.

Try Something New

4 Distinct species that are able to interbreed in nature are said to "hybridize," and their offspring are called "hybrids." The gray oak and the Gambel oak can mate to produce fertile hybrids in regions where they co-occur. However, the gene flow in nature is sufficiently limited that, overall, the two species remain phenotypically distinct. If the hybrid offspring survive well and reproduce to the extent that there is a large population of hybrid individuals that breed between themselves but do not interbreed with either of the two original parent species (the gray and the Gambel), which of the following would you say most likely led to the new hybrid species?

(a) prezygotic reproductive barriers

(b) sympatric speciation

(c) allopatric speciation

(d) postzygotic reproductive barriers

(e) none of the above

5 The four-eyed fish, *Anableps anableps*, really has only two eyes, which function as four, enabling the fish to see clearly through both air and water. *A. anableps* is a surface feeder, so the ability to see above water helps it locate prey such as insects. Its unique eyes also enable it to scan simultaneously for predators attacking from above (such as birds) or below (such as other fish). This species of fish _____ crafted by natural selection.

Anableps anableps

(a) is an example of coevolution

(b) has an evolutionary adaptation

(c) is an example of sympatric speciation

(d) is an example of allopatric speciation

6 Give an example of a barrier that might bring about allopatric speciation. In addition to this barrier, what would be necessary for speciation to occur?

7 Researchers from the Smithsonian Institution were startled to discover that the 3 species of *Starksia* blennies they had been studying in the Caribbean islands were really 10 different species. How could these researchers have thought that these 10 different reef fish species were only 3 species?

Leveling Up

8 *Write Now* **biology: adaptations** Select an organism (other than humans) that you find interesting. Research two adaptations of your organism and describe them in detail. Explain carefully why each of these features is considered an important adaptation for its species. Discuss how a change in the environment might change the usefulness of these adaptations.

9 **Doing science** The eastern and western meadowlarks are very similar in morphology. Both are grassland species with some overlap in their range in the upper Midwest, and each sings a distinctly different song. Create a hypothesis as to whether these are two different species or two populations of the same species, and describe an experiment that would enable you to test your hypothesis. What result would support your hypothesis, and what result would cause you to reject your hypothesis?

10 **What do *you* think?** According to the Defenders of Wildlife organization, there were 20–30 million American range bison in the old West. During the late 1800s the American range bison were hunted to the brink of extinction, leaving behind a bottleneck population of only 1,091 individuals. The population has since rebounded to about 500,000 bison. Unfortunately, almost all of these bison are the descendants of these few individuals crossbred with domestic cattle by ranchers. Scientists and conservationists want to genetically test bison to find those of pure bison origin to preserve the species. Only these, they argue, should be called American range bison and be allowed to roam free in the national parks as bison. They think hybrids should be confined to farms and ranches, should be called "beefalo" rather than bison, and should not be afforded the protection that pure bison currently have. What do you think? Should bison tainted with cattle genes be removed from free-range parks? Should the government spend scarce conservation monies on genetic testing and breeding efforts to preserve the pure bison population? Investigate conservation efforts and the costs of genetically testing and relocating bison to help you with your decision. Is speciation at the hands of human beings now part of evolution as we know it?

Thinking Multidimensionally

11 In biology, the "Red Queen hypothesis" is based on the idea that organisms need to continually change in order to keep up with the evolving competition and that variety among offspring, which comes about from the shuffling of genes during sexual reproduction, is necessary to ensure survival during adversity.

Consider a common microscopic lab worm called *Caenorhabditis elegans*. This worm usually reproduces on its own since it's a hermaphrodite (an organism with both male and female reproductive parts), but it can also reproduce sexually.

a. When pathogenic bacteria were introduced to a population of *C. elegans*, what form of reproduction do you think this worm mostly employed, and why?

b. Create a chart and list the biological advantages and disadvantages to reproducing sexually as opposed to asexually.

12 When horses and donkeys interbreed, they produce mules that are sterile. The diploid chromosome number in horses is 64; donkeys, 62; and mules, 63.

a. Given your understanding of meiosis, why is it not surprising that interspecies' hybrids, such as mules, are usually infertile?

b. What type of reproductive barrier is this an example of?

c. Research and list other examples of sterile hybrid organisms.

Donkey

Horse

Mule

For more, visit digital.wwnorton.com/bionow2 for access to:

The First Bird

A fresh wave of fossils challenges the identity of the first dino-bird.

After reading this chapter you should be able to:

- Identify critical events on a time line of life on Earth and explain their evolutionary importance.
- Interpret an evolutionary tree of a group of organisms.
- Explain how new scientific data may change an evolutionary tree without challenging our understanding of evolution.
- Classify a given species at each level of the Linnaean system.
- Give an example of an adaptation that enabled a group of organisms to transition from an aquatic existence to life on land.
- Describe the impact of prehistoric mass extinctions on Earth's biodiversity, and infer the likely impact of the current extinction rate on biodiversity.

CHAPTER
14
THE HISTORY OF LIFE

DRIVING QUESTION

How are newly discovered fossils impacting our understanding of bird evolution?

- HS-LS4.C: Adaptation
- HS-LS4.D: Biodiversity and Humans

In 1861, quarry workers in Germany unearthed the fossil of a crow-sized bird. The 150-million-year-old preserved skeleton looked bizarre: it had feathers and clawed hands like a bird, but teeth and a long, bony tail like a reptile.

After close examination by paleontologists, the fossil was hailed as a transitional form between birds and reptiles and labeled the oldest known bird. Named *Archaeopteryx*, the feathered dinosaur became famous around the world as the first solid evidence that birds descended from dinosaurs (**Figure 14.1**). Charles Darwin, who proposed the theory of descent with modification, called the discovery "a grand case for me."

Over 100 years later, a young fossil hunter named Xu Xing bent over the dirt at a dig site in Liaoning province in northern China. It was Xu's first time in the field since finishing college. An excellent student, Xu had wanted to study economics at Peking University in Beijing, but at the time, students in China did not pick their majors, and Xu was assigned paleontology. Luckily, farmers in Liaoning had recently begun to uncover huge numbers of fossils, and research was booming.

So, as a grudging paleontologist, Xu began searching for dinosaurs in Liaoning. He had a particular interest in feathered dinosaurs. By the late twentieth century, the idea that birds descended from dinosaurs was no longer a hypothesis, but a well-established scientific theory backed by a mountain of data. And atop that mountain of data sat *Archaeopteryx*, the earliest known bird.

Little did Xu know, as he dug through the dirt that day, that he would soon be the one to knock *Archaeopteryx* from its perch.

Dinosaurs and Domains

Although the *Archaeopteryx* fossil is old—150 million years old, in fact—much older fossils have been found.

Our solar system and Earth formed 4.6 billion years ago. Some of the oldest known rocks on Earth are 3.8 billion years old and contain carbon deposits that hint at life. Cell-like structures have been found in layered mounds of sedimentary rock called stromatolites that formed 3.7 billion years ago, and projections based on DNA analysis also support the idea that life appeared on Earth at that time.

The question of how life arose from nonlife is one of the greatest riddles in biology, but scientists have little doubt that all life on Earth is related. As noted in Chapter 1, all living organisms are united by a basic set of characteristics.

Figure 14.1

The original *Archaeopteryx* fossil

There are 11 complete *Archaeopteryx* fossils, but this one from Germany was the first to be found. Note the faint impressions of feathers on the wings and tail.

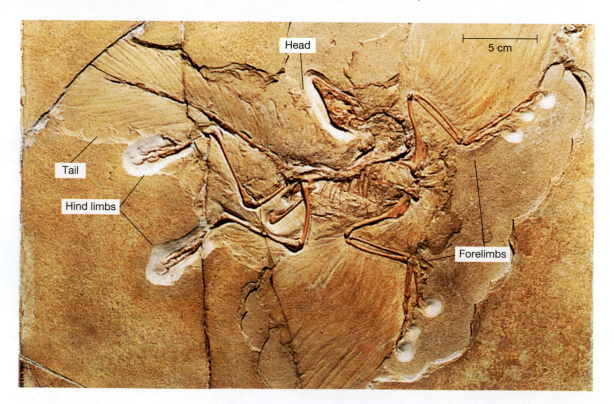

Life shares this set of common properties because all living organisms descended from a common ancestor, known as the *universal ancestor*. This hypothetical ancestral cell is placed at the base of the tree of life. From that cell, all life emerged. Biological diversity, or **biodiversity**, embraces the variety of all the world's living things, as well as their interactions with each other and the ecosystems they inhabit. Biodiversity can be described at the level of genes, or species, or entire ecosystems.

In spite of intense worldwide interest, scientists do not know the exact number of species alive today. Most traditional estimates have fallen in the range of 3–30 million species, but in 2016, scientists combined microbial, plant, and animal data sets from around the world, scaled up, and estimated that Earth could contain nearly a trillion species. So far, about 1.5 million species have been collected, named, and placed into an evolutionary tree, meaning it is possible that only one ten-thousandth of 1 percent of species are known to us. In other words, 99.9999 percent of species are yet undiscovered.

Life throughout history and today is so diverse that biologists created a classification system to organize it into categories. The **domains** form the highest, most inclusive hierarchical level in the organization of life, describing the most basic and ancient divisions among living organisms. There are three domains of life (**Figure 14.2**):

- **Bacteria**, which includes familiar disease-causing bacteria such as *E. coli*

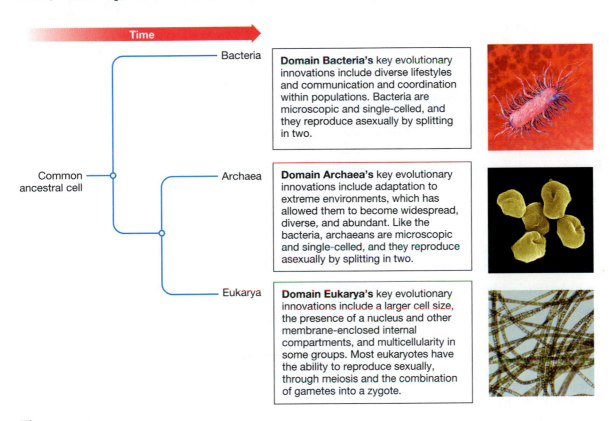

Figure 14.2

Three domains of life

This "tree of life" shows the relationships of the three domains of life.

Q1: Why is there a shared line from the universal ancestor for Archaea and Eukarya?

Q2: Where would birds be found within this figure? What about humans?

Q3: To which domain would you expect a disease-causing organism to belong? What if the organism was multicellular?

Billions of years ago (bya)	Millions of years ago (mya)					
4.6 3.8	540	490	445	415	360	

Precambrian	Paleozoic

Geologic period

Precambrian	Cambrian	Ordovician	Silurian	Devonian	Carboniferous

Major events

Origin of life; photosynthesis causes oxygen content of Earth's atmosphere to increase; first eukaryotes; first multicellular organisms	Large and relatively sudden increase in the diversity of animal life; increase in diversity of algae; first vertebrates	Further increases in diversity of marine invertebrates and vertebrates; plants and fungi begin to colonize land; mass extinction at end of period	Increase in diversity of fishes; first hints of colonization of land by insects and other invertebrates	Increase in diversity of land plants; first amphibians colonize land; mass extinction late in period	Extensive forests; amphibians dominate life on land; increase in diversity of insects; first reptiles
Life begins	**Invertebrates fill the seas**	**Plants begin to colonize land**	**Fish diversity increases**	**Amphibians appear**	**Earth is covered with forests**

Figure 14.3

The geologic timescale and major events in the history of life

The history of life can be divided into 12 major geologic time periods, beginning with the Precambrian (4.6 bya to 540 mya) and extending to the Quaternary (2.6 mya to the present). This time line is not drawn to scale; to do so would require extending the diagram off the book page to the left by more than 5 feet (1.5 meters).

- **Archaea**, which consists of single-celled organisms best known for living in extremely harsh environments
- **Eukarya**, which includes all other living organisms, from amoebas to plants to fungi to animals

Humans, dinosaurs, and birds are all part of the Eukarya domain. They are *eukaryotes*. Bacteria and Archaea are two different domains—Archaea are more closely related and in some ways more similar to Eukarya than to Bacteria—yet because neither Bacteria nor Archaea are eukaryotes, the two have traditionally been lumped under a common label: *prokaryotes*.

Prokaryotes first appear in the fossil record at about 3.7 billion years ago (**Figure 14.3**), but the first eukaryotes did not evolve until over a billion years later. Luckily for us, and all other eukaryotes, roughly 2.8 billion years ago a group of bacteria evolved a type of photosynthesis that releases oxygen as a by-product. As a result, the oxygen concentration in the atmosphere increased over time, and about 2.1 billion years ago the first single-celled eukaryotes evolved. When the oxygen concentration reached its current level, by about 650 million years ago (mya), the evolution of larger, more complex multicellular organisms became possible, including fish, then land plants, then insects, amphibians, and reptiles. One group of reptiles, which would eventually dominate most

XU XING

Xu Xing is a paleontologist at the Chinese Academy of Sciences in Beijing. He has discovered more than 60 species of dinosaurs and specializes in feathered dinosaurs and the origins of flight.

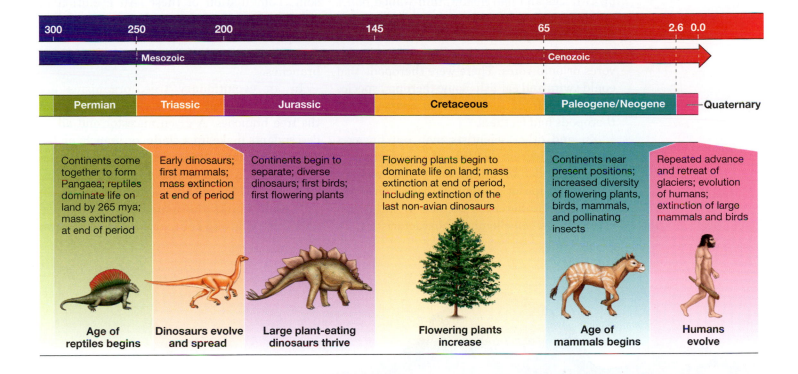

300	250	200	145	65	2.6 0.0

Mesozoic — Cenozoic

Permian	Triassic	Jurassic	Cretaceous	Paleogene/Neogene	Quaternary

Continents come together to form Pangaea; reptiles dominate life on land by 265 mya; mass extinction at end of period

Early dinosaurs; first mammals; mass extinction at end of period

Continents begin to separate; diverse dinosaurs; first birds; first flowering plants

Flowering plants begin to dominate life on land; mass extinction at end of period, including extinction of the last non-avian dinosaurs

Continents near present positions; increased diversity of flowering plants, birds, mammals, and pollinating insects

Repeated advance and retreat of glaciers; evolution of humans; extinction of large mammals and birds

Age of reptiles begins | **Dinosaurs evolve and spread** | **Large plant-eating dinosaurs thrive** | **Flowering plants increase** | **Age of mammals begins** | **Humans evolve**

Q1: During what geologic period did life on Earth begin?

Q2: How long ago did species begin to move from water to land? What period was this?

Q3: In what period would *Archaeopteryx* have been alive?

---

other species, was the dinosaurs. Dinosaurs first appeared about 230 mya, during the Triassic period, and they took over the planet.

Feathered Friends

Xu Xing may not have wanted to be a paleontologist when he went to college, but by the time he graduated, he was hooked on dinosaurs. Over the next 20 years, Xu became one of the most productive researchers in his field. To date, he has discovered and named more than 60 extinct species—mostly dinosaurs, but also a reptile and a salamander. And the majority of those dinosaur fossils have feathers (**Figure 14.4**).

As scientists traced back the **lineage**, or line of descent, from birds to dinosaurs, it became clear that birds are most closely related to theropods—fast-moving dinosaurs that ran

Figure 14.4

A feathered dinosaur tail, trapped in amber

In 2016, scientists discovered a 99-million-year-old dinosaur tail, complete with its feathers, in a piece of amber.

on two legs and had hollow, thin-walled bones (as birds do). Theropods were a diverse group of dinosaurs (**Figure 14.5**). Most were carnivores, including insectivores, but a few were herbivores or omnivores. There were theropods that could swim and eat fish. Some theropod species boasted enlarged scales, and of course, many theropods had feathers.

Xu and other scientists map out lineages using a diagram called an **evolutionary tree**, a model of evolutionary relationships among groups of organisms based on similarities and differences in their DNA, physical features, biochemical characteristics, or some combination of these. An evolutionary tree maps the relationships between ancestral groups and their descendants, and it clusters the most closely related groups on neighboring branches.

In an evolutionary tree, organisms are depicted as if they were leaves at the tips of the tree branches. A given ancestor and all its descendants make up a **clade**, or branch, on the evolutionary tree. *Archaeopteryx* and all subsequent animals that evolved from it are considered a clade (**Figure 14.6**).

A **node** marks the moment in time when an ancestral group split, or diverged, into two

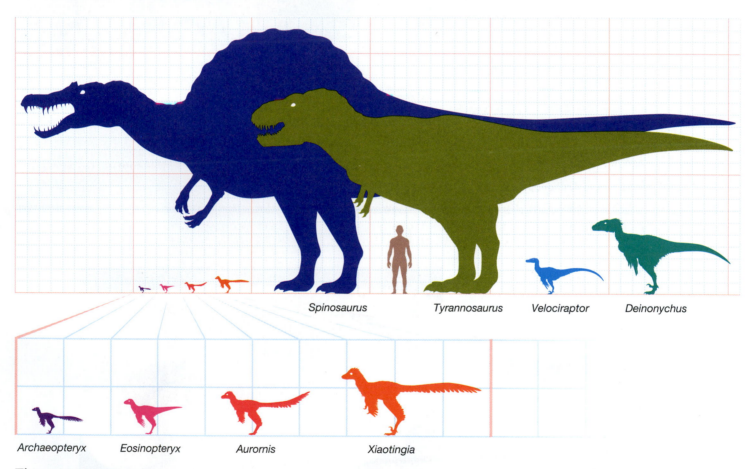

Spinosaurus   Tyrannosaurus   Velociraptor   Deinonychus

Archaeopteryx   Eosinopteryx   Aurornis   Xiaotingia

Figure 14.5

Dinosaurs large and small

Theropods ranged in size from tiny, like a chicken, to huge, like the group's most famous member, *Tyrannosaurus rex*.

Q1: In what ways were theropods the same as modern birds? Give at least two similarities.

Q2: In what ways did theropods differ from modern birds? Give at least two differences.

Q3: Birds are often referred to as "living dinosaurs." Is this accurate? Why or why not?

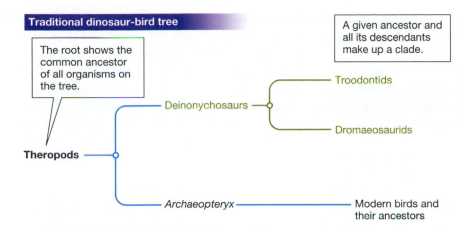

Traditional dinosaur-bird tree

The root shows the common ancestor of all organisms on the tree.

A given ancestor and all its descendants make up a clade.

Theropods

Deinonychosaurs

Troodontids

Dromaeosaurids

Archaeopteryx — Modern birds and their ancestors

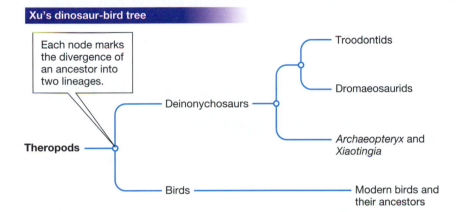

Xu's dinosaur-bird tree

Each node marks the divergence of an ancestor into two lineages.

Theropods

Deinonychosaurs

Troodontids

Dromaeosaurids

Archaeopteryx and Xiaotingia

Birds — Modern birds and their ancestors

Figure 14.6

The evolutionary origins of birds

(*Top*) The traditional evolutionary tree, showing *Archaeopteryx* as an early bird that split off from the deinonychosaurs (birdlike, carnivorous dinosaur). (*Bottom*) The evolutionary tree proposed by Xu after the discovery of *Xiaotingia*, with *Archaeopteryx* as a deinonychosaur rather than an early bird.

Q1: In the traditional tree, identify the node showing the common ancestor for early birds and dinosaurs.

Q2: What do both the traditional tree and Xu's tree suggest about troodontids and dromaeosaurids?

Q3: In both trees, identify the node for the common ancestor of *Archaeopteryx* and other birds. In what way are the nodes different in the two trees?

separate lineages. A node represents the **most recent common ancestor** of two lineages in question—that is, the most *immediate* ancestor that *both* lineages share. For over 100 years, researchers considered *Archaeopteryx* the most recent common ancestor of both birds and dinosaurs and thus placed it at the root of the avian clade—the first bird (**Figure 14.6**, top).

The first bird, that is, until Xu stumbled across a fossil that threw the field into controversy. In 2008, Xu visited the Shandong Tianyu Museum of Nature, a dinosaur museum in eastern China. There, he happened upon a unique fossil that had been collected by a farmer in Liaoning and sold through a dealer to the museum. The fossil, entombed in yellowish rock, shows a small, birdlike dinosaur seemingly craning its neck forward and spreading its short wings (**Figure 14.7**). "I saw it and said, 'Oh, this is an important species,'" recalls Xu. He asked the museum to let him study it.

As Xu examined the fossil, which he later named *Xiaotingia zhengi*, he wondered where it belonged on the dinosaur-bird evolutionary tree.

To find out, he analyzed *shared derived traits* of the fossil and similar early birds. **Shared derived traits** are unique features common to all members of a group that originated in the group's most recent common ancestor and then were passed down in the group (but not in groups that are not direct descendants of that ancestor). In this case, the original ancestor

5 cm

Head

Forelimbs

Hind limb

Tail

Figure 14.7

Xiaotingia zhengi: a controversial new leaf on the dinosaur-bird evolutionary tree

in question was assumed to be *Archaeopteryx*, and the shared traits included feathers, clawed hands, and a long, bony tail.

By comparing *Xiaotingia* to *Archaeopteryx* and other related species, Xu created a new evolutionary tree of early birds (**Figure 14.6**, bottom). Suddenly, *Archaeopteryx* wasn't in the avian clade. Instead, *Archaeopteryx* and *Xiaotingia* were in a different clade, grouped with deinonychosaurs—the small, birdlike, carnivorous dinosaurs commonly called raptors. (The term "raptor" refers to carnivorous modern birds like hawks and owls, but it is also used informally to describe this group of dinosaurs.) *Archaeopteryx*, Xu believed, was not the first bird, but a raptor. It had feathers, but its descendants did not evolve into birds.

"It was a big change," says Xu. Yet, he adds, it wasn't entirely unexpected. In the last 20 years, more and more fossils of early avian species have been discovered—and the more Xu compared *Archaeopteryx*, discovered in Europe, to other early birds discovered in China, the less it looked like a bird and the more it looked like a raptor. Early birds have small, thick skulls and two toes on each foot; *Archaeopteryx*, on the other hand, has a long, almost pointy skull and three toes on its feet. "*Archaeopteryx* is just so different from other early birds," says Xu.

Xu published his revised evolutionary tree in 2011. It took the scientific world by storm. Some researchers embraced the idea: "Perhaps the time has come to finally accept that *Archaeopteryx* was just another small, feathered bird-like theropod fluttering around in the Jurassic," wrote one paleontologist in the journal *Nature*. Others disagreed, arguing that Xu's analysis was not convincing. For them, *Archaeopteryx* remained the first bird.

The History of Life on Earth

Xu knew that one new fossil was but a single shred of evidence and therefore unlikely, by itself, to convince the scientific community. "The evidence is not very strong," he admits, "but it was a question we wanted to discuss and add more analyses to, including old and new fossils." The identity of the first bird is important because our understanding of how flight evolved is based on the classification of *Archaeopteryx*. If *Archaeopteryx* is not the species from which flight evolved, then our understanding of flight is wrong.

Xu's redesign of the early bird evolutionary tree raised another important question: Did flight evolve more than once? If *Archaeopteryx*, which looked capable of flight, was a bird, then flight probably evolved just once among dinosaurs, in the avian lineage. If *Archaeopteryx* was not a bird, and was instead simply a raptor that could fly, then flight evolved at least twice among dinosaurs, once in raptors and once in avians—but only avians went on to evolve into modern birds.

Biological classification helps us answer important evolutionary questions like these. In addition to recognizing three broad domains of life, biologists group the Eukarya into four distinct **kingdoms**, the second-highest level in the hierarchical classification of life (**Figure 14.8**).

Figure 14.8

The tree of life

The domains Bacteria and Archaea do not have separate kingdoms within them. The domain Eukarya encompasses four kingdoms: Protista (protists, an artificial grouping that includes organisms such as amoebas and algae), Plantae (plants), Fungi (including yeasts and mushroom-producing species), and Animalia (animals).

These kingdoms are **Protista**, a diverse group that includes amoebas and algae; **Plantae**, which encompasses all plants; **Fungi**, which includes mushrooms, molds, and yeasts; and **Animalia**, which encompasses all animals, including dinosaurs, birds, and humans. The members of each of these kingdoms share evolutionary innovations that adapted the organisms to their environments, enabling them to live and reproduce successfully (**Figure 14.9**).

Below the level of kingdom, biological classification can get even more specific using the **Linnaean hierarchy**, a system of biological classification devised in the eighteenth century by a Swedish naturalist named Carolus Linnaeus. The smallest unit of classification in the Linnaean hierarchy—the **species**—reflects individuals that are the most related to each other. The most closely related species are grouped together to form a **genus** (plural "genera"). Using these two categories in the hierarchy, every species is given a unique, two-word Latin name, called its **scientific name**. The first word of the name identifies the genus to which the organism belongs; the second word defines the species. Comparing our own species name, *Homo sapiens*, to that of the Neanderthals, *Homo neanderthalensis*, shows that the two are classified as separate species but belong to the same genus, *Homo* (**Figure 14.10**).

In the Linnaean hierarchy, each species is placed in successively larger and more inclusive categories beyond genus. Closely related genera are grouped into a **family**. Closely related families are grouped into an **order**. Closely related orders are grouped into a **class**. Closely related classes are grouped into a **phylum** (plural "phyla"). And, you guessed it, closely related phyla are grouped together into a kingdom.

The earliest forms of life evolved in water. About 650 mya (during the Precambrian period), the number of organisms appearing in the fossil record increased. At that time, much of Earth was covered by shallow seas filled with small, mostly single-celled organisms floating freely in the water.

Then, about 540 mya, the world experienced an astonishing burst of evolutionary activity, with a dramatic increase in biodiversity. Most of the major living animal groups first appear in the fossil record during this time, popularly known as the **Cambrian explosion**. The Cambrian explosion changed the face of life on Earth from a world of relatively simple, slow-moving, soft-bodied scavengers and herbivores to a world dominated by large, fast predators. The presence of predators sped up the evolution of Cambrian herbivores, judging by the variety of scales and shells and other protective body coverings typical of many Cambrian, but not Precambrian, fossils (**Figure 14.11**).

The Cambrian explosion, however, occurred primarily in the oceans. Because life first evolved in water, the colonization of land by living organisms posed enormous challenges. Indeed, many of the functions basic to life, including support, movement, reproduction, and the regulation of heat, had to be handled very differently on land than in water. About 480 mya, near the beginning of the Ordovician period, plants were the first organisms to meet these challenges. These early terrestrial colonists were single-celled or had just a few cells.

Fungi are thought to have made their way onto land next, according to fossil evidence. For example, scientists have found fossils of terrestrial fungi that are 455–460 million years old. Reconstructing the evolutionary history of eukaryotes from DNA data, scientists estimate that the common ancestor of fungi and animals diverged from all other eukaryotes about 1.5 billion years ago, and fungi diverged from their closest cousins, the animals, about 10 million years after that.

Next, land plants evolved and diversified from the original green algae that made it to land. By 360 mya, at the end of the Devonian period, Earth was covered with plants. The first land animals likely emerged about 400 mya. Many of the early animal colonists on land were carnivores; others fed on living plants or decaying plant material.

The first vertebrates to colonize land were amphibians, the earliest fossils of which date to about 365 mya (see Chapter 17 for more on vertebrates). Early amphibians descended from lobe-finned fishes. Amphibians were the most abundant large organisms on land for about 100 million years. Then, in the late Permian period, reptiles took over as the most common vertebrate group. Reptiles were the first group

Protista

The kingdom Protista is an artificial grouping defined by what members of this group are not: protists are not plants, animals, fungi, bacteria, or archaeans.

Although most protists are harmless, the best-known ones are pathogenic, like *Plasmodium vivax*, the protist that causes malaria.

Animal-like protists are consumers.

Fungus-like protists are decomposers.

Plant-like protists are photosynthetic.

Most protists are single-celled and microscopic, and they can swim with the help of one or more flagella, or by waving a carpet of tiny hairs called cilia.

Protista

| Diplomonads, others | Euglenoids, others | Ciliates, others | Brown algae, diatoms, others | Forams, others | Red algae | Green algae |

The evolutionary history of the protists remains unresolved.

Common ancestor of plants and the green algae

Common ancestor of the Eukarya

Plantae

Plants are multicellular autotrophs and mostly terrestrial. Because plants are producers, they form the basis of essentially all food webs on land.

Bryophytes (mosses and liverworts)

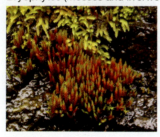

Plants have a waxy covering, known as the *cuticle*, that covers their above-ground parts. A waxy cuticle holds in moisture—an important adaptation to life on land.

Ferns

Gymnosperms

Gymnosperms were the first plants to evolve pollen, a microscopic structure that contains sperm cells, which freed them from a dependence on water for fertilization. Gymnosperms were also the first to evolve seeds, which can be disseminated so they will not compete with the mother plant for sunlight, or for water and nutrients in the soil.

Angiosperms

Flowering plants, or angiosperms, are dominant and the most diverse group of plants on our planet. The keys to the success of angiosperms are the flower, a structure that evolved through modification of early plant reproductive organs, and the fruit, a fleshy ovary wall that protects and helps disperse the seeds inside it.

Figure 14.9

The four kingdoms of the domain Eukarya

Q1: What group of organisms shares the most recent common ancestor with plants?

Fungi

Zygomycetes (molds)

Fungi digest organic material outside the body and absorb the molecules released as breakdown products.

Ascomycetes (sac fungi)

Basidiomycetes (club fungi)

Fungi play several roles in terrestrial ecosystems. Many are decomposers, acting as garbage processors and recyclers by speeding the return of the nutrients in dead and dying organisms to the ecosystem.

Fungi are similar to animals in that they store surplus food energy in the form of glycogen. Like some animals, such as insects and lobsters, fungi produce a tough material called *chitin* that strengthens and protects the body. Unlike animals, fungal cells have a protective cell wall that wraps around the plasma membrane and encases the cells.

| Plantae | Protista | Animalia | Fungi |

Plants

Amoebas, others

Animals

Fungi

Common ancestor of animals and fungi

Animalia

Animals are multicellular ingestive heterotrophs, obtaining energy and carbon by ingesting food. All animals are consumers, and some are important decomposers in the ecosystems they inhabit.

Animal cells differ from those of plants and fungi in that they lack cell walls. Instead, many cells in the animal body are enveloped in, or attached to, a felt-like layer known as the *extracellular matrix*. An important evolutionary innovation of animals is the development of true tissues. Most animals have two or three main tissue layers that give rise to a structurally complex body.

Sponges

Cnidarians

Mollusks

Arthropods

Echinoderms

Chordates

The sponges are the most ancient animal lineage. Cnidarians, a group that includes jellyfish and corals, evolved next. The remaining animal phyla fall into two groups: the *protostomes* and *deuterostomes*, distinguished by different patterns of embryonic development. Protostomes comprise more than 20 separate subgroups, including mollusks (such as snails), annelids (segmented worms), and arthropods (including spiders and insects). Deuterostomes include echinoderms (sea stars and their relatives) and the chordates. The chordates are a large group that includes all animals with backbones, such as fish, birds, and humans.

Q2: Are fungi more closely related to plants or to animals? Does the answer surprise you? Why or why not?

Q3: If you were to create an evolutionary tree in which amoebas were included within the kingdom of organisms to which they were the most closely related (rather than with protists, where they are currently placed), where would you put them?

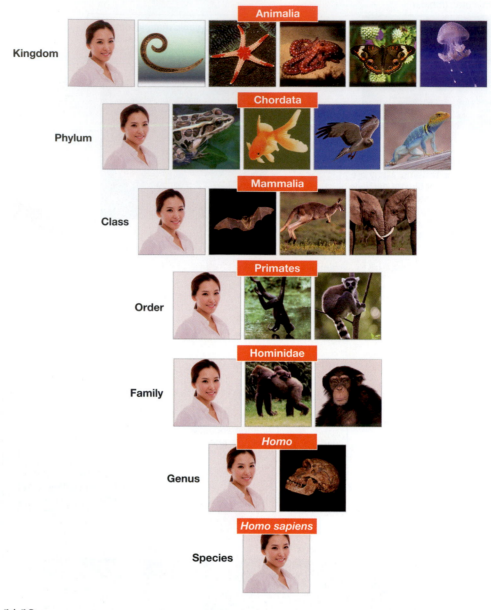

Figure 14.10

The Linnaean hierarchy of classification

Q1: Within which category are individuals most closely related to one another?

Q2: Within which category are individual species most distantly related?

Q3: Are individual species more closely related within the same order or within the same family?

of vertebrates that could reproduce without returning to open water, because they lay amniotic eggs that have a built-in food source and are protected from drying out by a hardened shell, compared to the jellylike sac that encloses the eggs of other vertebrates. The evolution of the amniotic egg was a major event in the history of life because it established a new evolutionary branch, the amniotes, which later included all reptiles, birds, and mammals.

And so, with the rise of reptiles 230 mya, the age of dinosaurs began.

mya
4,600   545   490   445

| Precambrian | Cambrian | Ordovician | |

Cambrian explosion

Figure 14.11

Cambrian biodiversity

During the Cambrian period, animal diversity increased dramatically. The remains of many of these species have been found in Canada (the Burgess Shale), China (the Maotianshan Shale), and fossil beds in Greenland and Sweden. Some of the fossils look familiar, resembling sponges and brachiopods, but do not appear to be related to any living groups of animals.

Tussling with Trees

In January 2013, a year and a half after Xu published his controversial paper about *Xiaotingia* and *Archaeopteryx*, his work received unexpected support. Pascal Godefroit, a paleontologist at the Royal Belgian Institute of Natural Sciences, reported the discovery of a feathered dinosaur called *Eosinopteryx brevipenna*. A commercial collector in northeastern China had dug up *Eosinopteryx* in the same area where *Xiaotingia* was discovered. The tiny 161-million-year-old dinosaur was preserved as a virtually complete skeleton, with its legs bent and arms out, as if about to jump (**Figure 14.12**).

When Godefroit and his team added *Eosinopteryx* to the evolutionary tree of feathered dinosaurs, they came to the same surprising conclusion that Xu had: *Archaeopteryx* was not a bird. Instead, *Archaeopteryx* was a raptor along with *Eosinopteryx* and *Xiaotingia*. All three species share traits that early birds did not have, such as arms longer than their legs, reduced tail plumage, and primitive feather development.

But science is a continuously changing process, and only 4 months later, Godefroit published data about another birdlike fossil that caused him to revise his hypothesis once again. This time it was a feathered dinosaur that Godefroit found collecting dust in the archives of a Chinese museum. The 18-inch-long fossil had small, sharp teeth and long forelimbs. Godefroit believed the dinosaur, which his team named *Aurornis xui* to honor Xu's work, probably couldn't fly, but instead used its wings to glide from tree to tree. But its other features, including the hip bones, were clearly shared by modern birds (**Figure 14.13**).

Using *Aurornis*, Godefroit constructed another evolutionary tree. This time he started from scratch, compiling data on almost a

PASCAL GODEFROIT

Pascal Godefroit is a paleontologist at the Royal Belgian Institute of Natural Sciences in Brussels. He has discovered numerous feathered dinosaurs.

Figure 14.12

Another feathered dinosaur, *Eosinopteryx brevipenna*, supports Xu's new tree

Figure 14.13

A namesake fossil, *Aurornis xui*, illuminates the dinosaur-bird transition

thousand characteristics of skeletons of 101 species of dinosaurs and birds. "It's very impressive," paleontologist Mike Lee at the South Australian Museum told *National Geographic*. "They considered more than twice as much anatomical information as even the best previous analyses."

Contrary to the tree that arose from his first study, Godefroit's second tree places *Archaeopteryx* back on its roost in the bird family, although no longer as the oldest bird. That place belongs to *Aurornis*, says Godefroit (**Figure 14.14**).

Still, the debate is not over. "Of course we need more evidence and more work," says Xu. "Many of these new species are a possible candidate for

Figure 14.14

The early birds

Godefroit's 2013 study places *Archaeopteryx* and *Xiaotingia* with birds rather than dinosaurs, as in the traditional dinosaur-bird evolutionary tree (see Figure 14.6, top). But it places *Aurornis* as the earliest known bird on the evolutionary tree.

Q1: Is *Xiaotingia* an earlier or later bird than *Archaeopteryx* in this tree?

Q2: If a future study, based on more fossils or new measurements, placed *Archaeopteryx* back with dinosaurs, would this suggest that birds are not related to dinosaurs? Why or why not?

Q3: If you were to create an evolutionary tree of modern birds, where would you expect to place the roadrunner (judging by its appearance in this figure) as compared to a house sparrow or pigeon?

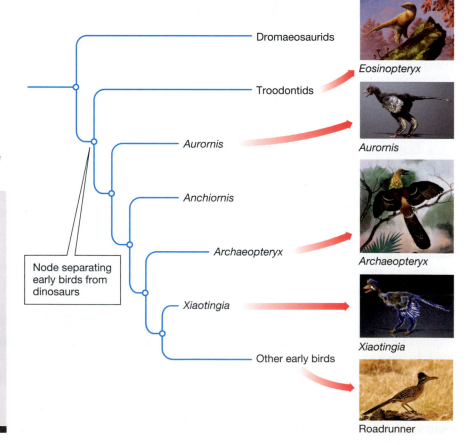

the earliest bird." There are likely to be more fossils that will shake up the bird tree, says Xu, but that's a good problem to have in science. "There are so many new species, it just makes it difficult for us all to agree."

Xu, Godefroit, and others are sure to continue to dig up dinosaur fossils for some time yet. Dinosaurs arose about 230 mya and dominated the planet from about 200 mya to about 65 mya. Then the majority of them went extinct, except for those that evolved into birds. As the fossil record shows, species have regularly gone extinct throughout the history of life. The rate at which

this has happened—that is, the number of species that have gone extinct during a given period—has varied over time, from low to very high. At the upper end of this scale, the fossil record shows that there have been five **mass extinctions**, periods of time during which great numbers of species go extinct (**Figure 14.15**).

Although difficult to determine, the causes of the five mass extinctions are thought to include such factors as climate change, massive volcanic eruptions, changes in the composition of marine and atmospheric gases, and sea level changes. The Cretaceous extinction event occurred

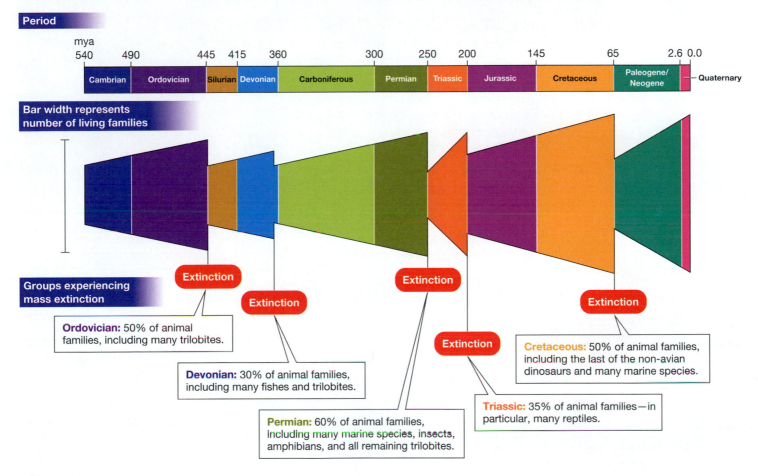

Figure 14.15

Mass extinctions and biodiversity

In addition to the marine and terrestrial animal groups shown here, plant groups were severely affected by the five mass extinctions that have occurred in Earth's history. After each extinction, life again diversified.

Q1: What extinction event occurred about 200 mya? What animal groups were most affected by this event?

Q2: Which of the mass extinctions appears to have removed the most animal groups? How long ago did this extinction occur?

Q3: The best studied of the mass extinctions is the Cretaceous extinction. Why do you think it has been better studied than the other extinctions?

The Sixth Extinction

On at least five occasions, mass extinctions have occurred across the globe, caused at different times by climate change, volcanic eruptions, and possible asteroids. Today, scientists agree we are in the midst of a sixth extinction, and this time, we, the human race, are the cause.

Assessment available in smartwork**5**

Ghosts of species past

Passenger Pigeon

The passenger pigeon, the most common bird in North America 200 years ago, was hunted to extinction in the nineteenth century. The last wild bird was shot in 1900, and the last captive bird died in 1914.

Xerces Blue Butterfly

The Xerces blue butterfly once lived on sand dunes around San Francisco—until its habitat was destroyed by urban development. It was last seen in 1943.

Caribbean Monk Seal

The Caribbean monk seal was the only seal native to the Caribbean Sea and Gulf of Mexico. It was overhunted for oil, and its food sources were overfished. It was last sighted in 1952.

Golden Toad

Once common in the cloud forests of Monteverde, Costa Rica, the golden toad has not been seen since 1989. Pollution and global warming likely contributed to its extinction.

Number of species known to be extinct or extinct in the wild since 1500

Group	Count
Arachnids	9
Crustaceans	12
Reptiles	22
● Amphibians	36
● Insects	58
Fishes	71
● Mammals	79
Plants	134
● Birds	145
Mollusks	324

about 65 mya and wiped out three-fourths of plant and animal species on Earth, including non-avian dinosaurs. Researchers suspect that a massive comet or asteroid slammed into the Gulf of Mexico, choking the skies around the planet with debris and decreasing the ability of plants to photosynthesize. As the plants died, so, too, did animals further up the food chain.

Mass extinctions affect the diversity of life in two main ways: First, entire groups of organisms perish, changing the history of life forever. Second, the extinction of one or more dominant groups of organisms can provide new opportunities for groups that previously were of relatively minor importance, thereby dramatically altering the course of evolution. When a group of organisms expands to take on new ecological roles and to form new species and higher taxonomic groups, that group is said to have undergone an **adaptive radiation**. Some of the great adaptive radiations in the history of life occurred after mass extinctions, such as when the mammals diversified after the extinction of the dinosaurs.

Today, one species of mammal, *Homo sapiens*, dominates life on land, and our impact on biodiversity is unprecedented. Because of human activities, the world is losing species at an alarming rate. In July 2016, researchers at the United Nations Environment Programme World Conservation Monitoring Centre in the United Kingdom announced that worldwide biodiversity had fallen below predetermined "safe" levels, thresholds below which ecological function is likely to be negatively affected. An estimated 58 percent of the world's land has lost more than 10 percent of its biodiversity, the researchers found, with grasslands and biodiversity hot spots such as the Amazon rainforest hit the hardest.

That is alarming news, because nonhuman organisms provide even the most basic requirements for human life. Plants and plantlike protists produce the oxygen we breathe and provide us with food. Whole ecosystems provide so-called "ecosystem services," environmental benefits that humans rely on. For example, coast redwood trees in northern California intercept fog, mist, and rain, channeling the water onto and into the ground.

Biologists today assert that we are on our way toward a new mass extinction, and the cause is clear: the activities of the ever-increasing number of humans living on, and exploiting, Earth.

REVIEWING THE SCIENCE

- The first single-celled organisms resembled bacteria and probably evolved about 3.7 billion years ago.

- The most basic and ancient branches of the tree of life define three **domains**: **Bacteria**, **Archaea**, and **Eukarya**. All life-forms fall into one of these three domains. The Eukarya are further divided into four kingdoms: **Protista**, **Fungi**, **Plantae**, and **Animalia**. The variety of these life-forms, as well as their interactions with each other and the ecosystems they inhabit, is **biodiversity**.

- Scientists use **evolutionary trees** to model ancestor-descendant relationships among different organisms. The tips of branches represent existing groups of organisms, and each **node** represents the moment when an ancestor split into two descendant groups. A **clade** is an ancestral species and all its descendants.

- Closely related groups of organisms share distinctive features that originated in their **most recent common ancestor**. These **shared derived traits** are used to identify the **lineages** of a species.

- The **Linnaean hierarchy** places each **species** in successively larger and more inclusive categories. Closely related species are grouped together into a **genus**, related genera are grouped into a **family**, related families into an **order**, related orders into a **class**, related classes into a **phylum**, and finally, related phyla into a **kingdom**.

- Organisms are identified by their genus and species names, together referred to as their **scientific name**.

- The release of oxygen by photosynthetic bacteria caused oxygen concentrations in the atmosphere to increase. Rising oxygen concentrations made possible the evolution of single-celled eukaryotes about 2.1 billion years ago. Multicellular eukaryotes followed about 650 mya.

- Life in the oceans changed dramatically during the **Cambrian explosion**, when large predators and well-defended herbivores suddenly appear in the fossil record. The Cambrian explosion is an example of **adaptive radiation**, in which a group of organisms take on new ecological roles and form new species and higher taxonomic groups.

- The land was first colonized by plants (about 480 mya), fungi (about 460 mya), and invertebrates (insects about 400 mya), which were followed later by vertebrates (about 365 mya). Each group evolved unique adaptations that enabled life to flourish on land.

- There have been five **mass extinctions** during the history of life on Earth. The extinction of a dominant group of organisms may provide new opportunities for other groups.

THE QUESTIONS

Challenge Yourself

1 *Archaeopteryx*

(a) belongs to the domain Prokarya.

(b) belongs to the kingdom Protista.

(c) lived during the Precambrian period.

(d) was much larger than modern birds.

(e) is an early example of the evolution of birds.

2 Identify the domain and kingdom for each of the following organisms.

(a) *Archaeopteryx*

(b) chanterelle mushrooms

(c) palm tree

(d) green algae

(e) you

3 "King Philip came over for good soup" is a mnemonic to remember each level of the Linnaean hierarchy.

a. List each level of the Linnaean hierarchy, in the order indicated by the mnemonic.

b. What level of the Linnaean hierarchy is missing from the mnemonic?

c. Create your own mnemonic to remember the Linnaean hierarchy.

4 Place the following evolutionary events in the correct order from earliest to most recent by numbering them from 1 to 5.

_____ a. Cambrian explosion

_____ b. origin of life on Earth

_____ c. plants' transition to land

_____ d. oxygen-rich environment created by bacteria

_____ e. evolution of birds from dinosaurs

5 Fill in the tree below to show the evolutionary relationships among the domains and kingdoms of life.

Try Something New

6 In each of the following cases, which domain(s) or kingdom(s) might contain organisms with the traits described? (You may need to review Figure 14.9 to answer.)

a. motile, with flagella or cilia

b. All organisms within the domain or kingdom are single-celled.

c. found in extreme environments (for example, high temperature, low oxygen, high salt)

d. multicellular with organ systems

e. photosynthetic

7 The traditional dinosaur-bird tree (Figure 14.6, top) can be restated as a hypothesis: "We hypothesize that *Archaeopteryx* is an early bird, and that birds split off from the closely related dinosaur groups Troodontidae and Dromaeosauridae." Restate Xu's tree (Figure 14.6, bottom) and Godefroit's tree (Figure 14.14) as hypotheses.

Leveling Up

8 *Write Now* **biology: early life on Earth** Read the *New York Times* article "World's Oldest Fossils Found in Greenland" (http://www.nytimes.com/2016/09/01/science/oldest-fossils-on-earth.html). Answer the following questions, writing one paragraph for each.

a. Summarize the findings of the study reported in the *Times* in your own words.

b. If the scientific community reached a consensus that these are *not* fossils, how would this conclusion affect our understanding of early life on Earth? Explain why such a conclusion would not be a failure of the scientific method, but rather a natural result of it.

c. Define the Late Heavy Bombardment (LHB). Describe how our understanding of the LHB has changed, and how this change has influenced our hypotheses about early life on Earth.

9 **Is it science?** Some people became quite upset when scientists agreed that the dinosaur *Brontosaurus* was in fact *Apatosaurus*, that *Triceratops* was only a juvenile *Torosaurus*, and that Pluto was not a planet. They clung to their *Brontosaurus*, or *Triceratops*, or planet Pluto, dismissing the scientists' arguments by saying, "Scientists are always changing their minds." Explain how scientists' "changing their minds" is not a problem but instead a necessary and useful feature of science, using the science described in this chapter as an example.

Thinking Multidimensionally

10 When one envisions the animals of the Galápagos Islands that a young Charles Darwin encountered in 1835, images of giant tortoises, marine iguanas, and finches probably come to mind, not an estimated 180 million black rats that are found there today! How did these invasive species arrive on the islands?

Rats eat the eggs of the tortoises and birds and are the most serious threat to the biodiversity of the Galápagos. To deal with this infestation, conservationists drop tons of poisons by helicopter onto the islands. What other issues might this solution cause?

11 If you go back in time to the Cambrian period when an explosion of new life-forms evolved, you would encounter a slender lancelet-shaped critter name *Pikaia*. This 2-inch-long creature may have been the ancestor of all vertebrates.

Look at the image above depicting *Pikaia*. What evidence has led scientists to think *Pikaia* may be our ancestor?

For more, visit digital.wwnorton.com/bionow2 for access to:

Navel Gazing

Scientists discover hundreds of new species of microbes—burrowed deep in our belly buttons.

After reading this chapter you should be able to:

- Differentiate between bacteria and archaea, and explain why they are together referred to as prokaryotes.
- Explain why prokaryotes can reproduce more rapidly than eukaryotes and why rapid reproduction can be advantageous.
- Give examples of the environments in which archaeans are found.
- Illustrate through example the breadth of metabolic diversity found in bacteria.
- Explain the function of at least one cellular structure that is unique to prokaryotes.
- Explain how citizen science contributes to scientific research.

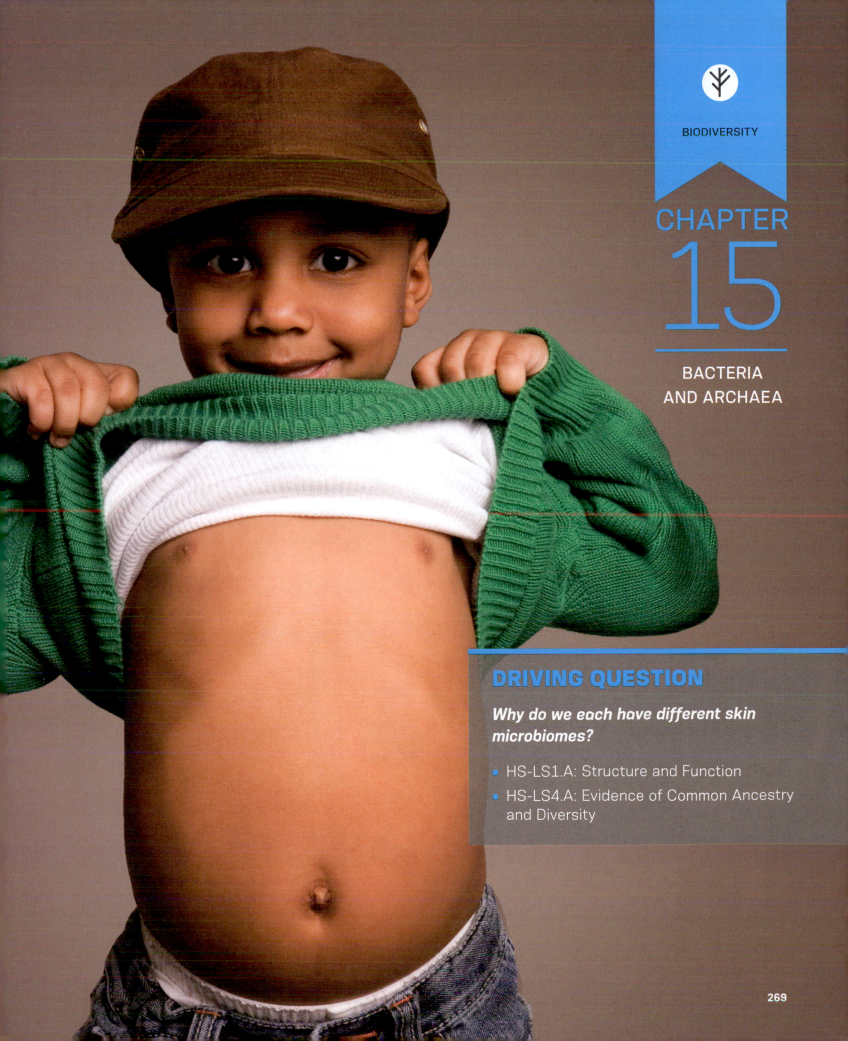

CHAPTER

15

BACTERIA
AND ARCHAEA

DRIVING QUESTION

Why do we each have different skin microbiomes?

- HS-LS1.A: Structure and Function
- HS-LS4.A: Evidence of Common Ancestry and Diversity

The gathering crowd jostled for position around the table. The scientists behind the table smiled as they handed out long, thin cotton swabs. Holly Menninger thanked them for the swab, then stared down at her shirt. She had been told what to do: lift your shirt, insert the swab into your belly button, swirl it around, place it into a test tube, seal the top. "Oh, the things we do for science," Holly thought. Then she swirled.

After returning her tube, Menninger—an entomologist by training and, at the time, coordinator of the New York Invasive Species Research Institute at Cornell University—struck up a conversation with the researchers manning the table, including Rob Dunn of North Carolina State University. Dunn, an applied ecologist and writer with an infectious passion for science, had initiated the project to answer a simple question: What lives on our skin? The belly button, as small and strange as it is, was an excellent place to look.

If you pick away the lint, the leftover schmutz in your belly button isn't dirt; it's alive. Our bodies are ecosystems, home to an estimated 39 trillion resident **microbes** (microscopic organisms), primarily bacteria. Considering that the average human body is made up of 30 trillion human cells, that's about a 1:1 ratio of microbial cells to human cells. (The old 10:1 ratio is a miscalculation that has unfortunately become enshrined in popular culture.) The **human microbiome**—the complete collection of microbes that live in and on our cells and bodies—affects human gut health, brains, and even body odor.

Microbes live in our guts, on our eyelashes, and yes, in our belly "holes," as Dunn calls them. In fact, the "belly hole" is an ideal place for a scientific study of resident microbes because it is a protected, moist patch of skin and one of the few areas that individuals don't regularly wash. Dunn knew that his students, if they could determine which microbes were swimming around in individual volunteers' navels, could then dig into the more burning question: *Why* do we each have the microbes we do?

"We know that which microbes you have on your skin influences your risk of infection, how attractive you are to other people, and how attractive you are to mosquitoes," says Dunn. "So this question of 'What determines which microbes are on your skin?' is super intriguing."

Of the three domains of life on Earth, Bacteria was the first to split off from the shared ancestor of the Archaea and Eukarya (**Figure 15.1**). Some fossil evidence places that split at about 3.48 billion years ago, yet a 2016 discovery in Greenland suggests that bacteria existed as early as 3.7 billion years ago, close to a period of time when Earth was being bombarded by asteroids (**Figure 15.2**).

Figure 15.1

Two of the three domains of life

Bacteria and Archaea are two distinct domains. They share many characteristics that are not found in the Eukarya.

Figure 15.2

Earth's first life left its mark in Greenland rocks

In 2016, Australian scientists found stromatolites, the fossilized secretions of bacteria, in rocks exposed by receding glaciers. These rocks formed 3.7 million years ago in a part of the world that is now Greenland.

Archaea likely split from Eukarya much later, around 2.7 billion years ago. The microbes that make up Bacteria and Archaea display many small but significant differences in their DNA, plasma membrane structure, and metabolism. They also share several important characteristics and so they are traditionally lumped under a common label: *prokaryotes*.

Prokaryotes are single-celled organisms, with a single loop of DNA floating free in the cytoplasm of the cell (review Figure 4.8 for a comparison of prokaryotic and eukaryotic cells). Unlike cells found in the Eukarya, prokaryotes do not have membrane-enclosed organelles. Prokaryotic cells are not only simpler than eukaryotic cells; they are smaller—almost exclusively microscopic, and invisible to the naked eye because of their diminutive size. The simple structure and single loop of DNA enable prokaryotes to reproduce at a much more rapid rate than eukaryotes, doubling in number every 10–30 minutes (**Figure 15.3**).

Prokaryotes in both domains are widespread and extremely abundant, and they display an astonishing diversity in metabolism. The vast majority of life on Earth is single-celled and prokaryotic. Scientists estimate that the number of prokaryotes on Earth is about 5,000,000,000,000,000,000,000,000,000,000 (5 nonillion, or 5×10^{30}). For example, prokaryotes—not fishes or algae—are the most abundant organisms in the open ocean, where they play a crucial role in the ecology of our biosphere.

The success of prokaryotes is due in part to how quickly a prokaryotic population reproduces, and also to the fact that they can live practically anywhere, including many places where few other forms of life can, such as deep, hot thermal vents or the acidic environment of the human intestine. Humans are saturated in microbes. We are each a zoo.

Menninger was so intrigued by Dunn's project that when a job became available in his lab a few months later, she immediately applied for and got the position. By then, the team had gathered about 60 swabs at local events, including the aforementioned conference, and now they had bigger ambitions: Dunn hired Menninger to spearhead a massive public outreach effort to acquire belly button swabs from all around the country. In **citizen science** projects such as this, the public participates in research by collecting

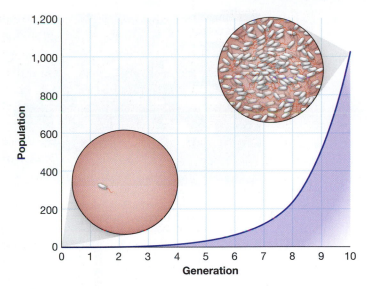

Figure 15.3

Prokaryotes are capable of extremely rapid population growth

An individual prokaryote is able to divide in two within 10–30 minutes. Those two can each divide in two in the same amount of time. This means that even a single bacterium or archaean can become a large population in a very short period of time.

Q1: If an individual prokaryote divides every 20 minutes, how many individuals will there be after an hour?

Q2: If the generation time is 20 minutes, how much time will have gone by when the final generation shown has doubled?

Q3: Many bacteria are able to reproduce more quickly in warmer conditions. What does this suggest to you about the importance of refrigerating foods?

and sometimes even analyzing data in cooperation with professional scientists (**Figure 15.4**).

"At the time, we were right at the forefront of an explosion in research, learning about how important the microorganisms that live on and in our bodies are to our health and well-being," says Menninger. "And this was a really great project to get people talking about the skin microbiome."

ROB DUNN

Rob Dunn is an applied ecologist at North Carolina State University. His fascination with species around us in our everyday lives has led him to research projects on microbes in our belly buttons, armpits, homes, and more.

Data collection

Citizen scientists from the Republic of the Congo are learning to map the forest with GPS.

Data analysis

A citizen scientist in North America uses an app to document plant budding dates.

Participation as an experimental subject

Sebastian Seung, a professor at Princeton University, works with citizen scientists around the world to map the brain through a video game.

Figure 15.4

Citizen scientists work alongside professional scientists to increase our knowledge about the natural world

There are many ways you can contribute to science—by collecting data, analyzing data, or even being an experimental subject like the video game players who work with Sebastian Seung. And the advantages are not one-way: citizen scientists learn new things, often have fun, and report feeling a sense of purpose and contribution.

Q1: In which of the three ways did the navel microbiome participants contribute?

Q2: Which of the advantages listed above do you think the navel microbiome citizen scientists received?

Q3: Would you be willing to contribute to the navel microbiome project? Why or why not?

Dutch tradesman Antonie van Leeuwenhoek first opened our eyes to the microbial world back in 1668, when he used simple, handcrafted microscopes to discover bacteria swimming around in pond water. In the intervening 350 years, our knowledge of the microbial world has come a long way.

Today, we know that bacteria and archaeans make up more than two-thirds of the species found on Earth. In 2016, microbiologists at UC Berkeley published a revised evolutionary tree of life that, for the first time, included the vast diversity of bacteria and archaeans lurking in Earth's nooks and crannies. With the data used to generate that tree, they showed that life is clearly dominated by prokaryotes (**Figure 15.5**).

Merry Microbes

In 2010, as a gag, an undergraduate student in Dunn's lab decided to make a Christmas card with streaks of microbes taken from people in the lab. She asked for belly button swabs from her coworkers, spread each sample on a petri dish, and grew the bacteria into colonies. As she grew bacteria from multiple individuals, it quickly became clear that there was more variety among the microbes growing on people in the lab than had been expected.

"Then it went from being a fun lab project to a serious question," says Dunn. The Christmas gag had raised an important question: Why do one person's microbes differ from another's? From his experience studying ecology, Dunn knew that microbes are critical in our lives. But neither Dunn nor others knew which factors determine which particular skin microbes a person has, or lacks.

So Dunn and his team began to study the locally collected swabs. An early part of the experiment engaged participants by visually depicting each person's microbial menagerie. The team plated each sample onto a petri dish and then took a picture of it for the microbes' owner. Although prokaryotes are single-celled organisms, some form colonies or long chains of identical cells, produced by repeated splitting begun by one original cell. This bacterial pattern of reproduction resulted in bright, dramatic patterns on the plates (**Figure 15.6**).

Under the microscope, the individual microbes were even more diverse and elaborate. Bacteria and Archaea cells are quite variable in shape, ranging from rods to spheres to spirals (**Figure 15.7**). Still, they all have a basic structural plan. Most bacteria and many archaeans have a protective cell wall that surrounds the plasma membrane. Some have an additional wrapping around that cell wall called a **capsule**. The capsule, made of slippery biomolecules, works like an invisibility cloak: it helps disease-causing bacteria evade the immune system that protects organisms like us from foreign invaders.

Dunn's team observed many bacteria whose surface was covered in short, hairlike projections called **pili** (singular "pilus"), a common bacterial

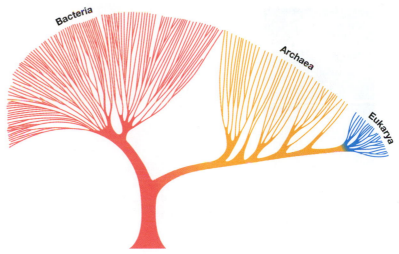

Figure 15.5

The most recent tree of life

Researchers sequenced DNA from individuals in 3,083 genera across all three domains. This information was combined by phylum, and the placement of each phylum in the tree was based on its relatedness to other phyla, as measured by DNA similarity.

Q1: Where in the figure would you place the first life found on Earth?

Q2: Identify where in the figure the Bacteria split off from the ancestor of Archaea and Eukarya.

Q3: The figure (and thus the study) demonstrates that Archaea and Eukarya are more closely related to each other than to Bacteria. How is that illustrated?

Bacillus *Staphylococcus*

Micrococcus Clostridiales

Figure 15.6

Belly button bacterial biodiversity

Each petri dish contains colonies of the bacteria collected from one person's belly button sample.

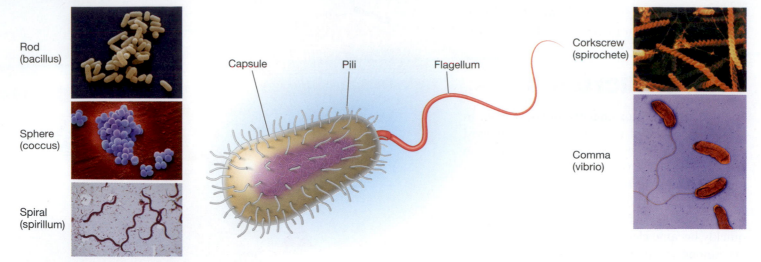

Rod
(bacillus)

Sphere
(coccus)

Spiral
(spirillum)

Capsule   Pili   Flagellum

Corkscrew
(spirochete)

Comma
(vibrio)

Figure 15.7

A simple structure, a diversity of forms

Bacteria and Archaea share a simple cell structure relative to eukaryotes, but they display a striking diversity of shapes and may have additional structures that perform special functions.

Q1: Which of these shapes do you think *Streptococcus* would take?

Q2: From the micrographs here, does it appear that all prokaryotes have a flagellum?

Q3: Which one of these shapes is most clearly capable of self-motility? Why?

feature. Bacteria use pili to link together to form bacterial mats, or to attach to surfaces in their environment, such as the cells of the human intestine. Some bacteria have one or more long, whiplike structures called **flagella** (singular "flagellum"), which spin like a propeller to push the bacterium through liquid.

Under the microscope, one obvious cellular feature is missing: prokaryotes do not have a true nucleus. They typically have much less DNA than eukaryotic cells have. Prokaryotes have less DNA because they have far fewer genes and because they contain relatively little non-coding DNA (DNA that is not used to construct proteins but may have regulatory or other functions). In contrast, eukaryotes generally have many more genes and far more non-coding DNA.

Dunn's lab partnered with Noah Fierer, an ecologist at the University of Colorado, to sequence the DNA from each belly button sample to determine the identity of the resident microbes, especially those that don't grow in a laboratory culture. To do so, lab members first had to access that DNA by mechanically crushing

the microbes and chemically dissolving their cell membranes with detergent, similar to how dish soap dissolves grease clinging to a dirty dish.

Next, they ran the sample through a silica strainer, called a column, to capture the DNA while allowing other debris to pass through. Once the DNA was isolated, the team amplified it using the polymerase chain reaction technique (see Figure 9.9) and then sequenced one short section of it, about 250 nucleotides of a well-studied gene called 16S ribosomal RNA (16S rRNA). Although 16S rRNA is found in all bacteria and archaeans, it is highly variable from one species to another, providing an identifying tag, like a fingerprint or barcode, to distinguish one species of microbe from another.

In 2012, the team published results from the initial set of swabs: among 92 belly buttons, the team identified 1,400 species of bacteria. "About 600 or so don't match up in obvious ways with known species, which is to say either they are new to science or we don't know them well enough," Dunn said in a media interview after the paper was published.

Archaea found in mineral hot springs are able to withstand extreme acidity *and* high temperatures.

A new archaean species was recently found in an abandoned copper mine. Others have been found living in acidic drainage from mines.

Hydrogen-eating, methane-producing archaeans have been found in deep-sea thermal vents.

Figure 15.8

Archaeans are everywhere

Archaeans, and bacteria to a lesser extent, thrive in environments that humans are incapable of surviving in.

Q1: Which shape in Figure 15.7 corresponds to the archaeans from deep-sea thermal vents?

Q2: Why are many archaeans referred to as "extremophiles"?

Q3: Is there anywhere you think archaeans could *not* survive? Justify your answer.

They had discovered new species of bacteria playing house in our navels. Still, Dunn was most interested in finding patterns hiding in the data. The first pattern they found was that the microbes in belly buttons are somewhat predictable. Six species showed up in about 80 percent of people, and when present, those species tended to be most widespread. "If I go to a cocktail party, I can tell you which species will be most abundant in the room," says Menninger with a laugh. That distribution pattern reminded Dunn of rainforest ecology: in any given forest, the types of plants might vary, but an ecologist can depend on a certain few tree types dominating the landscape.

Likewise, infrequent species tended to be present only in small quantities in belly buttons. Some were familiar: *Bacillus subtilis*, known to cause foot odor, also lives in some belly buttons (where it likely causes the same smell). Others were quite rare: one participant hosted a species previously found only in the soil of Japan, yet the man had never been to Japan. Dunn found a

species on himself known solely for eating pesticides. Why it was on his skin remains a mystery.

Amazingly, the team found more than just bacteria. On one particularly fragrant individual, who said he had not washed in years, they detected two species of archaeans. Although some bacteria thrive in unusual environments, the domain Archaea is well known for the extreme lifestyles of its members (**Figure 15.8**). Some are extreme **thermophiles** that live in geysers, hot springs, and hydrothermal vents—cracks in the seafloor that spew boiling water. The cells of most organisms cannot function at such high temperatures, but thermophiles have come up with evolutionary innovations that enable them to succeed where others cannot. Other archaeans, classified as **halophiles**, thrive in very salty, high-sodium environments where nothing else can live.

Yet archaeans have also been found in less exotic locations. In another citizen science project from Dunn and Fierer, students analyzed the microbial DNA swabs from home surfaces—kitchen counters, door frames, television screens, and more. Once again, they found several species of archaeans. "They are actually a lot more common than people tend to realize," says Menninger. "We're starting to have the molecular tools that allow us to find them." Dunn thinks the discovery is a bit ironic: "We spent so much time finding archaeans in the first place, in hard-to-reach, faraway places, and somebody could have just had some introspection with their own belly hole."

Yet back in the belly button, Dunn was unable to answer his main question about the bacteria: Why do individuals have the bacteria they do? The team looked at gender, ethnicity, how often participants washed their belly buttons, age, and more, yet they were unable to attribute any of the variation to biological or lifestyle factors.

So, as scientists tend to do, Dunn went looking for more samples. When Menninger joined the team, she initiated a program to have students and volunteers from around the country send in belly button swabs. Thanks to Menninger's outreach efforts, volunteers nationwide swabbed, swirled, and sealed. Soon the lab had over 600 samples to work with. That's a lot of microbes.

Talk, but No Sex

Like the plants and animals in a rainforest, microbes in an ecosystem communicate with each other, reproduce, and more. Researchers used to think prokaryotes were the ultimate loners—self-sufficient and maintaining strict single-celled lifestyles. Then they discovered that bacteria physically interact with each other in many ways.

Some bacteria (and Archaea) have a unique system of cell-to-cell communication called **quorum sensing** that enables them to sense and respond to other bacteria in the area in accordance with the density of the population. Disease-causing bacteria, for instance, may begin to multiply rapidly upon sensing that their numbers are high enough to overwhelm the host organism's immune system. Other bacteria coordinate their behavior by forming tough aggregates called *biofilms* that are made up of the same or different species (**Figure 15.9**).

It's strange to think about, but all these activities are happening on our skin: "Everything you can imagine life doing, happens in you," says Dunn, including reproduction. Prokaryotes

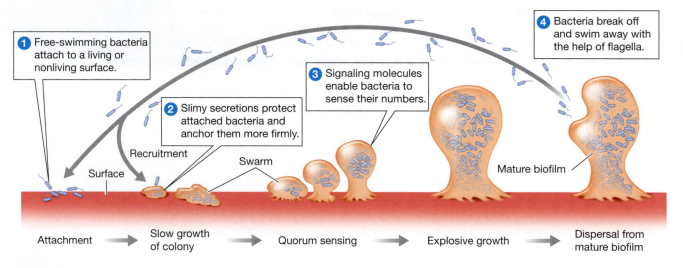

1 Free-swimming bacteria attach to a living or nonliving surface.

2 Slimy secretions protect attached bacteria and anchor them more firmly.

3 Signaling molecules enable bacteria to sense their numbers.

4 Bacteria break off and swim away with the help of flagella.

Recruitment

Surface

Swarm

Mature biofilm

Attachment → Slow growth of colony → Quorum sensing → Explosive growth → Dispersal from mature biofilm

Figure 15.9

Quorum sensing enables coordinated behavior in bacteria

With quorum sensing, a bacterial population can increase its virulence, reproductive rate, or antibiotic resistance under the appropriate conditions. Biofilms are produced via quorum sensing to protect bacteria from environmental hazards.

Q1: How do individual bacteria know that they have a "quorum"?

Q2: There is a well-known biofilm found in your mouth. What is it?

Q3: Under what conditions might bacteria want to coordinate (via quorum sensing) to increase their reproductive rate?

typically reproduce by splitting in two in a process called **binary fission**, a form of asexual reproduction (see Figure 6.3). The DNA in the parent cell is copied before fission, and one copy is transferred to each of the resulting daughter cells. The genetic information in the daughter cells is virtually identical to that of the parent cell, as is invariably the case in asexual reproduction.

Although sexual reproduction has not been seen in prokaryotes, they are adept DNA pickpockets: microbes can capture bits of DNA from their environment or other bacteria and incorporate them into their own genetic material. The transfer of genetic material between microbes is known as *horizontal gene transfer* and it involves plasmids, loops of extra DNA in the cytoplasm of prokaryotes (see Figure 12.12). (Horizontal gene transfer has been seen in some eukaryotes as well. Bdelloid rotifers, for example, a type of microscopic freshwater animal, stole about 8 percent of the DNA in their genome from bacteria.) A bacterium can directly trade DNA with another bacterium through a process known as *bacterial conjugation*. Alternatively, when a bacterium dies, the cell may burst open and another bacterium may simply take up the released DNA.

Some types of bacteria—but no archaeans—can undergo **sporulation**, the formation of thick-walled dormant structures called spores. Spores are the bomb shelters of microbes: they can survive boiling and freezing, thereby allowing the microbes to hang out for a long period of time until the conditions are again favorable to reproduce. In many cases, even antibiotics can't kill bacterial spores.

All Hands on Deck

Dunn, Menninger, and the team analyzed the additional samples. Still, they were unable to explain the differences among belly button microbes. They could not identify any factors affecting the number and types of species a person might host.

So they decided to take a step that more and more scientists are taking—they made all their data freely available online for other researchers to use (with all identifying information about the participants removed). "We asked other researchers to let us know what their insights are, because we're all in this together," says Menninger. "We need all hands on deck."

The gambit worked. Mathematician Sharon Berwick at the University of Maryland took a new approach to the data. Instead of focusing on characteristics of the participants, such as gender or location, she looked at the characteristics of the microbes and found that people tend to be dominated by one of two types of belly button bacteria: *aerobic* or *anaerobic*. **Aerobes** are prokaryotes that need oxygen gas to survive. *Micrococcus* species, which Dunn found in the navel, are aerobes; they need oxygen. Therefore, species of *Micrococcus* are unlikely to do well deep inside a belly button, says Dunn, but they appear to thrive on the surface.

Another type of navel resident, species of *Clostridia*, do not use oxygen. They are **anaerobes**, prokaryotes that survive without oxygen. In fact, some anaerobes may actually be poisoned by oxygen. Among the anaerobic archaeans are several species of **methanogens**, which feed on hydrogen and produce methane gas as a by-product of their metabolism. The ability to exist in both oxygen-rich and oxygen-free environments is another reason why prokaryotes can be found in most habitats.

Some prokaryotes can even switch between the two. One of the most common bacteria Dunn's team found in belly buttons was *Staphylococcus*. Although it is best known as a pathogen, staph on your skin is typically a good guy. On the skin, it is beneficial for your health, fighting off other pathogens that want to crowd in. Some species of staph, such as *Staphylococcus epidermidis*, are typically aerobes, but when oxygen is in short supply, they switch over to a special type of anaerobic metabolism known as **fermentation**, involving the breakdown of sugars (yes, the same process that is used to ferment beer and wine; to review fermentation, see Figure 5.12). In other words, as Dunn wrote in one description of *S. epidermidis*, "Right now you might be making a teeny tiny bit of navel wine."

Like staph, *Bacillus subtilis*, a rod-shaped, spore-forming bacterium, can grow in both aerobic and anaerobic conditions. *B. subtilis* is

HOLLY MENNINGER

Holly Menninger is the director of public science at North Carolina State University, where she works to engage the public in citizen science and science education.

one of the warrior clans in our belly buttons and other areas of the skin, producing antibiotics that kill off other bacteria and even foot fungi, says Dunn. Prokaryotes are either consumers, called **heterotrophs**, or producers, called **autotrophs** (**Figure 15.10**). Autotrophs make food on their own, but heterotrophs, like *B. subtilis*, obtain energy by taking it from other sources. Specifically, *B. subtilis* is a **chemoheterotroph**, an organism that consumes organic molecules to get energy (in the form of chemical bonds) and carbon (in the form of carbon-containing molecules). Some species of blue-green algae are **photoheterotrophs**, which acquire carbon from organic sources but their energy from sunlight.

Some autotrophs, called **photoautotrophs**, absorb the energy of sunlight and take in carbon dioxide to conduct photosynthesis. Others, called **chemoautotrophs**, get their energy from inorganic chemicals in their environment, including iron ore, hydrogen sulfide, and ammonia, instead of from sunlight.

In fact, the very first photoautotrophs on the planet were prokaryotes. One type of aquatic bacteria, the cyanobacteria, is believed to be responsible for changing Earth's chemistry by evolving photosynthesis and producing oxygen gas as a by-product. Oxygen gas accumulated in the air and water, and the levels rose from next to nothing to almost 10 percent about 2.1 billion years ago. At that time in the fossil record, eukaryotes appear, suggesting that the oxygen generated by cyanobacteria may have facilitated the evolution of eukaryotes, especially multicellular forms.

Today, prokaryotes continue to aid eukaryotes. Bacteria directly help plants through a process known as **nitrogen fixation**. Plants need nitrogen in the form of ammonia or nitrate, which they cannot make themselves. Bacteria, however, can take nitrogen gas from the air—our

		Source of Energy	
		Light	Chemicals
Source of Carbon	Carbon dioxide	Photoautotroph *Gloeocapsa*	Chemoautotroph *Acidithiobacillus*
	Organic matter	Photoheterotroph *Heliobacterium*	Chemoheterotroph *Escherichia*

Figure 15.10

How prokaryotes feed themselves

Prokaryotes have many more ways to feed themselves than do eukaryotes. They are categorized by their energy source, and by the source of carbon they use.

Q1: What source of energy would you expect a cave-dwelling prokaryote to use?

Q2: In which of these categories would you place the bacteria responsible for nitrogen fixing? Why?

Q3: In which of these categories do decomposers belong? Explain your reasoning.

atmosphere is 78 percent nitrogen—and convert it to ammonia, making it available for plants.

Autotrophs aren't the only microbes that enable life on Earth. Many heterotrophic bacteria and archaeans are decomposers, consumers that extract nutrients from the remains of dead organisms and from waste products such as urine and feces. Decomposers play a crucial role in **nutrient cycling**: by breaking down dead organisms or waste products, decomposers release the chemical elements locked in the biological material and return them to the environment. Those released elements, such as potassium or nitrogen or phosphorus, are used then by autotrophs and eventually by heterotrophs as well.

Healthy Balance

Since van Leeuwenhoek first looked through his microscope, people tended to regard microbes with suspicion and fear (**Figure 15.11**). We vilified

Figure 15.11

The first illustrations of microbes

Antonie van Leeuwenhoek's descriptions and illustrations of microscopic life led to his being considered the "father of microbiology."

Q1: From the prokaryotic structures shown in Figure 15.7, what shape would you assign to drawing number 8 in van Leeuwenhoek's illustration?

Q2: Which of the large prokaryote drawings has the coccus shape?

Q3: Do you think all of these "animalcules" drawn by van Leeuwenhoek are prokaryotes? Why or why not?

The Bugs in Your Belly Button

There's no way around it: Your belly button is full of bugs. According to research from Rob Dunn's lab at North Carolina State University, the most abundant species in a belly button—those with the biggest populations, such as *Corynebacterium* and *Staphylococcus*—are found in most people. Yet no two people have exactly the same set of microbes. In other words, our belly button microbiomes look alike, but we each have a unique navel fingerprint.

Assessment available in smartwork5

31%
Corynebacterium

28%
Staphylococcaceae

7% *Anaerococcus*

6% *Porphyromonas*

3% *Alicyclobacillus*

Prevotella 2%

Finegoldia 2%

Streptococcus 2%

Enterobacteriaceae 2%

Pseudomonadaceae 1%

Sphingomonas 1%

Acinetobacter 1%

Campylobacter 1%

Clostridiaceae 1%

Peptoniphilus 1%

Pseudomonas 1%

Clostridiales 1%

Cupriavidus 1%

The other 617 species

9%

microbes as the bad guys that our bodies were constantly at war with. But now we recognize that our ecosystems wouldn't function without microbes: we would have no oxygen to breathe, no plants to eat. And microbes in our gut and on our skin, when in healthy balance, promote human health.

We emphasize a healthy balance because it is possible for communities in the human microbiome to shift out of balance—a phenomenon called *dysbiosis*, which may cause illness. Alternatively, pathogenic bacteria can make us sick as well. Although the great majority of bacteria are harmless and many are actually beneficial to humans, some cause mild to deadly disease. Interestingly, archaeans are not known to be pathogens of any organism.

The biodiversity of microbes on our skin actually helps keep pathogens away. When a dangerous microbe lands on your skin, even before it meets the immune system it meets other microbes. If the skin has a diversity of microbes, odds are "one of them has the ability to kick the butt" of the pathogen, says Dunn.

After the overwhelming response to the belly button project, the lab kicked off a new citizen science spectacular: "Armpit-pa-looza" (**Figure 15.12**). As with the belly button, the team began collecting swabs of microbes living in people's armpits. In a preliminary study, the team found that the use of antiperspirants or deodorants dramatically inhibits bacterial growth, affecting the composition of the microbial populations on the skin more strongly than any other factor. "It now looks

Figure 15.12

Sampling armpits for science
Citizen scientists volunteer to have samples taken of the prokaryotes living under their arms.

like one of the biggest things affecting skin microbes in general is the use of antiperspirant or anti-odor products," says Dunn.

Now the team is taking a look at microbes on the skin of dogs. Skin wounds on pet pooches actually heal about four times faster than human skin wounds do, says Dunn, and he suspects that skin microbes are involved. "We're starting to figure out if we can predict wound healing rates as a function of which microbes are there to start with," says Dunn. "I'm pretty excited about that."

REVIEWING THE SCIENCE

- The non-Eukarya, or **prokaryotes**, fall into two domains: Bacteria and Archaea.

- All prokaryotes are **microbes**—microscopic, single-celled organisms—but the Bacteria and Archaea differ in significant ways, such as in their DNA, plasma membrane structure, and metabolism.

- Prokaryotes can reproduce extremely rapidly and are the most numerous life-forms on Earth. They also have the most widespread distribution.

- Many prokaryotes, particularly bacteria, have specialized structures. Bacteria with a **capsule** surrounding the cell wall can avoid detection by

organisms' immune systems. Short, hairlike **pili** help bacteria attach to surfaces and to each other. Whiplike **flagella** assist in locomotion.

- Some prokaryotes, particularly archaeans, thrive in extreme environments. **Thermophiles**, for example, live in extremely hot places, and **halophiles**, in very salty places.

- Prokaryotes typically reproduce by **binary fission**, and they may acquire DNA from each other or their environment via horizontal gene transfer.

- Some prokaryotes use **quorum sensing** to communicate with each other, and some undergo **sporulation** for protection.

- Prokaryotes exhibit unmatched diversity in methods of getting and using nutrients and energy. **Autotrophs** (producers) make their own food, whereas **heterotrophs** (consumers) obtain food from other sources. **Chemoautotrophs** and **chemoheterotrophs** use inorganic chemicals as their energy source; **photoautotrophs** and **photoheterotrophs** use sunlight.

- Prokaryotes perform key tasks in ecosystems, including photosynthesis, **nitrogen fixation** (providing nitrogen to plants), and **nutrient cycling** (decomposing dead organic matter).

- Although prokaryotes are useful to humanity in many ways, some cause deadly diseases.

THE QUESTIONS

Challenge Yourself

1 Which of the following can prokaryotes *not* do?

(a) communicate with each other about environmental conditions

(b) reproduce sexually

(c) live in extreme environments, including high salt, high temperature, and high atmospheric pressure

(d) share DNA with other prokaryotes

(e) double in population size two or more times an hour

2 Which of the following descriptions is true *only* of prokaryotes?

(a) They reproduce through binary fission.

(b) Their genetic material is DNA.

(c) They are unicellular.

(d) They contain organelles within the cell.

(e) none of the above

3 Place the following steps of the belly button research described in this chapter in the correct order by numbering them from 1 to 5.

_____ a. An observation was made that belly button bacteria seemed to differ across individuals.

_____ b. The DNA was sequenced.

_____ c. DNA was isolated from the samples.

_____ d. Volunteers had their navels swabbed to collect bacterial samples.

_____ e. The bacterial samples were grown in petri dishes.

Try Something New

4 Which of the following is citizen science?

(a) reporting when the first hummingbirds arrive at your feeders in the spring

(b) completing a Facebook quiz on your horoscope

(c) filling out your medical history at the doctor's office

(d) all of the above

(e) none of the above

5 There are more genera of bacteria and archaeans in the world than of eukaryotes, and more individuals. However, the total mass of prokaryotes is thought to be approximately equal to that of eukaryotes. How can this be?

6 Why are prokaryotes able to replicate so much more quickly than eukaryotes? (*Hint:* What cellular components must be copied before a cell splits in two?) Why is this difference in replication rate an important part of our vulnerability to bacterial pathogens?

Leveling Up

7 **Doing science** The research discussed in this chapter would not have been possible without the many citizen scientists who shared samples of their belly button microbes with the research team. Choose a citizen science project: Perform an Internet search on the term "citizen science project," or find regularly updated lists of projects at Scitable (http://www.scitable.com) and Your Wild Life (http://www.yourwildlife.org). Write a one-paragraph summary of the goals of the project you have chosen. Would you participate in this project? Why or why not? Describe a citizen science project that you *would* want to contribute to.

8 **Looking at data** The table here is based on the published results of the study discussed in the chapter. It shows the number of bacterial phylotypes (kinds of bacteria) that were found on a particular percentage of people.

a. How many bacterial phylotypes were found on only 10 percent or fewer of the people sampled? How many were found on more than 90 percent of people?

b. How many phylotypes were found on more than half of the people sampled?

c. Describe in one or two sentences the frequency of phylotypes found in the study. Were most phylotypes rare or common?

Phylotypes	Percentage of Human Samples Where Found
2,188	1–10
97	11–20
31	21–30
12	31–40
17	41–50
11	51–60
4	61–70
2	71–80
2	81–90
4	91–100

Thinking Multidimensionally

9 The 2-inch Hawaiian bobtail squid is a nocturnal predator, hunting shrimp at night. Its bioluminescence is a result of a symbiosis with light-producing bacteria called *Vibrio fischeri*.

Moonlight shining down on this predator casts a shadow on the seafloor and alerts the shrimp to its presence.

a. Explain how the squid's partnership with *V. fischeri* counteracts this barrier, enabling this squid to be a successful hunter.

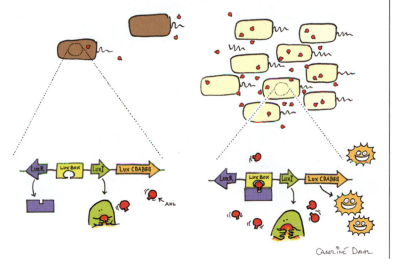

The squid flushes out most of the glowing bacteria from its body each morning and has to reestablish a fresh new supply before the next night's hunt. Looking at the accompanying diagram, explain the following:

b. How do the bacteria communicate their presence to one another?

c. How do they reproduce?

d. What action occurs when they form a large quorum?

10 Over a century ago, French pediatrician Henry Tissier stated that during development a fetus grows inside a microbe-free womb.

a. According to Tissier, how would a newborn baby then become inoculated with microbes in and on its body?

More recent studies indicate the placenta, amniotic fluid, and umbilical cord blood are not sterile, and therefore babies are born with microbes already living on and inside them.

b. If a mother is put on antibiotics during pregnancy, how might this impact the microbiome of her fetus?

c. If one expecting mother goes on the meat-heavy Atkins diet and another mother opts for a vegan diet, how would you expect the gut microbes to differ between these pregnant mothers?

d. Would their diets impact the microbiomes of their developing fetuses? Explain your answer.

For more, visit digital.wwnorton.com/bionow2 for access to:

The Dirt on Black-Market Plants

Nighttime poaching and illegal trade threaten to drive rare plants and fungi to extinction.

After reading this chapter you should be able to:

- Compare and contrast prokaryotes and eukaryotes.
- Explain what is distinctive about the kingdom Protista.
- Outline the key evolutionary innovations of plants.
- Connect structure and function in how fungi obtain energy from the environment.
- Identify key characteristics and give an example organism for each kingdom covered in this chapter.

DRIVING QUESTION

Why are wild plants and fungi being poached, and what are the ecological consequences?

- HS-LS1.A: Structure and Function

Jacob Phelps spent a lot of his time in Thailand trying not to arouse suspicion. Walking among vendor stalls at wildlife markets in Bangkok, the tall, thin graduate student stopped occasionally to help a trader trim dead leaves off his plants, or to chat about the weather. Wherever he walked, plants surrounded him—hanging from the ceiling, stuffed into boxes, piled into mounds.

Phelps was careful about the questions he asked, because he was there to document illegal activity. The market traders had tens of thousands of plants for sale, most of which weren't supposed to be there. But when Phelps cautiously asked whether the plants were wild or rare, the traders were surprisingly relaxed. Concern for plant conservation, and enforcement of trading laws, is so limited that illegal sales occur openly at public markets across Southeast Asia.

Even Thailand's wildlife trade management authority at the time claimed that illegal trade in ornamental flowers was limited, "found in small case [sic] in some parties." But Phelps was familiar enough with plants to know that most of what he saw in the Thailand markets were wild, protected plants that were illegal to sell: the charismatic pink blossoms of *Dendrobium* orchids; the dense, fragrant plumes of *Rhynchostylis* flowers; the delicate, wavy petals of the coveted lady slipper orchid, *Paphiopedilum*. Plant enthusiasts prize such species for their beauty, fragrance, and rarity. Coveted wild orchids make up more than 80 percent of the plants traded at these unregulated markets (**Figure 16.1**).

As a graduate student at the National University of Singapore, Phelps received the blessing of his PhD supervisor, plant ecologist Ted Webb, to conduct an extensive survey of the illegal plant trade at the four largest plant markets in Thailand. It was the first time such a survey had been conducted of plant markets anywhere in

Southeast Asia. In fact, the illegal trade of plants is often called the "invisible wildlife trade" because it is so rarely discussed or documented, in contrast to the widely publicized illegal trade of animals and animal products such as elephant ivory and rhinoceros horns.

Plants often take second-tier status behind animals—it's easier to get the public to care about a fuzzy baby tiger than a sprouting redwood tree—but their biology is no less wondrous. In Chapter 15 we explored the microscopic worlds of the Bacteria and Archaea domains of life. Here, we begin to meet the Eukarya, consisting of four kingdoms: Protista, Plantae, Fungi, and Animalia. Animalia will be covered in detail in Chapter 17.

The defining feature of the Eukarya is a true nucleus: instead of floating free in the cytoplasm, eukaryotic DNA is enclosed in two concentric layers of cell membranes that form a nuclear envelope. In addition to the nucleus, eukaryotes have a great variety of membrane-enclosed subcellular compartments, many of which are specialized for various tasks—such as sending messages, producing energy, or

Figure 16.1

A plant market in Thailand
Jacob Phelps visited plant markets like this one to conduct research on illegal trade in endangered orchid species.

JACOB PHELPS

Jacob Phelps is an environmental scientist at Lancaster University in the United Kingdom, studying illegal wildlife trade and tropical deforestation.

cleaning up—so that the cell can function efficiently (review Figure 4.8 for a comparison of prokaryotic and eukaryotic cells, and Figure 4.9 for an overview of eukaryotic organelles). All of these organelles take up space, so the diameter of a eukaryotic cell is, on average, 10 times larger than that of a prokaryotic cell, and the eukaryotic cell volume is a thousandfold greater.

This compartmentalization of the cell interior enables eukaryotic cells to do things that most prokaryotic cells cannot do. For example, some eukaryotes engulf their prey and digest them internally. That's the way many single-celled eukaryotes, such as the blob-like amoebas, eat: they extend gooey cytoplasmic arms to engulf other cells whole, then digest their prey with an elaborate system of internal compartments for digestion, ridding the cell of waste and storing surplus food (**Figure 16.2**).

Figure 16.2

Eukaryotes have a true nucleus and compartmentalized cells
The more complex cell structure of eukaryotes enables them to perform functions that are impossible for prokaryotes. Intracellular compartments such as food vacuoles enable this amoeba to digest its food, a single-celled alga. The amoeba expels excess water with the help of its contractile vacuole, another type of intracellular organelle.

Fungi Play Well with Others

Plants probably would not have been as successful on land if they had not entered into a mutualistic relationship with fungi almost immediately on their arrival. Today, the vast majority of plants in the wild have mutualistic fungi, known as **mycorrhizal fungi**, associated with their root systems. (The mutualistic relationships between plants and mycorrhizal fungi are called "mycorrhizae.") Truffles, morels, and chanterelles, all beloved by gourmets, are the reproductive structures of mycorrhizal fungi.

Mycorrhizal fungi form thick, spongy mats of mycelium on and in the roots of their plant hosts and also extend into the surroundings, sometimes permeating several acres of the soil around the root. Mycelia are thinner, more extensively branched, and in closer contact with the soil than even the thinnest branches on a plant root. As a result, a mycelial mat plumbs far more water and mineral nutrients, such as phosphorus and nitrogen, than the plant's root system could absorb on its own. In return for sharing absorbed water and mineral nutrients, the fungus obtains sugars that the plant manufactures through photosynthesis.

Mycorrhizal fungi assist in providing nutrients to orchid seeds, which are tiny and lack stored food. The embryo within a newly sprouted orchid seed could not survive without the mycorrhizal network linking the seedling to mature photosynthesizing plants, from which the seedling draws nourishment until it can photosynthesize on its own.

Fungal mutualisms are not found only with other eukaryotes. A **lichen** is a mutualistic association between

Plants that have no mycorrhizal fungi associated with their roots do not grow as well…

…as those that host the beneficial fungi.

a photosynthetic organism and a fungus. The fungus receives sugars and other carbon compounds from its photosynthetic partner, usually a green alga or a cyanobacterium. In return, the fungus produces lichen acids, a mixture of chemicals that scientists believe may function to protect both the fungus and its partner from being eaten by predators.

Lichens grow very slowly and are often pioneers in barren environments. Lichen acids wear down a rocky surface, facilitating soil formation. Soil particles build up from the slow weathering of rock, and over time, other life-forms, including plants, gain a toehold in the newly made soil.

Peculiar Protists

Amoebas are single-celled eukaryotes. Most single-celled eukaryotes are **protists**. Protista is a strange kingdom that is defined by exclusion instead of inclusion: although all protists are eukaryotes, they are grouped together in this kingdom simply because they are *not* plants, animals, or fungi (**Figure 16.3**). Researchers have proposed several classification schemes to split the protists into multiple kingdoms; however, no consensus exists on the best way to do so. For now, protists remain divided into two traditional, broad categories: the **protozoans**, which are nonphotosynthetic and motile (capable of moving); and the **algae**, which are photosynthetic and may or may not be motile.

Protists are diverse in size, shape, cellular organization, and mode of nutrition. Most protists are single-celled and microscopic, but some protists have evolved from free-living single cells into multicellular associations, such as slime molds and kelp, large seaweed such as kelp. Certain single-celled protists are bound by nothing more than a flexible plasma membrane, while others are covered in protective sheets, heavy coats, or other types of armor.

Most protists are motile and can swim with the help of one or more flagella, or by waving a carpet of tiny hairs called cilia. Others can crawl on a solid surface with the help of cellular projections called *pseudopodia* (false feet).

Many protists are heterotrophs and eat other organisms. Some of these heterotrophs function as decomposers, breaking down waste material and releasing nutrients into the environment to be taken up by producers and cycled back into the food chain. Other protists are nutritional

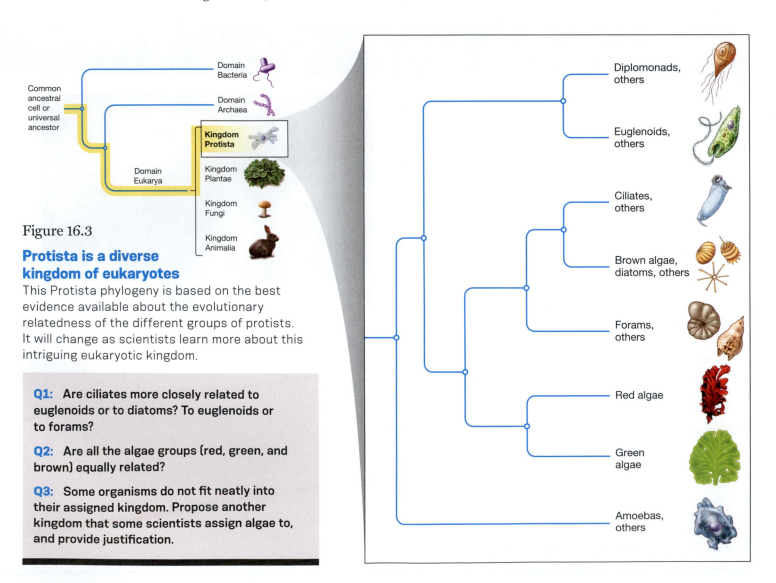

Figure 16.3

Protista is a diverse kingdom of eukaryotes

This Protista phylogeny is based on the best evidence available about the evolutionary relatedness of the different groups of protists. It will change as scientists learn more about this intriguing eukaryotic kingdom.

Q1: Are ciliates more closely related to euglenoids or to diatoms? To euglenoids or to forams?

Q2: Are all the algae groups (red, green, and brown) equally related?

Q3: Some organisms do not fit neatly into their assigned kingdom. Propose another kingdom that some scientists assign algae to, and provide justification.

opportunists, or *mixotrophs*, organisms that use energy and carbon from a variety of sources to fuel their growth and reproduction.

Although most protists are harmless, many of the best-known protists are pathogens (disease-causing agents), such as *Plasmodium*, which is transmitted by mosquitoes and causes malaria, and *Toxoplasma gondii*, passed to humans through poorly cooked food or cat feces. Toxoplasmosis typically causes mild flu-like symptoms, but can be passed from mother to child in utero and cause more severe symptoms.

Two Cells Are Better Than One

Multicellular forms evolved several times among different lineages of the eukaryotes. A multicellular organism is a well-integrated assemblage of genetically identical cells, in which different groups of cells perform distinctly specialized functions. This functional compartmentalization makes it easier for multicellular eukaryotes to sense and respond to the external environment, through the use of complex structures such as leaves, fruit, eyes and wings.

Multicellularity also enables an individual organism to grow large, which can be advantageous for evading potential predators. A bigger individual can also gather resources from its environment more effectively than can a smaller individual. Having more resources, such as light or food, may translate into producing more surviving offspring, the ultimate measure of biological success.

To make those offspring, eukaryotes reproduce via either asexual or sexual reproduction. Asexual reproduction is common among species in this domain and generates genetically identical offspring. Protists, for example, can split into two in a process similar to binary fission in prokaryotes. Many plants reproduce asexually by fragmenting into pieces, each piece developing into a new individual.

But it is through the second type of reproduction that eukaryotes have made an indelible mark on our planet. By combining genetic information from two parents, sexual reproduction produces offspring that are genetically different from each other and from both parents. Although their life cycle is distinctly different from that of animals, plants similarly produce embryos: the fusion of

egg and sperm produces a single cell, called a zygote, which then divides to produce a multicellular structure called an embryo. Sexual reproduction is one means by which natural populations become genetically diverse; it is why there are an estimated 400,000 species of plants on Earth.

Through these unique traits—a true nucleus, cellular compartmentalization, multicellularity, and sexual reproduction—eukaryotes have evolved into amazingly diverse and dynamic species. Of these species, Phelps was particularly concerned about orchids. In addition to spending time in Bangkok, the budding ecologist devoted months traveling along Thailand's borders with Laos and Myanmar, wandering hours upon hours through wildlife markets. Over time, he built local networks that allowed him to interview more than 150 plant harvesters and intermediaries, including market traders, online traders, nursery owners, and more.

Orchids are, on paper, one of the most heavily protected families, representing almost 75 percent of all species, plant or animal, for which international trade is regulated. Artificially propagated orchids can be traded legally, but wild orchid species in Southeast Asia are protected, says Phelps: "None should be collected from the wild, unless you have specialized permissions." Yet, as he came to discover, there is no enforcement of those protections in Thailand or across the region. Orchids and other ornamental and medicinal plants were not only being traded, but also poached from nearby countries, threatening the very existence of certain species and the overall biodiversity of the region.

Green-Fingered Thieves

Under the dark of night, the thieves worked swiftly. Armed with shovels, they snuck past the closed gift shop and made a beeline for their quarry. They knew exactly what to take.

The next morning, the staff at the Quail Botanical Gardens in Encinitas, California, discovered the theft. In a frantic effort, they spread details of the heist to local newspapers, television, and online community message boards. Their quick action paid off; the loot became too hot. Within days, an anonymous tip led authorities to a rural road where the thieves

Figure 16.4

Cycads: ancient, rare, and hard to move

Several workers struggle to move a cycad. In spite of their size, many cycads have been stolen from public and private gardens, and then sold on the black market.

had dumped their haul in a sorry-looking pile: 21 stocky plants, specimens with thick trunks covered in woody scales and a crown of long, green palm-like leaves, like a pineapple.

The thieves had attempted to steal a group of rare African cycads, an ancient type of plant that once lived alongside dinosaurs in the Jurassic period (**Figure 16.4**). As with the Thai orchids, horticulturalists around the world highly prize cycads as ornamental plants. Today, a rare, mature cycad can fetch $20,000 or more on the international black market. That's right—enough to fund a year of college. There are about 300 species of cycads, most threatened with extinction.

Ironically, 4 of the 21 cycads stolen from the Quail Botanical Gardens (now the San Diego Botanic Garden) were part of a rescue program; the plants had been illegally brought into the United States, where they were seized upon import by authorities, and the botanical garden had taken over their care. Some of the plants survived the ordeal and were replanted. But they weren't the only cycads stolen that year; a nursery in San Diego lost almost 40 large cycads to thieves, and many homes in the Long Beach area had plants swiped right out of their front yards.

Plants are multicellular autotrophs that are mostly terrestrial (**Figure 16.5**). Like the green

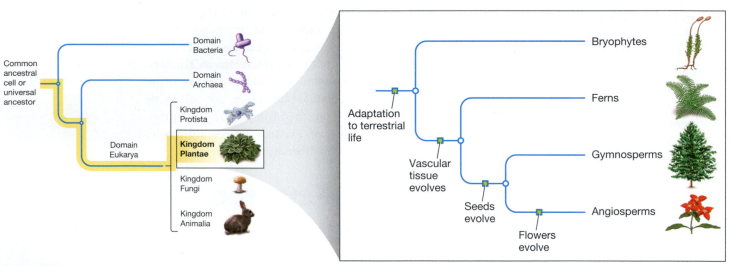

Figure 16.5

Plants are eukaryotic autotrophs

Plants are photosynthetic, making their own food from sunlight, and are almost exclusively found on land. They are at the base of almost all terrestrial food webs.

Q1: What evolutionary innovation separates all land plants from their aquatic ancestors?

Q2: How do ferns differ from bryophytes? Do they share this difference with other plant groups? (You will need to read ahead to answer this question.)

Q3: What group(s) might a plant with seeds belong to? What about a plant with flowers?

algae from which they evolved, plants use chloroplasts in order to photosynthesize. Most photosynthesis in plants takes place in their leaves, which typically have a broad, flat surface—a design that maximizes light interception. Because plants are producers, they form the basis of essentially all food webs on land. They may not be as cute, fuzzy, or exciting as animals, but animals wouldn't exist without them: nearly all organisms on land ultimately depend on plants for food, either directly by eating plants or indirectly by eating other organisms that eat plants.

In addition to being food, plants are valuable for many other reasons. Many organisms live on or in plants, or in soils largely made up of decomposed plants. By soaking up rainwater in their roots and other tissues, plants prevent runoff and erosion that can contaminate streams. Plants (and algae) also recycle carbon dioxide and produce the oxygen we breathe.

Many of these traits are adaptations to the challenges of living on land (**Figure 16.6**). The greatest of those challenges is how to obtain and conserve water. The first plants, the ancestors of

Algae (aquatic)

- Photosynthesis occurs and CO_2 is absorbed throughout the organism.

- Water and minerals are absorbed by the whole organism.

- Water supports the whole organism.

- Dehydration is not an issue in an aquatic environment.

Plants (terrestrial)

- Photosynthesis occurs and CO_2 is absorbed primarily in the leaves. Air enters leaf cells through minute openings in the cuticle. Many plants have more elaborate pores (*stomata*) that open and close to regulate the flow of gases into and out of the leaf.

- Roots absorb water and minerals from the soil.

- Roots anchor and support the plant within the ground. Lignin and vascular tissues help support the plant aboveground.

- The waxy cuticle holds in moisture to keep plant tissues from drying out, even when exposed to sunlight and air throughout the day.

Figure 16.6

Moving to land brought unique challenges to plants

Terrestrial plants evolved in response to challenges that their aquatic ancestors had not faced, which resulted in adaptations designed to slow dehydration, provide support and anchoring, and enable photosynthesis and nutrient uptake.

Q1: In what ways are terrestrial plants and their aquatic ancestors the same? Give at least two similarities.

Q2: In what ways do terrestrial plants and their aquatic ancestors differ? Give at least two differences.

Q3: Would you predict that aquatic plants (which have secondarily evolved to live in water) would be more like plants in a rainforest or more like desert plants? Explain your reasoning.

present-day **bryophytes** (liverworts and mosses), grew as ground-hugging carpets of greenery. These simple plants had relatively thin bodies, often just a few cells thick, and absorbed water through a wicking action. However, absorption by direct contact cannot transport fluids effectively in a plant that rises a foot or more aboveground, so more complex plants, including cycads, have a network of tissues, called the **vascular system**, that is made up of tubelike structures specialized for transporting fluids. Roots, found in most plants, also have an extensive vascular system. The first land plant with a vascular system was the ancestor of present-day **ferns**.

Like green algae from the protist kingdom, plant cells have strong but flexible cell walls composed of the substance known as **cellulose**, which gives structural strength to the cells, including those of the low-growing bryophytes. Beginning with the ferns, plants evolved yet another type of strengthening material, **lignin**, that allowed them to grow even taller. One of the strongest materials in nature, lignin links together cellulose fibers in the cell wall to create a rigid network. Lignin is the reason that some cycads in Japan can grow over 20 feet (7 meters) tall, though it takes them about 50–100 years to achieve such heights.

Figure 16.7

A flower attracts many pollinators

Pollination in the angiosperms is more efficient than in the wind-pollinated gymnosperms. Flowers attract animal pollinators that carry pollen to and from individual plants of the same species.

One reason rare cycads are so prized is that it can take a long time to raise and grow them. Cycads are **gymnosperms**, the first plants to evolve **pollen**, a microscopic structure containing sperm cells that can be lofted into the air in massive quantities. The evolution of pollen freed gymnosperms from a dependence on water for fertilization: instead of having to transport sperm cells through water, they could do it through air.

Gymnosperms were also the first plants to evolve the **seed**, which consists of the plant embryo and a supply of stored food, all encased in a protective covering. The embryo uses this stored food to grow until it is able to make its own food via photosynthesis. Seeds also provide embryos with protection from drying and from attack by predators. Unfortunately, they don't provide protection from poachers. Today, the San Diego Botanic Garden keeps not only its prized cycads, but also their valuable seeds, in a greenhouse under lock and key.

Searching for Flowers

Botanical gardens also go to extreme measures to keep flowering plants, or **angiosperms**, safe for long-term conservation. Orchids, as we have seen, are the most common illegally traded angiosperms. Others include *Galanthus*, or "snowdrop" plants, with delicate white flowers. In 2012, a single bulb of a rare variety of snowdrop fetched $945 at auction.

Compared to gymnosperms, angiosperms are a relatively recent development in the history of life. With about 250,000 species today, angiosperms are the most dominant and diverse group of plants on our planet. Nearly all agricultural crops are flowering plants. These plants also provide humans with materials such as cotton and pharmaceuticals.

Angiosperms' key evolutionary innovation is the flower, a structure that evolved through modification of the cone-like reproductive organs of gymnosperms. **Flowers** are structures that enhance sexual reproduction in angiosperms by bringing male gametes (sperm cells) to the female gametes (egg cells) in highly efficient ways, by attracting animal pollinators through scent, shape, and color (**Figure 16.7**).

Gymnosperms produce "naked" seeds that sit bare, unwrapped in any additional layers, on the modified leaf (the scale in a pine cone). In angiosperms, the modified leaf evolved into the ovary wall, which consists of tissue layers that enclose and protect the egg-bearing structures, or **ovules**. After fertilization, the ovules develop into seeds, and the ovary wall that enclosed them becomes the fruit wall.

At the Thailand plant markets, Phelps had trouble identifying orchid species until they flowered, so he went back again and again—four times per year to four different markets—looking for newly opened flowers, listing species he recognized, taking photos of others, and even occasionally asking for a flower off a plant he did not recognize, which he quickly stored in a vial of alcohol to take back for identification. In the end, Phelps gathered evidence of 348 orchid species in 93 genera, representing 13–22 percent of the area's known orchid flora, and tens of thousands of individual plants, including several new species.

Phelps's results were shockingly different from those published in preexisting government reports on plant trades among Thailand, Laos, and Myanmar. The Convention on International Trade in Endangered Species of Wild Fauna and Flora, or CITES, is an international treaty of over 175 countries that monitors and regulates international trade of plants and animals. Member countries are required to produce permits for wildlife trade of species protected under the agreement, including all wild orchids. These permits are used to guarantee that plants are legally harvested in a sustainable way that does not endanger either the species or the environment. "CITES is about conservation and sustainable use," says Anne St. John, a biologist with the U.S. Fish and Wildlife Service's Division of Management Authority, which implements CITES in the United States. "The goal is to ensure that these species are around for our grandchildren. That includes not only tigers and elephants, but also bigleaf mahogany and Brazilian rosewood."

And the orchids of Southeast Asia. Over nine years, Laos reported permits for the export of just 20 wild-collected orchids into Thailand; Myanmar reported none. Yet during just one day with a single market trader at the border between Laos and Thailand, Phelps documented that the woman sold at least 168 plants of eight different genera. In just one day, she sold eight times more plants than the government reported as sold over

9 years. "It's totally anecdotal, but incredibly illustrative of the problem," says Phelps. "This trade is completely unacknowledged. It's an open secret."

Other countries are working hard to crack down on the plant black market. American ginseng, a short leafy plant with a tan, gnarled root commonly used in Chinese herbal medicine, is the largest CITES-regulated plant export of the United States (**Figure 16.8**). A pound of quality, dried ginseng can sell for up to $900, so some people try to bypass CITES permits, harvesting plants that are too young (legally

Figure 16.8

Ginseng plants carpet a forest floor

After pollination, the ginseng flowers develop a bright-red seed head that helps ginseng hunters (inset) find it in the filtered light of the forest.

Q1: What feature(s) of the ginseng plant tell you that it is not a bryophyte?

Q2: What feature(s) of the ginseng plant tell you that it is not a fern or gymnosperm?

Q3: Because of the CITES classification of ginseng, you are not allowed to sell plants younger than 5 years even if they grew on your own land. Do you agree with that law? Why or why not?

traded roots must be 5 years or older), out of season, or poached from federal lands.

"As you would expect, species that are high value and in high demand are the subject of illegal trade," says St. John. The illegal trade of ginseng has even spawned a reality show on the History television channel: *Appalachian Outlaws*. To combat the criminal activity, some states have taken to spraying ginseng plants on state and federal lands with paint, so if a marked plant shows up in a batch of ginseng to be exported to China, authorities will know where it came from. Those caught selling protected plants face fines and jail time, and violating CITES is a federal crime.

Truffle Trouble

The illegal wildlife trade is a problem not just for animals and plants, but also for **fungi**, none of which have been listed as CITES-protected species so far. Fungi are absorptive heterotrophs:

they digest organic material outside the body and absorb the breakdown products. The majority of fungal species fall into three main groups: **zygomycetes**, which contains many species of molds; **ascomycetes**, a diverse group informally known as sac fungi; and the more familiar **basidiomycetes**, or club fungi (**Figure 16.9**). Each of these groups differs in—and is named for—its unique reproductive structures.

Fungi have properties in common with both plants and animals. Like plant cells, all fungal cells have a protective cell wall that wraps around the plasma membrane and encases the cell. However, fungi are similar to animals in that they store surplus food energy in the form of glycogen.

In 2012, two bandits broke through security gates and stole an estimated $60,000 of fungi from a locked warehouse. The loot was truffles, the fruiting body of a particular group of ascomycetes. Certain white European truffles can sell for as much as $3,600 per pound, making them the most expensive food in the world. That high price also makes them a target for thieves, especially since

Figure 16.9

Fungi are eukaryotic absorptive heterotrophs

Fungi must take their food from other organisms, and they do this in a unique way. Instead of ingesting their food, as most heterotrophs do, fungi use chemicals to break down their food outside of their bodies and then absorb the nutrients.

Q1: What group of fungi most resembles the mushrooms you buy in a grocery store?

Q2: Are sac fungi more closely related to molds or to club fungi?

Q3: How do we know that fungi are eukaryotes rather than prokaryotes?

Figure 16.10

A truffle hunter and his dog search for a fungus beloved by gourmets

While farmers continue attempts to grow truffles commercially, their success has been limited. For now, we must rely on truffle hunters and their dogs (or pigs, which can also be trained to sniff out truffles) for this fungal delicacy.

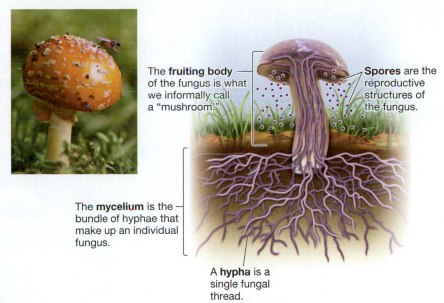

The **fruiting body** of the fungus is what we informally call a "mushroom."

Spores are the reproductive structures of the fungus.

The **mycelium** is the bundle of hyphae that make up an individual fungus.

A **hypha** is a single fungal thread.

Figure 16.11

A fungus lives both belowground and aboveground

The main body of a fungus is unseen, belowground. To reproduce, the fungus generates a fruiting body that is usually aboveground and releases spores. Spores travel by wind and develop into new fungi.

Q1: Why is it important that the fruiting body is aboveground?

Q2: What part of a fungus is the mushroom that you can buy in the grocery store?

Q3: Write a sentence in your own words that uses the terms "mycelium," "fruiting body," and "spore" correctly.

white truffles cannot be cultivated in greenhouses as other fungi can.

There's danger in the forests too. Truffle hunters use dogs and pigs to sniff out wild truffles (**Figure 16.10**), and some have reported competitors planting spiked traps or poisoned meatballs in forests to eliminate trained dogs. Other hounds have been stolen from their owners, never to be seen again.

That's a lot of drama for some fungi. Most fungi are multicellular, but there are some single-celled species collectively known as **yeasts**. Many of us are familiar with yeasts thanks to two important products they produce: alcohol and carbon dioxide, crucial to the rising of bread, the brewing of beer, and the fermenting of wine.

Fungi's key evolutionary innovation is their body form, which is well suited for absorptive heterotrophy (**Figure 16.11**). They are made up of a network of fine, colorless, branching, hair-like threads called **hyphae** (singular "hypha"), which absorb nutrients from the environment. The entire bundle of hyphae, composing the main body of the fungus, is called the **mycelium** (plural "mycelia").

Because of this unique body form, many fungi are decomposers and consume nonliving organic material. As they eat, these fungi release back into the environment inorganic chemicals that had previously been trapped in the bodies of dead organisms. Once these chemicals are back in the environment, plants and algae scoop them up and use them to manufacture food. Fungi interact with plants in other important ways too: the vast majority of wild plants have mutualistic fungi in their root systems that help the plants

Food Banks

Over millennia, farmers have saved seeds from crops that were easiest to grow, process, or store, and have used those seeds and their offspring year after year. The results of this domestication process are staple crops we rely on today: corn, soybeans, wheat. Yet to feed a growing human population, we're going to have to look beyond what we're familiar with, to the smorgasbord of edible plants that pepper the planet.

Assessment available in smartwork5

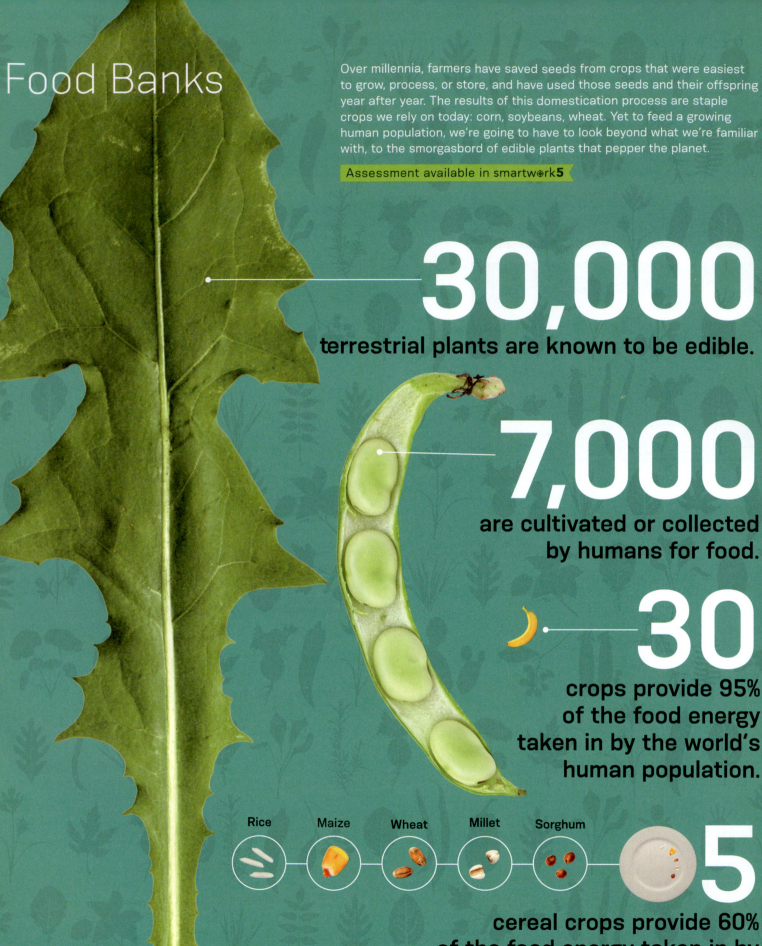

30,000 terrestrial plants are known to be edible.

7,000 are cultivated or collected by humans for food.

30 crops provide 95% of the food energy taken in by the world's human population.

Rice   Maize   Wheat   Millet   Sorghum

5 cereal crops provide 60% of the food energy taken in by the world's human population.

absorb nutrients from the soil (see "Fungi Play Well with Others" on page 287).

Despite that assistance, fungi and plants are not always close friends. Fungi are the most significant parasite of plants: they are responsible for two-thirds of all plant diseases, causing more crop damage than bacteria, viruses, and insect pests combined.

Like many plants (and animals and protists), fungi can reproduce both asexually and sexually. Some species appear to multiply only asexually, and most multicellular fungi can reproduce asexually through fragmentation—that is, by simply breaking off from the mother colony. When fungi do reproduce sexually, they do not have distinct male and female individuals. Instead, a sexually reproducing mycelium belongs to one of two (or more!) mating types. Each mating type can mate successfully with only one of the other types. After mating, a **fruiting body** is formed that may be large enough to be readily observed.

Fungal fruiting bodies release offspring as sexual spores. A **spore** is a reproductive structure that can survive for long periods of time in a dormant state and will sprout under favorable conditions to produce the body of the organism. Spores released from a fruiting body that is raised up in the air are better able to catch a ride on wind currents or to attract animals that can carry them far and wide. Many fungi also produce asexual spores.

Fighting for the Future

The illegal sale of plants and fungi could drive species to extinction. In 2015, for example, the International Union for Conservation of Nature (IUCN), a global environmental authority, announced that 31 percent of cactus species—renowned for their unique forms and beautiful flowers—are in danger of extinction, and that the greatest threat to these plants is illegal trade (**Figure 16.12**).

During his studies, Phelps observed traders selling plants that are highly endangered in the wild. He published results from his thesis in 2015 in the journal *Biological Conservation*, pleading with scientific and policy communities for a greater focus on these plants and calling

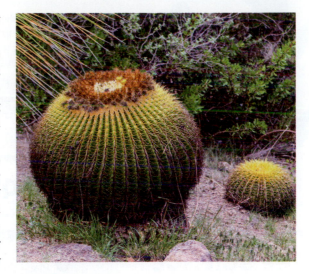

Figure 16.12

The greatest threat to endangered cacti is illegal trade

Like orchids and cycads, the main threat to most species of cacti is illegal trade. Collectors pay high prices to smugglers who are willing to risk fines and prison time to meet the demand.

the illegal plant trade "a major conservation challenge that has been almost completely overlooked."

"It's hard to get people to care about plants, and nobody can care if we don't have data to show it's a problem," says Phelps, who is now trying to engage governments and experts to raise awareness and initiate policies that will better regulate plant trade. At the CITES conference in September 2016, country delegates increased protections for a large number of plant species, including ponytail palms, which are in demand as houseplants, and all rosewoods, whose wood is used to manufacture fine musical instruments.

"All these plants are part of a larger ecosystem," says St. John. "They serve important purposes in temperature regulation, in creating oxygen, and more. We need to recognize that plants are an integral part of a healthy ecosystem and key to the diversity of life."

ANNE ST. JOHN

Anne St. John is a biologist with the Division of Management Authority for the U.S. Fish and Wildlife Service.

and **basidiomycetes**—and each group is characterized by distinctive reproductive structures (fruiting bodies).

- Most plant roots in natural habitats form close associations with beneficial fungi, called **mycorrhizal fungi**. A **lichen** is a mutually beneficial association between a fungus and a photosynthetic microbe, usually a green alga or a cyanobacterium.

REVIEWING THE SCIENCE

- The domain Eukarya is traditionally divided into four major kingdoms: Protista, Plantae, Fungi, and Animalia. Eukaryotes possess a true nucleus; they have complex subcellular compartments, which enable larger cell size. Sexual reproduction and multicellularity are key evolutionary innovations of the Eukarya.

- Kingdom Protista, the **protists**, lumps together many evolutionarily distinct lineages under one highly diverse grouping. Most protists are single-celled and microscopic.

- Protists are traditionally divided into two categories: **protozoans**, which are nonphotosynthetic and motile; and **algae**, which are photosynthetic and may or may not be motile.

- **Plants** are descended from green algae and have evolved numerous evolutionary innovations to adapt to life on land. All plants photosynthesize and use **cellulose** to strengthen their cell walls.

- The first plants, the ancestors of present-day **bryophytes** (liverworts and mosses), had relatively thin bodies and absorbed water through a wicking action.

- **Ferns** were the earliest plant group to grow larger and taller than the bryophytes. This extended growth required both cellulose and **lignin** to provide structure to plant bodies, as well as a network of fluid-transporting tissues called the **vascular system**.

- **Pollen** and **seeds** first evolved among the **gymnosperms**, the plant group that includes cycads.

- **Angiosperms** evolved **flowers**, and they enclose their seeds in the fruit. Many angiosperms recruit animals to deliver pollen and also to disperse their seeds.

- The **fungi** are distinguished by their mode of nutrition: they acquire their nutrients by absorption, digesting their food outside of their bodies. To accomplish this, their bodies consist of a **mycelium** composed of hairlike threads called **hyphae**.

- Fungal reproduction may be sexual or asexual. During reproduction, a **fruiting body** is formed that releases **spores** into the environment. Much as seeds do, these spores then develop into new individuals.

- There are at least three main groups of fungi— **zygomycetes**, **ascomycetes**,

THE QUESTIONS

Challenge Yourself

1 Place the following adaptations in the correct order from earliest to most recent by numbering them from 1 to 5.

_____ a. vascular tissue

_____ b. flowers

_____ c. seeds

_____ d. multicellularity

_____ e. movement to land

2 Which of the following groups contains only multicellular species?

(a) algae

(b) protists

(c) eukaryotes

(d) fungi

(e) angiosperms

3 Of the kingdoms covered in this chapter, which, if any, are composed of only autotrophic species? Which, if any, have only heterotrophs?

Try Something New

4 Symbiosis is a long-term and intimate association between two different types of organisms. A symbiotic organism may live on or inside another species. For each of the following symbiotic relationships, (1) define the relationship in one to three sentences, (2) identify the domain—and for Eukarya, the kingdom—of each partner in the relationship, and (3) discuss whether the relationship evolved as a mutualism (both benefit), commensalism (one benefits, while the other is not affected), or parasitism (one benefits to the detriment of the other). You may need to read more than your textbook to answer this question.

a. mycorrhizae

b. lichens

c. hermit crabs/shells

d. malaria

5 Which plant groups produce pollen, and what is the adaptive value of pollen?

6 Plants are not the only organisms susceptible to fungal infections. Give an example of a fungal infection to which humans are susceptible.

Leveling Up

7 **Life choices** If fungi are more closely related to animals than to plants, should vegetarians and vegans refrain from eating fungi? Why or why not?

8 *Write Now* **biology: is a mass extinction under way?** The International Union for Conservation of Nature maintains what it calls its Red List, which identifies the world's threatened species. To be defined as such, a species must face a high to extremely high risk of extinction in the wild. The 2016 Red List contains almost 25,000 species threatened with extinction out of the approximately 79,000 species assessed.

Because this assessment accounts for only a small percentage of the world's 1.7 million described species, the total number of species threatened with extinction worldwide may actually be much larger. For example, only 6,051 of 1 million described insect species have been assessed in terms of their survival risk.

The Red List is based on an easy-to-understand system for categorizing extinction risk. It is also objective, yielding consistent results when used by different people. These two attributes have earned the Red List international recognition as an effective tool for assessing extinction risk.

a. If the threatened species listed by the IUCN do become extinct and the percentages of species under threat in other taxonomic groups turn out to be similar to those listed, then the percentages of species that will go extinct will approach the proportions lost in some of the previous mass extinctions. Does this mean that a mass extinction is under way? Why or why not? Explain your reasoning, using information about mass extinctions from Chapter 14 (in particular, Figure 14.15) to support your argument.

b. What are the causes of the current high extinction rates? Compare the causes of the current situation with those of previous mass extinction events. In what ways are they similar, and in what ways do they differ?

c. Why are some groups in more danger of extinction than others? Choose one group that has a large proportion of species threatened with extinction, and find out more about it (you can begin at www.iucnredlist.org):
   - What kinds of habitat are they found in?
   - Does a particular aspect of their ecology—for example, feeding or reproduction—make them more vulnerable to extinction?
   - Is anything being done to protect them and/or their habitat?
   - Has there been an increase or decrease in the number of species within the group that have been identified as vulnerable to extinction?

Thinking Multidimensionally

9 You've probably heard of the stereotypical cat-loving, eccentric woman dubbed a crazy cat lady, although you may not know that there is actually a "crazy cat lady syndrome" called toxoplasmosis. Caused by the parasite *Toxoplasma gondii*, infected individuals may exhibit flu-like symptoms or more severe reactions that impact the brain. It is estimated that 500 million people worldwide are infected but most are asymptomatic.

a. Aside from eating raw meat containing cysts of the parasite, how can a cat become infected?

b. How does the cat continue the cycle and spread the parasite to humans or other mammals?

c. Rodents infected with the parasite become more reckless, are attracted to feline urine, and even lose their innate fear of cats. How do these behaviors impact the cycle of the parasite?

d. In some countries, cats are consumed by humans as food. Would you expect higher instances of toxoplasmosis in those nations? Why or why not?

e. Investigate why pregnant women especially are advised to not change cat litter boxes.

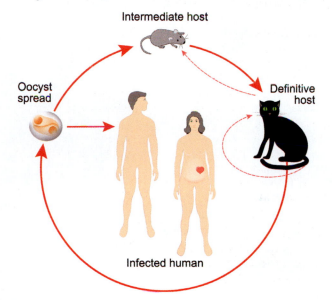

Intermediate host

Oocyst spread

Definitive host

Infected human

10 In 1692, two young girls, ages 9 and 11 in Salem, Massachusetts, began having violent muscle spasms, body contortions, outbursts of uncontrollable screaming, and hallucinations. A local doctor diagnosed them with bewitchment, and soon afterward other young girls in the community began experiencing similar symptoms. In the end, trials of the accused witches resulted in the deaths of 19 people, mostly by hanging.

a. Compare how the strange symptoms reported by the Salem accusers resemble the effects of a person on LSD, a drug that is manufactured from an ergot fungus that grows on grains.

b. Research the environmental conditions necessary to grow ergot fungus, and comment on the conditions and staple food crops in Salem during that time period.

c. In your opinion, is the evidence connecting the 1692–93 events of Salem to a fungus substantiated, or do you think other factors were involved? Explain your reasoning.

For more, visit digital.wwnorton.com/bionow2 for access to:

Neanderthal Sex

The relationship between modern humans and Neanderthals just got a whole lot spicier.

After reading this chapter you should be able to:

- Identify the key characteristics of animals, and give an example organism for each group within the chordates.
- Explain the significance of symmetry and segmentation for some animal groups.
- Compare and contrast the three types of mammals.
- Interpret an evolutionary tree of the primates.
- Illustrate the inheritance of mitochondrial DNA as compared to nuclear DNA.
- Describe the evidence suggesting that prehistoric hominin species reproduced with our direct ancestors.

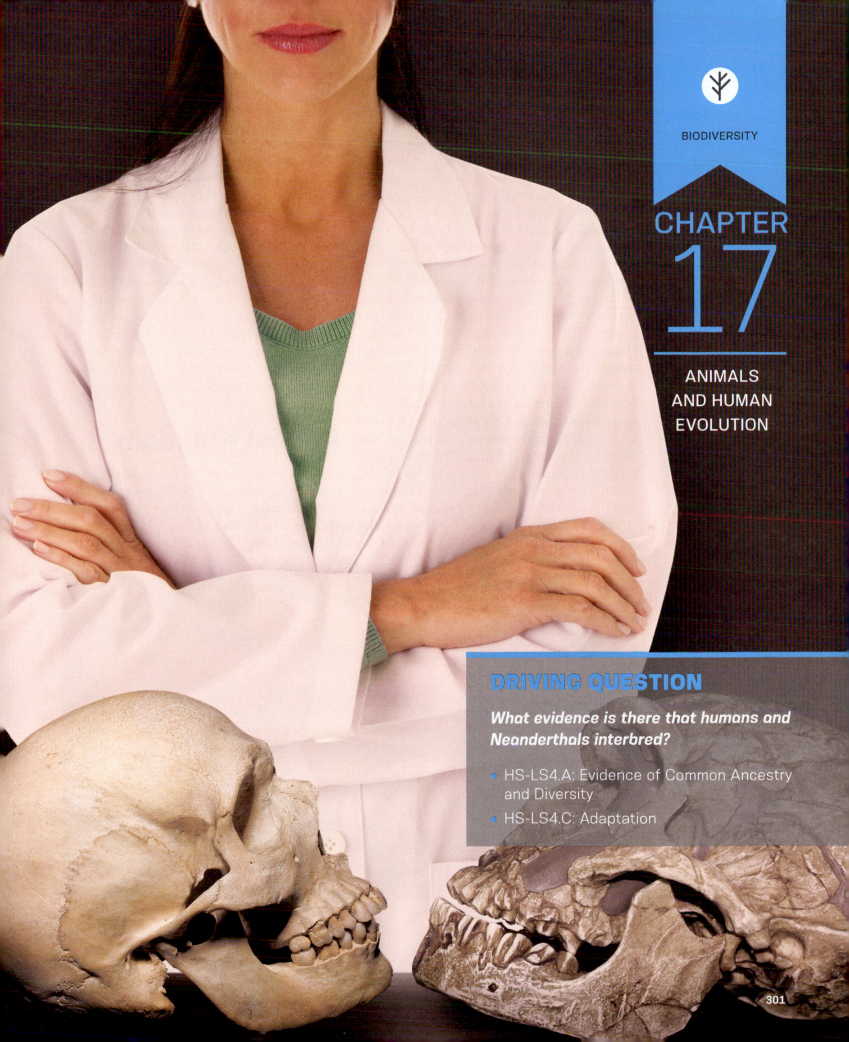

CHAPTER
17

ANIMALS AND HUMAN EVOLUTION

DRIVING QUESTION

What evidence is there that humans and Neanderthals interbred?

- HS-LS4.A: Evidence of Common Ancestry and Diversity
- HS-LS4.C: Adaptation

In popular culture, Neanderthals are ugly, hairy cavemen with big brains but no wits. The first Neanderthal bones were discovered in Germany in 1856, and since then we've cultivated an image of our closest extinct human relatives as hulking brutes who communicated by grunting, walked like chimps, and hit each other over the head with clubs. Yet, over the years, paleontologists have dug up fossilized vocal bones, sophisticated tools, and other evidence suggesting that Neanderthals were a fairly advanced group—and not as different from our own species as we once believed.

The big revelation came in 2010, when scientists sequenced the Neanderthal genome and compared it to the modern human genome: Humans have some Neanderthal DNA. Our two species may have been a whole lot closer than we thought. "There could have been interbreeding between modern humans and Neanderthals," says Silvana Condemi, a researcher at the National Center for Scientific Research (CNRS) at Aix-Marseille University in France. "We can imagine they not only exchanged culture, but exchanged genes." That's right: mounting evidence suggests that Neanderthals and modern humans had sex.

This story begins where many stories end—with a pile of bones. In the early years of the twenty-first century, Laura Longo, a curator at the Civic Natural History Museum of Verona, Italy, decided to take a second look at a group of fossils excavated from a rock shelter called Riparo Mezzena. Riparo Mezzena is nestled in the Lessini Mountains in northern Italy, a wide-open landscape speckled with large rocks and evergreen trees. The region is snowy and silent in the winter but green and thriving in the summer, when paleontologists come to work.

In the 1950s, paleontologists had carefully collected fossils of Neanderthals that lived around Riparo Mezzena about 35,000 years ago, late in the history of Neanderthals. But the fossils had sat at the museum, untouched, for over 50 years (**Figure 17.1**). Now, Longo believed, the fossils could help answer a hotly debated question about human history: How closely did modern humans and Neanderthals interact?

Animal Kingdom

Modern humans and Neanderthals lived in some of the same areas of Europe at the same times. Because of this proximity, some paleontologists hypothesize that as modern humans expanded their territory, Neanderthals were

Figure 17.1

Fossil remains

A jawbone like this one was found at the Riparo Mezzena rock shelter in the Lessini Mountains in Italy. The individual it belonged to lived between 40,000 and 30,000 years ago.

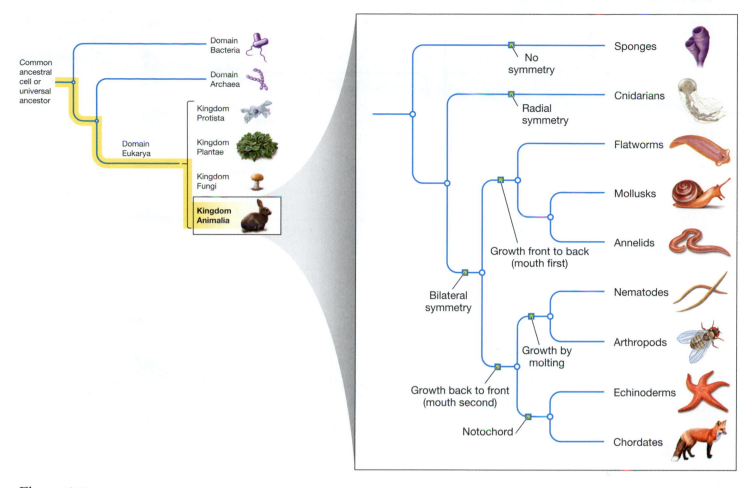

Figure 17.2

Animals are heterotrophic eukaryotes

Animals ingest their food, which can be organisms from any of the other domains and kingdoms, and then digest it internally.

Q1: Are mollusks more closely related to flatworms or to annelids? Explain your reasoning.

Q2: If you found an animal with no symmetry, to what group would it belong? Give an example of an animal that is radially symmetrical, and one that is bilaterally symmetrical.

Q3: How do we know that a sponge is an animal, and not a plant or an alga?

quickly driven to extinction, and therefore the two did not live side by side. Others claim the opposite: that Neanderthals, *Homo neanderthalensis*, were slowly incorporated into the population of newly incoming humans, *Homo sapiens*. But to know how humans evolved, it is important first to realize where we came from.

All species in the *Homo* genus, of which *H. sapiens* is the only one alive today, are in the Animalia kingdom. **Animals** are multicellular ingestive heterotrophs; that is, we obtain energy and carbon by ingesting food into our bodies and digesting it internally. Animals first evolved

some 700 million years ago, descended from flagellated protists, single-celled organisms with whiplike tails that thrived in wet habitats and began to use oxygen to break down food into energy. From those humble origins, countless animal species evolved—beginning with sponges and including mollusks such as snails and clams, the annelids (segmented worms), and arthropods like crustaceans, spiders, and insects (**Figure 17.2**).

Chordates make up a large phylum that includes all animals with backbones, such as fishes, birds, and mammals (see **Figure 17.3** and

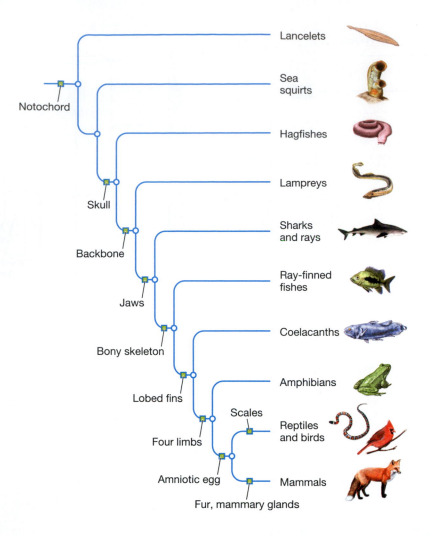

Figure 17.3

Not all chordates have a backbone

As with all eukaryotes except the protists, the evolutionary relatedness of the chordates is very well known. Not all chordates have a backbone, but all have a notochord, which is the precursor to part of the vertebrate backbone.

Q1: Do amphibians have amniotic eggs?

Q2: What group of animals has jaws but not a bony skeleton?

Q3: When people talk about animals, they are sometimes referring only to mammals. How would you explain to them their error?

"Get a Backbone!" on page 305). The phylum also includes several subgroups of less familiar animals, such as sea squirts and lancelets, that have a nerve cord along the back of the body but no backbone. Chordates that do possess a backbone are known as **vertebrates**. All other phyla of animals are informally lumped together as invertebrates. But keep in mind that "invertebrates" is not an evolutionarily meaningful label; it is an artificial grouping (like that of protists) of animals with varied evolutionary histories and different sets of evolutionary adaptations.

Body plan is another key factor that distinguishes one group of animals from another. All animals except sponges, the most ancient of the animal lineages, have a distinct body symmetry. The asymmetric sponges aside, animals can be divided into two main groups: those with radial

Get a Backbone!

Within the kingdom Animalia, humans belong to the phylum of chordates and the subphylum of vertebrates. All chordates have a dorsal **notochord**, a flexible rod along the center of the body that is critical for development. In vertebrates, the dorsal notochord has evolved to become the cushioning discs between **vertebrae** (singular "vertebra"), which are strong, hollow sections of the backbone, or vertebral column. Vertebrates include fishes, amphibians (frogs and salamanders), reptiles (snakes, lizards, turtles, and crocodiles), birds, and mammals.

The jawless fishes were the first vertebrates to evolve. Their skeletons—including the backbone—were made from a strong but flexible tissue called **cartilage**. Only a few groups of jawless fishes have survived to the present day, most notably the lampreys. The next great leap in vertebrate evolution was hinged jaws, which enabled predators to grab and swallow prey efficiently. The evolution of teeth made jaws even more effective because teeth enabled animals to seize and tear food.

Another major step in the evolution of vertebrates was the replacement of the cartilage-based skeleton with a denser tissue strengthened by calcium salts: bone. Although the descendants of cartilaginous fishes—sharks, skates, and rays—are still with us today, bony fishes are far more diversified and widespread in both marine and freshwater environments. With more than 30,000 species, bony fishes are the most diverse vertebrates today.

The advent of lungs was a crucial milestone in the transition of vertebrates onto land. Amphibians made this transition only partially; they can live on land but must return to the water to lay eggs and breed. The several thousand species of amphibians include frogs and salamanders.

Reptiles were the first vertebrates to head into drier environments, and they evolved a number of adaptive traits to deal with the risk of dehydration. These adaptations included skin covered in waterproof scales, a water-conserving excretory system, and the amniotic egg with its calcium-rich protective shell, which retards moisture loss while allowing the entry of life-giving oxygen and the release of waste carbon dioxide for the developing embryo. Reptiles dominated Earth during the age of the dinosaurs, and the dinosaurs' descendants (as we saw in Chapter 14) remain with us today in the form of birds. Like mammals, birds are warm-blooded, but they have feathers instead of fur for insulation. At least 10,000 different species of birds are living today.

symmetry and those with bilateral symmetry. The body of an animal with **radial symmetry** (**Figure 17.4**, left)—including cnidarians like jellyfish, sea anemones, and corals—can be sliced symmetrically along any number of vertical planes that pass through the center of the animal, like cutting a pie. Radial symmetry gives an animal sweeping, 360-degree access to its environment. The animal can snare food drifting in from any direction of the compass, and it can also sense and respond to danger from any side.

Animals with **bilateral symmetry** (**Figure 17.4**, right), on the other hand, can be divided by just one plane passing vertically from the top to the bottom of the animal into two halves that mirror each other. Bilateral animals, which include all chordates, have distinct right and left sides, with near-identical body parts on each side. The symmetrical arrangement of body parts on either side of a central body facilitates movement in bilateral animals. The paired arrangement of limbs or fins, for example, enables quick and efficient movement on land

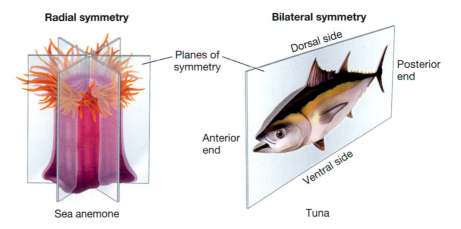

Figure 17.4

Body symmetry in animals

All animals other than sponges have symmetrical bodies that enable them to better sense and respond to the world around them.

Q1: Is a sea star radially or bilaterally symmetrical?

Q2: What advantage might a bilaterally symmetrical animal have over one that is radially symmetrical, and vice versa?

Q3: What kind of symmetry do you (a human) have?

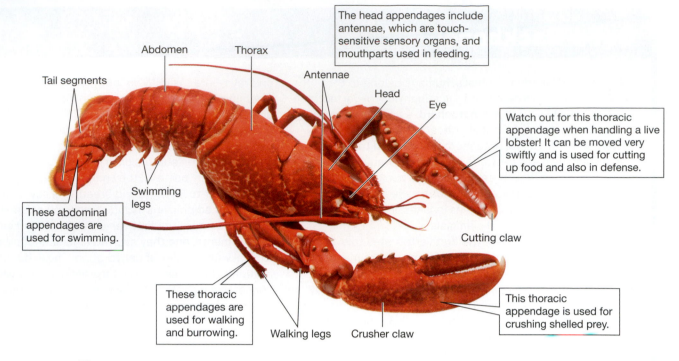

Tail segments

Abdomen

Thorax

Antennae

Head

Eye

The head appendages include antennae, which are touch-sensitive sensory organs, and mouthparts used in feeding.

Watch out for this thoracic appendage when handling a live lobster! It can be moved very swiftly and is used for cutting up food and also in defense.

Cutting claw

Swimming legs

These abdominal appendages are used for swimming.

These thoracic appendages are used for walking and burrowing.

Walking legs

Crusher claw

This thoracic appendage is used for crushing shelled prey.

Figure 17.5

Segmentation in animals

Segmentation, a body plan in which segments repeat, enabled the evolution of diverse uses of appendages, shown here in a lobster.

Q1: List all of the lobster's thoracic appendages.

Q2: Which of the lobster's appendages is most important for sensing the environment?

Q3: What body segments do you (a human) have?

or in water. Locomotion is a key evolutionary innovation of animals and one that has sparked a wide range of behaviors, including varied ways of capturing prey, eating prey, avoiding being captured, attracting mates, caring for young, and migrating to new habitats.

Many animals have segmented bodies; that is, their body plan consists of repeated units known as segments (**Figure 17.5**). Specialized body parts, known as appendages, often originate in pairs from specific segments of the body, such as fins or limbs. Over evolutionary time, the segments and the appendages that spring from them have evolved diverse form and function, enabling the animal body to adapt to new habitats or acquire new modes of life.

The evolution of just the posterior segments of arthropods illustrates how evolution can take a basic body plan and modify it to produce many variations over time: The last segment in the body of arthropods has evolved into the delicate abdomen of the butterfly, the piercing abdomen of the wasp, and the delicious tail of the lobster. The front appendage of vertebrates has evolved as an arm in humans, a wing in birds, a flipper in whales, an almost nonexistent nub in snakes, and a front leg in salamanders and lizards.

Mammals R Us

To get back to our egocentric focus—that is, humans—we are part of the kingdom Animalia and the class Mammalia. Humans share specific features with all other mammals, including body hair, sweat glands, and milk produced by mammary glands. And at the risk of tooting our own horn, **mammals** have been a highly successful class of animals, with over 5,000 species living in a variety of habitats. This is largely thanks to

Eutherian

Marsupial

Monotreme

Figure 17.6

Three kinds of mammals

All mammals have milk-producing mammary glands for suckling their young. Eutherians like the polar bear give birth to well-developed young. Kangaroos are marsupials, and give birth to immature young that finish developing in a pouch. Monotremes like the platypus lay eggs instead of giving birth to live young.

Q1: The Virginia opossum, or possum, is the only North American marsupial. Describe how its young are born and develop.

Q2: If you were a monotreme, would you still produce milk and nurse your young?

Q3: What kind of mammal is a cow? How about a human?

the extinction of dinosaurs; if dinosaurs still roamed Earth, it is likely that mammals would be nothing but dinner. Lucky for us, mammals have replaced dinosaurs as the top predators in most terrestrial habitats, and they thrive in both marine and freshwater environments. Only one type of mammal can fly—bats—although a few others can glide through the air.

Mammals can be divided into three broad categories, all of which feed their offspring with milk (**Figure 17.6**). More than 95 percent of mammals alive today are **eutherians**, including humans. A unifying characteristic of eutherians is that the offspring are nourished inside the mother's body through a special organ called the *placenta* and are therefore born in a relatively well-developed state. **Marsupials**, the second category of mammals, have a simple placenta, resulting in offspring born early, which then complete development in an external pocket or pouch. Marsupials are found mainly in Australia and New Zealand, with a few species in the Americas.

Making up the third category, **monotremes** are egg-laying mammals that lack a placenta altogether. The only living species of monotremes are just one platypus species and several echidna species, all confined to Australia and New Guinea.

Moving now from class to order, humans and Neanderthals are in the order of the **primates** (**Figure 17.7**). Like all other primates, we have flexible shoulder and elbow joints, five functional fingers and toes, thumbs that are **opposable** (that is, they can be placed opposite each of the other four fingers), flat nails (instead of claws), and brains that are large in relation to our body size.

Within the primates we are members of the ape family, the **hominids**. We are not just closely related to apes; we *are* apes. As such, we share many characteristics with other apes, especially chimpanzees, including the use of tools, a capacity for symbolic language, and the performance of deliberate acts of deception. But we are part of a distinct branch of apes called **hominins**—the "human" branch of the ape family that includes

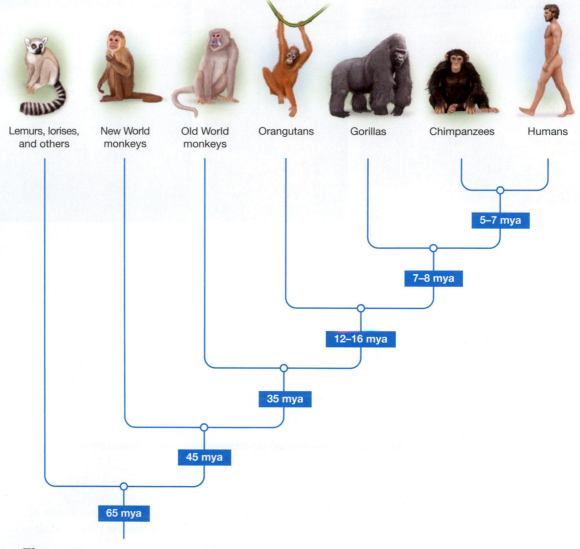

Figure 17.7

The primates include lemurs, monkeys, and apes

Genetic analyses and a series of spectacular fossil discoveries have led scientists to propose that the human lineage diverged from that of chimpanzees about 5–7 million years ago (mya). Similar evidence suggests that the evolutionary lineage leading to humans diverged from the lineage leading to gorillas about 7–8 mya, and from the lineage leading to orangutans about 12–16 mya.

Q1: According to this evolutionary tree, which primate group is most closely related to humans?

Q2: According to this evolutionary tree, which primate group is most distantly related to humans?

Q3: What characteristics are common to all the primates, including humans?

our extinct relatives, such as Neanderthals. The members of the hominin lineage have one or more humanlike features—for example, thick tooth enamel or upright posture—that set them apart from the other apes, like gorillas and chimpanzees.

Rise of the Apes

A major step in hominin evolution, and the main feature that distinguishes hominins from other hominids, was the shift from moving on four legs to being **bipedal**, walking upright on

Hereditary Heirlooms

Because of a shared common ancestor 1.6 billion years ago, humans share DNA with all animals, plants, and fungi. But how much? Take a look to see how much genetic material you have in common with other organisms.

Percentage of genes shared with humans

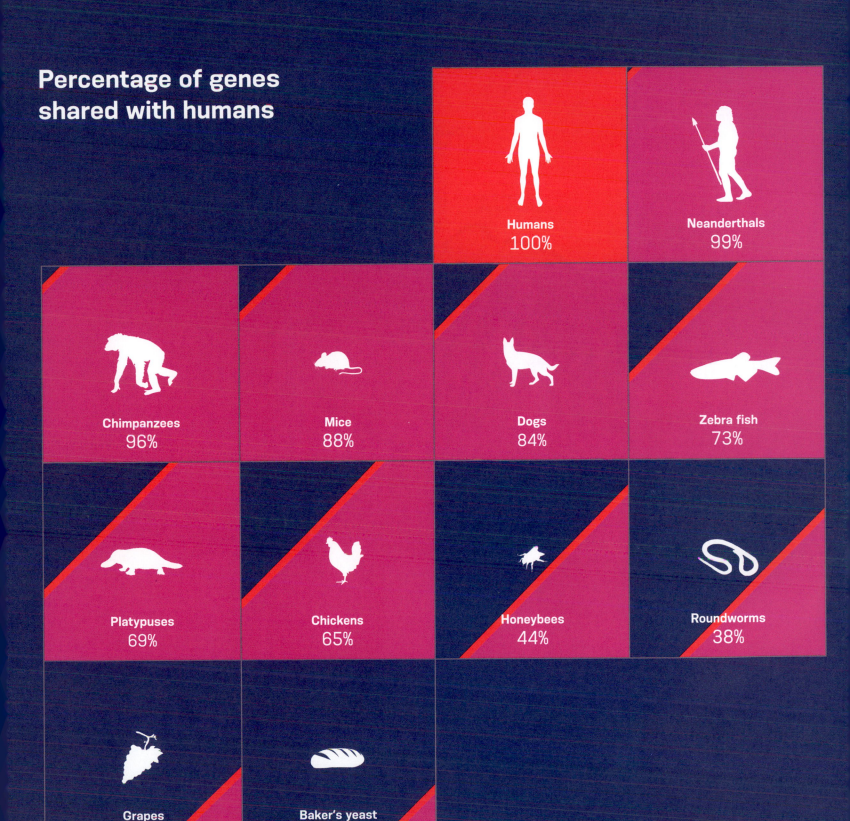

Humans
100%

Neanderthals
99%

Chimpanzees
96%

Mice
88%

Dogs
84%

Zebra fish
73%

Platypuses
69%

Chickens
65%

Honeybees
44%

Roundworms
38%

Grapes
24%

Baker's yeast
18%

Other apes

The skull and its relation to the spinal cord

The face is pulled forward.

The spinal cord enters from the rear of the skull.

The pelvis—frontal view

Shifting weight support

The angle of weight support in other apes...

Structure of foot and function of big toe

The big toe is opposable.

There is no arch.

Humans

The face is relatively flat.

The spinal cord enters from the center of the skull.

The hip bones in humans allow an upright posture.

...differs from that in humans.

The big toe is not opposable.

The arch and heel are pronounced.

Overall skeletal structure

Other apes have a posture supported by four limbs.

Humans have an upright posture.

Pelvis

Figure 17.8

Evolutionary differences between humans and other apes

The switch to walking upright required a drastic reorganization of primate anatomy, especially of the hip bones. ▶▶

two legs (**Figure 17.8**). Many skeletal changes accompanied the switch to walking upright, including the loss of opposable toes, as you will notice if you try touching your little toe with the big toe on the same foot.

The loss of opposable toes that accompanied walking upright would have been a handicap in trees, since opposable toes help grasp branches during climbing. It is therefore likely that bipedalism was an adaptation for living on the ground. Walking on two feet freed the hands to carry food, tools, and weapons, and it also elevated the head, enabling the walker to see farther and over more things.

The shift to life on the ground was probably not sudden or complete. The skeletal structure of some of the oldest fossil hominins (3–3.5 million years old) indicates that they walked upright. However, foot bones and fossilized footprints show that the hominins living at that time still had partially opposable big toes (**Figure 17.9**). Perhaps they still occasionally climbed trees.

The earliest known hominin is *Sahelanthropus tchadensis*, identified from a 6- to 7-million-year-old skull discovered in 2002. Other early hominins include *Ardipithecus ramidus*, who lived 4.4 million years ago, and several *Australopithecus* species that are 3–4.2 million years old, including the first full-time walker with the first modern foot: *Australopithecus afarensis*. All of these hominins are thought to have walked upright. Their brains were still relatively small (less than 400 cubic centimeters in volume), and their skulls and teeth were more similar to those of other apes than to those of humans (**Figure 17.10**). A typical modern human has a brain volume of about 1,400 cubic centimeters, about the same volume as a 1.5-liter soda bottle.

Within the hominin branch is the *Homo* genus. The fossils identified at Riparo Mezzena were believed to be *Homo neanderthalensis* bones, but they had never been closely studied. So, Longo, the curator from Verona, assembled a team of

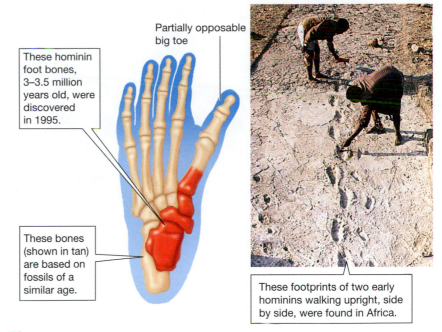

Partially opposable big toe

These hominin foot bones, 3–3.5 million years old, were discovered in 1995.

These bones (shown in tan) are based on fossils of a similar age.

These footprints of two early hominins walking upright, side by side, were found in Africa.

Figure 17.9

Early hominins had an upright stance and partially opposable big toes

Fossilized foot bones show that some hominins living between 3.5 and 3 million years ago walked upright but had partially opposable big toes.

Q1: What other reason besides continuing to use trees might explain why early hominins had partially opposable big toes?

Q2: In what way does the pattern of footprints in this figure suggest that the print makers were walking upright?

Q3: Why do you think we no longer have partially opposable big toes?

researchers to analyze them, including Condemi, the anthropologist. Condemi had long been interested in the movement of Neanderthals across Europe and how their populations overlapped with modern *Homo sapiens* populations, and she wanted to compare the Riparo Mezzena fossils to fossils from Neanderthal groups that had been dug up elsewhere around Europe, hoping to see what the fossils might say about how Neanderthals and modern humans interacted.

Q1: Through natural selection, deleterious traits will tend to disappear from a population over time. Which traits might have been deleterious for ground-dwelling early hominins?

Q2: Through natural selection, advantageous traits will tend to persist in a population over time. Which traits might have been advantageous for ground-dwelling early hominins?

Q3: Adaptations to upright walking also mean that human females have more difficulty giving birth than do females of other species. What adaptation would you predict has had the greatest impact on this result?

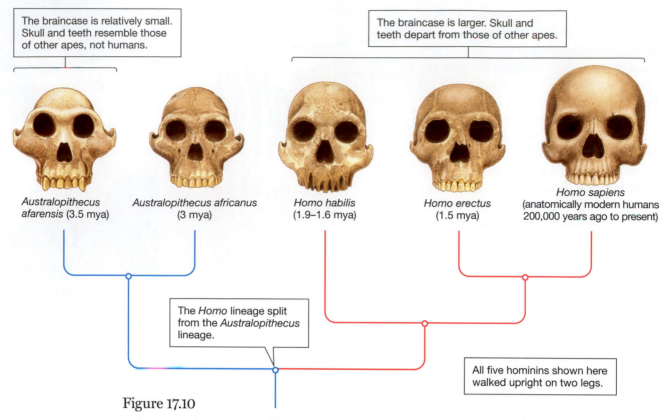

The braincase is relatively small. Skull and teeth resemble those of other apes, not humans.

The braincase is larger. Skull and teeth depart from those of other apes.

Australopithecus afarensis (3.5 mya)

Australopithecus africanus (3 mya)

Homo habilis (1.9–1.6 mya)

Homo erectus (1.5 mya)

Homo sapiens (anatomically modern humans 200,000 years ago to present)

The *Homo* lineage split from the *Australopithecus* lineage.

All five hominins shown here walked upright on two legs.

Figure 17.10

A gallery of hominin skulls

This tree shows the evolutionary relationships and the skulls of five hominin species. A complete evolutionary tree of hominins would be "bushier," with multiple side branches emerging at different times.

Q1: Where would the Neanderthal species branch be on this tree?

Q2: How would the Neanderthal skull differ from the *Homo erectus* skull?

Q3: How would the Neanderthal skull differ from the *Homo sapiens* skull?

Hominins United

"For years, it was a very simple story: Neanderthals either disappeared quickly when modern humans came or they integrated with humans," says Condemi. But she suspected that the dynamic was more complex. "In some regions, Neanderthals disappeared very quickly. In other regions, we have evidence the two [species] could have overlapped." Researchers, including Condemi, had often suggested that during the overlap, Neanderthals and humans interbred. But there was little physical evidence.

Even DNA evidence initially suggested that there was no interbreeding. DNA from the mitochondria of Neanderthals (isolated first from a single Neanderthal fossil in 1997 and then from four Neanderthal fossils in 2004) was compared to modern human mitochondrial DNA, and the tests showed that there was no genetic overlap between the species. Mitochondrial DNA (mtDNA) is unique because it is passed down virtually unchanged from mother to child, so it can be tracked from one generation, or one species, to another (**Figure 17.11**). But modern *Homo sapiens* did not have Neanderthal mitochondrial DNA, so it appeared there was no interbreeding, at least not involving female Neanderthals.

That mitochondrial work was performed by Svante Pääbo and researchers at the Max Planck Institute for Evolutionary Anthropology in Leipzig, Germany. Pääbo is one of the founders of the effort to use genetics to study early

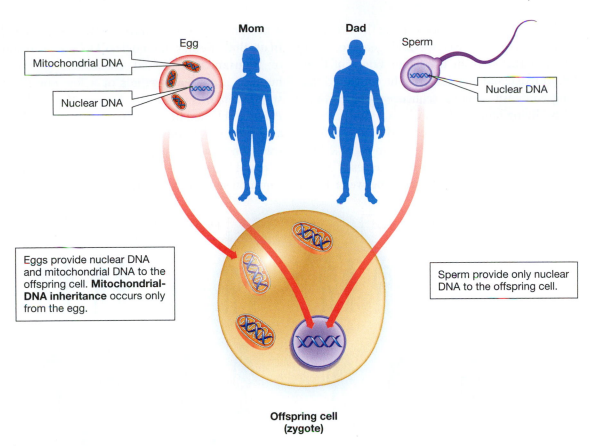

Mom **Dad**

Egg

Mitochondrial DNA

Nuclear DNA

Sperm

Nuclear DNA

Eggs provide nuclear DNA and mitochondrial DNA to the offspring cell. **Mitochondrial-DNA inheritance** occurs only from the egg.

Sperm provide only nuclear DNA to the offspring cell.

Offspring cell
(zygote)

Figure 17.11

Mitochondrial DNA comes only from your mom (not from your dad)

In addition to the chromosomes in the nucleus (nuclear DNA), cells also have DNA in their mitochondria, called mitochondrial DNA. Mitochondria and their resident DNA all come from Mom.

Q1: Why does mitochondrial DNA come only from your mother?

Q2: If a Neanderthal-human hybrid was born to a human mother and a Neanderthal father, could you tell by mitochondrial-DNA sequencing?

Q3: If a Neanderthal-human hybrid was born to a human father and a Neanderthal mother, could you tell by mitochondrial-DNA sequencing?

humans and other ancient populations, and he has pioneered numerous techniques to extract delicate DNA from even the tiniest slivers of fossilized bones. At the time that he began his work, it was easier to find and extract mitochondrial DNA because cells contain hundreds of copies of this type of DNA, and only one copy of nuclear DNA. In addition, mitochondrial DNA can be isolated from cells and tissues that aren't so well preserved and from damaged DNA. Whole-genome sequencing, by contrast, requires well-preserved cells or tissues with fully intact nuclear DNA. But despite his initial findings that modern humans and Neanderthals did not share

mtDNA, Pääbo still thought there might be room for some small contribution. Being a diligent scientist, he decided to look even deeper—this time at the whole Neanderthal genome.

SVANTE PÄÄBO

Svante Pääbo is a Swedish geneticist who directs the Department of Genetics at the Max Planck Institute for Evolutionary Anthropology in Leipzig, Germany. He specializes in using genetics to study early humans and other ancient populations.

Pääbo spent 4 years sequencing the 1.5 billion base pairs in the Neanderthal genome, using DNA extracted from the femur bones of three 38,000-year-old females. Then he compared that long, composite genome sequence to the genomes of five living humans from China, France, Papua New Guinea, southern Africa, and western Africa.

According to the results, all modern ethnic groups, other than Africans, carry traces of Neanderthal DNA in their genomes—between 1 and 4 percent. Most of us have a little Neanderthal in us. But whether that nuclear DNA is the result of thousands of sexual encounters between humans and Neanderthals or a few one-night stands remains unknown (**Figure 17.12**). When Pääbo published his work in 2010, he and others admitted it was possible that the shared DNA wasn't necessarily a product of interbreeding. It could have been a remnant of DNA from a common shared ancestor.

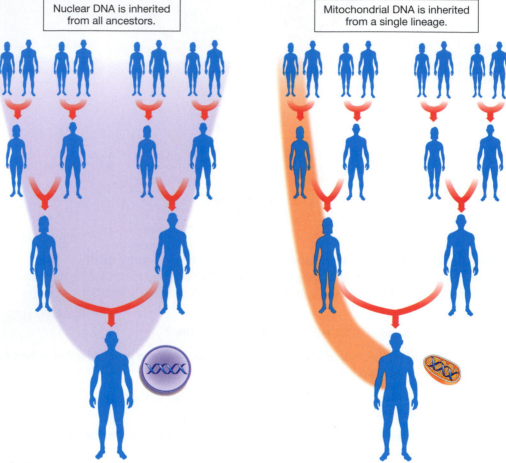

Nuclear DNA is inherited from all ancestors.

Mitochondrial DNA is inherited from a single lineage.

Figure 17.12

Nuclear-DNA inheritance

By sequencing the bases of nuclear DNA, scientists can determine how related an individual is to its ancestors, both male and female (see Chapter 11). Mitochondrial-DNA sequencing can determine only how related an individual is to the female ancestors on its mother's side.

Q1: If a human-Neanderthal hybrid was born to a human mother and a Neanderthal father, could you tell by whole-genome DNA sequencing that it was a hybrid?

Q2: If a human-Neanderthal hybrid was born to a Neanderthal mother and a human father, could you tell by whole-genome DNA sequencing that it was a hybrid?

Q3: Under what circumstances are scientists able to do whole-genome sequencing, and when are they restricted to mitochondrial-DNA sequencing?

Who would that ancestor have been? The oldest *Homo* fossil fragments were found in Ethiopia in 2015 and date between 2.80 and 2.75 million years ago (mya), suggesting that the earliest members of the genus *Homo* originated in Africa 2–3 mya. More complete early *Homo* fossils exist from the period 1.9–1.6 mya; these fossils have been given the species name *Homo habilis*. The oldest *H. habilis* fossils resemble those of *Australopithecus africanus*, yet more recent *H. habilis* fossils show a more rounded skull and a face that isn't pulled as far forward. *H. habilis* fossils therefore provide an excellent record of the evolutionary shift from ancestral hominins (in *Australopithecus*) to more recent species, such as *Homo erectus*, the most likely candidate for a shared common ancestor of Neanderthals and modern humans (**Figure 17.13**).

Figure 17.13

Meet the folks

Cartoon representations of seven hominin species depicting their average height in centimeters and their presumed features. *Homo sapiens* is represented as a 6-foot-tall male for reference.

Q1: Are you surprised by the interpretations of the hominins in this picture? Why or why not?

Q2: Describe the main differences that distinguish the hominin species.

Q3: From what you've learned about these species, do you think these representations are accurate? How can you find more information about each species to help you answer this question?

Taller and more robust than *H. habilis*, *H. erectus* also had a larger brain and a skull more like that of modern humans. It is likely that by 500,000 years ago, *H. erectus* could use, but not necessarily make, fire. In addition, *H. erectus* probably hunted large animals, as suggested by a remarkable 2010 discovery in Germany of three 400,000-year-old spears, each about 2 meters long and designed for throwing with a forward center of gravity (like a modern javelin). *H. erectus*, or one of the other *Homo* ancestors, migrated from Africa about 2 million years ago. From there, this ancestor species spread around the Middle East and into Asia. *Homo* fossils dating from the period 1.9–1.7 mya have been found in the central Asian republic of Georgia, in China, and in Indonesia.

Overall, current research on *H. habilis*, *H. erectus*, and other early *Homo* species indicates that there were more species of *Homo* than was once thought, and that several of these species existed in the same places and times. More research and evidence will be necessary before general agreement is reached regarding the exact number of early *Homo* species and their evolutionary relationships.

So, was the Neanderthal DNA found in modern human genomes simply a remnant of a common ancestor? In 2012, Pääbo's team and others were able to determine the age of the pieces of Neanderthal DNA in the human genome. They found that the DNA had been introduced into our genome between 90,000 and 40,000 years ago, around the same time that modern humans spread out of Africa and met the Neanderthals. A remnant of DNA from a common ancestor would have been 10 times older.

All in the Family

The fossil record indicates that the first *Homo sapiens*, called archaic *H. sapiens*, originated between 400,000 and 300,000 years ago.

Uniquely Human?

Yes, the frontal lobe of the brain is unique to human beings, and it enables us to reason like no other animal on our planet. But what about the rest of the attributes that we so commonly consider unique to us? Humans pride themselves on their intelligence and deep emotional connections to others, but are these really only human traits?

- **Language.** Researchers once believed language was an exclusively human trait. We now know that chimpanzees in the wild use sign language, with approximately 70 different signs for distinct words. Meanwhile, other primates, birds, whales, and bats have distinct, learned vocalizations that they use to communicate.

- **Memory.** Some have suggested humans alone possess the ability to store memories. But dogs easily learn and remember many commands, while crows can learn and remember shapes better than human adults can, and they can use causal reasoning, not trial and error, to unlock doors and find hidden objects.

- **Social culture.** Once thought to be strictly human, social culture is a learned trait that chimpanzees, Japanese macaques, and killer whales pass on throughout their populations. Tool use by dolphins, elephants, and octopi varies in its specifics from population to population—a sure sign of learned behavior.

- **Emotions.** Our emotions make us human, right? Others in the animal kingdom have been documented expressing empathy (elephants), grief (dolphins, elephants), jealousy (apes), curiosity (cats, lizards), altruism (apes), and gratitude (whales). Apes have been seen laughing at a clumsy fellow ape and using deception to outwit a family member.

- **Self-Awareness.** The ability to recognize oneself in the mirror, or show self-awareness, was once thought to be ours alone. As it turns out, all the apes, some gibbons, elephants, magpies, and some whales pass the mirror test of self-awareness.

- **Morality.** Finally, what about a sense of morality or an understanding of social norms? Monkeys and rats will not accept offered food if, in doing so, a fellow member of their species receives an electric shock.

To be sure, there *is* something unique about humans that lies at the intersection of all these abilities. However, our expanding knowledge of animal behavior can't help but make us feel more closely connected to the other species with whom we share this planet.

Figure 17.14

Anatomically modern humans evolved in Africa

The earliest known archaeological specimens of modern humans (*Homo sapiens*) come from Africa. The dates provided give the age of the earliest evidence that anatomically modern humans lived in different regions of the world. These dates are continually challenged by new fossil evidence that scientists must then work to confirm.

Q1: What evidence suggests that Neanderthals never lived in Africa?

Q2: How does the hypothesized origin of modern humans (*Homo sapiens*) differ from the hypothesized origin of Neanderthals (*Homo neanderthalensis*)?

Q3: What species of hominins other than the Neanderthals may have commingled with modern humans?

Archaic *H. sapiens* bore features intermediate between those of *H. erectus* and those of "anatomically modern" *H. sapiens*—our species—which arose some 195,000–200,000 years ago. These ancestors of anatomically modern humans developed new tools and new ways of making tools, used new foods, and built complex shelters. (But humans are not the only organisms to do many of these things; see "Uniquely Human?" on page 316.)

Early populations of archaic humans eventually gave rise to both the Neanderthals (who lived from 300,000 to 28,000 years ago) and us—that is, anatomically modern humans. There is some debate as to whether Neanderthals are simply an odd form of archaic *H. sapiens*, or their own distinct species. That question has yet to be resolved.

According to the out-of-Africa hypothesis, anatomically modern humans first evolved in Africa about 195,000–200,000 years ago from a unique population of archaic *H. sapiens*, and then spread into other continents to live alongside other hominins (**Figure 17.14**). Evidence from the fossil record indicates that anatomically modern humans overlapped in time with *H. erectus* and Neanderthal populations, yet remained distinct from them. Neanderthals and modern humans coexisted in western Asia for about 80,000 years, and in Europe for some 10,000 years, until modern humans completely replaced all other *Homo* populations.

But what happened in that intervening time? Were modern humans and Neanderthals friendly neighbors, or were the latter quickly wiped out by the former?

Figure 17.15

Neanderthal and human skulls

The skull on the right is from a Neanderthal; the one on the left, from a modern human. Notice the differences in the eyebrow ridge, forehead, and lower jaw.

Q1: Describe the difference you observe between the modern human skull's chin and the Neanderthal skull's lower jaw.

Q2: What other differences do you observe between the two skulls?

Q3: Why would you expect a hybrid of Neanderthals and modern humans to have intermediate features?

With a previous team, Condemi had used fossil evidence from southern Italy to determine that modern humans arrived on the Italian peninsula between 45,000 and 43,000 years ago, *before* the disappearance of Neanderthals. So the two populations likely made contact in Italy. During this period, it's possible that "there was a kind of interbreeding," says Condemi. But if the two species interbred and had children, what did those children look like? And why hadn't Condemi's team found their bones?

Then Condemi examined the Riparo Mezzena bones. One caught her attention. It was a jawbone, a mandible, from a late Neanderthal living in Italy at the same time that modern humans had already made their way into Europe (see **Figure 17.1**). But the jawbone didn't look like a Neanderthal's, which has no chin. Instead, the face of the Riparo Mezzena individual, when reconstructed with three-dimensional imaging, had an intermediate jaw, something between no chin and a strongly projected chin. Because chins are a feature unique to modern humans (**Figure 17.15**),

SILVANA CONDEMI

Silvana Condemi is an anthropologist and research director at the National Center for Scientific Research (CNRS) at Aix-Marseille University in France. She studies the movement of Neanderthals across Europe and how their populations overlapped with modern humans.

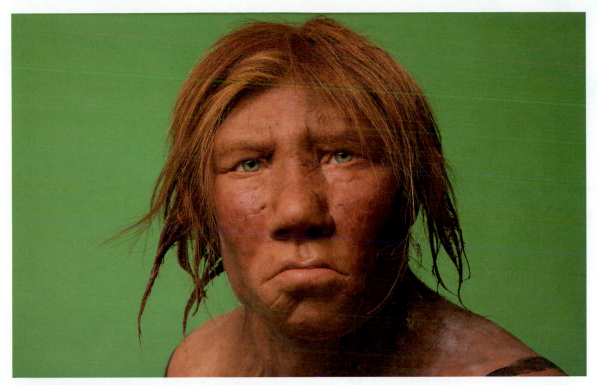

Figure 17.16

Neanderthal-human hybrid
Paleontological reconstruction of a Neanderthal-human hybrid.

the jaw appeared to be something of a hybrid between a Neanderthal and a modern human. "This, in my view, could only be a sign of inter-breeding," says Condemi.

To back up her hypothesis, Condemi's team analyzed the fossil's DNA. The fossil had Neanderthal mitochondrial DNA, confirming that at least the individual's mother was a Neanderthal. From the DNA and imaging evidence, Condemi and her team concluded that it was the child of a "female Neanderthal who mated with a male *Homo sapiens*" (**Figure 17.16**). What's more, she added, this evidence supports the idea of a slow transition from Neanderthals to anatomically modern humans, in which the two species inter-mingled in both culture and sex, rather than the abrupt extinction of Neanderthals when modern humans arrived.

It is unlikely that Condemi's finding will be the last word on Neanderthal-human inter-breeding. In 2016, the first in-depth genetic analysis of a Neanderthal Y chromosome revealed mutations in genes on the Neanderthal chromosome that would have been incompatible with comparable genes in the modern human Y chromosome. In other words, these Y chromosome mutations could have discouraged or prevented the formation of a hybrid Neanderthal–modern human species.

An even bigger question lingers for future scientists: While modern humans continued to develop their culture and populate the planet, Neanderthals went extinct. Why? Perhaps another pile of bones will someday reveal the answer.

REVIEWING THE SCIENCE

- **Animals** are multicellular ingestive heterotrophs. The **chordates** include all animals with a dorsal nerve cord. Chordates with a backbone are **vertebrates**, and all other phyla of animals are informally designated as invertebrates.

- Sponges lack distinct symmetry; some invertebrates have **radial symmetry**; all other animals, including all chordates, have **bilateral symmetry**.

- Many animal groups have a body plan consisting of repeated units known as **segments. Appendages,** body parts with specialized functions, develop in pairs from particular segments of the body. The evolution of segmentation and appendages has allowed animals to exploit new habitat and food sources in novel ways.

- **Mammals** are grouped into three categories: **Eutherians** are placental and born in a well-developed state. **Marsupials** have a simple placenta and finish development in a pouch. **Monotremes** lack a placenta and lay eggs.

- Humans are mammals (class), and more specifically **primates** (order), having flexible shoulder and elbow joints, five functional fingers and toes, **opposable** thumbs, flat nails (instead of claws), and large brains.

- **Hominins** are characterized by **bipedalism**, the ability to walk upright on two legs. They are a branch of the apes, the **hominids** (family).

- **Mitochondrial-DNA inheritance** occurs exclusively through the maternal egg; the sperm contributes essentially no mitochondria and none of the mitochondrial DNA. **Nuclear-DNA inheritance** occurs equally through both the eggs and sperm; all individuals have half maternal and half paternal nuclear DNA.

THE QUESTIONS

Challenge Yourself

1 _____ specimens have features that are intermediate between those of *Australopithecus africanus* and *Homo erectus* and provide an amazing record of the evolutionary shift from ancestral hominin characteristics seen in *Australopithecus* fossils to more recent ones seen in *H. erectus* fossils.

(a) *Homo sapiens*

(b) *Homo neanderthalensis*

(c) *Homo habilis*

(d) *Ardipithecus ramidus*

2 Briefly describe the key differences that distinguish monotreme, marsupial, and eutherian mammals.

3 Place the following hominids in the correct order from earliest to most recent by numbering them from 1 to 5.

_____a. archaic *Homo sapiens*

_____b. *Australopithecus afarensis*

_____c. modern *Homo sapiens*

_____d. *Homo habilis*

_____e. *Homo erectus*

Try Something New

4 You visit your local museum of natural history and come upon an exhibit showcasing hominin fossils that date back 300,000–400,000 years ago. You notice that these fossils have features intermediate between those of *Homo erectus* and those of "anatomically modern" *Homo sapiens*. Who were these fossils?

(a) archaic *Homo erectus*

(b) archaic *Homo sapiens*

(c) *Homo neanderthalensis*

(d) *Homo habilis*

5 In 2004, scientists discovered the fossilized remains of an extinct species they named *Tiktaalik roseae*. The fossil appears to be a transitional species between (aquatic) fishes and (terrestrial) four-legged amphibians. In what environments do you predict this animal lived?

6 Give an example animal for some of the key adaptations of chordates: (a) backbone, (b) skull, (c) amniotic egg, (d) bony skeleton. Choose a different animal for each adaptation.

7 How does the fact that all ethnic groups *except* Africans contain some Neanderthal DNA (1–4 percent of their DNA) support the out-of-Africa hypothesis for the origin of modern humans (*Homo sapiens*)?

Leveling Up

8 *Write Now* biology: if we were not alone Fossil evidence indicates that in the relatively recent past (about 30,000 years ago), anatomically modern humans, or *Homo sapiens*, may have shared the planet with at least three other distinct hominins: *H. erectus*, *H. neanderthalensis*, and *H. floresiensis*. If one or more of these species were alive today, how would their existence affect the world as we know it?

9 **Is it science?** Watch the original 1968 movie *Planet of the Apes*. Document as many scientific problems with the movie as you can. Which of the apes' adaptations would be biologically possible, and which ones would be impossible, from your understanding of apes on Earth today? If this species did evolve to have the adaptive traits of *Homo*, would its members still be called apes?

Thinking Multidimensionally

10 Chimpanzees are our closest living relatives in that we share 96 percent of our DNA.

a. With a diploid number of 48 and a haploid number of 24, how do these chimpanzee chromosome numbers compare to humans?

b. Using the evidence from the image below, explain what brought about these differences in chromosome numbers between the two primates.

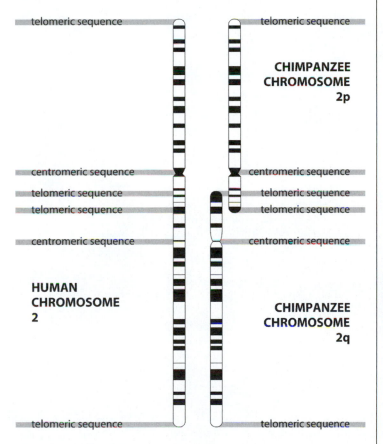

telomeric sequence — telomeric sequence

CHIMPANZEE CHROMOSOME 2p

centromeric sequence — centromeric sequence
telomeric sequence — telomeric sequence
telomeric sequence — telomeric sequence
centromeric sequence — centromeric sequence

HUMAN CHROMOSOME 2

CHIMPANZEE CHROMOSOME 2q

telomeric sequence — telomeric sequence

In 1960, at the age of 23, Jane Goodall traveled to Africa to study the behavior of chimpanzees in the wild. At that time, scientists thought that humans were the only organisms capable of problem solving with the aid of tools.

c. Look at the image above to witness what Goodall referred to as "termite fishing." Describe what the chimps are doing next to the termite mound and their motivation for these actions.

11 Variation in human skin color is the result of different kinds and amounts of the pigment melanin.

a. To understand why humans have a range of different skin tones, look for overall patterns in the map above, and comment on the role geography (i.e., latitude and elevation) played in the evolution of different skin tones.

b. Where is UV radiation most intense on the map?

c. How does UV light affect DNA in our cells?

SKIN COLOR INFOGRAPHIC

For more, visit digital.wwnorton.com/bionow2 for access to:

Amazon on Fire

Can the world's largest rainforest survive a vicious cycle of wildfires and climate change?

After reading this chapter you should be able to:

- Define the biosphere and the role that humans play in it.
- Articulate the difference between a biotic factor and an abiotic factor in the study of ecology.
- Distinguish between climate and weather.
- Describe the greenhouse effect.
- Compare and contrast the water cycle and the carbon cycle.
- Explain how global warming contributes to climate change.
- List the consequences of climate change, and describe one consequence in detail.

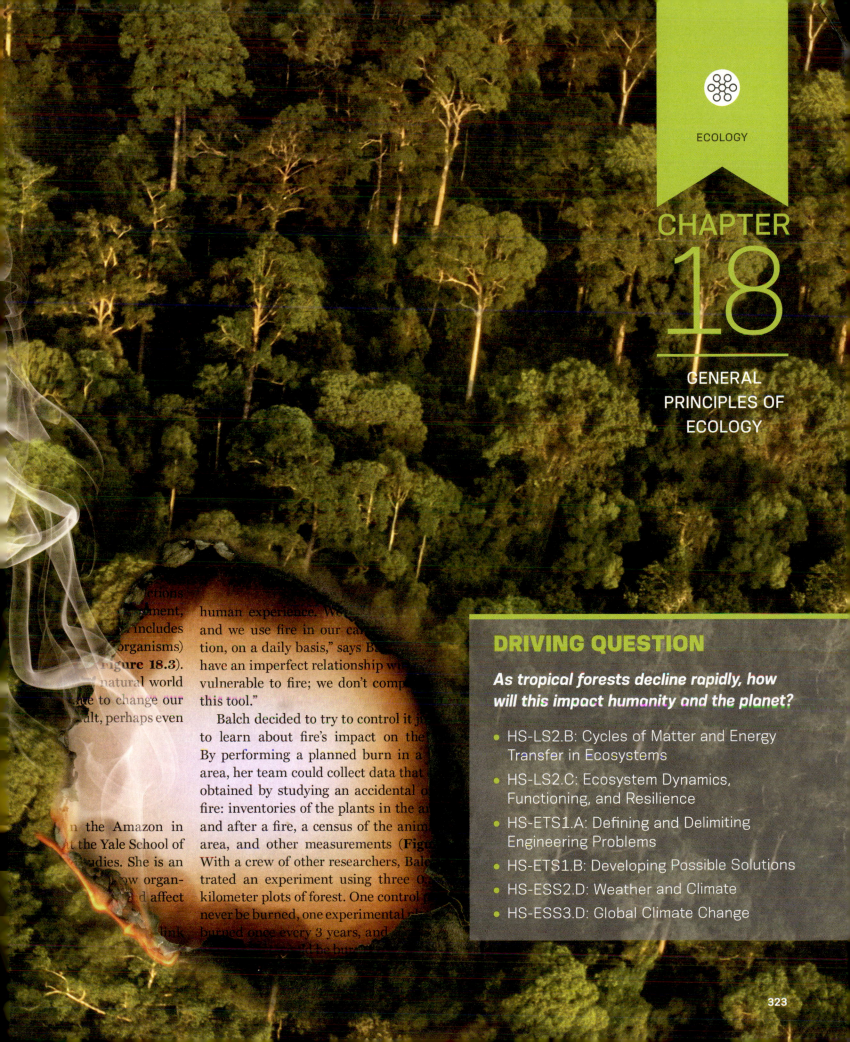

CHAPTER
18

GENERAL
PRINCIPLES OF
ECOLOGY

...tions
...ment,
...includes
...organisms)
...gure 18.3).
...natural world
...e to change our
...ult, perhaps even

human experience. We
and we use fire in our ca
tion, on a daily basis," says Ba
have an imperfect relationship wi
vulnerable to fire; we don't comp
this tool."

Balch decided to try to control it ju
to learn about fire's impact on the
By performing a planned burn in a
area, her team could collect data that
obtained by studying an accidental o
fire: inventories of the plants in the a
and after a fire, a census of the anim
area, and other measurements (Figu
With a crew of other researchers, Bal
trated an experiment using three 0.
kilometer plots of forest. One control
never be burned, one experimental
...burned once every 3 years, and

...n the Amazon in
...t the Yale School of
...udies. She is an
...ow organ-
... d affect

...link,

DRIVING QUESTION

*As tropical forests decline rapidly, how
will this impact humanity and the planet?*

- HS-LS2.B: Cycles of Matter and Energy
  Transfer in Ecosystems
- HS-LS2.C: Ecosystem Dynamics,
  Functioning, and Resilience
- HS-ETS1.A: Defining and Delimiting
  Engineering Problems
- HS-ETS1.B: Developing Possible Solutions
- HS-ESS2.D: Weather and Climate
- HS-ESS3.D: Global Climate Change

Flames rise from the forest floor, licking at Jennifer Balch's heels. Balch walks carefully ahead of the heat, stoking it, encouraging it. She tips a large metal container, dripping flaming kerosene onto another pile of dead branches on the ground (**Figure 18.1**). The forest burns behind her.

Balch steps back to look at her work—the destruction of a small patch of the Amazon rainforest. Balch, an ecologist at the University of Colorado Boulder, is in the southern part of the Amazon basin in Brazil, in the state called Mato Grosso. There, on a small square kilometer and a half of land belonging to a soybean farmer, Balch is experimenting to see what happens when the Amazon burns.

The Amazon is the largest rainforest in the world and a critical part of our planet's biosphere. The **biosphere** consists of all of Earth's organisms, plus the physical space we all inhabit. It includes inorganic chemicals like water, our nitrogen-rich atmosphere, every living organism, and more (**Figure 18.2**). Put more simply, the biosphere is the integration of all the environments on Earth. It is crucial to our survival and well-being because humans depend on the biosphere for food and raw materials.

Within the biosphere, the Amazon is home to more than half of the world's millions of species of plants and animals. It also contains one-fifth of the world's freshwater. Yet the Amazon is under threat: since 1960, the human population of the Brazilian Amazon region has increased from 6 million to 25 million people, leading to a dramatic expansion of agriculture, including major deforestation as humans cut down trees to make room for roads, cattle, and fields. The forest cover in the Amazon has declined by 20 percent during this period, and the grasslands and pastures that have taken its place—including soybean fields like the one Balch works at—are threatening the rainforest that is left.

JENNIFER BALCH

Jennifer Balch is an ecologist at the University of Colorado Boulder who studies how fire disturbance affects ecosystems. In 2004, she initiated a one-of-a-kind experimental burn study in the Amazon to see how wildfires affect the tropical rainforest.

Figure 18.1

Jennifer Balch ignites a controlled burn in an experimental plot in the rainforest

It's traditional for farmers in Brazil to clear land for agriculture by setting fires to burn away trees and brush. In addition, fire is a useful tool for clearing out old grasses to encourage the growth of new grasses for cattle to feed on. Yet when these fires are not kept under control, they can spread to nearby rainforest and destroy it. Historically, this has not been a major concern, because the Amazon is wet and humid, so forest fires burned slowly, extinguishing themselves before spreading too far. Yet a dramatic shift in the planet's climate is making the Amazon drier and more vulnerable to wildfires.

Our planet is warming. As a result of global climate change, the temperature in Earth's atmosphere is increasing, causing less rainfall and more drought in the Amazon, leaving the forests primed for fire. And they've begun to burn. Even worse, that loss of trees adds to the warming, resulting in a vicious cycle in which the Amazon becomes hotter and drier, burns again, and contributes to further climate change.

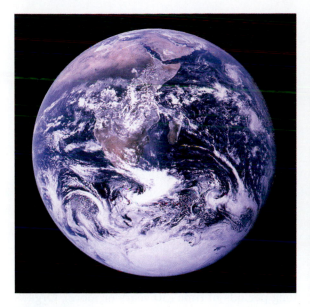

Figure 18.2

The biosphere is Earth and all of its inhabitants

This is a view of Earth from space. The atmosphere, Earth's surface, and all of the organisms make up the biosphere.

Abiotic factors: rocks, water, air

Biotic factors: plants, animals, microbes

Figure 18.3

Amazon rainforest ecology

Ecology is the study of how living organisms (biotic factors) interact with other organisms, and how they all interact with their nonliving (abiotic) environment.

Q1: List as many biotic and abiotic factors in this photograph as you can.

Q2: Is the forest part of the biotic or abiotic environment? Explain.

Q3: Is the river part of the biotic or abiotic environment? Explain.

Balch and other scientists have been investigating whether it might be possible to intervene in that cycle to slow or stop the loss of the Amazon rainforest. Their latest experiment, burning plots of Amazon rainforest to see what happens afterward, is an experiment in ecology. **Ecology** is the scientific study of interactions between organisms and their environment, where the environment of an organism includes both **biotic** factors (other living organisms) and **abiotic** (nonliving) factors (**Figure 18.3**). Ecology helps us understand the natural world we live in, but humans continue to change our biosphere in ways that are difficult, perhaps even impossible, to fix.

Hot and Dry

Balch began setting fires in the Amazon in 2004 as a graduate student at the Yale School of Forestry and Environmental Studies. She is an ecologist, a scientist fascinated by how organisms and environments interact with and affect each other.

Balch has always been interested in the link between humans and fire. "Fire is integral to the human experience. We play with fire as kids, and we use fire in our cars, through combustion, on a daily basis," says Balch. "But humans have an imperfect relationship with fire. We are vulnerable to fire; we don't completely control this tool."

Balch decided to try to control it just enough to learn about fire's impact on the Amazon. By performing a planned burn in a restricted area, her team could collect data that cannot be obtained by studying an accidental or escaped fire: inventories of the plants in the area before and after a fire, a census of the animals in the area, and other measurements (**Figure 18.4**). With a crew of other researchers, Balch orchestrated an experiment using three 0.5-square-kilometer plots of forest. One control plot would never be burned, one experimental plot would be burned once every 3 years, and a second experimental plot would be burned once every year.

Figure 18.4

Scientists take measurements before a planned burn in the Amazon

To prevent fires in their two experimental plots from spreading, each plot was protected by a firebreak—a perimeter around the plot, several meters wide, cleared of all plants, sticks, grasses, and debris. Once the plots were prepared, it was time to set the fires. The team walked along 10 kilometers of trail cutting through the plots, dripping flaming kerosene from special tanks and watching the forest burn.

The experiment began not a moment too soon. In 2005, the Amazon experienced an extreme drought. Experts called it a "hundred-year drought," meaning that such a drought is expected to occur, on average, only once in a century. But just five years later, in 2010, there was another hundred-year drought, this one even more widespread and severe. And in both cases, extensive wildfires followed the drought, destroying over

MICHAEL COE

Michael Coe is a hydrologist and leader of the Amazon Program at the Woods Hole Research Center in Massachusetts. He investigates how humans affect the water and energy balance in tropical South America.

30,000 square miles of rainforest. In 2005 alone, fires increased 20-fold in the year following the drought. "It was a huge increase," says Michael Coe, a researcher at the Woods Hole Research Center in Massachusetts and one of Balch's collaborators. "And if humans are increasing drought frequency, that's going to be a big problem."

Humans, researchers suspect, are increasing the number and intensity of droughts in the Amazon through climate change. It's important to distinguish between "climate" and "weather." **Weather** refers to short-term atmospheric conditions in a limited geographic area, such as today's temperature, precipitation, wind, humidity, and cloud cover. **Climate** describes the prevailing weather of a region over relatively long periods of time (30 years or more). Organisms are more strongly influenced by climate than by any other feature of their environment. On land, for example, features of climate such as temperature and precipitation determine whether a particular region is desert, grassland, or tropical forest.

Climate change, then, is a large-scale and long-term *alteration* in Earth's climate, and it includes such phenomena as global warming, change in rainfall patterns, and increased frequency of violent storms. Although Earth has gone through many changes in its average climate over its 4.6-billion-year history, the speed of the change that has taken place in the past 100 years is without precedent in the climate record. Climate change in recent history has been caused to a large extent by human actions, and its consequences are likely to be negative for people and ecosystems around the world (**Figure 18.5**).

"There are a lot of indications that climate change causes increased drought events, causing increased fire events," says Balch. "It's hard to predict the future, but that's definitely the trend."

A Warmer World

The terms "global warming" and "climate change" are related but not synonymous. **Global warming** is a significant increase in the average surface temperature of Earth over decades or more. Temperature on Earth is generally determined by the angle at which sunlight strikes the planet. Sunlight strikes Earth most directly at the equator, but at a more slanted angle near the North and

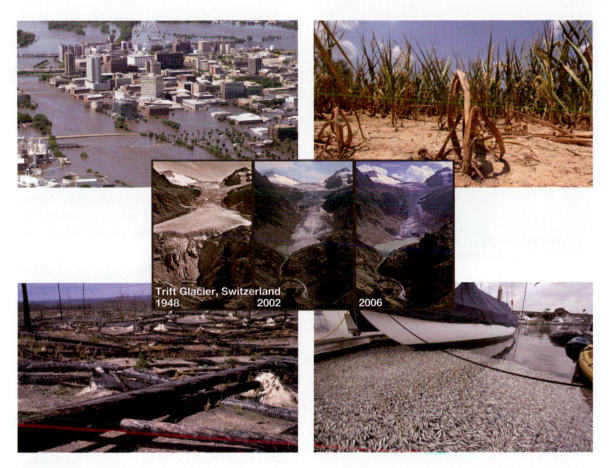

Figure 18.5

Consequences of climate change

Extreme weather and new rainfall patterns caused by climate change have led to increased flooding in some areas and drought in others, bringing about ecosystem and habitat destruction, and species extinction. The increase in global temperatures has led to melting of glaciers and polar ice, causing sea level rise, increase in ocean temperatures, and coral reef bleaching and death. Effects on our food sources include increased crop failure and depletion of important fisheries.

Q1: Name two ways in which climate change affects the frequency and severity of floods.

Q2: How has climate change caused a rise in sea level?

Q3: Give an example of an environmental effect of climate change in your state or region.

South Poles (**Figure 18.6**). Where sunlight hits the atmosphere more directly, its solar energy is more concentrated. For this reason, more solar energy reaches the equator, making it and neighboring tropical regions much warmer than the poles. Because Earth is tilted on its axis as it revolves around the sun on its annual orbit, sunlight hits different areas of the planet at different angles over the course of the year. This is the cause of our seasons: the Northern Hemisphere experiences summer when Earth is tilted toward the sun, while the Southern Hemisphere simultaneously experiences winter because it is tilted away.

Global warming, however, is not dependent on how sunlight strikes the Earth. Instead, it is caused by an increase in **greenhouse gases**. Some gases in Earth's atmosphere, such as carbon dioxide (CO_2), water vapor (H_2O), methane (CH_4), and nitrous oxide (N_2O), absorb heat that radiates away from Earth's surface, preventing it from being released into space. These gases are called greenhouse gases because they

The sun's rays hit Earth at an angle near the poles, resulting in cooler temperatures.

The sun's rays strike Earth directly near the equator, resulting in warmer temperatures.

Figure 18.6

Sunlight strikes Earth most directly at the equator

The angle at which the rays of the sun strike Earth determines how much energy or heat reaches Earth's surface. The more direct the rays are when they strike Earth, the more heat they deliver.

Q1: Why is it colder at the poles than at the equator?

Q2: Why is it warmer at the equator than at the poles?

Q3: During part of the year the Northern Hemisphere is tilted at a more direct angle to the sun than the Southern Hemisphere is, and for the other part of the year the opposite is true. How does this tilt explain temperature differences in summer and winter?

function much as the walls of a greenhouse or the windows of a car do: they let in sunlight and trap heat in a process known as the **greenhouse effect** (**Figure 18.7**).

Greenhouse gases are not inherently bad; in fact, they have existed in Earth's atmosphere for more than 4 billion years, and they play an important role in maintaining temperatures that are warm enough for life to thrive on Earth. Yet human activities, primarily the burning of fossil fuels, have released an excess of greenhouse gases into the atmosphere, especially the infamous king of greenhouse gases: **carbon dioxide** (CO_2).

Scientists have estimated atmospheric CO_2 levels for both the recent and the relatively distant past, up to hundreds of thousands of years ago, by measuring CO_2 concentrations in air bubbles trapped in ice. This evidence shows a near-perfect historical correlation between CO_2 levels and the surface temperature on Earth. Yet during the last 200 years, levels of atmospheric CO_2 have risen greatly—from roughly 280 to 400 parts per million (**Figure 18.8**). Measurements from ice bubbles show that this rate of increase is greater than even the most sudden increase that occurred naturally during the past 420,000 years. Carbon dioxide levels are now higher than those estimated for any time during that period.

The cause of the increasing levels of atmospheric CO_2 has been linked to the burning of fossil fuels like coal and oil, which releases CO_2 into the air. About 75 percent of the current yearly increase in atmospheric CO_2 is due to the burning of fossil fuels. Logging and burning of forests are responsible for most of the remaining 25 percent of the increase, but industrial processes also make a significant contribution. "Humans have a giant impact on the environment," says Coe. "With more than seven billion people on the planet, individual decisions add up to big global changes." (For more on the environmental impact of personal choices, see "How Big Is Your Ecological Footprint?" on page 332.)

All this is to say that human burning of fossil fuels has led to an increase in carbon dioxide in our atmosphere. That carbon dioxide acts as a greenhouse gas and traps additional heat as it leaves the atmosphere, causing temperatures on Earth to rise. Since the early twentieth century, Earth's mean surface temperature has increased by about 1.4°F (0.8°C), and it is estimated to rise another 2°F–11.5°F (1.1°C–6.4°C) in the future (**Figure 18.9**). As of the date this book went to publication, 2016 was the hottest year ever recorded; 2015 was previously the hottest year ever; 2014 was the winner before that. We keep beating our own record.

Our globe is warming. And the effects of that warming on the biosphere are now evident, especially in the Amazon, where trees dominate the landscape. Trees act as a layer connecting the atmosphere above them and the ground beneath them, absorbing water from the ground and releasing oxygen into the air. Any change in tree cover can change the local

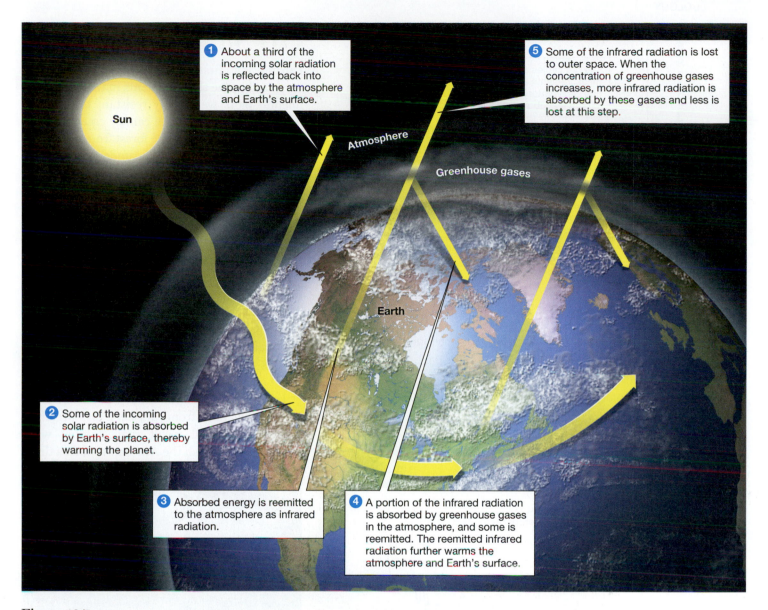

1 About a third of the incoming solar radiation is reflected back into space by the atmosphere and Earth's surface.

5 Some of the infrared radiation is lost to outer space. When the concentration of greenhouse gases increases, more infrared radiation is absorbed by these gases and less is lost at this step.

Sun

Atmosphere

Greenhouse gases

Earth

2 Some of the incoming solar radiation is absorbed by Earth's surface, thereby warming the planet.

3 Absorbed energy is reemitted to the atmosphere as infrared radiation.

4 A portion of the infrared radiation is absorbed by greenhouse gases in the atmosphere, and some is reemitted. The reemitted infrared radiation further warms the atmosphere and Earth's surface.

Figure 18.7

How greenhouse gases warm the surface of Earth

Carbon dioxide (CO_2), water vapor (H_2O), methane (CH_4), and nitrous oxide (N_2O) are known as greenhouse gases because they absorb and trap heat that would otherwise radiate away from Earth. ▶️

Q1: How much of the incoming solar energy is reflected back to outer space?

Q2: What kind of energy is reemitted to the atmosphere after being absorbed by Earth's surface?

Q3: How are greenhouse gases like a blanket on your bed at night?

climate, and increasing wildfires are devastating that tree cover.

"The big question is, What's the limit for these tropical forests? How much fire can they withstand before they convert to something else?" says Balch. That something else would be grasslands, also called savanna. If major parts of the Amazon rainforest were to convert to savanna, the climate would be significantly affected, and there would be no easy way—perhaps no way at all—to revert that land back to rainforest.

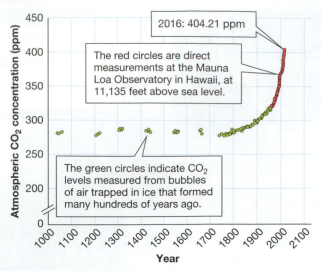

Figure 18.8

Atmospheric CO$_2$ levels are rising rapidly

Atmospheric CO$_2$ levels (measured in parts per million, or ppm) have increased greatly in the past 200 years.

Q1: What measurements do the green circles represent?

Q2: What measurements do the red circles represent?

Q3: For approximately how many years has the Mauna Loa Observatory been recording CO$_2$ levels?

Figure 18.9

Global temperatures are on the rise

Average global temperature has increased greatly over the 140 years since it has been recorded.

Q1: In what years were global temperatures the lowest?

Q2: In what years were global temperatures the highest?

Q3: What trend is apparent in this graph of actual global temperatures?

Fire and Water

During her first burn in 2004, Balch was surprised to see that the plants in the two experimental plots were fairly resistant to fire. The Amazon's tall, dense tree canopy creates humid air below it, protecting the forest from most fires, which tend to burn slowly under such conditions. For this reason, Balch's experimental plots initially suffered little damage.

That wasn't true in the following years. Remember that the Amazon saw a major drought in 2005, and droughts cause tree death, allowing more sunlight to reach the forest floor and dry the leaves and shrubs there, priming the forest for fire (**Figure 18.10**). During the planned burns after the drought in 2005, the two experimental plots experienced major damage. The majority of the trees on the plots died, and grasses took their place—particularly at the forest edge, where it was hottest and driest.

In both the 1-year and the 3-year burned plots, grasses invaded from adjacent pasturelands. This experimental result suggests that wildfires do indeed push the Amazon toward a drier,

Figure 18.10

Before and after the Amazon drought of 2005

The drought devastated the forest to this extent over much of the forest's range.

Forest Devastation

Agriculture is the main driver of tropical deforestation, and Brazil clears more land, by far, than any other country in the world. Since 1990, Brazil has cleared over 42 million hectares of forest, an area the size of California. Unless drastic measures are taken, deforestation will continue as farmers and companies clear land to meet rising global food demands.

Assessment available in smartwork5

Causes of deforestation
(Latin America)

- 67%
- 27%
- 3%
- 2%
- 1%

- ● Agriculture (commercial)
- ● Agriculture (local/subsistence)
- ● Infrastructure (e.g., roads)
- ● Mining
- ● Urban expansion

Drivers of forest degradation over time
(Latin America)

- 73%
- 16%
- 8%
- 3%

- ● Timber logging
- ● Uncontrolled fires
- ● Wood for fuel and charcoal
- ● Livestock grazing in forest

Top 5 countries that cleared the most forest between 2000 and 2005

Deforested area, in thousands of hectares

Brazil	Indonesia	Sudan	Myanmar	Zambia
15,515	9,357	2,945	2,332	2,224

How Big Is Your Ecological Footprint?

An action or process is **sustainable** if it can be continued indefinitely without causing serious damage to the environment. The current human impact on the biosphere is *not* sustainable.

Each of us can help build a more sustainable society. We can advocate legislation that fosters less destructive and more efficient use of natural resources, patronize businesses that take measures to lessen their negative impact on the planet, support sustainable agriculture, and modify our own lifestyles. For example, we can increase our use of renewable energy and energy-efficient appliances; reduce all unnecessary use of fossil fuels (for instance, by biking to work or using public transportation); buy seafood from sustainable fisheries; use "green" building materials; and reduce, reuse, and recycle waste. Experts estimate that more than 200 million women around the world wish to limit their family size but have no access to family planning. Those of us who live in developed countries can support aid efforts that provide education, health care, and family-planning services in developing countries.

One measure of sustainability is an **ecological footprint**, which is the area of biologically productive land and water that an individual or a population requires to produce the resources it consumes and to absorb the waste it produces. Scientists compute an ecological footprint using standardized mathematical procedures and express it in *global hectares* (*gha*). One gha is equivalent to one hectare (2.47 acres) of *biologically productive* space. Approximately one-fourth of Earth's surface is considered biologically productive; this definition excludes areas such as glaciers, deserts, and the open ocean.

According to recent estimates, the ecological footprint of the average person in the world is 2.7 gha, which is about 60 percent higher than the 1.7 gha that would be needed to support each of the world's 7.4 billion people in a sustainable manner. An ecological footprint can also be expressed in **Earth equivalents**, the number of planet Earths needed to provide the resources we use and absorb the wastes we produce. Currently, the global population uses 1.6 Earth equivalents each year (as shown at the bottom of the figure on the facing page).

Overall, such estimates suggest that, since the late 1970s, people have been using resources faster than they can be replenished—a pattern of resource use that, by definition, is not sustainable. As the world population grows, the amount of biologically productive land available per person continues to decline, increasing the speed at which Earth's resources are consumed.

The per capita consumption of Earth's resources by different countries is most directly related to energy demand, affluence, and a technology-driven lifestyle. As people in populous countries such as China and India become wealthier, their ecological footprints are growing rapidly.

What is *your* ecological footprint? If you are a typical American college student, your footprint is probably close to the U.S. average of 8.2 gha. It would take nearly five planet Earths to support the human population if everyone on Earth enjoyed the same lifestyle that you do (see the top row of the accompanying figure). Your ecological footprint depends on four main types of resource use:

1. *Carbon footprint*, or energy use

2. *Food footprint*, or the land and energy and water it takes to grow what you eat and drink

3. *Built-up land footprint*, which includes the building infrastructure (from schools to malls) that supports your lifestyle

4. *Goods-and-services footprint*, which includes your use of everything from home appliances to paper products

If you drive a gas guzzler, live in a large suburban house, routinely eat higher up on the food chain (more beef than grains or veggies/fruits), and do not recycle much, your footprint is likely to be higher than that of a person who uses public transportation, shares an apartment, eats mostly plant-based foods, and sends relatively little to the local landfill. Most of us can significantly reduce our ecological footprint with little reduction in our quality of life, while bestowing an outsized benefit on our planet.

---

grassier savanna-like ecosystem (**Figure 18.11**). "This could be the trigger for changing forest to savanna," says Balch. "The message is that repeated disturbances, such as multiple fires or drought and fire, cause forest to reach its tipping point, to shift into grassland."

While Balch and her team observed changes in plants after fire, ecologist Michael Coe investigated how fire affects its nemesis: water. "The forest is an incredibly important part of the **hydrologic cycle**," the circulation of water from the land to the sky and back again, says Coe (**Figure 18.12**).

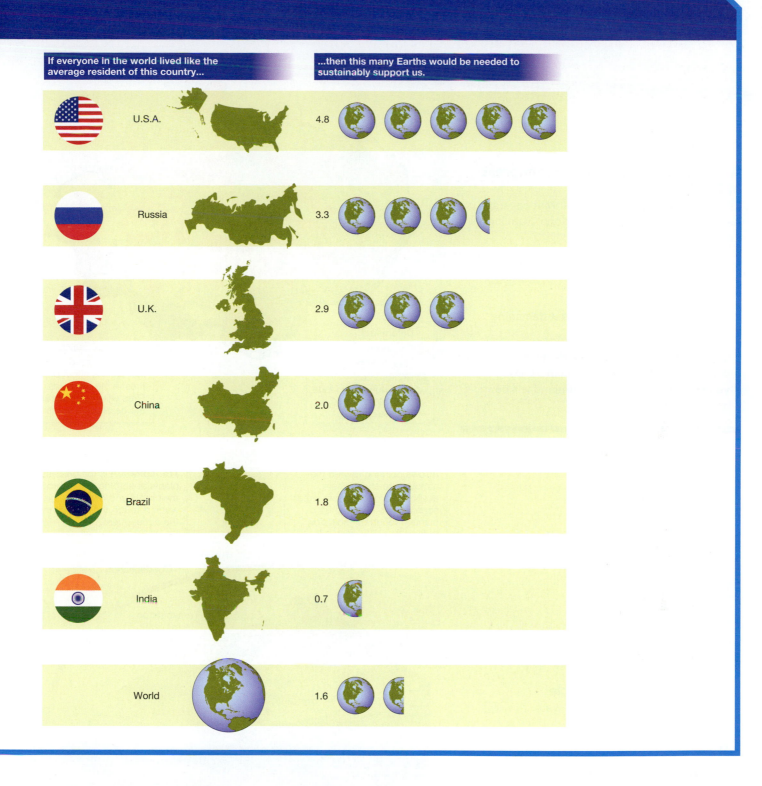

If everyone in the world lived like the average resident of this country...	...then this many Earths would be needed to sustainably support us.
U.S.A.	4.8
Russia	3.3
U.K.	2.9
China	2.0
Brazil	1.8
India	0.7
World	1.6

Near the equator, direct sunlight causes water to evaporate from Earth's surface. The warm, moist air rises because heat causes it to expand, making it less dense and lighter than air that has not been heated. But as it rises, the warm, moist air begins to cool. Because cool air cannot hold as much water as warm air can, much of the moisture from a cooling airmass is "wrung out" and falls as rain.

For this reason, most tropical regions, including the Amazon, receive ample rainfall. Earth has four giant **convection cells** in which warm,

Figure 18.11

A projection of the Amazon rainforest in 2030

This map of future Amazon rainforest coverage is based on the assumption that global warming will persist but not increase.

Q1: Where will fire most seriously affect the Amazon rainforest?

Q2: Where will fire be the least damaging to the Amazon rainforest?

Q3: This map does not include an increase in pasturelands for grazing animals. Do you think more or less pastureland will be needed in 2030? Explain.

The historical forest cover in this area is denoted by the dark black outline.

If global warming continues to increase, dry areas will likely increase as well.

- ☐ Forest
- ☐ Deforested
- ☐ Dry and/or logged forest
- ☐ Nonforested area (now grasslands and fields, due to climate change)

This depiction does not include the effect of fires in the dry areas. Fires will likely increase the deforested areas.

Figure 18.12

The hydrologic (water) cycle

Arrows indicate the direction of water flow. ▶

Q1: What is transpiration?

Q2: Why is transpiration important to the water cycle?

Q3: If there are fewer plants and therefore less transpiration in a given area, what will happen to the humidity or cloud cover in this area?

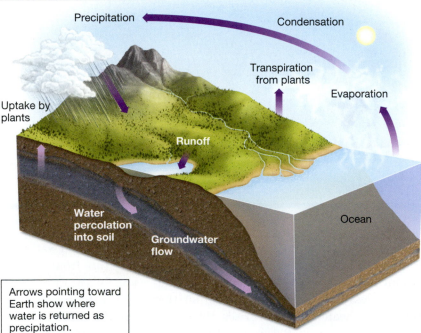

Arrows pointing toward the atmosphere show where evaporation occurs from water sources or is released from plants through transpiration to form humidity or clouds.

Transpiration is the process of plants absorbing water through their roots and releasing this water through their leaves into the atmosphere.

Precipitation   Condensation

Transpiration from plants

Evaporation

Uptake by plants

Runoff

Water percolation into soil

Groundwater flow

Ocean

Arrows pointing toward Earth show where water is returned as precipitation.

moist air rises and cools, releasing moisture as rain or snow depending on temperature, and then sinks back to the ground as dry air (**Figure 18.13**). These convection cells, in combination with the angle of sunlight striking the Earth, play a large role in the creation of regional environments on Earth, such as rainforests and deserts.

In the Amazon, trees are very important players in the water cycle, adds Coe. "The trees pull water out of the soil and evaporate it into the atmosphere in the process of photosynthesizing (through transpiration), so they're the mediators between the rainfall and the streams," he says. "Burning the trees greatly reduces the amount of water getting back into the atmosphere." (See Chapter 5 for more on photosynthesis.)

Each year of Balch's study, Coe joined Balch at the test site. There, his team dug 10-meter-deep soil pits—long, dark caverns in which they inserted instruments to measure the moisture content of the soil. In a healthy forest ecosystem, trees absorb a lot of the water in the soil, leaving it nice and dry, with only minimal water runoff into streams. This is what Coe observed in the control plot where nothing was burned. "A healthy forest uses up almost all the water [in the soil]," says Coe.

But in the other two plots—burned every year or every 3 years—he found that the soil in the pits was wet to the touch. When the forest burned, trees died, so nothing absorbed the moisture from the soil. Consequently, nearby streams were overflowing with water—up to four times the volume of water seen in healthy forests. "That's not a good thing," says Coe. "We're circumventing the natural cycle. Instead of this water going back into the atmosphere, creating more rain and driving vegetation, it's flushing the water out of the system."

And on the burned plots, the invasive grasses that took the place of the trees have very shallow roots, absorb less moisture from the ground, and evaporate less water into the air. In this way, deforestation of large areas—whether through unintentional wildfires or the intentional cutting down of trees—results in less rainfall and hotter temperatures. "If you deforest enough of it, you're going to really decrease the rainfall over a broad swath of this region," says Coe.

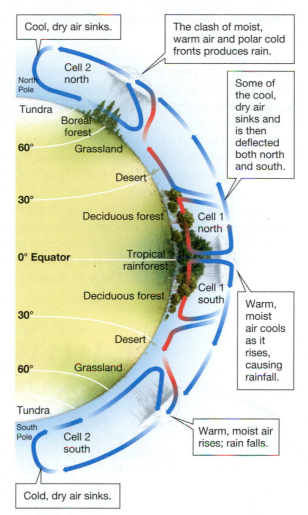

Figure 18.13

Earth has four giant convection cells

Two giant convection cells are located in the Northern Hemisphere and two in the Southern Hemisphere.

Q1: How do the patterns of rainfall in the Northern and Southern Hemispheres compare?

Q2: How do the patterns in the kinds of environments shown in the Northern and Southern Hemispheres compare?

Q3: What happens at the equator to make this region so wet?

The Carbon Games

In addition to measuring the hydrologic cycle and the expansion of grasses into the region, the teams led by Balch and Coe measured how much carbon was released into the atmosphere when the forests burned.

Carbon, in the form of CO_2 gas, makes up only about 0.04 percent of Earth's atmosphere, although that percentage has been creeping upward every year for the last 200 years and causing global warming, as we've seen. Carbon is also found in Earth's crust, where carbon-rich sediments and rocks formed from the remains of ancient marine and terrestrial organisms. Carbon is present in every living thing.

Living cells are built mostly from organic molecules—molecules that contain carbon atoms bonded to hydrogen atoms. After oxygen, carbon is the most abundant element in cells by weight; every one of the large biomolecules in an organism has a backbone of carbon atoms. Living organisms, in both aquatic and terrestrial ecosystems, acquire carbon mostly through photosynthesis. Aquatic producers, such as photosynthetic bacteria and algae, absorb dissolved CO_2 and convert it into organic molecules using sunlight as a source of energy. Plants, the most important producers in terrestrial ecosystems, absorb CO_2 from the atmosphere and transform it into food with the help of sunlight and water. This is how a plant builds itself—its leaves, stems and branches, flowers, and other structures.

The transfer of carbon within biotic communities and their physical surroundings, the abiotic world, is known as the global **carbon cycle** (**Figure 18.14**). (For more on nutrient

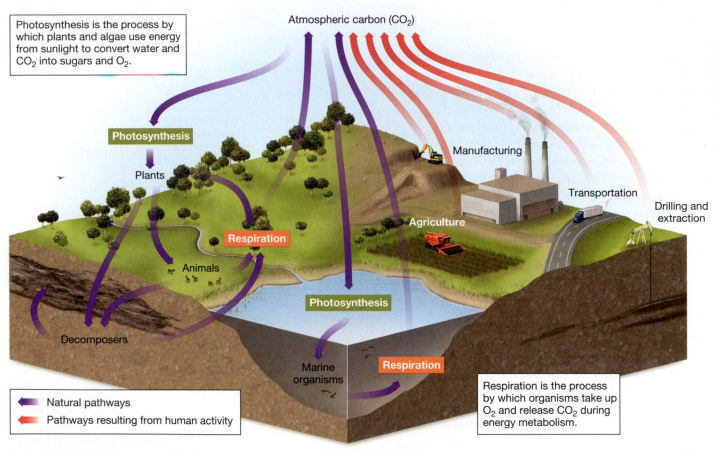

Figure 18.14

The carbon cycle

Arrows indicate the direction of carbon flow.

Q1: What are three ways that carbon is released into the atmosphere?

Q2: Are all of the pathways you listed for question 1 affected by human activity?

Q3: What are two biotic reservoirs of carbon?

cycling, see Chapter 21.) One way that carbon is transferred between the biotic and abiotic worlds is through combustion—the burning of carbon-rich materials, living or not.

Some of the organic matter from ancient organisms has been transformed by geologic processes into deposits of fossil fuels such as petroleum, coal, and natural gas. When we extract these fossil fuels and burn them to meet our energy needs, the carbon that was locked in these deposits for hundreds of millions of years is released into the atmosphere as carbon dioxide.

Plants also release carbon back into the atmosphere when they are burned. To assess the amount of carbon released by a burning rainforest, Balch and her team measured the amount of leaf litter and branches—the biomass—on the forest floor both before and after a burn, as well as the number of trees affected. Half of that biomass is carbon, so half of the difference between the biomass before and after the burn is the amount of carbon that was released into the atmosphere. It was a lot.

The very first burn, for example, released "about 20 tons of carbon per hectare," says Balch. That's approximately equal to the carbon emitted by a passenger car driven 864,000 miles (burning about 40,000 gallons of gasoline). And burns have another, indirect, detrimental effect that Balch wasn't able to directly measure: they destroy trees that normally absorb carbon dioxide from the air. In a normal year, trees in the Amazon absorb approximately 1.5 billion tons of carbon dioxide, helping to slow climate change. In this way, they act as a **carbon sink**, a natural or artificial reservoir that absorbs more carbon than it releases.

But during the 2005 drought, as trees died and rotted, the rainforest stopped absorbing more carbon dioxide than it gave off, and it actually released carbon *into* the atmosphere. In 2005, 5 billion tons of carbon were released. Instead of acting as a carbon sink, the Amazon became a **carbon source**, a reservoir that releases more carbon than it absorbs (**Figure 18.15**). And with increasing temperatures, decreasing rainfall, and increasing wildfires, ecologists fear that the Amazon could transition more and more into a carbon source instead of a carbon sink. If that happens, our planet will lose one of its greatest buffers against future climate change.

Figure 18.15

Carbon sinks and sources

Q1: How does a carbon source contribute to global warming?

Q2: How does a carbon sink protect against global warming?

Q3: How can trees act as both a source and a sink?

Waiting and Watching

Balch and her crew have stopped burning the plots and have begun to record what happens to them as they recover. In addition to making direct observations of which plants grow back and how their growth affects the water and carbon cycles, the team is conducting new experiments on the burned plots to see whether it is possible to help prevent the invasion of grasses and encourage a return to rainforest.

In one area, the scientists are planning to plant different types of trees to see whether particular species can successfully reestablish themselves in a mat of grasses and retake that land. Another experiment, also still in the planning phase, will test different amounts of watering to see whether the lack of precipitation is preventing the growth of new trees. If any strategies work well on the small plots, perhaps they can be adapted to save larger areas of rainforest affected by fire.

"We're hoping we find ways to help the forest recover," says Balch. Still, the researchers worry about the future of the region. "Have we crossed a threshold where this part of the forest isn't going to grow back as rainforest?" asks Coe, worry in his voice. If that is true for the small experimental plots, it bodes danger for large regions of the Amazon that border agricultural land and are threatened by wildfires.

REVIEWING THE SCIENCE

- **Ecology** is the study of interactions between organisms and their environment. All ecological interactions occur in the **biosphere**, which consists of all living organisms on Earth (**biotic** factors), together with the physical environment they inhabit (**abiotic** factors).

- **Weather** consists of the short-term atmospheric conditions in a given area, such as temperature, precipitation, wind, humidity, and cloud cover. **Climate**, the prevailing weather of a specific region over relatively long periods of time, has a major effect on the biosphere. Climate is determined by incoming solar radiation, global movements of air and water, and major features of Earth's surface.

- **Climate change** is a large-scale and long-term alteration in Earth's climate, and it includes such phenomena as more extreme temperatures, change in rainfall patterns, and increased frequency of violent storms.

- Carbon dioxide acts as a **greenhouse gas** and contributes to the **greenhouse effect** by absorbing heat released at Earth's surface, rather than allowing it to leave the atmosphere. The greenhouse effect has led to a rapidly rising average global temperature, a phenomenon known as **global warming**.

- The concentration of **carbon dioxide** (**CO_2**) gas in the atmosphere is increasing at a dramatic rate because of an increase in **carbon sources**, such as the release of CO_2 through the burning of fossil fuels, and a loss of **carbon sinks**, such as the absorption of CO_2 through photosynthesis in large forests.

- Both the **hydrologic cycle** and the **carbon cycle** are affected by global warming, which disrupts the natural cycling of water and carbon molecules in the biosphere.

- The area of the biosphere required to produce the resources and to absorb the waste produced by an individual or population is known as the **ecological footprint**. The ecological footprint is often expressed in **Earth equivalents**, the number of planet Earths needed to provide the resources and absorb the waste of an individual or population.

THE QUESTIONS

Challenge Yourself

1 The carbon cycle and the water (hydrologic) cycle are similar in that

(a) both cycle molecules between abiotic and biotic components of the environment.

(b) photosynthesis is a critical process in both.

(c) both are involved in global warming and climate change.

(d) all of the above

2 An ecologist is studying an area of forested land that experiences a forest fire every 5 years. In the years between forest fires, the area of study absorbs 300 million tons of CO_2. If a forest fire releases 200 million tons of CO_2 in this area, does this forested area act as a carbon sink or a carbon source? Why?

3 Place the following events in the correct order of their onset by numbering them from 1 to 5, beginning with the earliest.

_____ a. Climate change, including more extreme weather

_____ b. Increase in human population and activities

_____ c. Increase in release of CO_2 and other greenhouse gases into the atmosphere

_____ d. Increase in calls to control the production and release of greenhouse gases

_____ e. Global warming

Try Something New

4 Which of the following behaviors could *decrease* the production of greenhouse gases, and thus perhaps curb global warming?

(a) using ethanol or biodiesel rather than fossil fuels to power autos

(b) using coal rather than gas to warm homes

(c) using solar or wind power rather than fossil fuels to produce electricity

(d) all of the above

(e) none of the above

5 Reviewing Figure 18.13, identify the environment at your latitude, and explain how it is influenced by Earth's convection cells.

6 Discuss briefly how the more extreme weather caused by global warming may affect where you live.

Leveling Up

7 **Looking at data** Review Figures 18.8 and 18.9. Describe the change in atmospheric CO_2 levels from 1880 to the present, and the change in average global temperature for the same time period. What were the CO_2 level and the average temperature in 1880? in 1960? in 2016? Is there a relationship between the two variables? How would you explain the pattern you see here?

8 **Life choices** You can estimate your impact on the planet by using one of the many "footprint calculators" on the Internet. Take three online quizzes available from organizations such as the Global Footprint Network, the Nature Conservancy, and the U.S. Environmental Protection Agency. Each site calculates your ecological footprint a little differently. Compare and contrast the different sites you used. Which one do you think is the most accurate? Why? Which one do you think is the most superficial? Why? Write a list of ways you can decrease your ecological footprint, and try to adhere to your new lifestyle!

9 ***Write Now* biology: climate denial** Watch the documentary *Frontline's Climate of Doubt* (2012), produced by PBS. Write an essay reflecting on what you learned from the film and from this chapter about global warming. Use your knowledge of your ecological footprint to reflect on how important it is for all citizens to be aware of this global crisis.

Thinking Multidimensionally

10 Agricultural practices are one of the largest sources of greenhouse gases next to carbon emissions from the burning of fossil fuels.

a. Which farm animals are the biggest contributors?

b. Ruminant animals have multichambered stomachs filled with bacteria to digest tough cellulose from plants. How does the structure of this digestive tract contribute to the release of greenhouse gases, and which gas are they primarily releasing?

c. Which greenhouse gas traps more heat around the planet, carbon dioxide or methane?

d. While the reality of major U.S. dietary shifts is unlikely, what would happen to greenhouse gas emissions if beans replaced beef in the United States?

11 Imagine you are scuba diving in the waters around Queensland, Australia, and you see a scene similar to the image drawn above.

a. Describe the realistic colors of the corals in the image.

b. Corals exist as a colony of smaller organisms called polyps. The polyps live in a symbiotic relationship with colorful algae called zooxanthellae. What role do the algae perform?

c. The colors of coral reefs are changing due to climate change and other factors. What color are the corals turning, and what is this process called?

For more, visit digital.wwnorton.com/bionow2 for access to:

Zika-Busting Mosquitoes

Genetically engineered insects help stop the spread of dengue, malaria, and Zika.

After reading this chapter you should be able to:

- Articulate the difference between population size and population density.
- Interpret graphical population data to determine whether the population exhibits logistic or exponential growth.
- Explain the concept of carrying capacity, and give examples of factors that can increase or decrease carrying capacity for a population.
- Differentiate between density-dependent and density-independent changes in population size.
- Illustrate population cycles of predators and prey.

CHAPTER
19

GROWTH OF POPULATIONS

DRIVING QUESTION

How can genetically modified mosquitoes be used to conquer the spread of mosquito-borne diseases?

- HS-LS2.A: Interdependent Relationships in Ecosystems
- HS-ETS1.A: Defining and Delimiting Engineering Problems
- HS-ETS1.B: Developing Possible Solutions

t is impossible to know when an infected mosquito bit the first person in Miami-Dade County, Florida. But on Friday, July 29, 2016, health officials confirmed that a woman and three men in a small area north of downtown Miami had been infected with the Zika virus through the bite of a local mosquito. It was the moment everyone had been dreading: the first known cases of the virus being transmitted by mosquitoes in the United States. "As we have anticipated, Zika is now here," Tom Frieden, the director of the U.S. Centers for Disease Control and Prevention (CDC), told reporters that afternoon (**Figure 19.1**).

By July 2016, Zika had become a household name. Zika fever, the illness that results from infection with Zika virus, is a mild sickness that typically lasts less than a week and involves fever, red eyes, headache, and a possible rash. But Zika has another, much more dire effect on health. A year before the Miami outbreak, doctors in Brazil had begun noticing an increasing number of infants born with a serious birth defect known as microcephaly, in which the head and brain are smaller than expected; they have not developed properly. Microcephaly can cause many problems in growing infants, including seizures, speech and intellectual delays, feeding and movement problems, and hearing and vision loss. Brazil's new cases of microcephaly were soon linked to a rapidly spreading virus passed through the bite of a mosquito: Zika.

The scientific evidence linking Zika virus and microcephaly is strong: infected mothers pass the virus to their fetuses, where it may stunt brain growth (**Figure 19.2**). Studies suggest that infants exposed to the Zika virus in utero are more than 50 times more likely to be born with microcephaly than those who are not exposed.

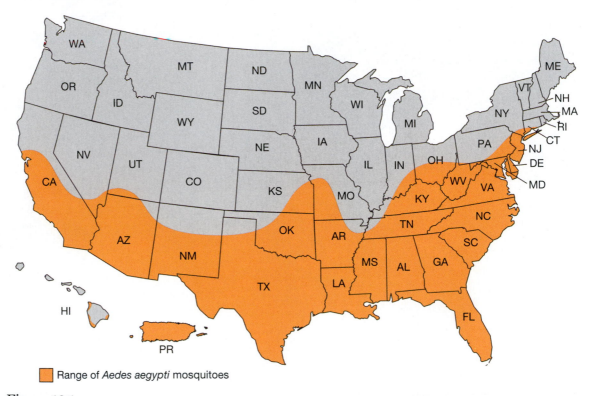

Range of *Aedes aegypti* mosquitoes

Figure 19.1

Range of the mosquito species most likely to carry the Zika virus

Aedes aegypti mosquitoes spread viruses like Zika, dengue, and chikungunya more often than other mosquito species do.

Q1: What parts of the United States are within the range of the mosquito that carries Zika?

Q2: What areas are *not* within the mosquito's range? Why do you think that is?

Q3: Find your own state, and where you are in the state. Are you at risk of contracting Zika?

In February 2016, the World Health Organization declared the spread of Zika virus to be a "public health emergency of international concern."

Since the outbreak in Brazil, governments around the world have begun looking for ways to stop Zika, and the local transmission in Florida galvanized U.S. officials and scientists. "It's scary," says Matthew DeGennaro, a biologist who studies mosquito genetics and behavior at Florida International University in Miami. "But people in other countries face these things all the time, and not much is being done about it. It is possible this [outbreak] will focus Americans' attention on mosquito-borne illness."

The human species has a long history of battling mosquito-borne viruses, including dengue virus and West Nile virus. Along the way, we've learned that when a vaccine is lacking, the best way to stop the virus is to stop the insect. Yet despite efforts to eradicate mosquito populations around the world, the pests continue to roar back no matter what we throw at them. Mosquitoes quickly develop resistance to common insecticides, for example, and in at least one case they showed an adapted behavior, attacking farmers outdoors in the early morning rather than indoors during the night when the farmers were protected by mosquito nets.

But now there's a new weapon in the arsenal against mosquito-borne diseases. Taking a technological approach to the problem, scientists have begun to target mosquito populations by using the insects themselves as weapons. It's a new kind of war—mosquito versus mosquito.

Population Control

One month after the first four reported cases of Zika transmission in Florida, officials captured Zika-infected mosquitoes at seven different locations on Miami Beach—the smoking gun to show that local disease transmission was indeed occurring. By October 2016, mosquitoes bearing Zika virus had spread another 3 miles north.

Zika is spread by females of both the *Aedes aegypti* species and the less common *Aedes albopictus* species, although the latter is less likely to bite humans because it breeds in rural areas and feeds on other animals in addition to humans. These two types of mosquitoes also

Figure 19.2

A Brazilian mother holds her microcephalic baby

The incidence of microcephaly in newborns has skyrocketed, and the increase has been linked to Zika infection of the mother during pregnancy.

serve as carriers, or *vectors*, for dengue, chikungunya, and other viruses. (Malaria, which is caused by parasitic protists, is transmitted among humans by female mosquitoes of the genus *Anopheles*.) Active primarily during the day, a female mosquito feeds on human blood via a tubelike mouthpart (which male mosquitoes lack) that pierces the skin of a host. During that moment, virus particles in the mosquito's saliva can be transferred to the person's skin. Once on the skin, the virus is able to replicate in skin cells and then can spread to the lymph nodes and bloodstream.

In addition to being transmitted by mosquitoes, Zika can spread from person to person via body fluids: blood, tears, semen, and saliva. In some cases, the virus can remain in those fluids for months after

MATTHEW DEGENNARO

Matthew DeGennaro is a neurogeneticist at Florida International University in Miami, studying mosquito genetics and behavior. As a Florida resident, he witnessed the outbreak of Zika in Miami and advocates the use of genetically modified mosquitoes to control mosquito populations and stop the spread of disease.

infection (**Figure 19.3**). Slowing the spread of Zika, therefore, involves the screening of donated blood and the proper use of condoms during sex. But true prevention ultimately requires decreasing the incidence of mosquito bites, which can be accomplished by getting rid of the mosquitoes.

Scientists have taken many approaches to trying to reduce the mosquitoes' **population size**, the total number of individuals in the population. A **population** is a group of organisms of the same species in a defined area. Population size tends to change over time—sometimes increasing, sometimes decreasing. Whether the size of a population increases or decreases depends on the number of births and deaths in the population, as well as on the number of individuals that enter (immigrate) or

PROTECT YOUR FAMILY AND COMMUNITY:
HOW ZIKA SPREADS

Most people get Zika from a mosquito bite

A mosquito bites a person infected with Zika virus

The mosquito becomes infected

A mosquito will often live in a single house during its lifetime

The infected mosquito bites a family member or neighbor and infects them

More mosquitoes get infected and spread the virus

More members in the community become infected

Other, less common ways, people get Zika:

During pregnancy
A pregnant woman can pass Zika virus to her fetus during pregnancy. Zika causes microcephaly, a severe birth defect that is a sign of incomplete brain development

Through sex
Zika virus can be passed through sex from a person who has Zika to his or her sex partners

Through blood transfusion
There is a strong possibility that Zika virus can be spread through blood transfusions

CDC

Figure 19.3

How Zika is transmitted
The CDC produces posters such as this for public health issues including Zika, sexually transmitted diseases, and the flu.

Q1: What is the main way by which someone is infected with the Zika virus?

Q2: Judging by the poster and your knowledge of mosquito behavior, what can you do to decrease your risk of being infected with the Zika virus?

Q3: Besides the transmission methods shown on the poster, what are some other ways you could become infected with the Zika virus?

leave (emigrate) the population. Birth and immigration increase population size; death and emigration reduce it. Environmental factors also have a strong impact on population size. Mosquito populations, for example, increase in warmer, wetter weather when the conditions are ripe for reproduction.

To target mosquito populations, scientists rely on **population ecology**, the study of the size and structure of populations and how they change over time and space. Population ecology can help us determine where mosquitoes live, eat, and breed. Mosquitoes congregate and lay their eggs at water sources, and many prefer urban environments with trash receptacles and concentrated groups of people. At these locations we find the highest **population density**, or number of individuals per unit of area.

To calculate population density, total population size is divided by the corresponding area of interest. For example, in an urban neighborhood of Rio de Janeiro, Brazil, where people were being regularly infected with dengue virus, scientists captured mosquitoes in mosquito traps and calculated the neighborhood's population of female mosquitoes to be 3,505 pregnant females. The neighborhood has an area of 911 acres, so the female mosquito population density was 3.85 mosquitoes per acre (3,505 mosquitoes divided by 911 acres). Keep in mind that population density is often difficult to measure, because it depends on an accurate count of the population size. Individuals may be hard to detect, may move between populations, or may inhabit a complex, hard-to-define area.

Scientists can track population density to see how well eradication efforts work against mosquitoes. As mentioned earlier, one of the most popular ways of reducing mosquito populations is poisoning them with insecticides. But that's not ideal, scientists agree. "Insecticides affect the entire insect population usually, and that leads to overall imbalances in the environment," says DeGennaro. "So putting a chemical into the environment is something you need to do very carefully."

An alternative approach, thanks to advanced genetic technologies, is to tweak the genes of the mosquitoes to sabotage their reproduction. If the mosquitoes can't reproduce, population size will plummet, and fewer mosquitoes means fewer people being bitten and infected. This gene tweaking is called **genetic modification**, or **GM**—altering the genes of an organism for a specific purpose. It is also referred to as genetic

engineering, or GE. In contrast to insecticides, GM mosquitoes can directly target just one insect species, says DeGennaro. "You're only affecting one species of mosquito, the species that is the primary vector for Zika, dengue, and more."

Early field tests of GM mosquitoes in Brazil appeared to stop the spread of dengue. But could they do the same for Zika?

Rapid Spread

Before it hit Florida, Zika virus struck Brazil like a tsunami. In a span of 3 months in early 2015, Brazil documented nearly 7,000 cases of mild illness similar to dengue fever, though no one yet suspected Zika. That's the problem with a newly emerging infectious disease: no one is looking for it.

Then, in May, a national laboratory reported that Zika virus was circulating around the country. Two months later, Brazilian health workers began reporting neurological disorders associated with Zika infection, including brain inflammation. In October came the worst news of all: reports of an unusual spike in newborns with microcephaly.

That dramatic spread and link to microcephaly put Zika on the map, but it isn't actually a new virus. Zika virus—a round particle with a dense core packed with RNA (**Figure 19.4**)—was first

Zika virus particle

Figure 19.4

Zika is a spherical RNA virus

isolated in 1947 in the Zika Forest of Uganda, for which it is named. The first human cases of Zika were detected in Uganda in 1952, and the little-studied virus subsequently spread across central Asia, the South Pacific, and into South and Central America and the Caribbean.

Humans are no strangers to mosquito-borne diseases (see "Mosquito-Borne Diseases"). While Zika and other diseases were once isolated to specific areas of the globe, global travel and human population growth, along with climate change, have increased mosquitoes' ability to proliferate and spread disease on a grand scale. In the second half of the twentieth century, our planet saw the fastest rate of human population growth in its history. **Population doubling time**—the time it takes a population to double in size—is a good measure of how fast a population is growing. For example, the doubling time for the U.S. human population at our current reproductive rate is about 100 years. In stark contrast, the population doubling time for mosquitoes can be as little as 30 days in an optimal environment.

Reaching Capacity

Now back to mosquitoes, and how to get rid of them. Scientists argue strongly that understanding mosquito population ecology is a prerequisite for eliminating mosquito-borne diseases. For example, many mosquitoes lay their eggs directly on the surface of water. That knowledge led to one of the most important public health efforts to reduce mosquito-borne diseases: the removal of pools of standing water. Eliminating these breeding sites makes it possible to reduce the **carrying capacity** of the environment, the maximum population size that can be sustained in a given area.

In most species, the growth rate of a population decreases as the population size nears the carrying capacity, because resources such as food and water begin to run out. If there are only a few locations of standing water in a given area, for example, a limited number of eggs can be laid in that area, thereby limiting the size of the population. Any limiting resource needed for survival, such as habitat, food, or water, will determine an environment's carrying capacity for a specific population. Because different species have unique needs, the same environment can have different carrying capacities for different resident species. For the same reason, two different environments in which the same species lives will have different carrying capacities. At the carrying capacity, the population growth rate is zero.

If a population has no constraints on its resources, it will experience **exponential growth**, which occurs when a population increases by a constant proportion over a constant time interval, such as one year. Exponential growth is represented by a **J-shaped growth curve** (**Figure 19.5**). A population that approaches its carrying capacity because of constrained resources will experience **logistic growth**, in which the population grows nearly exponentially at first but then stabilizes at the maximum population size that can be supported indefinitely by the environment. Logistic growth is represented by an **S-shaped growth curve**.

Mosquito-Borne Diseases

Mosquitoes cause more human suffering than any other organism (with the exception of our own species). Here are just a few of the deadly or debilitating diseases they carry:

- *Dengue virus* is a huge problem for the population of Brazil. Spread by the same mosquitoes as Zika, dengue also has no vaccine or cure, and it causes a disease chillingly known as "breakbone fever." Symptoms include continuous joint pain, vomiting, high fever, and severe rashes. In 2015, Brazil chalked up a record 1.6 million cases and 863 deaths to dengue—an 82.5 percent increase over 2014.

- *Malaria*, an ancient disease that has been traced as far back as 2700 BCE, affects an estimated 40 percent of the world's population. Hundreds of thousands of deaths are still reported annually from malaria.

- *Yellow fever*, against which there is a vaccine, still leads to an estimated 30,000 deaths per year.

- *West Nile virus*, which emerged in the 1930s, can invade the nervous system and cause death.

- *Chikungunya* is found mainly in Africa and Asia, but cases have been reported in Florida. The main symptoms are fever, rash, and joint pain that can last for years. Only one in 1,000 cases results in death.

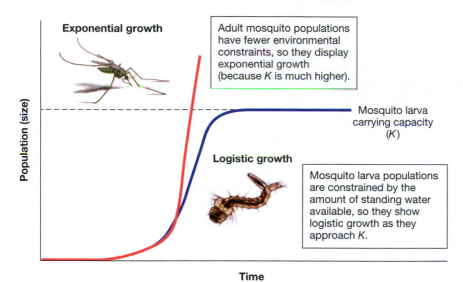

Figure 19.5

Populations can experience exponential growth or logistic growth

A population that is not constrained by resources or by the environment can grow exponentially, whereas one that is constrained by a set carrying capacity will show logistic growth.

Q1: Which form of population growth displays a J-shaped curve?

Q2: Which form of population growth displays an S-shaped curve?

Q3: Describe a situation in which a population initially shows exponential growth and later shows logistic growth.

Little research has been done on growth curves for mosquito populations, primarily because the insects are small and airborne, making them hard to track. But we do know how the human population has changed: Over the last 500 years, Earth's human population exhibited both logistic and exponential growth. At the end of the last ice age, in approximately 10,000 BCE, there were only 5 million people on Earth. With the advent of agriculture in about 8000 BCE, the world population began to rise logistically, until about 200 years ago. Then, alongside the use of fossil fuels and the industrial revolution, human population growth exploded exponentially. Modern populations have continued to show exponential growth to the detriment of the environment. Current estimates of the carrying capacity of Earth range from 2 billion to over 1,000 billion people, with the majority of studies insisting that 9–10 billion people is the maximum number that Earth can support. At current population growth rates, we will reach this number by the year 2050 (**Figure 19.6**).

Seeking Change

With human populations increasing in number and concentrating in urban areas, we can expect the continued spread of mosquito-borne illnesses. Thankfully, recent results from field

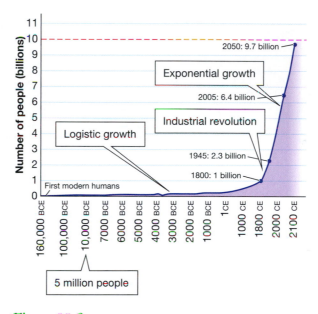

Figure 19.6

Curves of logistic and exponential growth in the world human population

The dashed line indicates the United Nations' estimated carrying capacity of Earth.

Q1: According to this graph, approximately when did exponential growth begin?

Q2: What milestone corresponds to the transition from logistic to exponential population growth?

Q3: What is the UN's projected carrying capacity of Earth, and when will we reach it?

tests of GM mosquitoes in Brazil suggest that we may have a new way to stop that spread.

Starting in 2002, British company Oxitec began producing a GM line of the *Aedes aegypti* mosquito. The company inserts a single gene into a line of lab-bred mosquitoes. This gene, designed to work only in insect cells, produces a protein that prevents the mosquito from transitioning between two of its life stages: larva and adult. In the lab, the mosquitoes are exposed to an antibiotic (tetracycline) that neutralizes the protein so that the mosquitoes can grow to be adults. Then, only males are released into the wild, because the males don't bite and so will not contribute to the spread of disease. Once free, the male GM mosquitoes mate with wild females and produce offspring with the lethal gene—and

no access to antibiotics—so the offspring die off before becoming adults and therefore do not reproduce. In this way the size of the population is reduced (**Figure 19.7**).

Populations—mosquito, human, or otherwise—can change in a density-dependent or density-independent manner. **Density-dependent population change** occurs when birth and death rates change as the population density changes. The number of offspring produced and the death rate are often density-dependent. Food shortages, lack of space, and habitat deterioration—all these factors influence a population more strongly as it increases in density (**Figure 19.8**).

In addition, when a population has many individuals, disease spreads more rapidly (because individuals tend to encounter one another more

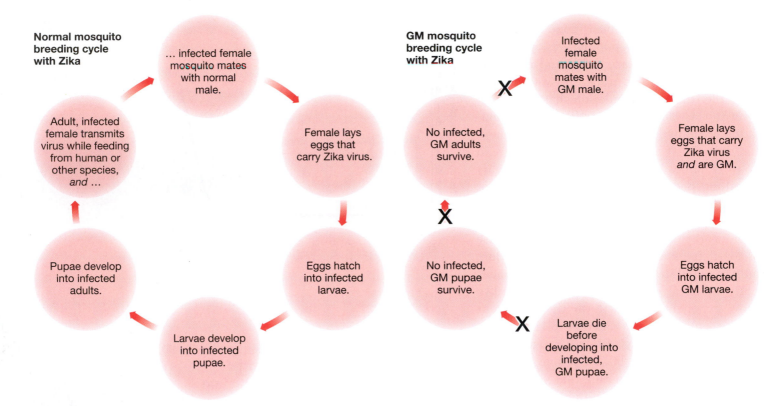

Figure 19.7

How mosquitoes are genetically modified to stop the Zika virus

Genetically modified (GM) male mosquitoes have been created and released in Brazil and Florida, where they mate with wild females and produce offspring that cannot grow to adulthood.

Q1: List all the stages of the mosquito life cycle.

Q2: Which life cycle stage is vulnerable to the GM treatment?

Q3: Why are only male GM mosquitoes released into the wild, rather than both males and females?

The number of seeds produced per plant drops dramatically under increasingly crowded conditions.

Figure 19.8

Overcrowded conditions result in density-dependent population change

Overcrowding affects many species. The plantain, shown here, is a small, herbaceous plant that has decreased reproduction in crowded conditions.

Q1: What factors may be limiting growth and reproduction in the plantain's crowded conditions?

Q2: Why are overcrowded conditions considered density-dependent population changes?

Q3: Relate this example of overcrowded conditions to the human population growth shown in Figure 19.6. How do you think the situations are similar? How are they different?

② The population rose rapidly to more than 2,000 individuals.

③ The population crashed because the reindeer heavily overgrazed their winter food source, lowering the carrying capacity of the environment.

① In 1911, 25 reindeer were introduced to St. Paul Island off the coast of Alaska.

④ By 1950, only 8 reindeer remained.

Figure 19.9

Habitat destruction results in density-dependent population change

Q1: In what year did the reindeer's numbers begin to rise exponentially?

Q2: In what years was the reindeer's population growth logistic?

Q3: How do you predict the graph (population size) would change if someone had begun bringing in supplemental food for the reindeer in 1940? Draw a sketch of your prediction.

often), and predators may pose a greater risk (because most predators prefer to hunt abundant sources of food). Disease and predators obviously increase the death rate. These changes are also density-dependent. A 2014 study found that the mosquito *Culex pipiens*, which spreads the West Nile virus, undergoes strong density-dependent population changes arising from competition for resources within the watery habitats where it lays its eggs. In other words, larvae compete with each other for food. But once they become adults, the population is no longer density-dependent, because the mosquitoes become airborne and have wider access to food.

For some organisms, if a population exceeds the carrying capacity of its environment by depleting its resources, it may damage that environment so badly that the carrying capacity is lowered for a long time. A drop in the carrying capacity means that the habitat cannot support as many individuals as it once could. Such habitat deterioration may result in widespread starvation and death, causing the population to decrease rapidly (**Figure 19.9**).

Not all population changes are due to density. **Density-independent population change** occurs when populations are held in check by factors unrelated to population density. Density-independent factors can prevent populations from reaching high densities in the first place. Year-to-year variation in weather, for example, may cause conditions to be suitable for rapid population growth. On the other hand, poor weather conditions may reduce the growth of a population directly (by freezing the eggs of a mosquito, for example) or indirectly (such as by decreasing the number of food plants available to an animal).

Other natural disturbances, such as fires and floods, can also limit the growth of populations in a density-independent way. Finally, the effects of environmental pollutants, such as the insecticide DDT, are density-independent; such pollutants can threaten natural populations with extinction (**Figure 19.10**).

Populations of many species rise and fall unpredictably over time. These **irregular fluctuations** in population size are far more common in nature than a smooth rise to a stable population size. In the 1950s, for example, Brazil mounted a massive antimosquito campaign to combat yellow fever—a disease also transmitted by *Aedes aegypti*—that included spraying insecticides and encouraging citizens to get rid of standing water. The success of the program led officials to declare in 1958 that *Aedes aegypti* had been eradicated. If only that had been true. By the 1970s the mosquitoes had come roaring back, breeding and spreading like crazy—an unwelcome irregular fluctuation. And now the pests are resistant to many chemical attacks.

Populations can also exhibit **cyclical fluctuations**, predictable patterns that occur with seasonal changes in temperature and precipitation or when at least one of two species is strongly influenced by the other. The Canadian lynx, for example, depends on the snowshoe hare for food, so lynx populations increase when hare populations rise, and they decrease when hare populations drop. In this example, the population cycles are also density-dependent population changes because each population is affected by the other's numbers. They cycle together in response to each other's density (**Figure 19.11**).

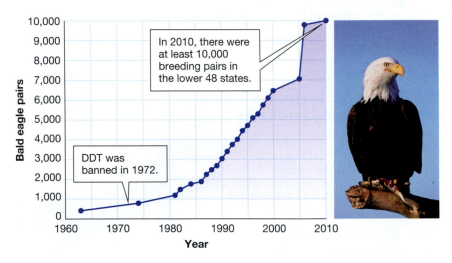

Figure 19.10

Banning the use of the pesticide DDT removed a density-independent population limit

DDT poisoning was directly responsible for declining eagle populations by the middle of the twentieth century. By the early 1960s, population counts indicated that only 417 breeding pairs of bald eagles remained in the lower 48 states—a huge drop from the estimated 100,000 breeding pairs present in 1800. Bald eagle populations increased dramatically after DDT was banned.

Q1: In what year did the bald eagle population rise to more than 2,000 breeding pairs?

Q2: Give some examples of possible density-dependent limits on bald eagle populations.

Q3: Is the population growth of bald eagles more like logistic or exponential growth? Explain why you think so.

Friendly Fight

The Oxitec GM mosquitoes—trademarked "Friendly" *Aedes aegypti*—were first released in Brazil in February 2011, under the direction of biochemist Margareth Capurro of the University of São Paulo. In a densely populated suburb in northeastern Brazil, the released mosquitoes reduced the wild population of *Aedes aegypti* by 85 percent, says Capurro. In 2013, a similar test in the village of Mandacaru resulted in a 96 percent reduction of the wild mosquito population in the area after only 6 months.

"For the first time, we demonstrated that transgenic mosquitoes can work," says Capurro, who is now developing alternate GM mosquito lines in her own laboratory, attempting to make the GM insects even more potent against their wild counterparts. For example, Capurro's lab is trying out different genetic mutations to create

Figure 19.11

Populations of two species may increase and decrease together

The Canadian lynx depends on the snowshoe hare for food, so the number of lynx is strongly influenced by the number of hare.

Q1: During which years did the hare likely have the greatest food supply?

Q2: Besides the number of hare, what other factors might contribute to the number of lynx?

Q3: Can you draw an average carrying-capacity line on these graphs? Why or why not?

The number of lynx rises and falls with the number of hare.

a line of mosquitoes that does not require the use of an antibiotic to be kept alive in the laboratory, which would be less expensive. "We are finishing the lines now. They are very promising," says Capurro. Other genetic modification techniques are being attempted in labs elsewhere to make *Anopheles* incapable of carrying the malaria parasite, for example.

More recently, Oxitec released Friendly *Aedes aegypti* in a neighborhood of 5,000 residents in the city of Piracicaba, Brazil, to combat the spread of dengue virus. The company released 3–4 million mosquitoes per month over a period of 10 months. The experiment reduced the number of wild *A. aegypti* larvae in the area by 82 percent, and the neighborhood experienced a

MARGARETH CAPURRO

Margareth Capurro is a biochemist at the University of São Paulo in Brazil. Starting in 2011, she led two field releases of Oxitec's genetically modified mosquitoes in Brazil, and today she continues to explore the best way to use GM mosquitoes to control mosquito-borne viruses.

91 percent reduction in cases of dengue, according to local officials.

In August 2016, shortly after the first cases of Zika infection from local mosquitoes were reported, the U.S. Food and Drug Administration (FDA) approved a field test of the Oxitec mosquitoes in Key Haven, Florida. Critics argue that the approach is too expensive, because mosquitoes have to be bred and released multiple times. There are also concerns that the engineered gene won't correctly insert in the genome or will become somehow deactivated in the wild. Release of such mosquitoes would simply exacerbate the problem, adding reproducing mosquitoes into the environment. After examining such claims, the FDA concluded that the proposed field trial in Florida "would not result in significant impacts on the environment." In a 3–2 vote in November 2016, the Florida Keys Mosquito Control District approved a trial of genetically modified mosquitoes, yet local opposition continues, and the district continues to seek a location for the trial.

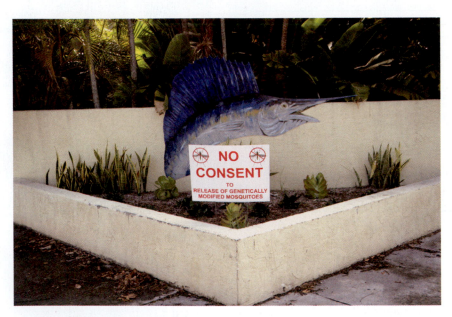

Figure 19.12

Some Florida residents oppose the release of GM mosquitoes in their community

Misinformation about genetically modified organisms, including mosquitoes, has caused concerns about their use.

Some reject the plan solely on the grounds that it involves genetic modification (**Figure 19.12**). "The public fears genetic engineering. Nearly all politicians don't understand it," said bioethicist Arthur Caplan at the New York University School of Medicine, as quoted in a 2016 article in the *Atlantic*. "I don't think the issue is economic. It is ignorance, distrust, fear of the unknown." Recent polls suggest that the tide may be changing: three independent polls found that between 53 and 78 percent of Americans support the release of GM mosquitoes in the United States. "I think it will work well," says DeGennaro. "It's time we use modern methods here in Florida like they are using in Brazil to reduce mosquito populations."

Just the Beginning?

As of October 2016, Zika infections had been reported in every U.S. state, with the exception of Alaska. These infections are due primarily to travel outside of the country. In Florida, however, local infections continue; as of this writing, 214 people have locally acquired the virus since 2015.

The CDC continues to warn pregnant women away from traveling to any area with Zika, which now includes dozens of countries; three U.S. territories; Brownsville, Texas; and, of course, parts of Florida. As of the date this book went to publication, there were no approved vaccines or therapies for Zika virus infection—but not for lack of trying. In June 2016, California-based Inovio Pharmaceuticals dosed the first humans with a prospective Zika vaccine. Even if the trials go well, however, it could take years for a vaccine to reach the market.

It remains too soon to tell whether GM mosquitoes will stop the spread of Zika. Success will depend on population ecology: on mosquito populations, the human population, and how much virus is circulating among both. But history suggests that in the fight against mosquitoes, it's time to try something new.

World's Deadliest Animals

You may have heard that humans are the deadliest animals on the planet. It's true that we, as a species, kill hundreds of thousands of humans. But there's one family of animals that has us beat: mosquitoes. Many species of these small, pesky insects transmit harmful infections, including Zika fever, malaria, West Nile disease, dengue fever, and many more.

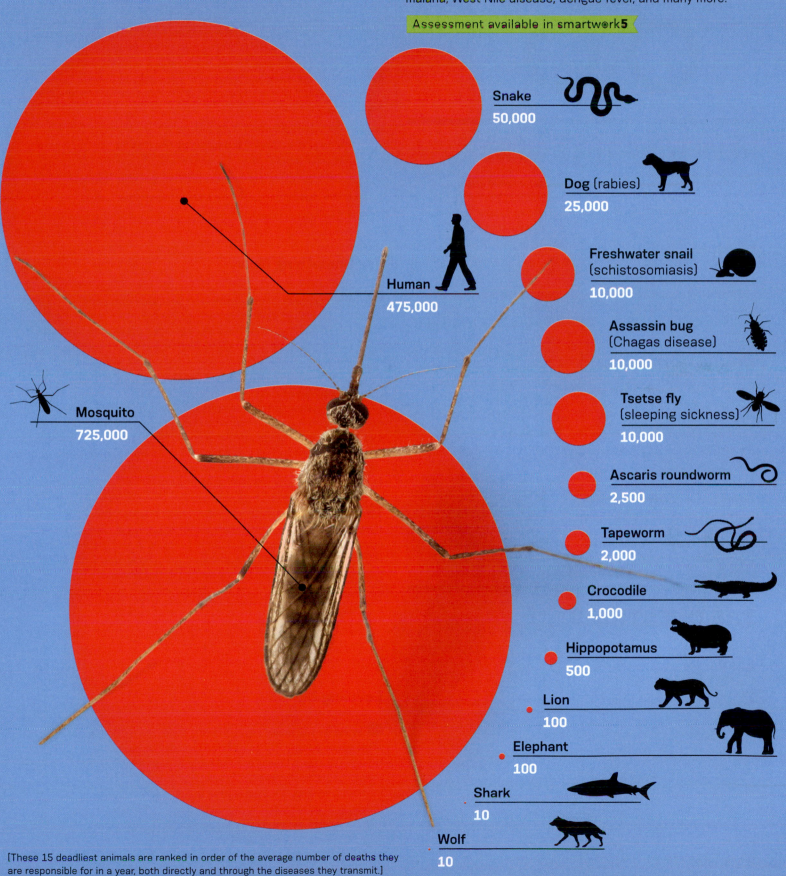

Snake
50,000

Dog (rabies)
25,000

Human
475,000

Freshwater snail
(schistosomiasis)
10,000

Assassin bug
(Chagas disease)
10,000

Tsetse fly
(sleeping sickness)
10,000

Ascaris roundworm
2,500

Tapeworm
2,000

Mosquito
725,000

Crocodile
1,000

Hippopotamus
500

Lion
100

Elephant
100

Shark
10

Wolf
10

[These 15 deadliest animals are ranked in order of the average number of deaths they are responsible for in a year, both directly and through the diseases they transmit.]

REVIEWING THE SCIENCE

- A **population** is a group of individuals of a single species located within a particular area.

- **Population size** is the total number of individuals in a population. **Population density** is the population size divided by the area covered by that population.

- Environmental factors such as lack of space, food shortages, predators, disease, and habitat deterioration limit populations. These factors affect the **carrying capacity**, the number of individuals that can live in an environment indefinitely.

- In **logistic growth**, a population grows exponentially at first but then stabilizes after reaching the carrying capacity. It is associated with an **S-shaped growth curve** when graphed.

- In **exponential growth**, a population increases by a constant proportion from one generation to the next generation. It is associated with a **J-shaped growth curve** when graphed.

- **Density-dependent population change** occurs when birth and death rates are affected by population density, which is the case when many individuals occupy the same space and therefore compete for resources.

- **Density-independent population change** occurs when population size is affected by factors that have nothing to do with the density of the population.

- Populations may rise and fall over time, exhibiting **irregular fluctuations** or **cyclical fluctuations**.

THE QUESTIONS

Challenge Yourself

1 A population of plants has a density of 12 plants per square meter and covers an area of 100 square meters. What is the population size?

(a) 120

(b) 1,200

(c) 12

(d) 0.12

(e) none of the above

2 Which of the following would *not* cause the carrying capacity of a mosquito population to change?

(a) lower availability of standing water in which to lay eggs

(b) fewer animals for female mosquitoes to feed on

(c) warmer weather that increases overwintering survivorship

(d) All of the above would change carrying capacity.

(e) None of the above would change carrying capacity.

3 Place the following elements in the strategy to control mosquito populations in the correct order from earliest to latest by numbering them from 1 to 5.

_____ a. Scientists genetically modify male mosquitoes in the lab.

_____ b. Mosquito offspring produced by the matings do not survive to adulthood.

_____ c. GM male mosquitoes mate with normal wild females.

_____ d. The mosquito population decreases dramatically.

_____ e. Scientists release GM male mosquitoes into the wild.

Try Something New

4 Draw a diagram that illustrates the possible fluctuating population cycles of a predator, such as the mountain lion, and its prey species, such as deer. The prey regularly experiences population crashes when it overgrazes its food source. Label your axes.

5 An artificial pond at a high-school campus was populated with several mating pairs of pond slider turtles. After many years of slow population growth, campus administrators installed a coin-operated food dispenser at the pond's edge. Students bought and fed the turtles an enormous amount of food each day, causing the population of turtles in the pond to increase exponentially. How did the installation of the food dispenser affect the carrying capacity of the turtle population at this pond?

6 Suppose that population ecologists at the high school in question 5 determined that the number of turtles in the pond had increased from 6 individuals to 24 individuals in just the first year after installation of the coin-operated food dispenser. What is the population doubling time for the turtles in this population?

(a) 6 years

(b) 6 months

(c) 1 year

(d) 1 month

Leveling Up

7 **What do *you* think?** Some Florida residents do not want to allow genetically modified (GM) mosquitoes to be released, even though they have been demonstrated to significantly reduce mosquito-borne illnesses in Brazil.

a. Should Floridians be allowed to decide not to have GM mosquitoes released in their neighborhoods? Should this question be decided by vote? Explain your reasoning.

b. Currently, the Florida Department of Health may require people to remove standing water from their yards in order to decrease breeding habitat for mosquitoes. Should people be allowed not to comply with this order, given that the yard is their private property? Explain your reasoning.

8 **Life choices** Do you feel that people should choose to limit the number of children they have, given the increasing human population? Alternatively, should people focus on decreasing their own ecological footprint—given that, for example, one American uses the same amount of resources as do four Chinese citizens? Explain your reasoning.

9 **Is it science?** Your best friend found a Reddit thread stating that GM mosquitoes do not suppress the population size of Zika-bearing mosquitoes, and instead claimed that Zika was *caused* by the release of GM mosquitoes. Putting aside the idea that Zika was caused by GM mosquitoes, which is refuted by multiple studies as discussed on many legitimate websites, you decide to simply show your friend some data comparing mosquito population sizes in areas with and without GM mosquitoes. In your own words, write a paragraph that describes the results shown in the figure here and explains to your friend how they relate to the scientific claims he read.

Mosquito larvae population numbers in areas with and without released GM mosquitoes

Wild mosquito larvae per trap in each area*

Source: Oxitec   *Mean *Aedes aegypti* Larvae/trap—4 week rolling average

- 🟥 Conventionally treated area
- 🟩 Area with released GM mosquitoes

Thinking Multidimensionally

10 In 1980, China's government enacted the one-child policy to curb runaway population growth in the most populous country on the planet.

a. Looking at the graph above, how did this impact the average birth rates for Chinese women?

b. How many people currently live in China? Will China remain the most populated country in the world based on future predictions? Explain.

c. Using Figure 19.6 on page 347, describe the worldwide growth rate trends between the years 1800 and 2000.

d. Approximately how many people are there on Earth today?

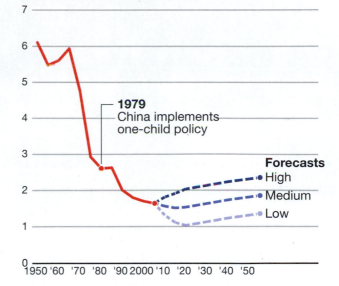

TOTAL FERTILITY RATE ESTIMATES
Number of children per woman

1979 China implements one-child policy

Forecasts
High
Medium
Low

11 Transgenic salmon are now being served on Canadian dinner tables. This 2017 landmark decision is the first time a genetically modified animal has been sold for human consumption.

a. How were these salmon produced, and how do they differ from their wild counterparts?

b. List some transgenic plants that are already on grocery store shelves. Describe changes that were made to them and why.

c. What do scientists say regarding the safety of genetically modified organisms (GMOs) on our health and the environment?

d. Do you think there should be a "GMO label" so consumers can make informed purchases? Why or why not?

e. Research the debate over scientists genetically engineering the DNA of plants and animals. What are the benefits and potential risks of these practices?

f. Create a poster that shows both sides of the argument and concludes with your opinion about the consumption of GMOs.

For more, visit digital.wwnorton.com/bionow2 for access to:

Of Wolves and Trees

The extermination of wolves in Yellowstone National Park had unforeseen effects on the park's ecosystem. Can the return of this top predator restore order?

After reading this chapter you should be able to:

- Explain the concept of an ecological community and how it relates to ecological diversity.
- Articulate the role of relative species abundance and species richness in defining ecological diversity.
- Distinguish between a food chain and a food web.
- Illustrate how the removal of a keystone species disrupts an ecological community.
- Describe the four major kinds of species interactions, and give an example of each.
- Compare and contrast primary and secondary succession.

CHAPTER 20

COMMUNITIES OF ORGANISMS

DRIVING QUESTION

How has the reintroduction of wolves into Yellowstone affected the entire ecosystem?

- HS-LS2.B: Cycles of Matter and Energy Transfer in Ecosystems
- HS-LS2.C: Ecosystem Dynamics, Functioning, and Resilience
- HS-LS2.D: Social Interactions and Group Behavior

357

ECOLOGY

Robert Beschta will always remember the day he visited Yellowstone National Park's Lamar Valley in 1996. Beschta, a hydrologist, was there to observe the Lamar River, which winds through the valley's lush lowland of grass and sage (**Figure 20.1**). But on that day, as he walked toward the waterway, he noticed something odd: all around the valley, there were not many trees. The few tall, white aspens he saw looked haggard, their bark eaten away. And there were no young saplings to be seen.

Beschta had studied forestry and rivers for decades and knew what a healthy valley was supposed to look like. This was not it. Beschta approached the river and observed that its banks were also devoid of trees. The leafy green cottonwoods and wide willows that had once arched gracefully from the riverbank over the water were absent. And with no tree roots to hold the soil in place, the riverbanks themselves

ROBERT BESCHTA

Robert Beschta is a professor emeritus at Oregon State University. He studies water processes in forest and rangeland ecosystems.

Figure 20.1

The Lamar River flows through Yellowstone National Park

were jagged and eroding. "I was dumbstruck," recalls Beschta. Something unprecedented was happening in Yellowstone.

Beschta returned to Oregon State University (OSU), where he worked, and described his observations in a seminar. He showed pictures of aspens with their bark stripped away and empty riverbanks where saplings should have been growing. In the audience, William Ripple sat up a little straighter. Also a scientist at OSU, Ripple studied forest ecology. He was particularly interested in aspen trees, which grow as tall as 70 feet and live up to 150 years. From what Beschta was saying, aspens were no longer growing in Yellowstone. Ripple wanted to know why.

"It was a scientific mystery as to what was the cause of the decline," he recalls. In that moment, listening to Beschta, Ripple knew exactly what his next research project would be: to document the extent of the aspen decline and determine why it was occurring.

Within a year, Ripple and one of his graduate students, Eric Larsen, traveled to the Lamar Valley. There, they drilled small holes into the centers of aspens and removed from each a plug of wood about the diameter of a pencil. Then they counted the growth rings in each plug—one for every year the tree has been in existence—to determine the age of the tree. They found that most of the aspens had begun to grow *prior* to 1920. After 1920, almost no new trees had begun growing.

"We started scratching our heads at that point," says Ripple. He, Beschta, and Larsen began to brainstorm reasons why the trees might have stopped regenerating. They looked for environmental changes in the 1920s that could have done it: a fire that killed off saplings, or a change in climate that reduced the trees' ability to reproduce. But nothing lined up—until an ordinary moment gave Ripple an extraordinary idea.

Standing in a gift shop in Grand Teton National Park, just south of Yellowstone, Ripple looked up at a poster on the wall. It featured a grove of tall, white aspen trees in the winter. In the middle of the trees, its paws covered in snow, stood a large gray wolf. "That was an 'aha' moment," says Ripple. "I thought, 'Maybe the wolf protects the aspen.'"

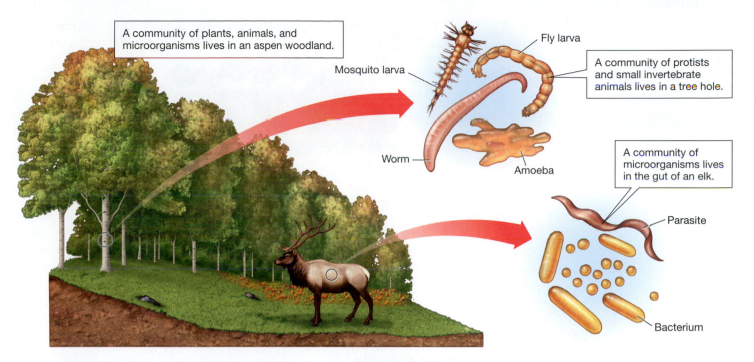

Figure 20.2

Ecological communities come in all sizes

Smaller communities can be nested within a larger community. This aspen woodland community contains the smaller communities of a temporary pool of water in a tree hole and an elk's gut, among others.

Q1: List another species that is part of this community.

Q2: Of which community could this aspen woodland be a smaller part?

Q3: Which other small communities could be found within this larger community?

The idea of wolves protecting aspens initially seems nonsensical. How would meat-eating predators protect trees? They wouldn't—at least not directly. Ripple surmised that wolves in Yellowstone might have had an indirect effect on the community. As an ecologist, Ripple had long studied **ecological communities**, associations of species that live in the same area. Communities vary in size and complexity, from a small group of microorganisms in a temporary pool of water to the whole of Yellowstone Park, home to an estimated 322 species of birds, 67 species of mammals, 1,349 species of plants, and an uncounted number of insects (**Figure 20.2**).

Ripple knew that an ecological community is characterized by the diversity of species that live there, and that diversity is governed by two things: the **relative species abundance** (how common one species is when compared to another) and the

species richness (the total number of different species that live in the community; **Figure 20.3**). Ripple also knew that communities are subject to constant change, and that something must have changed in Yellowstone.

Ecological communities change naturally as a result of interactions between and among species, and as a result of interactions between species and their physical environment. Ripple knew that both the relative abundance and the

WILLIAM RIPPLE

William Ripple is director of the Trophic Cascades Program at Oregon State University. There, he leads a research project investigating how gray wolves affect other species in the Yellowstone ecosystem.

This community has higher relative species abundance for white-barked trees than the community shown below.

This community has higher species richness than the community shown above.

Figure 20.3

Two measures of species diversity

The diversity of an ecological community is determined by relative species abundance and also species richness. High relative species abundance means that one species dominates the community (so the community is less diverse), and high species richness means that many species are present in the community (so it is more diverse).

Q1: If relative species abundance increased in the first community, how would the figure look different from now?

Q2: If species richness decreased in the second community, how would the figure look different from now?

Q3: How do relative species abundance and species richness define the species diversity of a forest community?

richness of species had changed in Yellowstone during the previous century. The relative species abundance had changed because aspen trees and willows were in decline, elk and coyote numbers had increased, and bison and beaver populations had decreased. Many of the common inhabitants of Yellowstone had changed in abundance; therefore, the relative species abundance of this community had changed.

The species richness had also changed because early in the twentieth century, a *keystone species*, the gray wolf, had gone missing. The total number of species in the community had gone down by only one, but that seemingly small change had a big impact. A **keystone species** is a species that has a disproportionately large effect on a community, relative to the species' abundance. There are few wolves compared to, say, rabbits, yet the wolves have a stronger effect on the community. Keystone species in other communities include prairie dogs in the American western plains, hummingbird pollinators in the Sonoran Desert, and sea stars in intertidal waters (**Figure 20.4**).

Keystone species are often recognized only when they go missing and their disappearance results in dramatic changes to the rest of the community. And that is exactly what happened in Yellowstone in the 1920s: the mighty gray wolf disappeared. Or, to be accurate, the mighty gray wolf was exterminated.

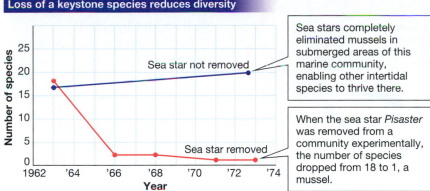

How a predator maintains diversity

A *Pisaster* sea star feeding on a mussel. Without this keystone species, the mussels crowd out other species in their community.

Loss of a keystone species reduces diversity

Sea star not removed

Sea star removed

Number of species

1962 '64 '66 '68 '70 '72 '74

Year

Sea stars completely eliminated mussels in submerged areas of this marine community, enabling other intertidal species to thrive there.

When the sea star *Pisaster* was removed from a community experimentally, the number of species dropped from 18 to 1, a mussel.

Figure 20.4

The star of the community

The sea star *Pisaster ochraceus* is an example of a keystone species. In a classic experiment conducted along the Pacific coast of Washington State in 1963, sea stars were removed from one site while an adjacent site was left undisturbed.

Q1: How many species were left in 1966 in the community where sea stars were *not* removed?

Q2: How many species were left in 1966 in the community where sea stars *were* removed?

Q3: How do your answers to questions 1 and 2 demonstrate the importance of a keystone species for the maintenance of diversity in a community?

A Key Loss

In the early 1900s, ranchers and homesteaders killed wolves all across the United States and eliminated them in many eastern states. Then, in 1915, the U.S. government began subsidizing wolf extermination programs all over the country, and a systematic slaughter began. Under the national program, states paid bounties of up to $150 for individual wolf pelts. Wolves, feared and hated by private landowners for killing livestock, were trapped, shot, and skinned. The extermination happened quickly: the last known wolf den in Yellowstone was destroyed in 1926. At least 136 wolves, maybe more, were killed during the eradication campaign in Yellowstone. Once it was done, the park was wolf-free.

Seven decades later, staring at the poster of a wolf standing among the aspens, Ripple wondered whether the loss of that keystone species had contributed to the decline of aspen trees. He immediately looked up the historical records to see whether the timing matched. Lo and behold, the last wolf had been killed at

about the same time the aspens stopped regenerating, in the mid-1920s. Suddenly, it seemed obvious why the aspens had declined: wolves kill elk and elk eat aspens—three species in the same food chain.

A **food chain** is a simple list of who eats whom. In scientific terms, it is the direct path by which nutrients are transferred through the community. A **food web**, on the other hand, is a more complex diagram of all the food chains in a single ecosystem and how they interact and overlap (**Figure 20.5**).

In the wolf-elk-aspen food chain, aspen are the **producers**, the organisms at the bottom of the food chain that use energy from the sun to produce their own food through photosynthesis. In Yellowstone and on land all over Earth, photosynthetic plants like trees, grasses, and shrubs are the major producers (**Figure 20.6**). In aquatic biomes, photosynthetic plankton are the major producers, as we'll see in Chapter 21.

Further up the food chain are **consumers**, organisms that obtain energy by eating all or parts of other organisms or their remains. Elk and wolves are both consumers: elk eat aspens, and wolves eat elk. In the Yellowstone food chain, elk are **primary consumers**: they eat producers. Wolves are **secondary consumers** because they eat primary consumers. This sequence of organisms eating organisms can continue: a bird that eats a spider that ate a beetle that ate a plant is a **tertiary consumer**; a killer whale that eats a leopard seal that ate a sea bass that ate a krill that ate a phytoplankton is a **quaternary consumer**. We will explore the flow of energy up the food chain in Chapter 21.

The more they discussed it, the more Ripple, Larsen, and Beschta believed that the loss of wolves in Yellowstone had allowed the elk population to flourish and eat so many young trees that the aspen population could not regenerate. "We developed a hypothesis that maybe the killing of wolves actually affected the reproduction of aspen trees," says Ripple. Any change in species diversity will have a ripple effect (no pun intended) throughout the community, and the wolves of Yellowstone were no exception. Ripple and Larsen published their hypothesis in 2000, suggesting that the loss of wolves had led to increased elk

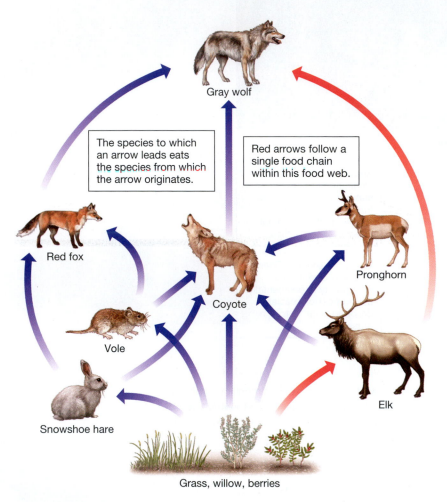

The species to which an arrow leads eats the species from which the arrow originates.

Red arrows follow a single food chain within this food web.

Gray wolf
Red fox
Coyote
Pronghorn
Vole
Snowshoe hare
Elk
Grass, willow, berries

Figure 20.5

Food webs show how energy moves through a community

Food webs are composed of many food chains that show one species eating another.

Q1: What species eats the coyote?

Q2: What species does the coyote eat?

Q3: What do you think would happen to a community that lost its coyotes?

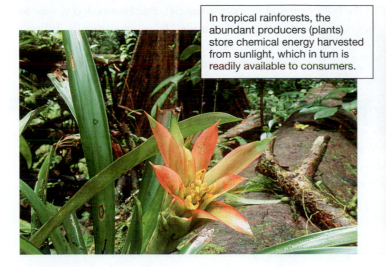

In tropical rainforests, the abundant producers (plants) store chemical energy harvested from sunlight, which in turn is readily available to consumers.

In deserts, the sparseness of plant life means that relatively little chemical energy is available for consumers.

Figure 20.6

Producers are the energy base of a food chain

All communities and ecosystems have producers, but they vary in abundance, depending on the circumstances.

Q1: Where do producers acquire the energy they need to perform their function in the food chain?

Q2: Why are producers necessary for life on Earth?

Q3: Given the lower abundance of producers in desert regions, compared to tropical rainforests, what would you predict about the abundance of consumers in the two environments?

populations and altered elk movements and feeding patterns. In other words, with wolves gone, elk were free to find and eat young aspen shoots whenever they wanted, with no fear of wolves.

In 2001, Beschta returned to the Lamar Valley and collected data on another species of tree, the cottonwood, which can live more than 200 years. He documented the same trend as the aspens showed: in the 1920s, cottonwoods suddenly stopped generating new, young trees. In fact, since the 1970s, not a single new cottonwood had been established. "This was dramatic," recalls Beschta. "It's a big deal when you can't have a single cottonwood in this large valley make it to a mature tree."

It is possible for a consumer to eat a species to extinction, so if elk populations had continued to grow unchecked, they could have reduced the aspen and cottonwood populations to zero, disrupting the ecological community permanently. But before that could happen, something dramatic occurred in Yellowstone: humans brought the wolves back.

A Second Ripple Effect

In 1973, wolves became the first animals to be protected under the Endangered Species Act. It was the dawn of the modern conservation movement, and the idea of returning wolves to Yellowstone grew in popularity. It took some time, but eventually, lawmakers agreed to the plan, and between 1995 and 1997, 41 wild wolves were captured in Canada and released in Yellowstone. It didn't take long for the wolf populations to recover: by 2007, an estimated 170 wolves lived in and around Yellowstone; today that estimate is 528. As of January 2016, at least 98 wolves in 10 packs were living within the park limits (**Figure 20.7**).

The loss, and then return, of the wolves had significant impacts on **species interactions** in the park. There are four central ways in which species in a community interact: *mutualism, commensalism, predation,* and *competition.*

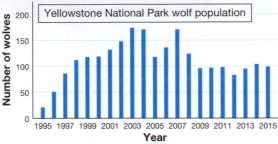

Figure 20.7

Wolves today in Yellowstone

Q1: When were wolves first reintroduced to Yellowstone?

Q2: What was the highest number of wolves observed in Yellowstone? In what year did that occur?

Q3: Why did it take a few years after the reintroduction of wolves for the aspen population to increase, as well as the beavers and bears?

The classification is based on whether the interaction is beneficial, harmful, or neutral to each species involved. These interactions affect where organisms live and how large their populations grow. Species interactions also drive natural selection and evolution, thereby changing the composition of communities over short and long periods of time.

With wolves back in the park, Ripple suddenly had a way to test his hypothesis about aspen decline. If the loss of wolves was responsible for aspen loss, then the return of the wolves should incite a revival of aspens (and other woody plants,

including cottonwoods). But he needed a way to quantify that change. "It's a scientifically difficult task to connect a wolf to a plant. Obviously wolves don't consume plants, so instead we had to connect the dots. There are data that wolves affect elk, and other data that elk affect plants," says Ripple.

Ripple and Beschta went into the field in 2006 and again in 2010 to take measurements. In addition to recording the ages of the trees, they looked for and documented signs of elk feeding on aspens—such as scars where branches or buds had been bitten off—and measured the heights of young aspens. Beschta did the same with the cottonwoods. They were eager to see which species interactions would occur now that wolves were back.

The first type of species interaction, **mutualism**, occurs when two species interact and both benefit. For example, Yellowstone is home to 4,600 bison, the largest land mammals in North America. Bison have a mutualistic relationship with the black-billed magpie. Pests such as ticks burrow into a bison's short, dense hair to suck the beast's blood, but hungry little magpies perch on top of the bison and eat those ticks. Thus, both the bison and the magpie benefit from close interaction with one another. Mutualism is common and important in ecosystems all over Earth: many species receive benefits from, and provide benefits to, other species, as the clownfish-anemone partnership in **Figure 20.8** illustrates. These benefits increase the survival and reproduction of both interacting species.

When they aren't perched atop bison, black-billed magpies can be found in large nests atop deciduous or evergreen trees, where they reproduce once a year. These trees, another member of the community, share a commensal relationship with the magpies. **Commensalism** happens when one partner benefits while the other is neither helped nor harmed—in this case, the magpie benefits from having a safe place to lay eggs, and the interaction has no effect on the tree. Another example of a commensal relationship is barnacles living on a whale (**Figure 20.9**).

As you might have guessed, not all species interactions are as pleasant as those among bison, birds, and trees. In two other types of interactions, at least one of the two species is harmed: *competition* (which we will return to in a moment) and *predation*.

The clownfish gains protection from predators by hiding within the sea anemone's stinging tentacles. The clownfish is not harmed, because a thick mucus covers its body.

The anemone is protected from grazing predators by the clownfish and absorbs the fish's nutrient-rich excrement.

Figure 20.8

Mutualism: friends in need

Both the clownfish and the sea anemone benefit from their relationship.

Q1: What might happen to an anemone without a resident clownfish?

Q2: What would happen to a clownfish that did not produce a mucous coating?

Q3: In the example of mutualism given in the text, what would happen if ticks were no longer able to feed on bison?

In **predation**, one species benefits and the other is harmed, and **predators** are defined as consumers that eat part or all of other organisms. A **parasite** is a kind of predator that lives in or on the organism it harms, its **host**. An important group of parasites is pathogens, which cause disease in their hosts. The bacteria that cause strep throat, tuberculosis, and pneumonia are pathogens, for example. Many organisms have evolved mechanisms to avoid being hosts, such as immune systems to help fight off parasitic diseases and infections.

Other types of predators are distinguished by what they eat. Elk are **herbivores**, animals that eat plants. Yellowstone elk feed on the shoots, saplings, and new branches of woody plants like aspen, cottonwood, and willow, especially in winter, when other plants are scarce. Wolves are **carnivores**, animals (and, in rare cases, plants) that kill other animals for food. Yellowstone wolves predominantly eat elk, especially in winter, but they also eat deer and any small mammals they can catch, notably beaver. Other animals, such as raccoons and coyotes, eat both animals and plants, so we call them **omnivores**. Both raccoons and coyotes are also **scavengers**, animals that eat dead or dying plants and animals. Of the species that are scavengers, **decomposers**, such as fungi, dissolve their food; and **detritivores**, including worms and millipedes, mechanically break apart and consume their food.

The barnacles on a whale's snout enjoy a continuous stream of nutrient-rich water flowing across them from which they can feed. The whale is neither harmed nor helped by these stowaways.

Figure 20.9

Commensalism: a whale of a ride!

This gray whale's snout is covered in barnacles.

Q1: How do barnacles benefit from living on a whale?

Q2: Do you think a whale could avoid being colonized by barnacles? Why or why not?

Q3: Explain why detritivores are considered commensal to the organisms they consume.

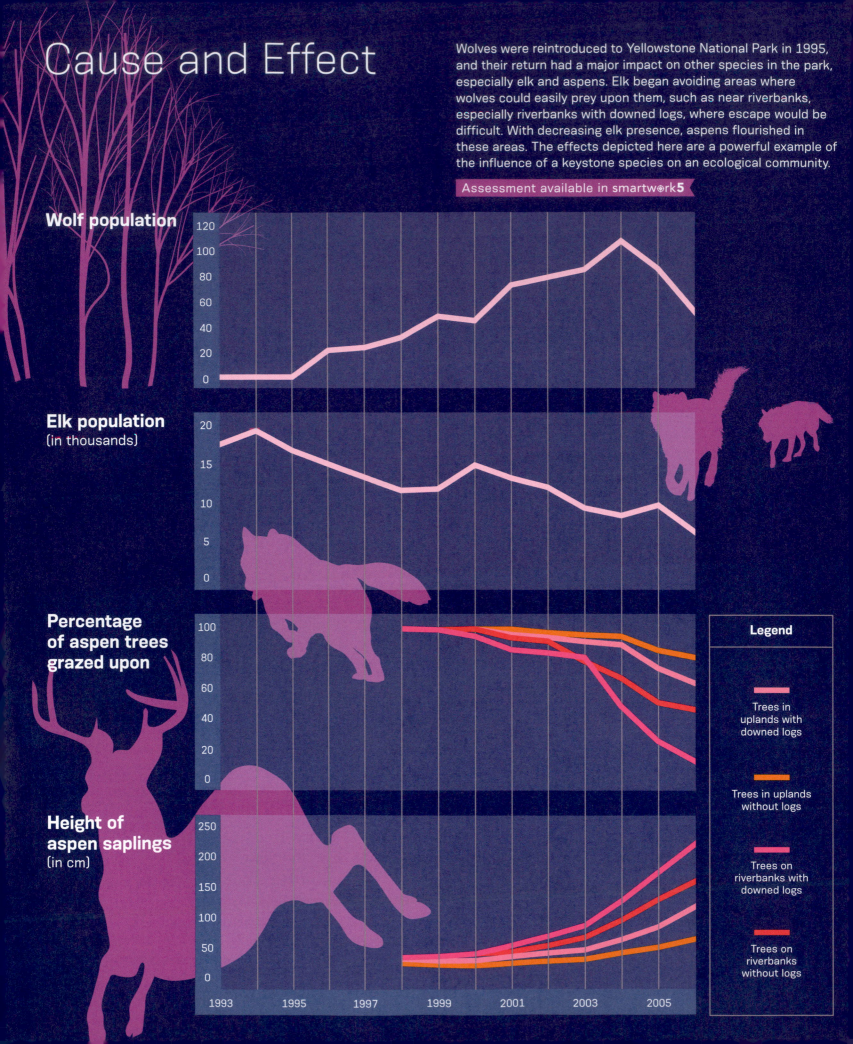

Cause and Effect

Wolves were reintroduced to Yellowstone National Park in 1995, and their return had a major impact on other species in the park, especially elk and aspens. Elk began avoiding areas where wolves could easily prey upon them, such as near riverbanks, especially riverbanks with downed logs, where escape would be difficult. With decreasing elk presence, aspens flourished in these areas. The effects depicted here are a powerful example of the influence of a keystone species on an ecological community.

Assessment available in smartwork**5**

Wolf population

Elk population
(in thousands)

Percentage of aspen trees grazed upon

Height of aspen saplings
(in cm)

Legend

Trees in uplands with downed logs

Trees in uplands without logs

Trees on riverbanks with downed logs

Trees on riverbanks without logs

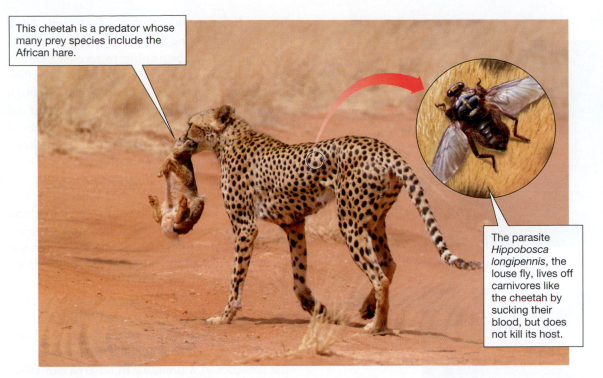

This cheetah is a predator whose many prey species include the African hare.

The parasite *Hippobosca longipennis*, the louse fly, lives off carnivores like the cheetah by sucking their blood, but does not kill its host.

Figure 20.10

Predators can also be prey

While the cheetah is what we traditionally think of as a predator, it can also be prey to parasites, and eventually to decomposers and detritivores.

Q1: What kind of predator is the cheetah in the figure?

Q2: What kind of predator is the louse in the figure?

Q3: What kind of predator are the elk that graze on aspen tree saplings?

The animals eaten by predators are called **prey**. All of the animal residents of Yellowstone (and most plants) are eaten by other species, except for grizzly bears, mountain lions, eagles, and gray wolves. These four animals are all at the top of the food chain. They are top predators, although they are themselves prey to parasites (**Figure 20.10**).

Back in the Park

Now that wolves, a top predator in the community, were back in Yellowstone, how would their prey, the elk, react? And how would that reaction affect the elk's food, the trees?

Using the plant measurements taken in the park and comparing those measurements to historical data, Beschta and Ripple found that between 1998 and 2010, as the wolf population in the park grew, the elk population decreased and therefore fewer aspens were eaten. In 1998, essentially 100 percent of the young aspen plants were being preyed upon, but by 2010, only 18–24 percent were being eaten. In addition, average aspen heights increased for all areas that the scientists observed. Cottonwoods experienced the same revival. In the 1970s, cottonwoods had entirely stopped adding new young saplings, but by 2012, some 4,660 young cottonwoods had grown to over 2 meters high.

Together, the aspen and cottonwood data sets convinced Ripple and Beschta that the reintroduction of wolves was responsible for a cascade of species interactions leading to the restoration of aspen and cottonwood populations. "With

Warning coloration

The bright colors of the poison dart frog warn potential predators of the deadly chemicals contained in its tissues.

Mimicry

The viceroy butterfly (left) mimics the color and pattern of the monarch butterfly (right), which contains toxic compounds.

Camouflage

With its long legs outstretched, this lichen-mimic katydid, a relative of the cricket, sits motionless on lichen-covered branches to escape detection by predators.

Figure 20.11

Adaptive coloration responses to predation

Prey species have adapted many elaborate strategies to avoid being eaten by predators, including warning coloration, mimicry, and camouflage.

Q1: How do predators know that brightly colored prey are usually toxic?

Q2: Do you think mimicry works if the toxic species is in low abundance? Why or why not?

Q3: Why is camouflage considered an adaptive response to predation?

wolves back, young, woody plants are doing better and growing taller," says Ripple. "Plant communities are beginning to recover." Different species are growing and spreading at different rates of recovery, he adds, "but there's enough new growth that we suggest it is in support of our basic hypothesis, that the presence or absence of the top predator—the wolf in this case—makes a difference in these plant populations."

Since Ripple and Beschta's discovery, the scientific community has been debating two potential reasons why the plants are flourishing with the return of the wolves. The most straightforward possibility is that the elk population has decreased: wolves kill elk, so there are fewer elk to consume the plants. Yet some of the tree populations seemed to recover faster than the drop in the elk populations would suggest. So a second possibility is that the presence of wolves led to a change in elk behavior called a *fear effect*. Often, the presence of a predator in an area can affect the behavior of its prey. In this case, it is possible that elk stopped grazing in areas where they could easily be seen by wolves, such as along the banks of the Lamar River.

"These two mechanisms, the population density and fear behavior, are difficult to tease apart, and we're working on that," says Ripple. "Many today believe it [the change in plant populations in Yellowstone] is due to a combination of the two." Beschta agrees: "In my opinion, they've both been going on."

Safety in Numbers and Colors

Elk may avoid lingering at streams as a way to evade their predators, but other prey have far more elaborate strategies to avoid being consumed. The poison dart frog, for example, is among the most toxic animals on Earth, and it evolved bright colors as **warning coloration** to alert potential predators to the dangerous chemicals in its tissues (**Figure 20.11**, top). Such warning coloration can be highly effective. Young blue jays, for example, quickly learn not to eat brightly colored monarch butterflies, which contain chemicals that cause nausea and, at high doses, death.

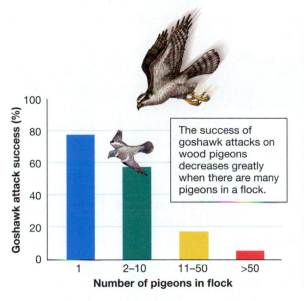

The success of goshawk attacks on wood pigeons decreases greatly when there are many pigeons in a flock.

Although a single musk ox may be vulnerable to predators such as wolves, a group forming a circle makes a difficult target.

Figure 20.12

Safety in numbers

Animals that live in groups are better able to warn each other and sometimes fend off attacking predators.

Q1: What percentage of pigeons are caught when they are alone and not in a flock?

Q2: For wood pigeons, what is the minimum number of individuals that provides protection from goshawks?

Q3: Why do you think a group of musk oxen versus a lone musk ox would be safer from a pack of wolves?

Then there are species that, though not poisonous, evolved coloration to make them look as if they were. Through **mimicry**, the viceroy butterfly, which is not poisonous, imitates the color and pattern of the monarch butterfly (**Figure 20.11**, middle). That "borrowed" coloration scares away blue jays and other birds that may have felt sick the last time they ate a monarch. Another mechanism to avoid being eaten is **camouflage**, any type of coloration or appearance that makes an organism hard to find or hard to catch (**Figure 20.11**, bottom). Finally, many prey, from musk oxen to wood pigeons, have evolved a different strategy to avoid becoming dinner: living together. By group living, these animals are able to act together to warn each other when a predator is about to attack and even to repel attacks as a united front (**Figure 20.12**).

In predation, the predator benefits. And each of the other two types of species interactions we've discussed—mutualism and commensalism—benefits at least one of the species involved. But in one final type of species interaction, **competition**, no one benefits. Instead, both interacting species are negatively affected.

Competition most often occurs when two species share an important but limited resource, such as food or space. In Yellowstone, both beavers and elk eat woody plants. For both species, woody plants are part of the **ecological niche**, the set of conditions and resources that a population needs in order to survive and reproduce in its habitat. Because the niches of beavers and elk overlap, these species compete. When two (or three or more) species compete, each has a negative effect on the other because one is using resources that the other then cannot access. (If resources are abundant, however, there may be no competition between species, even if their niches overlap.)

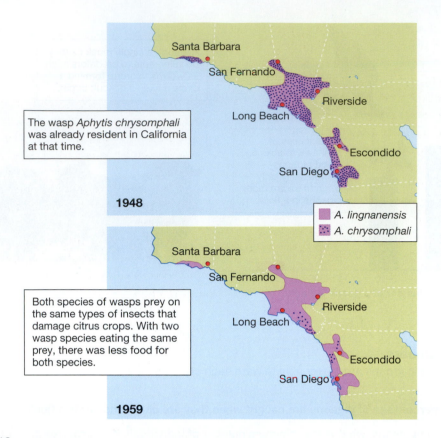

The wasp *Aphytis chrysomphali* was already resident in California at that time.

1948

A. lingnanensis
A. chrysomphali

Both species of wasps prey on the same types of insects that damage citrus crops. With two wasp species eating the same prey, there was less food for both species.

1959

Figure 20.13

Exploitative competition: a new species moves in

Two different species of wasps—*Aphytis lingnanensis* and *Aphytis chrysomphali*—feed on the same resource but do not directly compete for access to it.

Q1: Which species appears to be the superior competitor?

Q2: Why is this example considered exploitative competition?

Q3: What would you predict if these species had undergone competitive exclusion? (You will need to read the next paragraph to answer this question.)

There are two main categories of competition: *exploitative* and *interference*. In **exploitative competition**, species compete indirectly for shared resources, such as food. In this case, each species reduces the amount of the resource that is available for the other species, but they do not directly interact or come in contact with each other (**Figure 20.13**). When wolves returned and elk populations declined, beavers had less competition for food, especially willow trees, and the number of beaver colonies in Yellowstone rose from 1 in 1996 to 12 in 2009. By 2015, Yellowstone had an estimated 100 beaver colonies.

Elk and bears also interact through exploitative competition. Elk eat the leaves and branches on shrubs, resulting in a decrease in the number of berries the shrubs produce. That's not good for grizzly bears, which love to eat berries. Knowing of this relationship, Ripple hypothesized that a decrease in elk would result in an increase in berry-producing shrubs, and that bears would be eating more berries. To test this hypothesis, Ripple, Beschta, and others spent 2 years analyzing grizzly scat that had been collected in the park. They compared the percentage of fruit in current scat to that of scat data that had been collected and saved before 1995, prior to the wolf return. Over a 19-year period, they found that the percentage of berries in the grizzly diet went up as elk populations went down. This is an example of the **competitive exclusion principle**, which predicts that different species that use the same resource can coexist only if one of the species adapts to using other resources. In this case, grizzlies adapted to the presence of elk by finding food sources other than berries, and in the absence of elk they increased their consumption of berries.

Organisms also compete through **interference competition**, in which one organism directly excludes another from the use of a resource. In Yellowstone, for example, bears and wolves often fight over the carcass of an animal (**Figure 20.14**). In interference competition, one individual physically interferes with another individual that is trying to obtain a resource.

A Community Restored

Today, the return of the wolf is having a clear and significant impact on the ecological community of Yellowstone, from the rebirth of aspens and cottonwoods to growing beaver populations. These are signs of **succession**, the process by which the species in a community change over time. "We're on a very important upward trend," says Beschta.

All ecological communities change over time, sometimes because of human intervention, as in Yellowstone, but also because of natural changes in species composition—the number of individuals in a population often changes as the seasons change, for example—and natural disturbances such as fires, floods, and windstorms. In addition, communities

Figure 20.14

Interference competition can be dramatic

Stage 1: Bare sand is first colonized by dune-building grasses, which spread rapidly and stabilize the moving sand of the dunes.

Stage 2: Pines invade 50–100 years after the dunes are stabilized by the grasses.

Stage 3: The dominant species in the community, black oak, usually appears after 100–150 years.

Lake Michigan

Older sand dunes ⟶

Stage 4: Climax communities of black oak have lasted up to 12,000 years.

Figure 20.15

Primary succession: from nothing to climax community
Sand becomes woodland near Lake Michigan.

Q1: What species represents the first colonizers of the sand dunes?

Q2: What species is the intermediate species, and how does it become the dominant species?

Q3: What species is the mature, climax community species, and how does it become the dominant species?

can broadly change by the slow loss or gain of populations of species over long periods of time through natural selection.

In nature, ecologists have observed two major types of succession: primary and secondary. **Primary succession** occurs in newly created habitats, such as an island that just emerged from the sea, the soil left behind a retreating glacier, or new sand dunes (**Figure 20.15**). A new habitat begins with no species. Often the first species to colonize an area alters the habitat in ways that enable later-arriving species to thrive. A specific type of flowering plant may grow on a new island, for example, and then a species of bee that feeds on that flower subsequently joins the habitat.

Secondary succession occurs after a disturbance within a community—such as the loss of a keystone species like the wolves in Yellowstone or the plant loss caused by a forest fire—reduces the number of species in that community. During secondary succession, communities usually regain the successional state that existed before the disturbance. This type of succession does not take as long as primary succession, because some species still exist in the community.

Luckily, communities can and do bounce back from disturbances, but the time required to regain a previous state varies from years to decades to centuries. Yellowstone is currently experiencing a secondary succession as wolf, aspen, and beaver populations slowly return. It will likely still be a while before the park returns to its previous state, says Ripple. "We were 70 years without wolves, and now we're less than 20 years since wolves have returned, so this is going to take time." But he and others are hopeful that Yellowstone will return to its status as a **climax community**, a mature community whose species composition remains stable over long periods of time (**Figure 20.16**).

Yet even as Yellowstone's recovery is under way, ecological communities around the planet are being threatened by the loss of other keystone species, especially large carnivores. In 2013, Ripple and colleagues analyzed 31 carnivore species around the globe, including leopards, lions, cougars, and sea otters. They found that more than 75 percent of those 31 species are declining, and that 17 of them now occupy less than half of their former ranges. In 2016, they found that bushmeat hunting, in particular, is

In 1988, a large fire destroyed a portion of the mature lodgepole-pine forest in Yellowstone National Park.

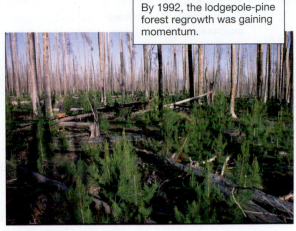

By 1992, the lodgepole-pine forest regrowth was gaining momentum.

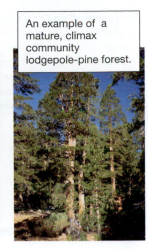

An example of a mature, climax community lodgepole-pine forest.

Figure 20.16

Secondary succession: from disturbance to climax community

Lodgepole-pine forest has been slowly but steadily regrowing in Yellowstone National Park.

Q1: What other types of disturbances could you imagine destroying a forest?

Q2: How is secondary succession different from primary succession?

Q3: What is a climax community?

driving many mammal species to extinction, including 113 species in Southeast Asia. Because of all the species interactions discussed here, it is clear that those changes will have major effects on ecological communities. "Humans are affecting predators around the globe in a major way," says Ripple. "It's a worldwide issue."

REVIEWING THE SCIENCE

- An **ecological community** can be characterized by its species composition, or diversity. This diversity has two components: **relative species abundance** (the number of individuals of each species that exist in the community) and **species richness** (the total number of different species that live in the community).

- **Keystone species** have a disproportionately large effect, relative to their own abundance, on the richness and abundances of the other species in a community. The removal or disappearance of these keystone species results in dramatic changes to the rest of the community.

- A **food chain** is a single direct line of who eats whom among species in a community. A **food web** depicts how overlapping food chains of a community are connected.

- **Producers**, organisms found at the bottom of a food chain that use light energy to produce their own food, are eaten by **consumers**.

- Consumers are classified as **carnivores**, **herbivores**, or **omnivores**, depending on whether they eat animals or plants or both, respectively.

- Dead and dying organisms are prey for many kinds of consumers. **Scavengers** are omnivores or carnivores that hunt for dead and dying prey. **Decomposers** dissolve their food, and **detritivores** mechanically break apart and consume their food.

- **Species interactions** in a community can be beneficial, harmful, or without benefit or harm to each of the interacting species.

  - In **mutualism**, both species benefit.

  - In **commensalism**, one species benefits at no cost to the other.

  - In **predation**, one species benefits and the other is harmed. **Parasites** are predators that live in or on their **hosts**.

In **competition**, both species are harmed. Competition occurs when **ecological niches** overlap and includes **exploitative competition** and **interference competition**. The **competitive exclusion principle** predicts that different species that use the same resource can coexist only if one of the species adapts to using other resources.

Succession establishes new communities (**primary succession**) and replaces disturbed communities (**secondary succession**). In stable environments without disturbances, called mature communities or **climax communities**, species composition remains stable over long periods of time.

THE QUESTIONS

Challenge Yourself

1 Wolves are considered a keystone species in Yellowstone because

(a) their removal in the early twentieth century caused many changes to the Yellowstone ecological community.

(b) their reintroduction in the late twentieth century caused many changes to the Yellowstone ecological community.

(c) when they were removed, elk populations increased, leading to increased competition with beavers and bears, which then declined.

(d) when they were reintroduced, aspen populations began to increase because of decreased predation from elk.

(e) all of the above

2 Link each species interaction with an example of the interaction.

MUTUALISM	1. Elk graze on aspen.
COMMENSALISM	2. Bison allow magpies to perch on them; the birds eat ticks they find on the bison's bodies.
PREDATION	3. Birds nest in aspen trees.
COMPETITION	4. Beavers and elk eat the same trees.

3 Place the following elements of the scientists' study of the relationship between aspen populations and wolf populations in the correct order from earliest to latest by numbering them from 1 to 5.

_____ a. Beschta hypothesized that the decimation of wolf populations in the early twentieth century allowed elk populations to grow, thus increasing grazing of the aspen by elk.

_____ b. Beschta observed there were few trees in the river valley.

_____ c. Ripple heard Beschta describe his observations in a talk.

_____ d. Ripple and his graduate student collected data showing that no new trees had grown in the valley since the 1920s.

_____ e. With the reintroduction of wolves into the park, elk populations declined and aspen populations rebounded.

4 Which ecological community would be more diverse: one with high relative species abundance or one with low relative species abundance? Which community would be more diverse: one with high species diversity or one with low species diversity?

Try Something New

5 When a female cat comes into heat and is ready to mate, she urinates more frequently and in a large number of places. Male cats from the neighborhood congregate near urine deposits and fight with each other for the female's attention and breeding rights. In what type of interaction are the male cats engaging?

(a) commensalism

(b) predation

(c) interference competition

(d) exploitative competition

6 Rabbits can eat many plants, but they prefer some plants over others. Assume that the rabbits in a grassland community containing many plant species prefer to eat a species of grass that happens to be a superior competitor. If the rabbits are removed from the region, predict whether relative species abundance will increase or decrease, and whether species richess will increase or decrease.

7 Identify whether each of the following is an example of primary succession (P) or secondary succession (S) in Rocky Mountain National Park.

_____ a. A mountain slope was cleared of evergreen trees and is now sprouting aspen trees.

_____ b. A mountainside with mature evergreen trees has dominated the landscape for generations.

_____ c. An area of soil, sand, and rocks remains after a dam burst and flooded the area.

_____ d. Lichens and mosses grow on bare rock at upper elevations because of an increase in average yearly temperatures.

8 Analyze the food web shown in the accompanying figure and answer the following questions:

a. Which species do *not* have a predator shown on the food web?

b. Which species shown on the food web performs photosynthesis (captures light energy to make its own food)?

c. Which species have only one predator and only one prey shown on the web?

d. Which species has the most predators shown on the web?

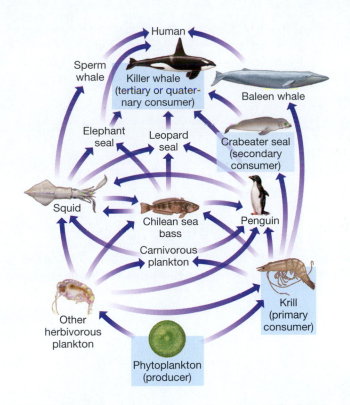

Human
Sperm whale
Killer whale (tertiary or quaternary consumer)
Baleen whale
Elephant seal
Leopard seal
Crabeater seal (secondary consumer)
Squid
Chilean sea bass
Penguin
Carnivorous plankton
Other herbivorous plankton
Krill (primary consumer)
Phytoplankton (producer)

Leveling Up

9 **Doing science** Citizen science is an amazing way to get involved in scientific research. Search the Internet for citizen science projects relevant to this chapter's topics, using keywords like "parasite," "predator," "prey," and "group living." Examples include Project Monarch Health (http://www.monarchparasites.org), in which volunteers sample wild monarch butterflies for a protozoan parasite to track its prevalence across North America. Participate in a project and, in writing, reflect on what you learned.

10 *Write Now* **biology: saving a species** Write a letter to your congressperson in which you use the concepts learned in this chapter to explain how an endangered species near you are beneficial to our wilderness ecosystems. Explain what role the species plays in the environment, what has caused it to be at risk, what the likely effect will be if it goes extinct (locally or globally), and what could be done to ensure its survival.

Thinking Multidimensionally

11 For nearly 60 years, scientists have studied the wolf packs on Isle Royale, a secluded island in Lake Superior.

a. Based on their location, what genetic issues would you expect these wolves to experience?

b. In 1997, after analyzing DNA found in the scat of a wolf nicknamed "Old Gray Guy," scientists noticed he had alleles unlike the other wolves on the island and that he looked and acted differently. From this information, they determined he migrated to Isle Royale to join the isolates. How could Old Gray Guy access the island and how could his presence impact the fitness of the gene pool?

c. In 2018, only two wolves, a male and female breeding pair, remained on the island, and the moose population grew to roughly 1,600. Predict the future of this food chain and the effects it will have on other organisms on the island.

d. On Isle Royale in 1980, canine parvovirus was accidentally introduced into the wolf population. How did this impact the wolf and moose populations, and how do you think the virus was brought to the island?

e. With only two wolves remaining, do you think scientists should intervene and add more wolves to Isle Royale? Why or why not?

12 Hidden beneath the soil of a forest are tree roots living in a mutually beneficial partnership with fungi. The trees provide sugars, and the fungi gather water and nutrients. The fungi form a network of white threads that connect the trees; in this way, the forest behaves like a single organism, allowing the trees to "talk" to one another.

a. What gases, nutrients, and other molecules do you think are the "words" that get passed among the trees?

b. If a tree is attacked by bugs and communicates this to neighboring trees via the fungal web, how might the other trees prepare for the intruders?

c. Pictured below is the phantom orchid. Based on its color, what pigment is it lacking, and how can it survive in a forest without this essential pigment?

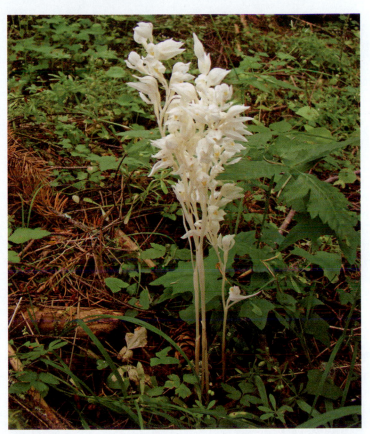

For more, visit digital.wwnorton.com/bionow2 for access to:

Here and Gone

Researchers discover an alarming decline of plankton in the ocean.

After reading this chapter you should be able to:

- Define an ecosystem and explain how it differs from an ecological community.
- Compare and contrast nutrient and energy cycles.
- Place the members of a specified ecosystem into an energy pyramid at the appropriate trophic level.
- Identify where on Earth a specified biome is found, and compare its relative productivity to other biomes.
- Define net primary productivity (NPP) and explain its significance to studies of ecosystems.

CHAPTER
21

ECOSYSTEMS

DRIVING QUESTION

Why are phytoplankton populations decreasing, and how will this change impact ocean ecosystems?

- HS-LS2.B: Cycles of Matter and Energy Transfer in Ecosystems
- HS-LS2.C: Ecosystem Dynamics, Functioning, and Resilience
- HS-LS4.D: Biodiversity and Humans

As a teenager, Daniel Boyce made extra cash working as a deckhand on fishing boats, so he was no stranger to the ocean. Eventually, Boyce turned that interest in marine environments into a career, and in 2007 he joined the lab of marine biologist Boris Worm at Dalhousie University in Nova Scotia, Canada.

Several years earlier, Worm had discovered that the industrialized fishing boom that began in the 1950s had decimated predatory fish communities. Boyce decided to follow up on his mentor's work by studying how that loss of predatory fish reverberated down the food chain. Had the disappearance of ocean predators affected plankton? Plankton, a diverse group of free-floating organisms, are the base of the ocean food web, supporting virtually all marine animals. Although they are tiny, they play a mighty role in marine environments.

When Boyce proposed this study, little did he know that it would uncover a profound shift in oceans around the world—a finding so shocking that just suggesting it would plunge him and his collaborators into a heated public controversy.

Going Green

Boyce's initial goal, similar to that of William Ripple's work studying wolves in Yellowstone National Park (see Chapter 20), was to show how the loss of a top predator affects the environment in which it lives. In Ripple's story, a variety of species living in Yellowstone interacted to form what we call an *ecological community*. A group of communities interacting with one another and with the physical environment they share is an **ecosystem** (**Figure 21.1**). To say it another way, an ecosystem is characterized by interactions of organisms in the biotic (living) world with each other, as well as with the abiotic (nonliving) world.

An ecosystem may be small or large; a puddle teeming with protists is an ecosystem, as is the Atlantic Ocean. And smaller ecosystems can be nested inside larger, more complex ecosystems. This variety means that ecosystems do not always have sharply defined physical boundaries. Instead, ecologists often define an ecosystem by the

Figure 21.1

Ecosystems in Nova Scotia, Canada

Overfishing has decimated fish populations in Nova Scotia, affecting the entire ecosystem in which these fish reside. Both biotic and abiotic elements of the ecosystem have been affected.

Figure 21.2

Phytoplankton bloom

The turquoise area in this aerial photo is a phytoplankton bloom occurring off the coast of Norway. When a population of phytoplankton (inset) increases rapidly, it discolors the water in which it resides.

distinctive ways in which it functions, especially the means by which *energy* and *nutrients* are acquired and distributed by the biotic community.

The activity of primary producers, in particular, profoundly influences the characteristics of an ecosystem. Ecologists often describe an ecosystem according to the types of producers it contains and the consumers that the producers support. A duckweed-covered pond, a tallgrass prairie, and a beech-maple woodland are all examples of ecosystems that can be defined by the specific types of producers that capture and supply energy to consumers.

To see how overfishing was affecting food chains in the ocean, Boyce first looked to the primary producers of the ocean ecosystem: phytoplankton. These small, floating microalgae come in a fantastic array of shapes and sizes, from smooth orbs to segmented spirals to pointy crescents (**Figure 21.2**). Phytoplankton are primarily microscopic, but in large groups they form the green color often seen in water. The more phytoplankton in the water, the greener the water is; the less phytoplankton, the bluer it is.

Phytoplankton are green because they are photosynthetic: they convert light energy from the sun into chemical energy using chlorophyll, the green pigment critical to the process

DANIEL BOYCE AND BORIS WORM

Daniel Boyce (left) is a postdoctoral researcher at Queen's University in Ontario, Canada. As a graduate student with Boris Worm (right), he studied the amount of phytoplankton biomass in the ocean. Boris Worm is a marine biologist at Dalhousie University in Nova Scotia, Canada, studying global marine biodiversity.

of photosynthesis. Because they are photosynthetic, these water-living organisms inhabit the top layer of water in the ocean (and almost every body of freshwater as well), a location that gives them easy access to sunlight. Thanks to their ability to photosynthesize using sunlight, phytoplankton are primary producers and the central means through which energy enters the ocean ecosystem. If overfishing was affecting phytoplankton levels, Boyce worried, the whole ocean ecosystem could be in danger.

Bottom of the Pyramid

Energy and nutrients flow through ecosystems in distinctive patterns. First, let's consider the path of energy. Producers, like phytoplankton, capture

energy from the sun and transform it into fuel energy. That fuel is passed up the food chain as one organism eats another. An **energy pyramid** represents the amount of energy available to organisms in an ecosystem. Each level of the pyramid corresponds to a step in a food chain and is called a **trophic level**. In the ocean, for instance, phytoplankton are on the first trophic level. Zooplankton, larger, multicellular plankton that feed on phytoplankton, are the second trophic level. Small fishes such as herring are the third level, and large fishes such as tuna are the fourth (**Figure 21.3**).

At each trophic level, a portion of the energy captured by producers is lost as **metabolic heat**, the heat released as a by-product of chemical reactions within a cell, especially during cellular respiration. Organisms lose a lot of energy as metabolic heat, as revealed by the fact that a small room crowded with people rapidly becomes hot; that warmth is the result of metabolic heat leaving our bodies. On average, roughly 10 percent of the energy at one trophic level is transferred to the next trophic level. The remaining 90 percent of the energy that is not transferred is either not consumed (for example, when we eat an apple, we eat only a small part of the apple tree), is not taken up by the consumer's body (for example, we cannot digest the cellulose contained in the apple), or is lost as metabolic heat.

Because of this steady loss of heat, energy flows in *only one direction* through ecosystems. It enters Earth's ecosystems from the sun (in most cases) and leaves them as metabolic heat. Every unit of energy captured by producers is eventually lost from the biotic world as heat. Therefore, energy cannot be recycled within an ecosystem. It travels up an energy pyramid, never down.

In contrast, **nutrients**—chemical elements required by living organisms—*are* recycled and reused within and across ecosystems. Although Earth receives a constant stream of light energy from the sun, our planet does not acquire more nutrients on a daily basis; rather, a constant and finite pool of nutrients cycles through the land, water, and air. If nutrients were not cycled between organisms and the physical environment, life on Earth would not exist.

Nutrients pass through the abiotic world, from rocks and mineral deposits into soil, water, and air, and then on to the biotic world via absorption by producers. Once in the biotic world, they are cycled among consumers for varying lengths

Trophic level   **Energy (Kilocalories)**

Only 10% of the energy captured and stored at each trophic level is transferred to the next trophic level. The remainder is lost through metabolic heat and inefficient consumption.

Tertiary consumer = fourth trophic level
10

Secondary consumer = third trophic level
100

Primary consumer = second trophic level
1,000

Of each 10,000 Kilocalories of energy captured from the sun by producers, primary consumers capture and store only about 10%.

Producer = first trophic level
10,000 Kcal

Figure 21.3

Energy pyramid

The levels of the energy pyramid correspond to steps in a food chain.

Q1: What percentage of the original 10,000 Kilocalories is available to a shark that might eat the tuna in this figure?

Q2: What trophic level and term would describe a predator of tuna?

Q3: Give an example of a primary consumer in a terrestrial environment.

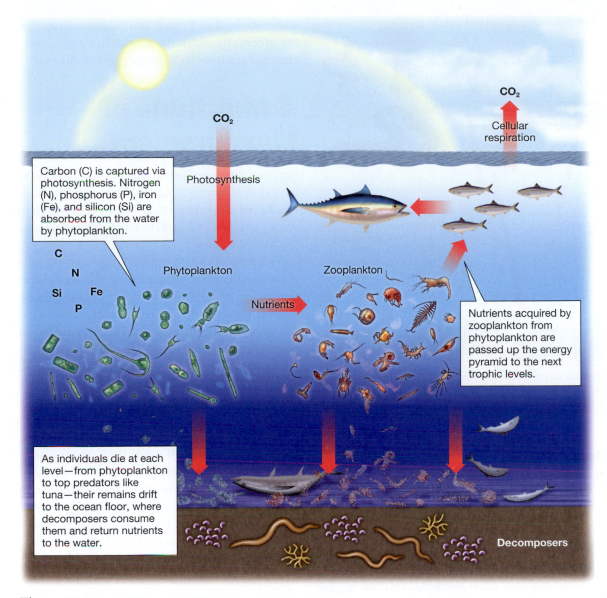

CO_2

CO_2

Cellular respiration

Photosynthesis

Carbon (C) is captured via photosynthesis. Nitrogen (N), phosphorus (P), iron (Fe), and silicon (Si) are absorbed from the water by phytoplankton.

C

N

Si   Fe

P

Phytoplankton

Zooplankton

Nutrients

Nutrients acquired by zooplankton from phytoplankton are passed up the energy pyramid to the next trophic levels.

As individuals die at each level—from phytoplankton to top predators like tuna—their remains drift to the ocean floor, where decomposers consume them and return nutrients to the water.

Decomposers

Figure 21.4

Nutrients cycle within and beyond the ocean ecosystem

In the ocean, phytoplankton absorb nutrients from the abiotic world. Within the biotic world, these nutrients are cycled among consumers, beginning with zooplankton and moving up through trophic levels. The nutrients then return to the abiotic world when dead or dying organisms are broken down by decomposers into their constituent elements.

Q1: Which organisms are the producers in this ecosystem?

Q2: How do nutrients flow from the abiotic to the biotic components of the ecosystem?

Q3: How do nutrients flow from the biotic to the abiotic components of the ecosystem?

of time. Phytoplankton, for example, require the nutrients nitrogen, phosphorus, iron, and silicon for growth. When zooplankton eat phytoplankton, they take up those nutrients, and so on up the food chain (**Figure 21.4**).

Nutrients are eventually returned to the abiotic world when **decomposers** break down the dead bodies of other organisms. In some ecosystems, decomposers break down 80 percent of the *biomass*, or biological material,

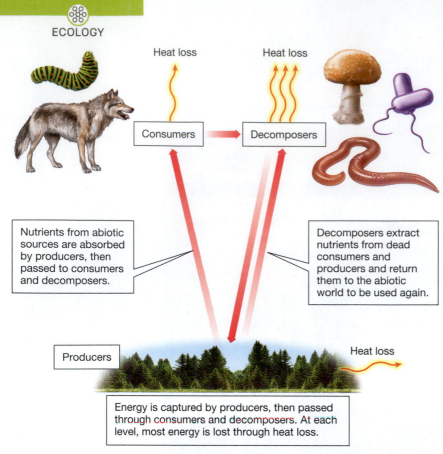

Heat loss   Heat loss

Consumers → Decomposers

Nutrients from abiotic sources are absorbed by producers, then passed to consumers and decomposers.

Decomposers extract nutrients from dead consumers and producers and return them to the abiotic world to be used again.

Producers   Heat loss

Energy is captured by producers, then passed through consumers and decomposers. At each level, most energy is lost through heat loss.

Figure 21.5

Energy flow and nutrient cycling

Unlike energy, which moves up and out of ecosystems, nutrients are constantly cycled between the abiotic and biotic worlds. Important nutrients for the biotic world include carbon (C), potassium (K), phosphorus (P), and nitrogen (N).

Q1: How is a decomposer different from a more typical consumer?

Q2: What is the difference between how carbon is brought into the biotic portion of the ecosystem, and how other nutrients, such as phosphorus, are brought in?

Q3: Describe all the points at which heat is lost in this figure.

made by producers. Without decomposers, nutrients could not be repeatedly reused, and life would cease because all essential nutrients would remain locked up in the bodies of dead organisms. In this way, decomposers are the "cleaners" of an ecosystem. Bacteria and fungi are important decomposers in the ocean, as are hagfishes, worms, and others.

Ecologists and earth scientists use the term "nutrient cycle" to describe the passage of a chemical element through an ecosystem (**Figure 21.5**). The nutrient cycle and the flow of energy are two of four processes that link the biotic and abiotic worlds in an ecosystem. These **ecosystem processes** also include the

water cycle (see Figure 18.12) and succession, the process by which the species in a community change over time (as discussed in Chapter 20).

A Multitude of Measurements

For over 100 years, researchers around the globe have studied ecosystems containing phytoplankton. Boyce tapped into that wealth of research to document past and present levels of phytoplankton in the oceans.

The amount of phytoplankton biomass in a given area can be estimated by the concentration of chlorophyll found there (**Figure 21.6**). For decades, nearly all ocean studies have used chlorophyll concentration as a reliable metric of phytoplankton biomass. Chlorophyll concentration is measured by detecting the color of water. Water takes on deeper shades of green as the amount of chlorophyll increases. When there is no chlorophyll, water appears clear.

Ideally, Boyce would have used satellite data to detect chlorophyll and thus phytoplankton concentrations, since satellites today take high-resolution color measurements of the ocean surface. Yet Boyce planned to review phytoplankton levels over the past 100 years, and high-quality satellite data have been available for only the last decade. He needed another source of data.

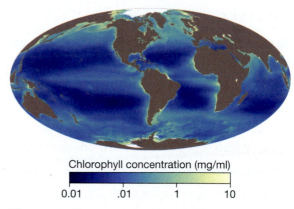

Chlorophyll concentration (mg/ml)

0.01   .01   1   10

Figure 21.6

Average chlorophyll concentration in the oceans

Phytoplankton are most abundant in high latitudes, along coastlines and continental shelves, and along the equator in the Pacific and Atlantic Oceans (yellow)—but are scarce in remote oceans (dark blue).

In a first-of-its-kind analysis, Boyce, together with his adviser Boris Worm and the oceanographer Marlon Lewis, combined two types of chlorophyll measurements. The first type, dating all the way back to 1899, were recorded with nothing more than a rope and a disk.

In 1865, the pope asked priest and astronomer Pietro Angelo Secchi to measure the clarity of water in the Mediterranean Sea for the purposes of the papal navy. Secchi designed one of the simplest measurement devices ever used: a dinner plate–sized disk painted with black and white stripes attached to a rope. The disk is lowered into water until the white stripes disappear (as they become obscured by chlorophyll from phytoplankton), and the depth at that point is recorded (**Figure 21.7**; see also the chapter-opening photo). Chlorophyll concentrations derived from Secchi disk measurements have been corroborated by satellite data, so scientists know they are reliable.

In addition to gathering Secchi disk data, scientists at sea use lab tools to directly measure the quantity of chlorophyll in the water (as opposed to observing its color and relating that to chlorophyll concentration). Boyce found hundreds of thousands of these direct chlorophyll measurements online in open-source databases. "There's been a huge increase in the amount of publicly available oceanographic data out there," says Boyce.

But to use the data, Boyce first had to separate the wheat from the chaff. "With any big database, there are bound to be measurements that are entered incorrectly, for whatever reason," says Boyce. He, Worm, and Lewis ruled out measurements that were inappropriate for their study, such as those taken where the ocean floor was less than 25 meters deep, because changes of water transparency in those cases could be caused by sediment or runoff from landmasses nearby rather than by phytoplankton.

The team analyzed each data set separately—Secchi disk measurements and direct chlorophyll measurements—and then together. To combine the two, they converted all the Secchi measurements into the same units as those used for direct chlorophyll concentrations. In total, the blended data set included 445,237 chlorophyll measurements collected between 1899 and 2008.

What they found made them pause. With those two different methods of analysis, they identified a significant decline in phytoplankton levels—a whopping 60–80 percent—in Earth's oceans where data were available during the

Figure 21.7

Secchi disks indirectly measure chlorophyll concentrations
The Secchi disk is lowered into water until its white stripes become obscured by the chlorophyll in phytoplankton.

last century. Overall, phytoplankton appeared to have declined by about 1 percent of the global average each year.

One percent sounds like a small number, but that amount every year just since 1950 translates into a staggering total phytoplankton decrease of 50 percent in the world's oceans.

The Precious 1 Percent

A 50 percent loss in the main producer in any ecosystem is a worrisome number, but especially with respect to phytoplankton. Phytoplankton support fisheries, produce half the oxygen we breathe, and take in carbon dioxide from the atmosphere, which helps offset the greenhouse effect and global warming.

An ocean with less phytoplankton will function differently because ecosystems depend on **energy capture**, the trapping and storing of solar energy by the producers at the base of the ecosystem's energy pyramid. Herbivores, carnivores, and detritivores all depend indirectly on energy capture. If an ecosystem has an abundance of producers, it can often support more consumers at higher trophic levels. In a tropical forest, for example, an abundance of plants capture energy from the sun, and the forest teems with life. On the flip side, relatively little energy is captured in an environment with few producers. In tundra or desert regions, for example, less food is available, and fewer animals can live there. These significant differences have prompted ecologists to categorize large areas of Earth's surface into distinct regions, called **biomes**, that are defined by their unique climatic and ecological features (**Figure 21.8**).

Assessing the overall amount of energy captured by producers is important in determining how an ecosystem works, because energy

The **boreal forest** is dominated by coniferous trees that grow in northern or high-altitude regions with cold, dry winters and mild summers. The soil is thin and nutrient-poor, so while plants generally receive adequate moisture during the growing season, plant diversity is relatively low. Large herbivores include elk and moose. Small carnivores, such as weasels, wolverines, and martens, are common. Larger carnivores include lynx and wolves.

The **tundra** biome is found at the poles and on mountaintops. Trees are absent or scarce because of the short growing season. The vegetation is dominated by low-growing flowering plants, and the boggy landscape is covered in mosses and lichens, important food sources of herbivores. Rodents provide food for carnivores like foxes and wolves. Bears and musk oxen are among the few large mammals.

The **chaparral** is a shrubland biome dominated by dense growths of drought-resistant plants in regions with cool, rainy winters and hot, dry summers. These conditions make the chaparral exceptionally susceptible to wildfires. Common vegetation in the California chaparral includes scrub oak, pines, mountain mahogany, manzanita, and the chemise bush. Small mammals such as jackrabbits and gophers are common, and there are many species of lizards and snakes.

Terrestrial biomes:
- Tundra
- Boreal forest
- Temperate forest
- Grassland
- Chaparral
- Desert
- Tropical forest

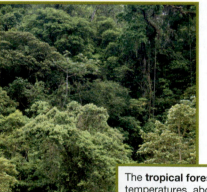

Ecosystems within the **freshwater biome** are heavily influenced by the terrestrial biomes that they border or through which their water flows. Lakes are landlocked bodies of standing freshwater. Rivers are bodies of freshwater that move continuously in a single direction. Wetlands are characterized by standing water that is shallow enough for rooted plants to emerge above the water surface. A bog is a freshwater wetland with stagnant, oxygen-poor water, and low productivity and species diversity. In contrast, grassy marshes and tree-filled swamps are highly productive wetlands with a high diversity of organisms.

The **tropical forest** biome is characterized by warm temperatures, about 12 hours of daylight each day, and either seasonally heavy or year-round rains, which tend to leach nutrients from the soil. Soils in this biome further tend to be nutrient-poor because a large percentage of nutrients are locked up in the living tissues (biomass) of organisms. Tropical rainforests, which may receive in excess of 80 inches (200 cm) of rain annually, are some of the most productive ecosystems on Earth, with a rich diversity of organisms.

The **grassland** biome cannot sustain vigorous tree growth, but its moisture levels are not as low as in deserts. Grasslands are found in both temperate and tropical latitudes and are dominated by grasses and herbaceous plants such as coneflower and shooting star. Scattered trees are found in some, such as the tropical grasslands known as the savanna. Burrowing rodents like voles and prairie dogs may aerate the soil, thereby improving growing conditions. Many grasslands have been converted to agriculture.

The **temperate forest** biome is dominated by trees and shrubs adapted to relatively rich soil, snowy winters, and moist, warm summers. These forests display greater species diversity than do the tundra and boreal forest biomes: oak, maple, hickory, beech, and elm are common. Herbivores include squirrels, rabbits, deer, raccoons, and beavers, while bobcats, mountain lions, and bears make up the carnivores. Amphibians and reptiles are common.

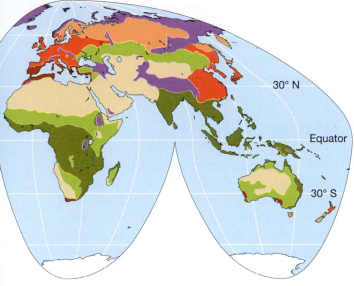

30° N

Equator

30° S

Because air in the **desert** biome lacks moisture, it does not retain heat well. As a result, temperatures can be above 113°F (45°C) in the daytime and then plunge to near freezing at night. Desert plants have small leaves to minimize water loss. Succulents, such as cacti, store water in their fleshy stems or leaves. Most animals in the desert are nocturnal, hiding in burrows during the heat of the day and emerging at night to feed.

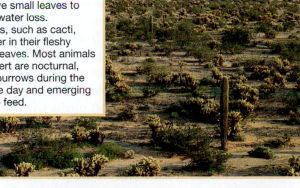

Estuaries are the shallowest but most productive of the aquatic biomes. They are tidal ecosystems where rivers flow into the ocean. They have a constant ebb and flow of fresh and salt water, and organisms must be able to tolerate daily changes in saltwater concentrations. The plentiful light, the abundant supply of nutrients delivered by the river system, and the regular stirring of nutrient-rich sediments by water flow create a rich and diverse community of photosynthesizers. Grasses and sedges are the dominant vegetation in most estuaries.

The **marine biome**, characterized by salt water, is the largest biome on our planet. The *coastal region* stretches from the shoreline to the edge of the continental shelf and is highly productive because of the availability of nutrients and oxygen. A majority of Earth's marine species live in the coastal region. The intertidal zone, closest to the shore, is a challenging environment where organisms such as seaweeds, worms, crabs, sea stars, sea anemones, and mussels are submerged and exposed to dry air on a twice-daily basis. The relatively nutrient-poor *open ocean* begins about 40 miles offshore and is much less productive than coastal waters.

Figure 21.8

Amazing biomes

Biomes do not begin and end abruptly, but rather transition into one another. Terrestrial biomes are categorized by temperature, precipitation, and altitude; aquatic biomes are determined by proximity to shorelines.

capture influences the amount of food available to other organisms. The **net primary productivity (NPP)** of an ecosystem is the energy, acquired through photosynthesis over a particular time period, that is available for the growth and reproduction of producers. NPP is the amount of energy captured by photosynthetic organisms minus the amount they expend on cellular respiration and other maintenance processes. NPP is typically determined by estimating the amount of carbon captured during photosynthesis. This can be done by measuring the amount of new biomass produced by the photosynthetic organisms in a given area during a specified period of time.

According to scientists' estimates, the NPP of all producers on Earth exceeds 100 billion tons of carbon biomass per year. Roughly half of this productivity comes from phytoplankton in the ocean. Therefore, phytoplankton capture approximately 50 billion tons of carbon per year. So if Boyce's calculations are right, a loss of 1 percent of that biomass is 500 million tons of organic matter lost from the oceans *each year*. That's a lot of biomass to lose.

Net primary productivity relies on four things: sunlight, water, temperature, and the availability of nutrients. The most productive ecosystems on land are tropical forests; the least productive are deserts and tundra (including some mountaintop communities). The most productive ecosystems in water are estuaries—regions where rivers empty into the sea—because nutrients drained off the land stimulate the growth and reproduction of phytoplankton and other producers, which in turn nourish large populations of consumers. The least productive aquatic biome is the deep ocean, where sunlight does not penetrate.

Despite similarities between the NPP requirements on land and in the ocean, the global pattern of NPP differs between the two. On land, the NPP is highest at the equator and decreases toward the poles. But in the ocean, the general pattern relates not to latitude but to distance from shore: the productivity of marine ecosystems is typically high in ocean regions close to land and relatively low in the open ocean (**Figure 21.9**). This is because nutrients needed by aquatic photosynthetic organisms are in better supply near land, thanks to delivery from streams and rivers. Wetlands such as swamps and marshes, which trap soil sediments rich in nutrients and organic matter, can be so productive that they match the productivity levels of tropical forests.

A loss of 500 million tons of phytoplankton each year could potentially affect the ocean's NPP and ocean life. "Almost all biological life in the ocean depends on phytoplankton. A reduction in the biomass of phytoplankton will result in less secondary production in the oceans," says Boyce. That means fewer sharks, whales, fishes, eels—you name it.

The team's discovery was shocking, to say the least. No one else had documented a global decline in phytoplankton before. So, to be confident in their results, Boyce and Worm did several more rounds of data analysis, checking again and again to make sure they were using the right numbers in ways that correctly represented what was happening in the natural world. And over and over, they came back with the same results: global phytoplankton declined over the last century. In 2010, they published that finding in the peer-reviewed scientific journal *Nature*.

Phyto-Fight

The scientific community reacted immediately. Some researchers doubted Boyce's conclusions; others were outright incredulous. Paul Falkowski at Rutgers University told a *New York Times* reporter that he had not found the same trend in a long-term analysis of the North Pacific (though Boyce contends that their trends were very similar), and another team had actually seen an increase in phytoplankton starting around 1978 in the central North Pacific. Falkowski called Boyce's paper "provocative" but said he would "wait another several years" to see whether satellite data would back up the finding.

Then, in 2011, three separate research teams went so far as to publish formal critiques of the work. One suggested that the declining trend was an error resulting from the use of two different types of measurements: Secchi disk readings and direct chlorophyll measurements. A second team echoed that idea, reanalyzed the data in a way that showed an increase in phytoplankton, and then bluntly concluded, "Our results indicate that much, if not all, of the century-long decline reported by [Boyce] is attributable to this [sampling bias] and not to a global decrease in phytoplankton biomass."

The third team noted that Boyce's finding conflicted with eight decades of data on phytoplankton biomass collected by a large project called the Continuous Plankton Recorder (CPR)

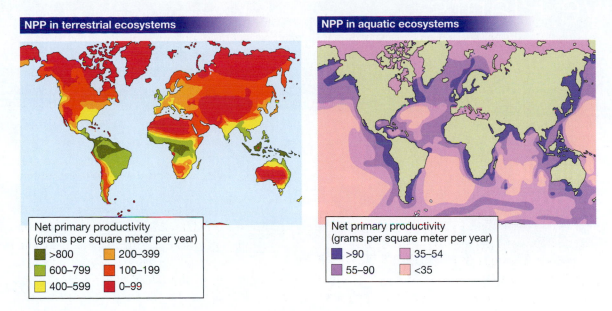

NPP in terrestrial ecosystems

Net primary productivity
(grams per square meter per year)

- >800
- 600–799
- 400–599
- 200–399
- 100–199
- 0–99

NPP in aquatic ecosystems

Net primary productivity
(grams per square meter per year)

- >90
- 55–90
- 35–54
- <35

Figure 21.9

Global variation in net primary productivity

NPP can be measured as the number of grams of new biomass made by producers each year in a square meter of each biome's area. NPP varies greatly across both terrestrial and aquatic ecosystems.

Q1: Which terrestrial biome has the lowest NPP? Which aquatic biome?

Q2: Where are the most productive terrestrial biomes located?

Q3: Give a possible reason for your answer to question 2.

survey, which monitors the Northeast Atlantic Ocean. The CPR survey, started in 1931, employs a unique instrument pulled through the ocean by commercial fishing vessels to collect millions of samples of plankton. The CPR survey found that over the last 20–50 years, phytoplankton biomass increased in the Northeast Atlantic, says Abigail McQuatters-Gollop, a former researcher at the Sir Alister Hardy Foundation for Ocean Science, which operates the survey.

In response, Boyce, Worm, and Lewis didn't get angry; they got focused. After reading the critiques, the three researchers went back to the data. First they applied a correction factor suggested by the critics, in the hopes of removing any bias between the two types of data. "We did that, and the trends remained similar," says Boyce. Next, they again estimated changes over time individually for the two data sources. "That didn't change the trends either," says Boyce. After that, they incorporated additional suggestions by their peers and created a new, expanded database of chlorophyll measurements to work from. Finally, they reestimated changes in chlorophyll

using this new database and their revised analysis methods, but the phytoplankton still seemed to be declining, independently of the type of data or how the data were analyzed. The three researchers published their reanalysis in a series of three papers in 2011, 2012, and 2014, demonstrating the same decline again and again.

But the additional work has not silenced the critics. "It's still pretty hotly debated," admits Boyce. "The story is not over by any means." In 2011, in fact, Worm traveled to an international plankton conference where he, McQuatters-Gollop, and others debated the topic in front of a live audience. "It was an amicable meeting that generated loads of discussion," says

ABIGAIL MCQUATTERS-GOLLOP

Abigail McQuatters-Gollop was a science and policy researcher at the Sir Alister Hardy Foundation for Ocean Science, home of the Continuous Plankton Recorder survey, from 2010 to 2015. She is now a lecturer in marine conservation at Plymouth University in the United Kingdom.

Productive Plants

Data on net primary productivity (NPP)—the total energy available in an ecosystem for the growth and reproduction of primary producers—is sparse for whole biomes, but researchers have been able to estimate the NPP of those listed here using information about different vegetation types and carbon sources in each area.

Assessment available in smartwork5

Total energy available from primary producers

(in grams per square meter per year)

Desert
150

Tundra
240

Lakes and streams
440

Grassland (temperate)
790

Grassland (tropical)
1,770

Wetlands
2,620

McQuatters-Gollop. Still, they could not agree on whether phytoplankton populations have increased or decreased in the ocean. At the meeting, the researchers agreed that the best thing to do was to combine as many data sets as possible.

Heating Up

In 2015, the long-awaited satellite data came in—and they weren't good. Ocean color measurements from two NASA satellites led scientists to conclude that diatoms, the largest type of phytoplankton, had declined more than 1 percent per year from 1998 to 2012 globally, with major losses in the North Pacific, North Indian, and Equatorial Indian Oceans.

It looks like Boyce, Worm, and Lewis may be right, but the important question of why a decline is occurring remains. As part of his research, Boyce investigated possible causes of the downward trend. In one study, he compared changes in sea surface temperature to changes in chlorophyll levels and noted a strong correlation: over the last 100 years, chlorophyll concentrations declined *and* ocean temperatures increased, in line with increases in global warning. This correlation has been closely followed in recent years (**Figure 21.10**).

But correlation does not prove causation, and much more work needs to be done to support the idea that global warming caused the decline. It is a logical hypothesis, however, because as the planet warms, water in the oceans mixes less, limiting the nutrients delivered to the surface from decomposers in the deep sea. As a result, phytoplankton do not receive the nutrients they need for growth and reproduction. In 2014, Boyce published experimental data supporting the hypothesis. Working with marine scientists in Germany, he found that warming the water in a controlled, experimental ocean water enclosure led to reduced phytoplankton biomass.

A 2014 study by a large, international group of researchers predicted that an increase in ocean temperature due to global warming would cause phytoplankton biomass to decrease by 6 percent by the end of this century. Consequences of continued phytoplankton decline could include altering the carbon cycle between the ocean and the atmosphere, changing heat distribution in the ocean, and causing a decrease in the supply of food in the ocean. Whatever the case, if phytoplankton populations are decreasing, says Boyce—and he is sure they are—there will be profound effects, which have only just begun.

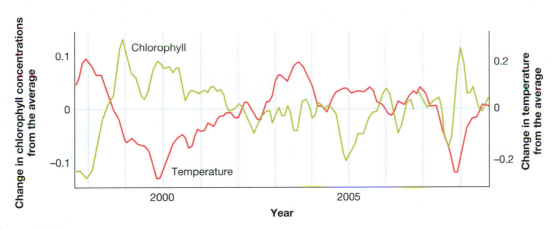

Figure 21.10

As ocean temperature increases, chlorophyll decreases
Between late 1997 and mid-2008, satellites observed that warmer-than-average temperatures (red line) were correlated with below-average chlorophyll concentrations (green line).

Q1: In what years were chlorophyll levels the highest?

Q2: In what years were the temperature changes from the average the greatest?

Q3: How do you predict this graph will look 10 years from now?

REVIEWING THE SCIENCE

- An **ecosystem** consists of communities of organisms and the physical environment in which those communities live. It is the sum of all biotic elements interacting with all abiotic elements. Energy, materials, and organisms can move from one ecosystem to another.

- An **energy pyramid** represents the amount of energy available to each **trophic level** of a food chain in an ecosystem.

- Energy enters an ecosystem when producers capture it from an external source, such as the sun. A portion of the energy captured by producers is lost as **metabolic heat** at each trophic level. As a result, energy flows in only one direction through ecosystems.

- **Nutrients** are the chemical elements required by living organisms. Unlike energy, nutrients are recycled and reused within and across ecosystems. Earth has a fixed amount of nutrients.

- **Decomposers** break down the dead bodies of other organisms, both consumers and producers, releasing the nutrients in the bodies of dead organisms back to the physical environment.

- Four **ecosystem processes** link the biotic and abiotic worlds in an ecosystem: nutrient cycling, energy flow, water cycling, and succession.

- Ecosystems depend on **energy capture**, the trapping of solar energy by producers via photosynthesis, and the storage of that energy as chemical compounds in their bodies.

- Earth is categorized into 10 major **biomes**, regions defined by their climatic and ecological features.

- **Net primary productivity (NPP)** is the energy acquired through photosynthesis that is available for growth and reproduction to producers in an ecosystem. NPP is estimated by the amount of biomass produced in a given area during a specified period of time.

THE QUESTIONS

Challenge Yourself

1 Select the correct terms:
The biome characterized by shrubs and nonwoody plants that grow in regions with cool, rainy winters and hot, dry summers is (**tundra / chaparral**). Another biome with few trees, but in this case dominated by grasses and nonwoody plants, is (**grassland / tundra**). The most productive aquatic biome is (**freshwater / estuaries**).

2 Which of the following is a component of an ecosystem but *not* of an ecological community?

(a) a producer

(b) water

(c) a secondary consumer

(d) a primary consumer

3 Place the following elements of the scientists' study of oceanic phytoplankton levels in the correct order from earliest to latest by numbering them from 1 to 5.

_____ a. Boyce analyzed data from multiple sources showing that ocean phytoplankton levels have declined dramatically over the last 100 years.

_____ b. Worm demonstrated that industrialized fishing had decimated predatory fish communities.

_____ c. Secchi designed a device to measure phytoplankton levels in water.

_____ d. Boyce's finding of declining phytoplankton levels was challenged by fellow scientists, because of his methodology.

_____ e. Boyce, a graduate student in Worm's lab, decided to study how the loss of predatory fish reverberated through the oceanic ecosystem—in particular, its phytoplankton populations.

4 How is an ecosystem different from an ecological community?

Try Something New

5 In the energy pyramid shown here, an owl, a cardinal, and a grasshopper are the fourth, third, and second trophic levels, respectively.

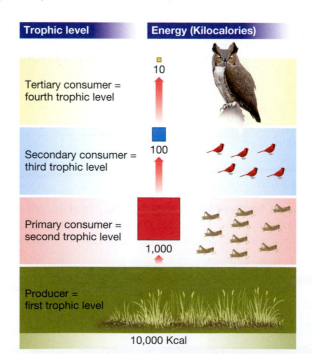

a. If each grasshopper passes 100 Kilocalories to the cardinal when eaten, how many grasshoppers would the cardinal have to eat to obtain 10,000 Kilocalories?

b. How many Kilocalories of grass are required to produce 10,000 Calories' worth of grasshopper?

c. Where does the lost energy go?

6 Is the water cycle (see Figure 18.12) more similar to the movement of energy or to the movement of nutrients through an ecosystem? Justify your answer.

7 While sloshing around in the swampy wetlands of a nearby forest preserve, a small child notices a rotting tree branch covered in fungi. What are these organisms doing on this tree branch?

(a) These are producers acquiring energy through photosynthesis.

(b) These are consumers acquiring energy from the wood and releasing nutrients back to the earth.

(c) These are decomposers acquiring energy from the wood and releasing nutrients back to the earth.

(d) These are decomposers releasing nutrients back to the earth, but acquiring no energy for themselves in the process.

Leveling Up

8 *Write Now* biology: human-caused biome shifts The location of Earth's different biomes depends on climate and altitude, for the most part. However, human activities play a role in the conversion of one biome to another, as has been seen many times in history. Research one major change in a biome category based on human activity, and describe how and why this change happened. Speculate on how this change could have been specifically avoided. (*Note:* Do not analyze a change via deforestation to agricultural land, since agricultural land is not a natural biome. *Hint:* Take a look at Easter Island as one example.)

9 **Doing science** Join forces with millions of others by classifying phytoplankton on your computer. Do an Internet search for "citizen science phytoplankton" and sign in as a citizen scientist, complete the tutorial, and start helping researchers quantify the phytoplankton in our oceans.

10 **What do *you* think?** Some people think the current U.S. Endangered Species Act should be replaced with a law designed to protect ecosystems, not species. The intent of such a law would be to focus conservation efforts on what its advocates think really matters in nature: whole ecosystems. Given how ecosystems are defined, do you think it would be easy or hard to determine the boundaries of what should and should not be protected if such a law was enacted? Give reasons for your answer.

Thinking Multidimensionally

11 Based on current agricultural production, some experts estimate there is enough food to feed 10 billion people worldwide, yet close to 1 billion people go to bed hungry each night.

a. While poverty and environmental problems are the largest contributing factors to world hunger, if we want to attempt to feed the world, from what trophic level(s) would most people need to eat? Explain.

b. Most food chains consist of four trophic or feeding levels. Explain why shorter food chains are more energetically stable.

c. As you might imagine, exceptions to short food chains exist. Look at the image on page 375 depicting an aquatic food web. Identify the chain of organisms in this community that has the largest number of trophic levels.

d. Calculate the energy transfers for this food chain based on the ~10% energy transfer rule.

e. What is the rationale for some humans to kill whales?

12 Have you ever thought about how a tiny acorn is able to grow into a mighty oak tree?

a. Referring back to Figure 5.2, which component(s) is/are responsible for a plant's gaining its biomass, and why?

b. Now consider the increased anthropogenic CO_2 on Earth. If plants are given higher concentrations of CO_2, would you expect higher or lower crop yields? Explain.

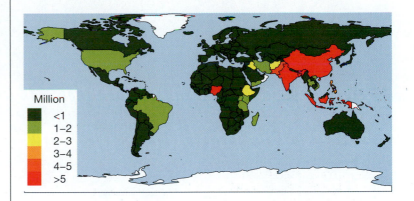

Million
- <1
- 1–2
- 2–3
- 3–4
- 4–5
- >5

The trade-off becomes a matter of quantity versus quality. Higher CO_2 levels increased carbohydrate levels in plants while decreasing the overall nutritional value of specific crops.

The most puzzling part of this experiment was, and still is, the effects that increased CO_2 has on protein quantities in some plants. Major food crops of the world, such as barley, wheat, rice, and potatoes exhibited 6%–15% lower protein concentrations. The map above shows the number of people estimated to be newly at risk of protein deficiency due to elevated carbon-dioxide levels.

c. With protein being an essential part of human dietary needs, which countries are predicted to be most at risk for protein deficiencies, and why not all nations?

d. What functions do proteins serve in our bodies?

For more, visit digital.wwnorton.com/bionow2 for access to:

Baby Bust

Facing dwindling births, Denmark searches to resolve problems of infertility.

PHYSIOLOGY

CHAPTER

22

HOMEOSTASIS,
REPRODUCTION,
AND
DEVELOPMENT

DRIVING QUESTION

Why are birth rates decreasing in Denmark, and should this decline be a cause for concern or celebration?

- HS-LS1.A: Structure and Function
- HS-LS1.B: Growth and Development of Organisms

A woman slides a bra strap off her naked shoulder. In Danish, a seductive man's voice asks, "Can sex save Denmark's future?"

The commercial, aired on Danish television, then switches to a view of an empty playground. Produced by a travel agency, the ad encourages responsible Danes to book a romantic holiday with the company's "ovulation discount" and try to conceive while on vacation. It concludes with a large banner proclaiming, "Do It for Denmark" (**Figure 22.1**).

Birth rates reached a 27-year low in the year 2014 in Denmark, a Nordic country in northern Europe. When asked, most Danish couples said they would like to have two or three children, yet the present fertility rate is only 1.69 children per family—not high enough to maintain Denmark's current population. Infertility is now considered an epidemic in the country. In fact, one in 10 children in Denmark is conceived using reproductive technologies. "We see more and more couples needing to get assisted fertility treatment," Bjarne Christensen, secretary general of Sex and Society, Denmark's leading family-planning association, told *Bloomberg News*. "We see a lot of people who don't succeed in having children."

Commercials like the one described here make a patriotic, if whimsical, appeal to Danish citizens to have more children, but experts debate whether falling fertility rates are a cause for concern or for celebration. Some applaud the decrease as a way to slow global population growth and reduce human consumption of limited natural resources. But others in countries facing declining fertility rates—including Denmark, France, Singapore, and recently the United States—worry about the burdens that will be placed on a generation smaller than the one before it (**Figure 22.2**). Government officials fear that this trend will produce a smaller workforce and fewer young people to care for retirees, reducing their countries' economic strength.

To counteract falling birth rates, some governments have begun taking profertility stances, including providing free postnatal care and subsidized day care. In Denmark, Sex and Society, a nongovernmental organization, provides sex education materials for most of Denmark's schools, and it recently unveiled a new series of lesson plans entitled "This Is How You Have Children!" Instead of focusing on contraception and how to avoid becoming pregnant, these new classes educate students about what fertility is, how aging affects fertility, and when may be the best times to have children.

"We have for many years addressed the very important issues of how to avoid becoming

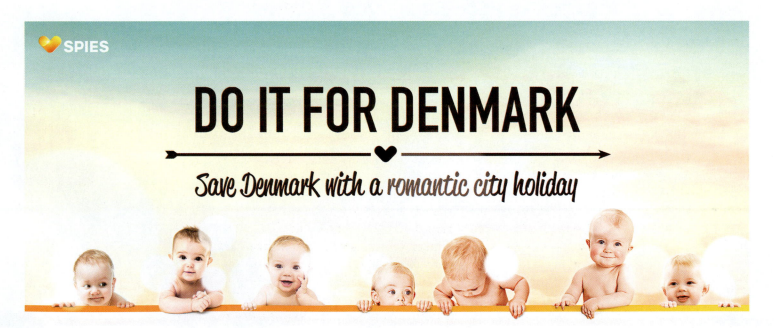

Figure 22.1

A print advertisement in the "Do It for Denmark" campaign

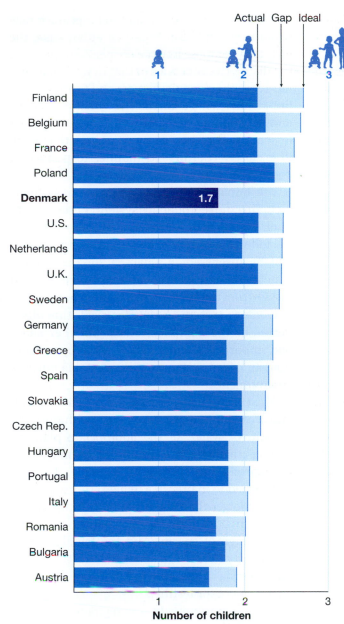

Actual Gap Ideal

Finland
Belgium
France
Poland
Denmark 1.7
U.S.
Netherlands
U.K.
Sweden
Germany
Greece
Spain
Slovakia
Czech Rep.
Hungary
Portugal
Italy
Romania
Bulgaria
Austria

1   2   3

Number of children

Sources: European data from 2011 Eurobarometer, U.S. data from 2006 and 2008 General Social Survey

Figure 22.2

Women in Europe and the United States want more children than they have

American and European women aged 40–54 were asked, "What do you think is the ideal number of children for a family to have?" Their answers were consistently higher than actual birth rates.

Q1: In which of the countries shown in the graph did women want the most children, on average? The least?

Q2: Which country has the largest gap between the number of children women would like to have and the number they actually have?

Q3: Which country has the highest average number of children? The lowest?

pregnant, how to avoid sexual diseases, how they have a right to their own bodies, but we totally forgot to tell the kids that we cannot have children forever," Søren Ziebe, head of Copenhagen University Hospital's fertility clinic, told reporters. "There is a biological limit."

Seeking Stability

That biological limit may be the cause of Denmark's falling birth rate. Population studies show that, on average, couples are waiting longer to have children, and conception can be more difficult after age 30.

But other causes are also possible. Laboratories are investigating whether chemicals or other environmental factors might be affecting the human **reproductive system**, the parts of the body responsible for producing offspring.

Our bodies are highly efficient and well-coordinated communities of 200 different specialized cell types. **Anatomy** is the study of the structures that make up a complex multicellular body, and **physiology** is the science of the functions of anatomical structures. Through physiological research, scientists hope to determine why fertility rates are falling in Denmark, and whether anything can be done about it.

Their first step has been to examine the reproductive system, which consists of cells, tissues, and organs. **Tissues** are made up of cells that act in an integrated manner to perform a common set of functions (**Figure 22.3**). A tissue

may be composed of just one cell type, or it may contain multiple cell types; in either case, the cells that compose a tissue cooperate to perform the distinctive functions of that tissue. An **organ** has more than one tissue type and forms a functional unit with a distinctive function, shape, and location in the body. An **organ system** is composed of two or more organs that work in a closely coordinated manner to perform a distinct set of functions in the body.

The human body has 11 major organ systems, including the reproductive system (**Figure 22.4**). Each system will be covered in more detail in the following chapters, and plant organ systems will be discussed in Chapter 26. Each organ system and its organs are unique in function and form—from the beating, blood-filled heart of the circulatory system to the electrical, threadlike nerves of the nervous system—but they share one important commonality: for proper function, an organ system requires a stable internal environment.

Most biological processes take place within only a certain temperature range, with the right amount of water, at the appropriate pH, and at a particular concentration of chemicals. In the reproductive system, for example, the male testes need to maintain a temperature about 1–2 degrees lower than the usual body temperature of 98.6°F (37°C) for the successful production and storage of sperm. And the female vagina requires a pH between about 3.8 and 4.5 to keep out bacteria while still maintaining a healthy environment for fertility.

These environments are regulated through **homeostasis**, the process of maintaining a relatively constant internal state despite changes in the external environment. Homeostasis enables an organism to continually sense its internal state and rapidly adjust. In this way, despite large fluctuations in the outside world, homeostasis maintains the internal conditions best suited for life processes.

Homeostasis occurs via **homeostatic pathways**, sequences of steps that reestablish homeostasis if there is any departure from the genetically determined normal state (also called the **set point**). Homeostatic pathways continually monitor the physical and chemical characteristics of the internal environment and trigger regulatory processes within the body if this monitoring system detects any deviation from the set point.

Nervous tissue communicates and processes information.

Brain

Spinal cord

Nerves

Epithelial tissue covers organs and lines body cavities.

Lining of digestive tracts and organs

Skin surface

Muscle tissue generates force by contracting.

Cardiac muscle

Smooth muscle

Skeletal muscle

Connective tissue binds and supports tissues and organs.

Fat and other soft tissues

Bone

Tendon

Figure 22.3

Tissues in animals

The different types of tissues found in the animal body can be placed into four broad categories: **epithelial**, **connective**, **muscle**, and **nervous tissue**.

Q1: Which tissue type is the primary component of skin?

Q2: Which tissue type is the primary component of bone?

Q3: Which tissue types does the hand contain?

The **urinary system** removes excess fluid from the body, along with waste products and toxins.

In the **digestive system**, large molecules of food are broken down in the mouth, stomach, and small intestine, and nutrients are absorbed in the small and large intestines.

The **circulatory system** diffuses oxygen from the lungs to the heart, which then pumps oxygen-rich blood to the rest of the body through a closed network of vessels.

The **respiratory system** brings in oxygen and expels carbon dioxide through the lungs.

The **endocrine system** works closely with the nervous system to regulate all other organ systems. It consists of a number of glands and secretory tissue.

The **integumentary system** is the largest organ system in the human body, covering and protecting the surface of the body.

The **nervous system** is a key player in sensing the external world and the body's internal state, and it communicates with all of the other organ systems.

The **skeletal system** provides an internal framework to support the body in vertebrates. It consists of bone, cartilage, and ligaments.

The **muscular system** produces the force that moves structures within the body. It works closely with the skeletal system.

The **immune system** defends the body from invaders such as viruses, bacteria, and fungi.

The **reproductive system** generates gametes and, depending on the animal group, may also support fertilization and prenatal development.

Figure 22.4

Organ systems

The 11 major organ systems of the human body work in an integrated manner.

Q1: Which organ system defends the body from infectious diseases such as the common cold or flu?

Q2: Which organ systems transport oxygen to cells?

Q3: Which organ systems regulate the activities of the other organ systems?

Homeostatic pathways depend on **feedback loops**. A **negative feedback** loop turns off or reduces the output of a process. For example, if a person drinks a large milk shake, the level of glucose in the blood rises. In response, cells in the pancreas produce insulin, which allows glucose to enter cells. So, blood glucose concentration declines. The regulation of body temperature also relies on negative feedback loops (**Figure 22.5**).

A **positive feedback** loop, on the other hand, increases the output of a process until an endpoint is reached. Blood clotting triggered by a broken blood vessel, for example, is a positive feedback loop: the process of clotting releases chemicals that lead to accelerated clotting, continuing until the clot plugs the break in the blood vessel wall. In this feedback loop, the chemicals increase the amount of clotting.

Overall, in homeostatic pathways negative feedback loops are more common than positive feedback loops. Examples of homeostatic pathways include *thermoregulation* (the control

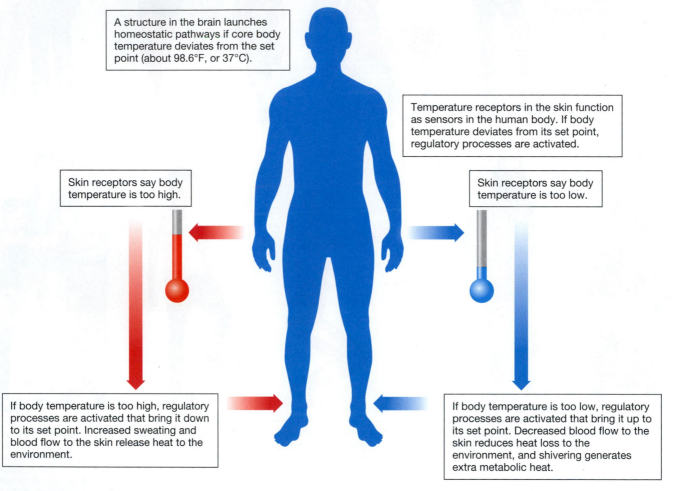

A structure in the brain launches homeostatic pathways if core body temperature deviates from the set point (about 98.6°F, or 37°C).

Temperature receptors in the skin function as sensors in the human body. If body temperature deviates from its set point, regulatory processes are activated.

Skin receptors say body temperature is too high.

Skin receptors say body temperature is too low.

If body temperature is too high, regulatory processes are activated that bring it down to its set point. Increased sweating and blood flow to the skin release heat to the environment.

If body temperature is too low, regulatory processes are activated that bring it up to its set point. Decreased blood flow to the skin reduces heat loss to the environment, and shivering generates extra metabolic heat.

Figure 22.5

Homeostasis maintains stable internal conditions

Because of homeostatic pathways, large fluctuations in external conditions produce little or no change in the overall state within the animal body. For example, even when the outside temperature is very hot or cold, the human body's internal temperature stays within the narrow range required for survival.

Q1: Why is it important to maintain a stable body temperature?

Q2: Which organ system is involved as a temperature sensor? (See Figure 22.4 for an overview of organ systems.)

Q3: Give another example of a homeostatic pathway in humans.

of heat gain and loss) and *osmoregulation* (the control of internal water content and solute concentration by an organism). Homeostasis is critical to the maintenance of our organ systems, including the reproductive system—which brings us back to doing it in Denmark.

All in the Timing

Danish researchers have long suspected that the primary reason for declining fertility rates is that women are waiting longer to have children. In the 1970s, on average, a Danish woman gave birth to her first child at 24 years old. But in 2014, that average age was 29, with more and more women waiting until they were over 35 to have a first child. And the older a woman gets, the more difficult it is for her to conceive.

Humans reproduce through **sexual reproduction**, in which haploid gametes from a male (the **sperm**) and a female (the **egg**, or ovum) combine to form a diploid **zygote**, which develops into a multicellular individual that is genetically unique and different from

either parent. Here, we will review primarily human sexual reproduction, but keep in mind that sex in other animals is more variable than our human perspective might lead us to expect (**Figure 22.6**). In particular, recall that some animals can also reproduce via **asexual reproduction**, in which cells from only one individual produce the offspring, so all of the offspring's genes come from that parent. (For more on the benefits and pitfalls of sexual and asexual reproduction, see "Why Sex?" on page 235 of Chapter 13.)

Human eggs develop through a series of cell divisions called **oogenesis**, the production of mature eggs capable of being fertilized (**Figure 22.7**, left). Oogenesis begins before birth, when germ line cells multiply and develop into immature, diploid egg cells called **primary oocytes**. At birth, the ovaries of a female already contain her entire lifetime supply of primary oocytes, about 1–2 million cells. These cells remain in a state of suspended development until the production of hormones at puberty stimulates one, or occasionally two, of them to mature each month in preparation for ovulation. By the time a female reaches puberty, at about

Sea stars can reproduce by breaking off an arm, which then regenerates into a new individual. Although some animals rely exclusively on asexual reproduction, most asexually reproducing species switch between sexual and asexual reproduction, depending on environmental conditions.

In frogs and other species, females release their eggs, which males then cover with sperm. External fertilization is common among aquatic animals; internal fertilization is more common among land animals.

Clownfish begin life as males, but the largest fish in a group will change to female. Other species begin as females and may then switch to males, while still other species are both male and female at the same time. Individuals that produce both functional testes and functional ovaries—and are therefore both male and female—are called **hermaphrodites**.

Figure 22.6

Animals display a rich variety of reproductive systems

Some animals reproduce by cloning themselves through asexual reproduction (left). Others fertilize gametes externally (middle), while still others are both male and female, either simultaneously or sequentially (right).

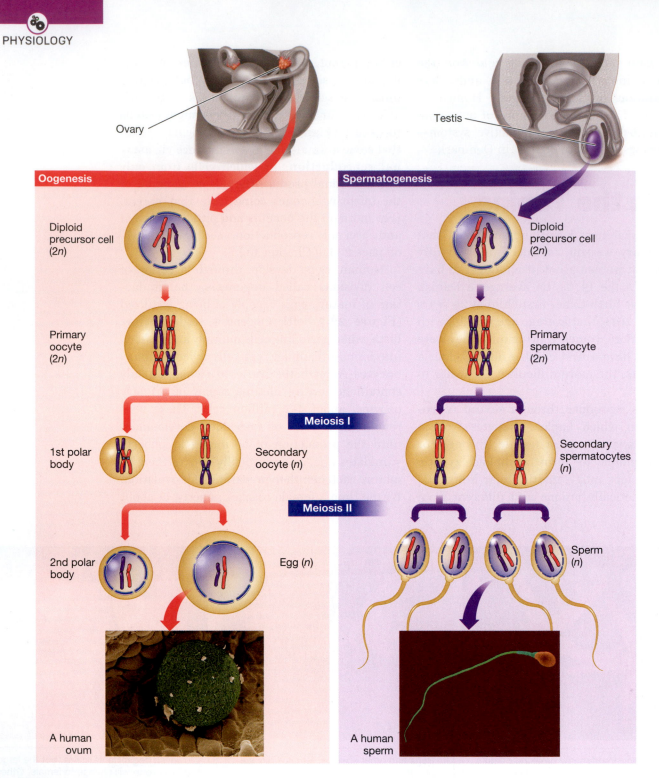

Figure 22.7

Sexual reproduction requires the production of haploid gametes

Oogenesis produces haploid eggs (ova), and spermatogenesis produces haploid sperm.

Q1: Identify one way in which oogenesis and spermatogenesis are the same.

Q2: How many eggs are produced from each precursor cell? How many sperm?

Q3: How much time elapses between the appearance of a precursor cell and the formation of an egg? How does this process differ for a sperm? (You will need to read ahead to answer this question.)

10–12 years of age, approximately 400,000 viable primary oocytes remain—still more than she will use in her lifetime.

Human females do not produce mature eggs continuously. Instead, individual eggs mature and are released in a hormone-driven sequence of events known as the **menstrual cycle (Figure 22.8)**. The menstrual cycle averages about 28 days, but cycle lengths from 21 to 35 days are considered normal.

A woman has more primary oocytes than she will use in her lifetime, but evidence suggests that those eggs decline in quality as a woman ages—a conclusion supported by the increased risk of birth defects in children born to older mothers. In addition, if eggs from younger women are implanted into women over 40, the pregnancy rate equals the rate associated with the younger women who donated the eggs. In other words, young eggs result in young pregnancy rates.

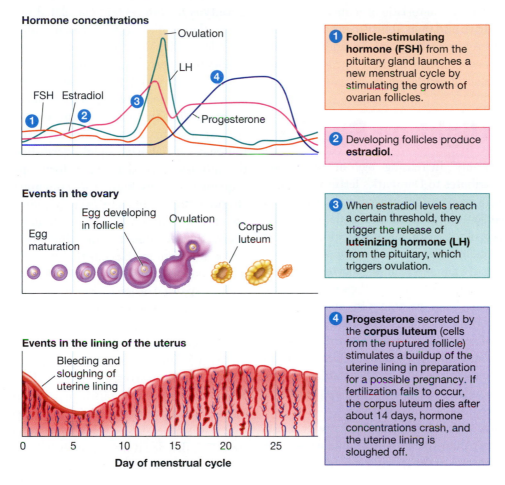

1 **Follicle-stimulating hormone (FSH)** from the pituitary gland launches a new menstrual cycle by stimulating the growth of ovarian follicles.

2 Developing follicles produce **estradiol**.

3 When estradiol levels reach a certain threshold, they trigger the release of **luteinizing hormone (LH)** from the pituitary, which triggers ovulation.

4 **Progesterone** secreted by the **corpus luteum** (cells from the ruptured follicle) stimulates a buildup of the uterine lining in preparation for a possible pregnancy. If fertilization fails to occur, the corpus luteum dies after about 14 days, hormone concentrations crash, and the uterine lining is sloughed off.

Figure 22.8

The human menstrual cycle

A menstrual cycle begins with the first day of bleeding, which marks the end of the previous cycle. Over the next few weeks, a succession of hormones stimulates the release of an egg and signals the uterine lining to grow and thicken in preparation for a potential pregnancy. If pregnancy does not occur, hormone levels plummet and the lining is sloughed off as menstrual flow, ending that menstrual cycle.

Q1: Which hormones important for the menstrual cycle are produced in the pituitary gland?

Q2: Which hormone is involved in producing the uterine lining?

Q3: How is the egg follicle involved in producing hormones?

When a woman passes 40 years of age, she shows a clear drop in her ability to produce normal eggs and bear children. Human females reach menopause—the end of their reproductive lives—around the age of 50, although this is highly variable among women.

A decline in egg quality is one suggested cause of Denmark's fertility problem: women today are waiting longer to have children, so their eggs are older and their fertility has decreased. Age also affects the fertility of men, although males do not undergo the clearly identifiable menopause that is characteristic of females. Instead, their sex drive and their ability to produce sperm slowly decrease as they age. In addition, men produce fewer sperm as they age, decreasing their chances of fertilizing an egg.

Spotlight on Sperm

While parents' increasing age at conception likely contributes to Denmark's fertility problem, researchers suspect that other factors are also at play—particularly, the quality of male sperm. The production of male gametes and the production of female gametes differ in several important ways. The supply of a female's primary oocytes is limited, and once a primary oocyte develops into a mature ovum, it is lost from the supply. In contrast, sperm precursor cells in males constantly replenish the pool of sperm. It is also noteworthy that in a normal menstrual cycle, a human female produces only one mature egg cell—in other words, only one egg per month—while a male produces hundreds of millions of sperm cells every day.

Another difference is that ova are typically much larger than sperm. The human egg is visible (just barely) to the naked eye, but individual sperm can be seen only under a microscope. Sperm contain little substance beyond chromosomes and the cellular machinery needed to move up the female reproductive tract, attach to an egg, and propel the sperm's chromosomes into the egg's cytoplasm. Sperm are simple

packages with valuable information inside. An ovum, on the other hand, is a plump, complex cell full of organelles.

The first reports of falling male sperm counts came in 1992 from Niels Skakkebæk at the University of Copenhagen. Looking at 61 different papers describing the semen quality of almost 15,000 men—both the density and volume of sperm in semen—Skakkebæk and his team found that sperm counts around the world had dropped by 50 percent from about 1940 to 1990. In a peer-reviewed paper they concluded, "As male fertility is to some extent correlated with sperm count, the results may reflect an overall reduction in male fertility." In other words, a decrease in the quality of semen could be lowering male fertility.

In human males, meiosis occurs inside structures called *seminiferous tubules*: twisty, spaghetti-like tubes that permeate the **testes**. In response to male hormones that surge at the onset of puberty, diploid germ line cells in the tubules start dividing to form sperm in a sequence of steps known as **spermatogenesis** (**Figure 22.7**, right). The average man produces about 300 million sperm each day. Surplus sperm that accumulate over time are degraded and reabsorbed by the cells that line the tubules.

Skakkebæk's 1992 study of declining semen quality sparked a lot of interest. Hundreds of studies followed, as researchers sought the cause for the decline. Many suspected an environmental cause, such as a toxin.

"We all wanted to study it in more detail," says Tina Kold Jensen, a researcher at the University of Southern Denmark. But instead of looking at past studies, Jensen and her collaborators determined that it was necessary to track these trends in real time. "We decided if we wanted to get closer to the truth, we had to collect our own data."

Jensen partnered with Skakkebæk and Niels Jørgensen at Copenhagen University Hospital. Starting around 1995, the scientists began recruiting young male volunteers at government-required physical exams, asking each to provide a sperm sample. By 2010 the team had amassed sperm from over 5,000 volunteers. When analyzed, the data seemed to contradict Skakkebæk's initial finding: over the 15 years of the study, semen quality did not decline; in fact, it improved slightly. Sperm concentrations rose from 43 million per milliliter in 1996–2000 to 48 million per milliliter in 2006–10.

TINA KOLD JENSEN

Tina Kold Jensen studies links between the environment and human reproduction as a professor of environmental medicine at the University of Southern Denmark.

Preventing Pregnancy

The Food and Drug Administration approved the first birth control pill for sale in the United States ▮ 1960. It became immediately popular as a way for women to control their fertility, and has been said to have contributed to a sharp increase in college attendance and graduation rates for women. Here is an overview of some of today's most popular forms of birth control, with data on just how effective each form is.

Assessment available in smartwork5

Pregnancies per 100 women in a year

Intrauterine Device
T-shaped plastic device must be positioned inside the uterus by a health care professional.

- ○ *Hormonal*
- ○ *Nonhormonal*

.2
.8

.5
.15

Female Sterilization
The oviducts are sealed with clamps or by other surgical means.

Male Sterilization
The tubes that carry sperm are sealed surgically by a health care professional.

The Pill
Hormones suppress ovulation, preventing the release of an egg, by mimicking the hormone levels of pregnancy.

9

12

Diaphragm
Dome-shaped latex cup filled with spermicide, inserted before intercourse, covers the cervix and keeps sperm out of the uterus.

Female Condom
Plastic pouch, inserted before intercourse, lines the vagina to prevent sperm from entering.

Male Condom
Plastic or latex pouch covers the penis to keep sperm from entering the vagina.

21
18

24
12

Sponge
A sponge containing spermicide is inserted deep in the vagina prior to intercourse.

- ○ *Women who have given birth before*
- ○ *Women who have never given birth*

Still, the rise was not enough to suggest that fertility improved during that period. "Although we see a slight rise, only 23 percent of the young men had optimal semen quality," said Jørgensen in a statement when the study was released. "In fact, the semen quality of 27 percent of the men was so poor, it will probably take these men longer to make their partner pregnant," he added. "Furthermore, for 15 percent of the men, semen quality was so poor, they are likely to need fertility treatment in order to conceive."

Driven by Hormones

Jørgensen and Jensen have gone on to study the factors that might have caused such low levels of semen quality. In May of 2014, their team found that 98 percent of 308 young men had detectable urinary levels of bisphenol A (BPA), a chemical found in plastics that disrupts the body's endocrine system (see Chapter 6 for more on BPA). Men with higher BPA levels, they discovered, also had higher levels of testosterone and other hormones. That correlation suggests that BPA could be affecting hormone feedback loops, but additional research is needed to identify possible mechanisms for such an effect.

Hormones regulate nearly all aspects of reproduction in animals, from mating behaviors to the development and birth of offspring. The emergence of sex-specific characteristics in the fetus and the maturation of reproductive organs during puberty are examples of long-term aspects of reproduction controlled by hormones. Hormones also regulate the recurring stimulation of sperm production in males and the monthly cycle of menstruation in females (see **Figure 22.8** for the role of hormones in the menstrual cycle, and Chapter 25 for more on hormones in general).

Testes and ovaries produce three major types of hormones: estrogens, progestogens, and androgens. Both males and females produce all three, but in different ratios; for example, males have more androgens than estrogens, and females have more estrogens than androgens.

- **Estrogens** play a role in determining female characteristics such as wide hips, a voice that is pitched higher than that of males, and the development of breast tissues. The primary estrogen is **estradiol**.

- **Progestogens** have a number of functions in the female body, including thickening the lining of the uterus and increasing its blood supply to create a suitable environment for a developing fetus. **Progesterone** is the most important of the progestogens.

- **Androgens** stimulate cells to develop characteristics of maleness, such as beard growth and the production of sperm. The primary androgen is **testosterone**. Together with another closely related androgen, testosterone directs the development of internal reproductive structures such as the sperm ducts and prostate gland. A third androgen directs the development of external structures such as the penis.

In addition to uncovering the link between BPA and hormone levels, the Danish research team recently found evidence that regular alcohol consumption may affect semen quality, possibly by changing testosterone levels. In particular, large amounts of alcohol significantly lowered semen quality: men who consumed 40 or more drinks in a week had a 33 percent reduction in sperm as compared with men who drank just 1–5 drinks per week. In their conclusion, the authors went so far as to warn young men to "avoid habitual alcohol intake."

"Do It for Denmark"

Jørgensen, Jensen, and others continue to seek the causes for declining semen quality, while demographers track social reasons for decreasing fertility rates, such as the increasing age of mothers. "Of course there are a lot of factors involved, including social factors," says Jensen. "But it's not all due to social factors. When we talk about fertility rates, we need to think about biology as well."

In the meantime, as we saw earlier, Danish officials are working to encourage young people to reproduce. After years of teaching how pregnancy can be prevented through abstinence—refraining from sexual intercourse—and the use of **contraceptives** such as the Pill, Danish teachers will now also teach how fertilization occurs (**Figure 22.9**).

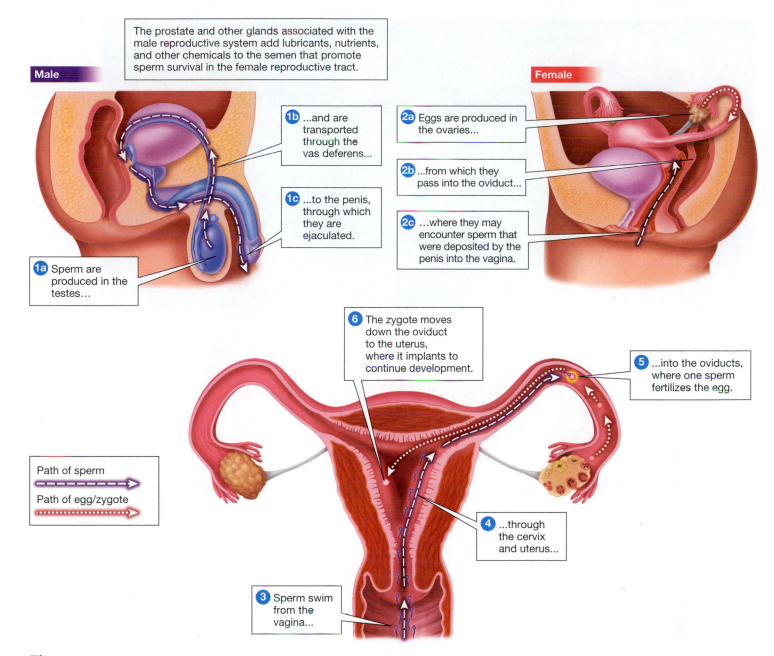

The prostate and other glands associated with the male reproductive system add lubricants, nutrients, and other chemicals to the semen that promote sperm survival in the female reproductive tract.

Male

1b ...and are transported through the vas deferens...

1c ...to the penis, through which they are ejaculated.

1a Sperm are produced in the testes...

Female

2a Eggs are produced in the ovaries...

2b ...from which they pass into the oviduct...

2c ...where they may encounter sperm that were deposited by the penis into the vagina.

6 The zygote moves down the oviduct to the uterus, where it implants to continue development.

5 ...into the oviducts, where one sperm fertilizes the egg.

Path of sperm

Path of egg/zygote

4 ...through the cervix and uterus...

3 Sperm swim from the vagina...

Figure 22.9

Fertilization takes place in the oviduct

Fertilization results in a zygote that can develop in the sheltered environment of the uterus.

Q1: If an egg is released but no sperm enter the oviduct, what is likely to occur?

Q2: If sperm enter the oviduct but no egg is present, what is likely to occur?

Q3: Sperm can live for up to 5 days inside a woman's body. Furthermore, eggs may be released at any point in the menstrual cycle, although midcycle ovulation is the norm. If you are trying *not* to become pregnant, when is it safe to have unprotected intercourse?

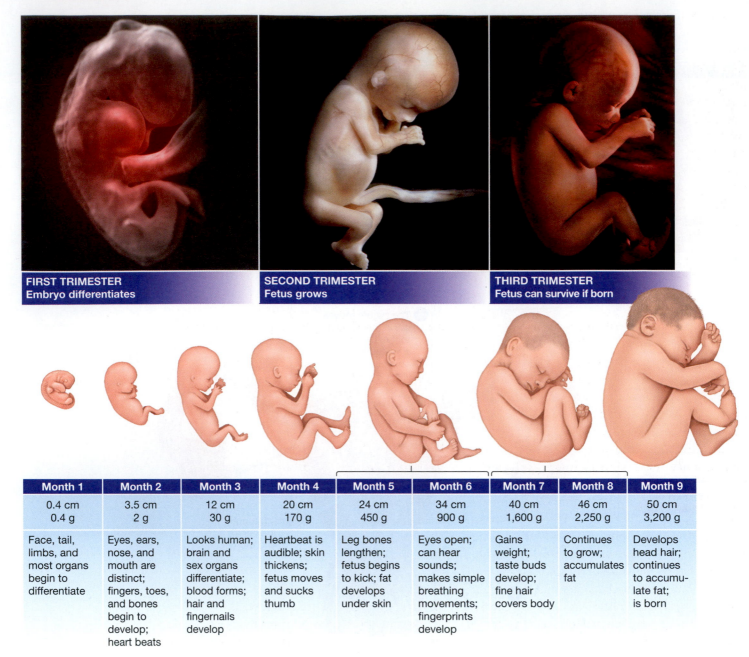

FIRST TRIMESTER
Embryo differentiates

SECOND TRIMESTER
Fetus grows

THIRD TRIMESTER
Fetus can survive if born

Month 1	Month 2	Month 3	Month 4	Month 5	Month 6	Month 7	Month 8	Month 9
0.4 cm 0.4 g	3.5 cm 2 g	12 cm 30 g	20 cm 170 g	24 cm 450 g	34 cm 900 g	40 cm 1,600 g	46 cm 2,250 g	50 cm 3,200 g
Face, tail, limbs, and most organs begin to differentiate	Eyes, ears, nose, and mouth are distinct; fingers, toes, and bones begin to develop; heart beats	Looks human; brain and sex organs differentiate; blood forms; hair and fingernails develop	Heartbeat is audible; skin thickens; fetus moves and sucks thumb	Leg bones lengthen; fetus begins to kick; fat develops under skin	Eyes open; can hear sounds; makes simple breathing movements; fingerprints develop	Gains weight; taste buds develop; fine hair covers body	Continues to grow; accumulates fat	Develops head hair; continues to accumulate fat; is born

Figure 22.10

Nine months in the womb

Although it is a critical step in human reproduction, fertilization marks only the beginning of a 9-month-long period of development within the mother's uterus.

Q1: Place these terms in the correct order of development: embryo, fetus, infant, zygote.

Q2: In what trimester is the fetus most likely to survive outside its mother's body?

Q3: The first trimester is the most sensitive time for exposure to mutagens. Why might that be?

A woman releases an egg from her ovaries, or *ovulates*, about once every 28 days. The egg moves down the **oviduct**, or *fallopian tube*, a 4-inch-long tube that connects the **ovary** to the **uterus**. If the woman has unprotected sexual intercourse, the man's **penis** ejaculates almost 300 million sperm into her **vagina**. The sperm swim from the vagina into the uterus through an opening called the **cervix**, and then up the oviduct in response to a chemical signal released by the ovary.

Only a few hundred sperm manage to reach the egg in the oviduct, and only one of those lucky sperm fertilizes the egg to create a zygote. Although both parents contribute equally to the genetic material of the zygote, its organelles and other cellular machinery come almost entirely from the female.

Human development in the uterus averages about 38 weeks and is divided into three stages known as **trimesters**, each about 3 months long (**Figure 22.10**). During the first trimester, the zygote develops from a single cell into an **embryo** possessing all the main tissue types. All organ systems are established by the third month, and the developing individual is now known as a **fetus**. It is during these first 3 months that most birth defects occur, when the fetus's organs are initially formed, many of them severe enough to cause miscarriage. Birth defects, the leading cause of infant mortality in the United States, are structural changes that affect how the body forms or works. Physicians know the cause of some birth defects, such as fetal alcohol syndrome resulting from heavy drinking by the mother during pregnancy. However, the causes of most birth defects remain unknown, though they are likely due to a mix of genetic and environmental factors.

During the next 3 months, the second trimester, the organs develop and the fetus increases in size. By the start of the third trimester, fetal development has progressed to the point that, with the help of modern technology, the fetus has reasonably good odds of surviving outside the mother's body. It gains a good deal of weight during the third trimester, and its circulatory and respiratory systems prepare for living in a gaseous atmosphere rather than the watery world of the amniotic fluid.

By the end of the third trimester, the fetus is ready for its sudden transition from the uterus to the outside world: childbirth (**Figure 22.11**). The last few weeks of pregnancy are marked by hormonal changes. Specifically, higher estrogen levels in the mother's blood make the muscles of her uterus more sensitive to **oxytocin**, a hormone secreted by the fetus and, later in the birth process, by the mother's pituitary gland. Oxytocin stimulates the uterine muscles and causes the placenta to secrete prostaglandins, which reinforce the contractions. Labor begins when the muscles of the uterus begin to contract in response to these hormones.

In a positive feedback loop, more contractions cause the production of more oxytocin, and the strength of the contractions increases as more oxytocin is produced. The cervix begins to open, and the increasingly strong contractions eventually push the fetus out of the mother's body. At this point the positive feedback ends, and the contractions subside as oxytocin levels decrease. The placenta, often referred to as the "afterbirth," is expelled during the last stage of childbirth.

At birth, a baby becomes physically independent of its mother. It no longer obtains its oxygen and nutrients directly from her blood, and it must eat, breathe, and maintain homeostasis on its own. Development does not end when an animal is born. Humans spend about a quarter of their lives reaching full adult size. Most of this growth occurs during childhood, before sexual maturity.

Amazingly, the "Do It for Denmark" ad blitz appeared to work, at least temporarily. In the summer of 2016, 9 months after the commercials premiered, the country was preparing for a baby boom, with 1,200 more babies due to be born in that summer than in the previous summer. But in most developed countries, including the United States, all signs point to birth rates continuing to fall. In August of 2016, the U.S. government reported that American fertility had fallen

NIELS JØRGENSEN

Niels Jørgensen is a member of the Department of Growth and Reproduction at Copenhagen University Hospital in Denmark. He studies male infertility.

2 Oxytocin and prostaglandins cause contractions.

Partially dilated cervix

Pelvis

3 The fetus is expelled from the uterus.

1 High estradiol levels cause muscles in the uterus to become sensitive to oxytocin.

4 The placenta and the umbilical cord are expelled.

Cervix   Vagina   Pubic bone

Figure 22.11

Childbirth is orchestrated by hormones

Childbirth occurs in stages, driven by the hormone oxytocin. Oxytocin signals uterine muscles to contract. The contractions become stronger as the amount of oxytocin increases. The mother's cervix opens, and the fetus is eventually expelled from the uterus, followed by the placenta soon after.

Q1: What is the role of estradiol in childbirth?

Q2: Explain how the involvement of hormones in childbirth is an example of a positive feedback loop.

Q3: If a woman has been pregnant for more than 40 weeks, her doctor might give her an injection of oxytocin to precipitate labor. How would that bring about labor?

to a record low of just 1.86 births per woman, putting us on the same page as Denmark, with a birth rate below the amount needed to keep the country's population from falling. As of yet, however, the United States has not adopted policies designed to encourage women to have more children—though the idea is not far-fetched. Is "Do It for America" in our future?

REVIEWING THE SCIENCE

- Cells that work in an integrated manner to perform a common set of functions constitute a **tissue**. Four main types of tissues are found in vertebrates: **epithelial**, **connective**, **muscle**, and **nervous**. An **organ** is made up of more than one tissue type and forms a functional unit with a distinctive function, shape, and location in the body. An **organ system** is composed of two or more organs that work in a closely coordinated manner to perform a distinct set of functions.

- **Homeostasis** is the process of monitoring the internal environment of an organism. **Homeostatic pathways** have two basic features: sensors that monitor the internal environment, and regulatory processes that attempt to restore the normal internal state when deviations from optimal conditions are detected.

- Many homeostatic pathways are controlled by **feedback loops**. In **negative feedback** loops, the results of a process cause that process to slow down or stop. In **positive feedback** loops, the results of a process cause it to speed up.

- **Sexual reproduction** involves the organs of the **reproductive system** and the union of male and female gametes (**sperm** and **eggs**). Most animals produce offspring in this way, although some can individually produce genetically identical offspring via **asexual reproduction**.

- **Oogenesis** and **spermatogenesis** are the production of eggs and sperm, respectively. Fertilization fuses the haploid sperm and haploid egg to produce a diploid **zygote**. Males produce sperm in **testes**, and females produce eggs in **ovaries**.

- Approximately monthly, one egg released from a woman's ovary moves into the **oviduct**, where it can be fertilized. During sexual intercourse, the man's **penis** releases into the woman's **vagina** nearly 300 million sperm, only one of which can fertilize the egg.

- During the first **trimester** of human development in the **uterus**, cells of the **embryo** rapidly differentiate into the various organs and structures present at birth. From the ninth week of development on, the developing human is called a **fetus**. During the second and third trimesters, the fetus grows rapidly.

- Childbirth occurs in stages. The hormone **oxytocin** signals uterine muscles to contract. The contractions become stronger as positive feedback increases the amount of oxytocin produced.

THE QUESTIONS

Challenge Yourself

1 Which of the following is an example of a positive feedback loop?

(a) the interaction of oxytocin and prostaglandins during labor

(b) the interaction of estradiol and progesterone during the menstrual cycle

(c) the interaction of glucose and insulin during eating

(d) both a and b

(e) both a and c

2 Sea stars (also called starfish) sometimes break off one of their "arms," which then develops into a separate sea star. Is this an example of asexual or sexual reproduction?

3 Select the correct terms:
Homeostasis maintains a constant internal state through (**homeopathic / homeostatic**) pathways that trigger regulatory processes when there is movement away from the body's (**set point / initiator**). For example, body temperature homeostasis is maintained through a (**negative / positive**) feedback loop—a process known as (**thermoregulation / osmoregulation**).

4 Place the following events of the human menstrual cycle, beginning with the final day of bleeding, in the correct order by numbering them from 1 to 5.

_____ a. The corpus luteum produces progesterone for about 14 days.

_____ b. Follicle-stimulating hormone (FSH) stimulates the growth of egg follicles.

_____ c. The egg is released from the follicle (now corpus luteum).

_____ d. The uterine lining sloughs off (menstruation occurs) with decreased progesterone.

_____ e. Estrogen levels increase and trigger luteinizing hormone (LH).

Try Something New

5 By producing leptin, fat cells signal to the body that it has sufficient energy stores. Higher leptin levels then decrease the feeling of being hungry. With fewer fat cells, less leptin is produced, and hunger levels are higher. This is an example of

(a) oogenesis.

(b) an organ system.

(c) spermatogenesis.

(d) homeostasis.

(e) contraception.

6 The actual effectiveness of birth control is often significantly lower than the theoretical effectiveness, or "effectiveness if used as directed." For example, there is a 9% annual pregnancy rate of women on the Pill, although its effectiveness is stated as over 99%. Similarly, although the condom is considered 98% effective, the associated annual pregnancy rate is between 10% and 18%. Would you expect a difference between theoretical and actual effectiveness for sterilization (female tubal ligation and male vasectomy)? If so, what would be the cause of that difference?

7 In maintaining homeostasis, small animals face more challenges than do large animals. Large animals tend to exchange water, solutes, and heat with their environment more slowly than small animals do, because a large animal has a larger volume relative to its surface area than a smaller animal has. The ratio between these two quantities—surface area and volume—determines how quickly or slowly an animal can gain or lose water, solutes, or heat. When the ratio of surface area to volume is relatively high, gains and losses are rapid; when the ratio is relatively low, gains and losses happen more slowly. Explain how this ratio indicates the need to be extra careful to protect newborns from temperature extremes. What else does the ratio of surface area to volume suggest that we need to consider when caring for a newborn?

Leveling Up

8 **Looking at data** To answer the following questions, refer to the accompanying graph, which shows data collected for the National Survey of Family Growth by the U.S. CDC (Centers for Disease Control and Prevention).

a. What does the *y*-axis show?

b. How many children, on average, does a woman with a bachelor's degree (or higher) have?

c. How many children, on average, does a man with a high school diploma or GED have?

d. Describe in your own words what the graph shows about the relationship between education and reproduction in the United States.

e. State a scientific hypothesis that might explain the relationship you described in part (d).

9 *Write Now* **biology: birth control and fertility rates** You are the aide to a U.S. senator, who has just sent you an e-mail including an attached article from a constituent in her (and your) home state. The constituent is concerned that people are having fewer children than in the past and thinks that the Senate should pass a law making it illegal to sell birth control to married couples. The senator asks you to write a two-page "white paper" (500 words) that she will use to respond to the e-mail and possibly also to propose legislation. The senator is extremely busy, and she neglected to take a biology class in college, so she is relying on you to give her a clear, concise, and accurate summary of the issue raised by the e-mail, as well as of possible actions to take.

Write a position paper addressing the following points (using about half a page, or 125 words, on each point). Begin by reading the article that was sent to the senator: http://national .deseretnews.com/article/1522/the-potential-impact-of-falling -fertility-rates-on-the-economy-and-culture.html. You may use other resources, including your textbook.

a. Summarize the main points of the article, defining terms (for example, "fertility rate," "replacement rate") as needed.

b. Find the fertility rate for your state, and compare it to the U.S. fertility rate. How do the two rates differ, and what is your best hypothesis for why they differ? Do the points made in the article hold for your state?

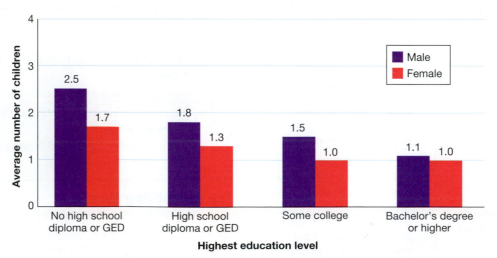

NOTE: GED is General Educational Development high school equivalency diploma.

c. What are some potential challenges to the legislation proposed in the e-mail? Is there alternative legislation—or another action—that might have the same effect with fewer challenges?

d. In your final paragraph, advise the senator on how she should respond to the letter and whether she should propose the legislation recommended in the e-mail. Also provide an alternative recommendation, which does not have to involve increasing reproduction. Justify your opinion with data and logic.

Thinking Multidimensionally

10 When it comes to most mammals, sex determination is a matter of inheriting XX or XY, where the sperm delivers either an X or Y chromosome to the egg. This familiar system is not the case for all organisms.

a. The XO system is found in crickets, grasshoppers, and roaches. Explain how this system differs from the XY system in humans.

b. The ZW system is found in birds, snakes, and crustaceans. Which parent determines the gender of the offspring of a chick?

c. In ants and honeybees, sex determination is a result of the egg either getting fertilized or not and is called the haplodiploid system. With honeybees, the female workers are sterile, while the queen bee is responsible for mating with the male drones and laying all the eggs. Research and explain the mechanism behind how the reproductive parts develop inside the queen yet fail to develop inside the workers.

d. Temperature can also play a role in sex determination, as is the case with turtles and alligators. Research and describe how male and female turtles are produced.

	XO SYSTEM	ZW SYSTEM	HAPLODIPLOID SYSTEM
FEMALE	XX	ZW	diploid
MALE	22 + X	76 + ZZ	16 haploid

11 Have you ever considered how the spots on a cheetah or the stripes on a tiger form? Alan Turing, better known as the founder of computer science, proposed that patterns could be explained by the interplay of the diffusion of activator and inhibitor chemicals and named this the reaction-diffusion model. These patterns are genetically controlled in the animal's skin during embryonic development.

a. What pigment molecule accounts for patterns and skin coloration?

b. What adaptive roles do spots and stripes have for an organism in nature?

c. Although still hypothetical, mathematical equations can often predict pattern formation for many animals during development. Larger domains or body parts are more prone to having more complex patterns, and smaller domains typically have simpler patterns. Compare the pattern on the back of a cheetah to the pattern on its tail. What do you notice?

d. Which pattern is more complex, spots or stripes?

e. Despite the mathematically predicted trend of more complex patterns forming on larger domains and simpler designs forming on smaller domains, exceptions exist. Name an animal whose markings defy the reaction-diffusion model in that the tail is striped and the body is solid-colored.

For more, visit digital.wwnorton.com/bionow2 for access to:

The Sunshine Vitamin

Is vitamin D the new supernutrient?

After reading this chapter you should be able to:

- Explain the roles of the major classes of nutrients in maintaining a healthy body.
- For a given vitamin or mineral, describe its main functions and dietary sources, and the possible effects of a deficiency.
- List the components of the integumentary system, and the layers of skin.
- Relate the structure to the function of one component of the digestive system.
- Demonstrate how flexible and rigid elements of a joint work together to produce controlled motion.
- Compare and contrast skeletal, smooth, and cardiac muscle.
- Compare and contrast cartilage and bone.

CHAPTER
23

DIGESTIVE, MUSCULAR, AND SKELETAL SYSTEMS

DRIVING QUESTION

What does the evidence reveal regarding vitamin D levels and a wide range of diseases?

- HS-LS1.A: Structure and Function

413

The babies' bones were not growing well. As a PhD student at McMaster University in Ontario, Canada, Hope Weiler was observing infants in a local hospital's neonatal intensive care unit, or NICU. They were being given dexamethasone, an anti-inflammatory steroid.

Prematurely born infants, or "preemies," often have chronic lung disease and cannot breathe on their own (**Figure 23.1**). To help these babies breathe, doctors commonly treated them with many doses of dexamethasone over the course of a month. But Weiler observed that the medication, while saving the infants' lives, also appeared to interfere with bone formation. Infants treated with dexamethasone had smaller heads, thinner bones, and shorter stature than those who didn't receive the medication.

"We wanted to know more about why that was happening," says Weiler, now the Canada Research Chair in Nutrition, Development, and Aging and a professor at McGill University in Quebec.

HOPE WEILER

Vitamin D₃

HO

Hope Weiler is an associate professor of nutrition, development, and aging at McGill University in Quebec, Canada, where she studies the effects of nutrition on bone health.

Figure 23.1

A premature baby receiving oxygen via a nose tube

Determined to find ways to help preemies grow, Weiler started a study tracking the bone mass of infants. During the study, she discovered that many newborn infants had low vitamin D levels in their cord blood.

Vitamins are small, organic nutrients needed by our bodies, but only in tiny amounts (**Figure 23.2**). They participate in a great variety of essential metabolic processes, such as helping blood cells form and maintaining brain function. Some vitamins bind to enzymes to speed up chemical reactions within a cell. Some act as a delivery service, supplying chemical groups needed in important metabolic reactions. Others act as signaling molecules. And some are even believed to work as antioxidants, substances that protect body tissues from destructive chemicals known as free radicals.

Most vitamins have multiple functions in the animal body, and vitamin D is no exception. Vitamin D, a fat-soluble vitamin, is required to absorb calcium from food and therefore is critical for bone growth and bone remodeling. Most vitamin D is naturally made by our bodies when ultraviolet (UV) rays are absorbed through the skin—an organ system in its own right, the *integumentary system* (see "The Skin We're In" on page 416).

Vitamins are a type of *nutrient*. **Nutrients** are components of foods that an organism needs to survive and grow. These can be micronutrients (vitamins and minerals) or macronutrients. Macronutrients are large organic molecules classified into three main categories: carbohydrates, lipids, and proteins. These biomolecules serve as sources of energy and furnish the body with chemical building blocks such as sugars, fatty acids, and amino acids. Although an adult human can synthesize some of the 20 amino acids needed to make proteins, we must get 9 of them, called **essential amino acids**, from food. Another type of macronutrient is dietary fiber, which does not contribute amino acids or energy to the body but is critical for survival because it affects how other nutrients are absorbed in the gut.

Minerals are inorganic chemicals that have critical biological functions. Carbon, hydrogen, oxygen, and nitrogen make up about 93 percent of the animal body, so, by convention, these four elements are excluded from the category of dietary minerals. But more than 20 other elements, such as fluoride, sodium, and

Vitamin C is abundant in fruits and vegetables, and it assists in the maintenance of teeth, bones, and other tissues. Deficiency of this vitamin leads to scurvy, in which teeth and bones degenerate.

Fish is the richest source of **vitamin D**; fortified foods (such as milk, soy milk, and breakfast cereals) are important sources for most people. A deficiency in vitamin D leads to poor formation of bones and teeth.

Folic acid, a **B vitamin**, is abundant in green vegetables, legumes, and whole grains. B_{12}, another B vitamin, is scarce in plant foods but abundant in milk, meat, fish, and poultry. It is important for the maintenance of teeth, bones, and other tissues.

Vitamin E is abundant in nuts, vegetable oils, whole grains, and egg yolk. It protects lipids in cell membranes and other cell components.

Leafy green vegetables and some fruits (e.g., avocado and kiwi) are rich in **vitamin K**, which is also manufactured by intestinal bacteria. A deficiency can cause prolonged bleeding and slow wound healing.

Carotene is responsible for the color of yellow and orange fruits and vegetables. It is converted into **vitamin A** within our bodies. Vitamin A aids in production of the visual pigment needed for good eyesight and is used in making bone.

Figure 23.2

Vitamins needed in the human diet

Humans need nine **water-soluble vitamins** (C and eight different B vitamins) and four **fat-soluble vitamins** (A, D, E, and K) from their diet. Because water-soluble vitamins are easily excreted in urine, they tend not to accumulate in body tissues, which means we must obtain these vitamins from food on a regular basis. Fat-soluble vitamins are not excreted as readily and tend to accumulate in body fat, so excessive consumption can lead to overdosing.

Q1: Which vitamins described in the figure are important for healthy bones?

Q2: Which vitamins are you more likely to overaccumulate?

Q3: In your own diet, are there any vitamins that you may not be eating enough of?

iodine, are essential for the normal function of most animals, and these are classified as dietary minerals. Calcium is the most abundant dietary mineral in the body; it makes up a large proportion of your bones.

Low levels of vitamin D have been identified not only in premature infants, but also in patients with ailments such as osteoporosis (a progressive bone disease), schizophrenia, erectile dysfunction, and more. If lack of vitamin D does contribute to these disorders, then supplementation with the vitamin should be protective. Many supplement manufacturers, in fact, now tout D as a "supervitamin" and "star supplement"

The Skin We're In

The **integumentary system** is the largest organ system in the human body, accounting for almost 15 percent of our weight. It covers the body and protects it from environmental hazards such as extreme temperatures and dangerous pathogens. It also prevents water loss and protects the body from physical damage. Sensory receptors are embedded in the skin (see Chapter 24 for more on the nervous system), and skin is the site of vitamin D synthesis.

The integumentary system consists of the skin and structures embedded in the skin, such as hair and nails in humans, or feathers, hooves, and scales in other vertebrates. The skin is made up of three layers. Moving inward from the outermost layer, the layers are the epidermis, dermis, and hypodermis.

As the figure shows, the skin contains multiple tissue types. The epidermis is an example of epithelial tissue. The nerve endings are examples of nervous tissue. The arrector pili muscle is composed of smooth muscle tissue. Much of the dermis is made up of connective tissue. And adipose tissue dominates in the hypodermis, a thick insulating sheet under the other layers of the skin.

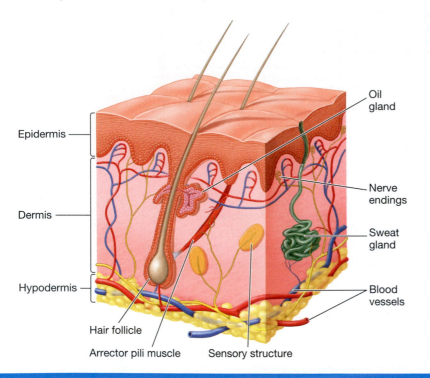

that will protect individuals from a wide range of diseases.

But has science shown that claim to be true? Just because low vitamin D and disease are correlated does not mean that one causes the other. So, scientists have long wondered whether low vitamin D actually causes disease or is involved in some other way.

You Are What You Eat

In her experiment studying premature infants, Weiler hypothesized that low vitamin D levels were an indirect result of treatment with dexamethasone.

Dexamethasone is a powerful steroid that helps a premature baby's lungs work better, but it has a side effect: it interferes with calcium absorption in the gut. When the drug is present, the intestines do not absorb calcium correctly. To counteract this effect, the babies' bodies began trying to find ways to compensate—by using up their vitamin D.

Vitamin D promotes calcium absorption in the intestines and in bones. The babies were using up all their vitamin D in an effort to absorb more calcium in the gut, so they didn't have enough left circulating in their blood to build bones.

Absorbing vitamins is one important function of the **digestive system**, which also processes food and eliminates unusable waste. The digestive system of most animals consists of a long, hollow passageway, known as the digestive tract, and a number of accessory organs, such as the pancreas and liver (**Figure 23.3**).

Eating, or **ingestion**, is the first stage in the processing of food by the digestive system. Digestion, the mechanical and chemical breakdown of food, begins almost immediately after ingestion in many species.

During ingestion, a bite of food—say, a spoonful of cornflakes—is deposited in the **oral cavity**, the mouth. There, an array of different types of teeth, which are shaped to cut, crush, or grind food into smaller pieces, begin to break apart the cornflakes. Many small pieces of food provide a greater surface area for digestive enzymes to work on than do fewer large pieces.

The muscular tongue mixes the crushed cereal particles with saliva. **Saliva** contains enzymes that start to break down starches— or any carbohydrate—into sugars. If you chew a piece of bread long enough, for example, its starches are digested to sugar and it will begin to taste sweet. Saliva is also important for turning the crunchy cereal into a moist mass that can slip easily down the throat.

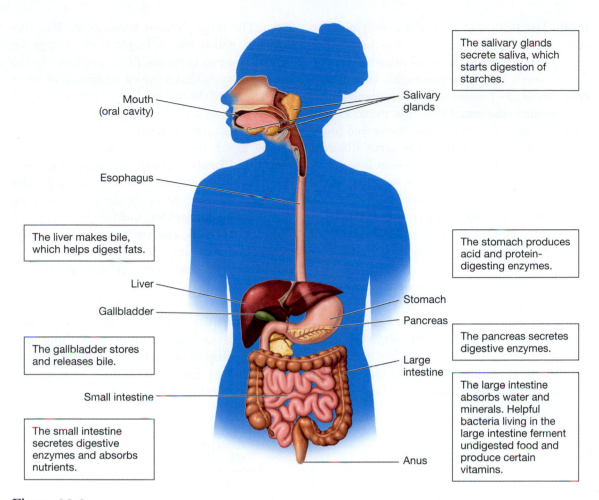

The salivary glands secrete saliva, which starts digestion of starches.

Mouth (oral cavity)

Salivary glands

Esophagus

The liver makes bile, which helps digest fats.

The stomach produces acid and protein-digesting enzymes.

Liver

Gallbladder

Stomach

Pancreas

The gallbladder stores and releases bile.

The pancreas secretes digestive enzymes.

Large intestine

Small intestine

The small intestine secretes digestive enzymes and absorbs nutrients.

The large intestine absorbs water and minerals. Helpful bacteria living in the large intestine ferment undigested food and produce certain vitamins.

Anus

Figure 23.3

The digestive system converts food into absorbable nutrients

As food moves through the digestive system, it is broken down into small molecules that can be absorbed by the lining of the intestine.

Q1: List, in order, the structures of the digestive system that a piece of swallowed food would pass through, beginning with the mouth.

Q2: What part of the digestive system hosts bacteria that produce vitamins?

Q3: What is the shared function of the liver and gallbladder?

The tongue assists in pushing the now-moist cereal into the throat, or **pharynx**, where the back of the mouth and the nasal cavity come together. The pharynx is the common entryway for both the air tube (the trachea) and the food tube (the **esophagus**). That is why, on occasion, you may cough up food or liquid that "went down the wrong tube"; it accidentally went down the trachea instead of the esophagus.

Normally, the pharynx is very good at separating air and food. When the mushy bite of cereal makes contact with the wall of the pharynx, it stimulates nerves that launch the **swallowing reflex**, in which a flap of tissue, called the epiglottis, seals off the entry into the trachea. The cereal is then pushed into the esophagus.

Waves of muscular contractions carry the cereal down the esophagus and into the **stomach**. Protein digestion begins in the stomach, which secretes acid and enzymes that break down complex protein molecules. Muscles in the wall of the stomach alternately contract and relax to mix the food particles with the acid and enzymes. The resulting watery mixture is stored in the stomach until it can move into the small intestine.

The **small intestine** is a highly coiled, thin tube about 3–4 centimeters in diameter. If straightened, the small intestine would extend about 20 feet (6 meters). The upper and lower regions of the small intestine serve different functions. The upper region, which lies nearer to the stomach, uses enzymes secreted by the **pancreas** and by the intestine itself to break down large molecules into simpler forms that the body can absorb. Here, the digestion of proteins, carbohydrates, and lipids, including fats, is completed.

The digestion of fats poses a particular problem because fats are not soluble in water, yet they need to be broken down and made to mix with the watery contents of the digestive tract. **Bile** is a fluid that helps digest fats by creating a coating that enables the fat globules to interact with water molecules to partially dissolve them. The large globules break down into tiny droplets, which offer a larger work surface for lipid-degrading enzymes. Bile is produced by the **liver**, an organ that serves a multitude of functions. Some of the bile made by the liver is stored in the **gallbladder**, which dispenses the bile into the small intestine as needed.

The lower region of the small intestine is specialized for **absorption**, the uptake of mineral ions—including calcium—and small molecules by cells lining the cavity of the digestive tract. These small molecules include broken-down sugars, fatty acids, and amino acids. The innermost lining of the small intestine presents a large surface area for that process (**Figure 23.4**).

Vitamins, including D, are also absorbed in the lower region of the small intestine. Only a few foods naturally contain vitamin D, including fatty fishes such as salmon and tuna, and egg yolks. Most of our ingested vitamin D comes from fortified foods—foods to which a vitamin or mineral is added. Milk, for example, is

Small intestine   Villus   Epithelial cell   Microvilli

Cavity

Extensive folding of the lining of the small intestine creates a large surface over which absorption can occur.

Blood vessels within each villus take up the nutrients absorbed through the epithelial cells.

Figure 23.4

The small intestine is specialized for absorption

Nutrients are absorbed in the small intestine by large numbers of fingerlike projections called **villi**. Each villus is about 1 millimeter long, with a surface consisting of cells specialized for nutrient absorption. The plasma membrane of each of these cells also has many tiny projections, called microvilli. This complex folding of the intestinal lining produces almost 300 square meters of surface area for absorption.

Q1: Why is a larger surface area important for absorption?

Q2: In what way are the epithelial cells lining the villi modified to increase absorption?

Q3: Where is the food coming from when it enters the small intestine? Where does it go next?

often fortified with vitamin D, voluntarily in the United States and by law in Canada. Both countries mandate that all baby formula be fortified with vitamin D. Breakfast cereals often contain added vitamin D as well.

Most of the nutrients absorbed by the digestive tract are sent to the bloodstream, which eventually delivers them to every cell in the body. Whether from the skin or the gut, vitamin D is sent to the liver and then to the kidney (**Figure 23.5**). In each organ, the nutrient is chemically modified to become an active form of the vitamin that our cells can use.

Our original bite of cereal contains very few nutrients by the time it arrives in the final segment of the digestive tract. This residual matter is prepared for **elimination**—the removal from the body of solid waste, consisting mostly of indigestible material and bacteria that inhabit the digestive tract—during

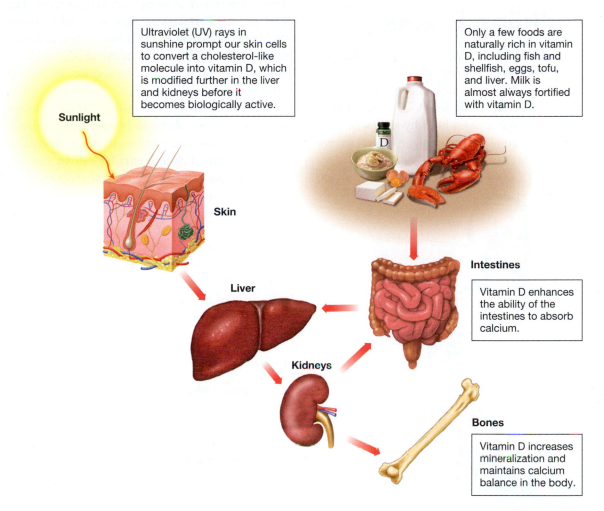

Ultraviolet (UV) rays in sunshine prompt our skin cells to convert a cholesterol-like molecule into vitamin D, which is modified further in the liver and kidneys before it becomes biologically active.

Only a few foods are naturally rich in vitamin D, including fish and shellfish, eggs, tofu, and liver. Milk is almost always fortified with vitamin D.

Sunlight

Skin

Liver

Intestines

Vitamin D enhances the ability of the intestines to absorb calcium.

Kidneys

Bones

Vitamin D increases mineralization and maintains calcium balance in the body.

Figure 23.5

Sources of vitamin D

Vitamin D is the only vitamin that we can manufacture entirely within our tissues, yet many Americans get inadequate amounts of vitamin D.

Q1: What organs of the digestive system are involved in processing vitamin D?

Q2: Do you get the majority of the vitamin D you need from the sun, from natural foods, or from dietary supplements?

Q3: It is true that you can increase your vitamin D levels by visiting a tanning booth (as described later in the chapter). Why is increasing vitamin D in this way considered a bad idea?

passage through the **large intestine**, or **colon**. The colon absorbs almost all remaining minerals and water from the waste. Then, large numbers of bacteria living in the colon break down the remaining waste to squeeze out the very last nutrients that the body can absorb. These bacteria also produce certain vitamins that are absorbed into the body from the colon. The waste, or **feces**, leaves the body through the **anus**, a muscle-lined opening.

Building Bones

Weiler's study became one of many that, over the years, have confirmed the cause-and-effect role of vitamin D on bone health in the **skeletal system**: with low levels of vitamin D, the babies did not grow a healthy skeleton. Today, dexamethasone is no longer used to treat premature babies, having been replaced with a safer drug.

Like most other vertebrates, humans have a bony internal skeleton that supports the body, gives it shape, and protects soft tissues and organs (**Figure 23.6**). The **axial skeleton**, made up of 80 bones, supports and protects the long axis of the body. It includes the skull, the ribs, and a long, bony spinal column. Although the axial skeleton plays a role in movement, its primary purpose is to protect vital organs. The 126 bones of the arms, legs, and pelvis make up the **appendicular skeleton** ("appendicular" means "relating to an appendage or limb"). These bones have more to do with motion than with protection.

Another important, though often passed over, part of our skeleton is **cartilage**, a dense tissue that combines strength with flexibility. In the human skeleton, cartilage gives form to the nose, the ears, and part of the rib cage. In addition, cartilage is found at nearly every point in the body where two bones would otherwise come into direct contact (see **Figure 23.6**); it creates a smooth surface that prevents the two bony surfaces from grinding against each other.

Cartilage does contain cells, but it consists primarily of nonliving, extracellular material—bundles of **collagen**, a tough but pliable protein

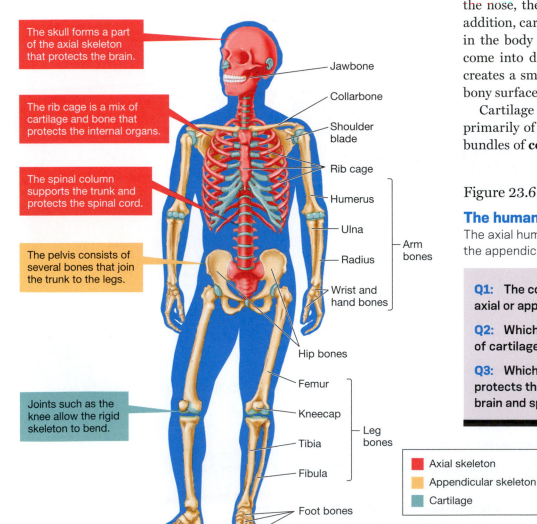

The skull forms a part of the axial skeleton that protects the brain.

The rib cage is a mix of cartilage and bone that protects the internal organs.

The spinal column supports the trunk and protects the spinal cord.

The pelvis consists of several bones that join the trunk to the legs.

Joints such as the knee allow the rigid skeleton to bend.

Jawbone

Collarbone

Shoulder blade

Rib cage

Humerus

Ulna

Radius — Arm bones

Wrist and hand bones

Hip bones

Femur

Kneecap — Leg bones

Tibia

Fibula

Foot bones

■ Axial skeleton
■ Appendicular skeleton
■ Cartilage

Figure 23.6

The human skeleton

The axial human skeleton protects vital organs; the appendicular skeleton facilitates movement.

Q1: The collarbone is part of which skeleton: axial or appendicular?

Q2: Which parts of the skeleton are made of cartilage?

Q3: Which skeleton—axial or appendicular—protects the central nervous system (the brain and spinal cord)?

found in a great variety of tissues, including skin, blood vessels, bones, teeth, and the lens of the eye.

Sharks belong to a small class of vertebrates that do not have a skeleton made of bones; their skeleton consists entirely of cartilage and connective tissue. Keep in mind, too, that not all animals have their skeleton on the inside. While humans and other vertebrates have an internal skeleton, or **endoskeleton**, many other animals, such as lobsters and insects, have an **exoskeleton**, an external skeleton that surrounds and encloses the soft tissues it supports (**Figure 23.7**).

Like cartilage, much of bone is made up of nonliving material. Still, it is a living tissue that has a blood and nerve supply. Specialized bone cells, called **osteocytes**, surround themselves with a hard, nonliving mineral matrix composed largely of calcium and phosphate. Vitamin D maintains calcium and phosphate concentrations in the body that are necessary to form that matrix. Although they are just single cells, osteocytes can live as long as the organism whose skeleton they belong to.

Bones are made of two major types of bone tissue (**Figure 23.8**): **Compact bone** forms the hard, white outer region. **Spongy bone**, honeycombed with numerous tiny cavities, lies inside the compact bone and is most abundant at the knobby ends of our long bones. Long bones and some others, such as the ribs and breastbone, have a hollow interior, which makes them light but strong. The cavities inside hollow bones contain **marrow**, a tissue that, depending on the type of bone, stores fat or produces blood cells.

Vitamin D, as previously noted, promotes calcium absorption in the gut and maintains calcium and phosphate concentrations. It is also needed for calcium absorption into bones, and thus is essential to bone growth and remodeling. Without it, bones can become thin, brittle, or misshapen—symptoms of a condition called rickets.

Rickets causes bones to soften, especially near joints, the junctions in the skeletal system that let the skeleton move in specific ways (**Figure 23.9**). Walking, for example, requires movement at the hips and knees, as well as at other joints. The lower jaw connects to the skull at a joint so that it can move relative to the rest of the skull, enabling us to chew and talk.

Figure 23.7

A newly molted cicada emerges from its exoskeleton

Exoskeletons provide a protective armor for many animals and also protect terrestrial invertebrates from excessive moisture loss. The rigidity of an exoskeleton means that immature animals that outgrow their exoskeletons must shed them periodically—a process known as molting.

Joints are held together by collagen-rich ligaments and tendons. **Ligaments** are specialized, flexible bands of connective tissue that join bone to bone, while **tendons** connect muscle to bone.

Wherever two moving parts rub against each other, as in a joint, wear can erode bone, and friction can waste energy. So, joints are lined with

Spongy bone

Bone marrow (produces red blood cells)

Compact bone

Fat-storing bone marrow

Figure 23.8

Bone has an intricate inner structure

The human humerus illustrates the complex structure of a bone's interior.

Muscle

Femur

Synovial sacs

Surface view

Tendons connect muscle to bone.

Kneecap

Back   Front

Flexible ligaments connect bone to bone.

Lubricating fluid in the synovial sacs allows the knee joint to bend easily.

Fibula

Tibia

Internal view

Cartilage cushions bone-to-bone connections.

Back   Front

■ Fluid-filled synovial sacs
■ Cartilage

Figure 23.9

The human knee illustrates how rigidity and flexibility together enable movement

Although our knees differ in detail from other joints in the body, they provide a good general model of the way flexible and rigid materials are combined in a joint to allow controlled motion.

Q1: What is the function of the synovial sacs?

Q2: Compare and contrast ligaments and tendons.

Q3: Knee injuries are some of the most common sports injuries. Why do you think that is?

sheets of tissue (synovial membranes) that form cavities called **synovial sacs**. The space inside each synovial sac is filled with a lubricating fluid that reduces friction between the two bony surfaces.

Altogether, these five components—bone, cartilage, ligaments, tendons, and synovial sacs—work to move a joint safely and with precise control.

Show of Strength

There is now no doubt that vitamin D is vital for bone health, but over decades, scientists have also been amassing data on the role of vitamin D in nonskeletal tissues. One of these is muscle tissue, which forms the body's **muscular system**.

Muscle tissue is unique to animals. The skeleton and its **joints** are the framework for motion, while muscles provide the power necessary for movement. Muscle tissue possesses a crucial property:

it can contract and relax. When we walk, run, or jump, we are using *voluntary* contractions.

But even when we are still, we're using *involuntary* muscles: the heart is pumping blood, the lungs are pumping air, and food is being moved along the digestive tract by muscular contractions. Involuntary muscles do their work without our having to think about them (**Figure 23.10**). Two specialized types of muscle engage in involuntary contraction: cardiac muscle and smooth muscle. The vertebrate heart is the only organ that contains **cardiac muscle. Smooth muscle** is found in the digestive tract, the walls of blood vessels, the respiratory tract, the uterus, and the urinary bladder. Some smooth muscles contract rapidly, while other types of smooth muscle can sustain a contraction for hours.

In November 2014, researchers in Belgium compiled all the available data on the role of vitamin D in **skeletal muscle**, the muscles involved in voluntary contractions. Skeletal muscle consists of many bundles of muscle fibers. A **muscle fiber** is a long, narrow cell that can span the length of an entire muscle because it is made up of several muscle cells that fused together during development (**Figure 23.11**). Each muscle fiber is packed with cylindrical structures containing proteins that contract by bracing against each other. Each such cylinder is known as a **myofibril**. Myofibrils are organized into series of contractile units called **sarcomeres**.

The simultaneous contraction of sarcomeres, usually taking no more than a tenth of a second, produces the contraction of a whole muscle. Sarcomeres are visible as bands when seen through a microscope. At extremely high magnification, additional details of their structure become evident (**Figure 23.12**). Two kinds of protein filaments are arranged in a specific manner inside each sarcomere: each **actin filament** is made up of molecules of the protein actin, and each **myosin filament** consists of the protein myosin. Under a microscope, each end of a sarcomere appears as a dark line called the **Z disc**. The two Z discs that mark the ends of the sarcomere contain a large protein that provides anchor points for actin filaments. Between the free ends of the actin filaments, attached to the middle of the sarcomere, lie thicker myosin filaments. The sliding of the myosin filaments against the actin filaments, which requires calcium, enables the sarcomere to contract.

To assess the impact of vitamin D supplementation on skeletal muscle health, the Belgian authors identified 30 randomized, controlled trials involving a total of 5,615 individuals. In each trial, an experimental group of people was given vitamin D supplements while a control group was not.

Looking at the data from all of those trials, the researchers found a "small but significant positive effect of vitamin D supplementation on global muscle strength" and saw that supplementation seemed to be more effective in people aged 65 and older than in younger people.

We've now seen that scientific evidence connects vitamin D to bone and muscle health. Yet some studies correlate low vitamin D levels with almost every other tissue and ailment under the sun, from heart disease to prostate cancer to

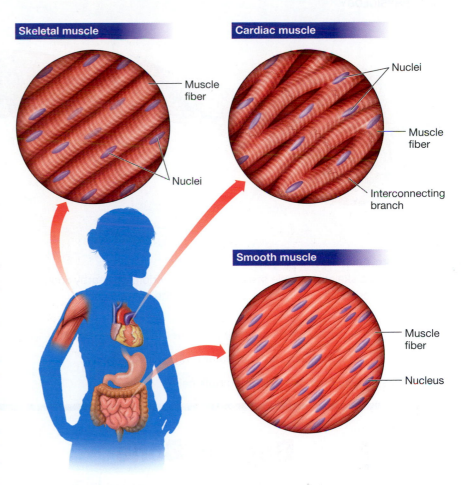

Figure 23.10

Specialized types of muscles for different types of movement

Skeletal muscle has a distinctive banded appearance, brought about by the sarcomeres that make it up (see Figure 23.12). Cardiac muscle is also banded, and its muscle fibers are branched, which helps produce the coordinated contractions known as heartbeats. Because it lacks sarcomeres, smooth muscle has no visible bands.

Q1: Which types of muscles can you voluntarily contract?

Q2: Do the muscles in the heart contract voluntarily or involuntarily?

Q3: You do not have to think about breathing (otherwise, sleeping would be dangerous!), but you *can* increase or decrease your rate of breathing. Are the muscles involved in breathing, then, voluntary or involuntary?

dementia. But, as we'll keep repeating, correlation isn't necessarily causation: a recent study has cast a large shadow of doubt over whether lack of vitamin D is actually the *cause* of these nonskeletal disorders.

Figure 23.11

The muscles that move you

Both ends of a skeletal muscle are anchored by tendons to nearby support structures, such as bones. A muscle contains bundles of muscle fibers, each one running the length of the entire muscle.

Q1: How is skeletal muscle attached to bones in the skeleton?

Q2: Do skeletal muscles contract voluntarily or involuntarily?

Q3: Is a muscle fiber one cell or multiple cells?

Beyond Bone

To one skin cancer researcher in France, the flurry of studies linking vitamin D to nonskeletal diseases was surprising. Epidemiologist Philippe Autier, vice president of population research at the International Prevention Research Institute (iPRI) in Lyon, France, had been following the vitamin D story from early on. He had grown skeptical of the health claims—that vitamin D could prevent a wide range of diseases—when the tanning industry began touting vitamin D as a benefit of tanning booths while conveniently glossing over the risks of skin cancer from the beds' UV rays (**Figure 23.13**).

Shortly after the tanning industry started promoting the health benefits of vitamin D, numerous studies began to emerge about the vitamin's role in a variety of nonskeletal diseases. "We were surprised by the fact that there were so many relationships between low vitamin D and diseases of all sorts, from brain disease to lung disease to infectious disease to cancer," says Autier. As he began reading the primary literature, he found that for virtually every disease, some correlation between vitamin D and the disease existed, and usually it followed a particular pattern.

"It was almost always that low vitamin D was associated with a greater risk of disease, and high vitamin D was protective," says Autier. But such conclusions were made from observational studies. Much rarer were randomized clinical trials, in which one group of people was given vitamin D supplements, another was not, and their likelihood of developing a particular ailment was compared. In those studies, Autier noticed, higher levels of vitamin D did not seem to protect the individuals taking supplements.

People began asking Autier whether they should take a vitamin D supplement to protect themselves. "The question was legitimate," says Autier, "so we had to address it."

With colleagues at iPRI, Autier analyzed data from 290 observational studies and 172 randomized trials examining the effects of vitamin D on nonskeletal disorders. They spent 6 years gathering and then analyzing the data from those

PHILIPPE AUTIER

Philippe Autier is an epidemiologist at the International Prevention Research Institute in Lyon, France. He researches cancer, specifically the role of UV light in skin cancer.

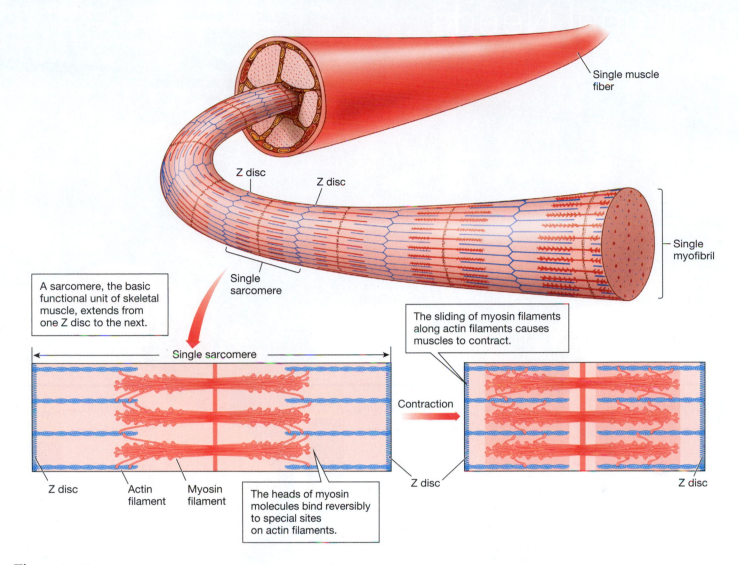

Figure 23.12

The microscopic structure of muscle

Each muscle fiber contains myofibrils made up of sarcomeres. Muscle contraction depends on the movement of actin and myosin filaments within each sarcomere. At each end of a sarcomere is a Z disc; each of the two Z discs contains a large protein that provides anchor points for the actin filaments.

Q1: List muscle structures from smallest to largest, beginning with sarcomeres.

Q2: What are the components of the sarcomere?

Q3: Across animal species, the microscopic structure of muscles is the same. Why, then, are there differences in strength among animals?

studies. "We had a huge database with many servers," says Autier. "You should have seen my office." From the data deluge, they found that high doses of vitamin D did not prevent any of the disorders they looked at. Not one.

Autier's team concluded, therefore, that low vitamin D levels are not a cause of ill health.

So why, then, do so many sick patients have low levels of vitamin D? One hypothesis is that diseases (especially cancer) are often associated with inflammation, and inflammation can reduce vitamin D concentrations in the body.

Three months after Autier's paper came out, a second, independent research team from

Nutritional Needs

Vitamins and minerals are essential nutrients that our bodies require in small amounts. Supplement manufacturers sell a range of different multivitamins, but the broad consensus from nutrition experts is that nothing substitutes for a healthy diet. Check out the data below to see how much of each nutrient you should be ingesting in a day, and some of the best foods in which to find them.

Assessment available in smartwork5

- 🟨 Men's multivitamin
- 🟪 Women's multivitamin
- ⬛ RDA male
- 🟧 RDA female

The recommended daily allowance (RDA) is the average daily dietary nutrient intake level that is sufficient to meet the nutrient requirements of nearly all healthy individuals (97%–98%) in a particular life stage and gender group.

Vitamin A (IU)

3500	🟨
2500	🟪
3420	⬛
2800	🟧

9189
1/2 cup raw carrots

3743
1 slice pumpkin pie

Vitamin C (mg)

80	🟨
60	🟪
90	⬛
75	🟧

95
1/2 cup raw red pepper

70
medium orange

Vitamin D (IU)

800	🟨
1000	🟪
200	⬛
200	🟧

447
3 oz sockeye salmon

41
1 large egg

Folic Acid (µg)

500	🟨
600	🟪
400	🟧
400	⬛

105
1/2 cup black-eyed peas

59
1/2 cup avocado

Calcium (mg)

200	🟨
400	🟪
1000	⬛
1000	🟧

415
8 oz plain yogurt

94
1 cup cooked kale

Iron (mg)*

27	🟪
8	⬛
18	🟧

8
1 cup white beans

7
3 oz dark chocolate

*0 mg in men's multivitamin

the University of Auckland in New Zealand published another meta-analysis that came to a similar conclusion: vitamin D supplementation does not reduce the risk of nonskeletal diseases.

All in all, these findings suggest that vitamin D supplements may not protect against nonskeletal disorders, though Autier emphasizes that their findings do not apply to pregnant women and young children, for whom vitamin D is critical during development.

Autier would be happy to see more large, randomized trials to confirm his findings. Weiler agrees. "Vitamin D doesn't have the same long history of people focusing on it, like heart disease," she says. "We know some relationships are there, but we haven't much looked at cause and effect."

In the meantime, consumers should not be nervous about taking vitamin D. The current recommended daily allowance from the Institute of Medicine, an impartial, nonprofit health and science policy institute, is 600 international units (IUs) of vitamin D per day to maintain health. (Seniors and breast-fed infants may need more.) Taking vitamin D for long periods of time in doses higher than 4,000 units per day is potentially unsafe, according to the National Institutes of Health, as it may cause excessively high levels of calcium in the blood. When blood levels of calcium are too high, the risk of heart attack and stroke rise dramatically.

Figure 23.13

Tanning beds are a risky source of vitamin D
Exposure to ultraviolet rays, either outdoors from the sun or indoors from tanning beds or lamps, greatly increases the risk of skin cancer. It can also cause eye cancer, cataracts, and premature wrinkling. About 400,000 Americans a year develop skin cancer as a result of indoor tanning.

Autier encourages other researchers to keep studying vitamin D. The story is not over. "It's quite good that we continue," he says, "because vitamin D levels could tell us a lot about the mechanisms of many diseases."

REVIEWING THE SCIENCE

- Animals rely on **nutrients** for chemical building blocks and energy. Macronutrients are carbohydrates, proteins, and lipids.

- Micronutrients include vitamins and minerals. **Vitamins** are organic compounds obtained from food that regulate metabolic processes in the animal body. **Minerals** are inorganic molecules needed by the body in small amounts.

- The **integumentary system** is the largest organ system in the human body, covering and protecting the surface of the body. It consists of skin and the structures embedded in the skin, such as hair and nails.

- The **digestive system** is a tubular passageway that, in conjunction with accessory organs, processes ingested food.

  - After **ingestion**, food is broken down into smaller pieces in the **oral cavity** through the grinding action of teeth.

- **Saliva** moistens the food and begins the chemical breakdown of starch.

- Food passes from the oral cavity to the **pharynx**, down the **esophagus**, and into the acidic environment of the **stomach**, where protein **digestion** begins.

- Partially digested food moves from the stomach into the upper region of the **small intestine**, where enzymes secreted by the **pancreas** and the intestine complete digestion of the food. The **liver** produces **bile** (stored and delivered by the **gallbladder**), which helps digest fats.

- In the lower region of the small intestine, digested nutrients are absorbed into the body. The lining of the small intestine is highly folded and bears fingerlike projections (**villi**) that present a large surface area for absorbing nutrients.

- From the small intestine, any unabsorbed material moves into the **colon**, where remaining water and minerals are absorbed. Here, bacteria break down the waste and release nutrients that the body can absorb.

- The **axial skeleton** supports and protects vital organs along the long axis of the body. The **appendicular skeleton** is composed of the bones in the arms, legs, and pelvis, and is primarily involved with movement.

- Most bones are made of **spongy bone** surrounded by harder **compact bone**. Hollow bones contain **marrow**.

- **Ligaments** connect bone to bone. **Tendons** connect muscle to bone. Tendons and ligaments are made of **collagen**.

- Muscles provide the power necessary for movement. Muscles consist of **muscle fibers**, each of which is packed with **myofibrils**. Myofibrils contain repeating units, called **sarcomeres**, that contract when actin and myosin filaments slide past one another.

- Movements are created by specialized muscle types. **Cardiac muscle** is found in the heart, where its contractions pump blood. **Smooth muscle**, which is found in the digestive tract and blood vessels, contracts in waves. **Skeletal muscle** is under conscious control and has a banded appearance.

THE QUESTIONS

Challenge Yourself

1 Which of the following are important for controlled movement of the knee?

(a) the axial and appendicular skeletons

(b) the ulna and the radius

(c) rigidity and flexibility

(d) the femur and the tibia

2 Select the correct terms:

Within a (**muscle / tendon**), sarcomeres shorten when their (**actin / myosin**) filaments slide over (**actin / myosin**) filaments joined to (**Z / X**) discs on either end of the (**muscle fiber / sarcomere**).

3 Place the following events of food movement through the digestive system in the correct order by numbering them from 1 to 5.

_____ a. Nutrients are absorbed through cells of the villi.

_____ b. Food is broken into smaller pieces through chewing.

_____ c. Digestive enzymes are released by accessory organs.

_____ d. Bacteria help digest food and produce some vitamins before sending wastes out of the body.

_____ e. Acids break down proteins for further digestion.

Try Something New

4 Which of the following is *not* involved in supporting movement of a joint?

(a) bone

(b) smooth muscle

(c) ligaments

(d) synovial sacs

(e) All of these are involved in joint movement.

5 Our bones are constantly changing in response to how we live. Physical activity builds stronger bones. The bones in a pitcher's throwing arm, for example, are stronger and have larger ridges to which muscles can attach than do the same bones in the other arm. Inactivity leads to weaker bones because special osteocytes step up the rate at which they remove tissue from the bone when the skeleton is not under physical stress; such weakening is seen, for example, in a person confined to bed by an injury or illness. Which of the following activities would be the least effective in increasing bone strength? Explain why.

(a) walking

(b) dancing

(c) weight lifting

(d) swimming

(e) running/jogging

6 A January 2013 Tumblr post that went viral on the Internet read as follows:

> eating is so bad*** i mean you put something in a cavity where you smash and destroy it with 32 protruding bones and then a meat tentacle pushes it into a vat of acid and after a few hours later you absorb its essence and transform it in[to] energy just wow.

Identify each element of the digestive system discussed in the post. Do you think this is an accurate description of digestion?

Leveling Up

7 **Looking at data** The recommended daily allowance (RDA) for nutrients is the average daily amount necessary to meet the nutrient requirements of most healthy adults (97%–98%). There are two kinds of necessary mineral nutrients: macrominerals, which your body needs in larger amounts; and trace minerals, which your body needs in smaller amounts. Macrominerals include calcium and phosphorus. Trace minerals include iron, copper, iodine, zinc, and fluoride.

a. According to the accompanying graph, what is the dividing line between macrominerals and trace minerals?

b. What does the *y*-axis show?

c. Describe in your own words what the graph shows.

d. How much potassium is needed by an average person?

e. Is the amount of potassium you stated in part (d) the same amount that we need to eat daily?

Average Mineral Content Required in the Body

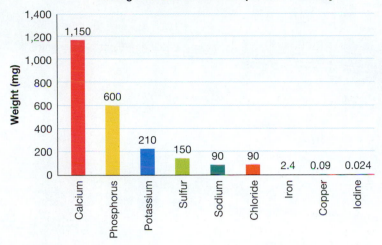

Mineral	Weight (mg)
Calcium	1,150
Phosphorus	600
Potassium	210
Sulfur	150
Sodium	90
Chloride	90
Iron	2.4
Copper	0.09
Iodine	0.024

8 **Life choices** Nutritionists agree that we can obtain all our necessary nutrients in optimal amounts from natural sources, provided we eat a well-balanced diet that includes a variety of foods. Recognizing that relatively few people meet this ideal in reality, some physicians advise their patients to take a multivitamin and mineral supplement as "added insurance." The use of this kind of dietary supplement raises the specter of toxicity from excessive intake, particularly of the fat-soluble vitamins. Most manufacturers, but by no means all, attempt to avoid high levels of the fat-soluble vitamins in their vitamin preparations, but the burden is largely on us to use supplements wisely, because this industry is not highly regulated by the government. Review the information provided on dietary supplements by the National Institutes of Health Office of Dietary Supplements (http://ods.od.nih.gov/HealthInformation/DS_WhatYouNeedToKnow.aspx) and the U.S. Food and Drug Administration, or FDA (http://www.fda.gov/Food/DietarySupplements/UsingDietarySupplements/ucm110567.htm). Take notes so that you can answer the following questions.

a. What aspects of the sale of dietary supplements does the government regulate?

b. Are you confident that this regulation is sufficient to make supplements safe to use as labeled? Why or why not?

c. List three recommendations that you found informative or helpful. Will they change your use of dietary supplements? Explain your reasoning.

d. Do you think a mobile app for dietary supplements would be helpful to have? Why or why not?

e. Identify one dietary supplement that you feel would be helpful for you to take. Will you take it? Why or why not?

Thinking Multidimensionally

9 The length of digestive tracts varies across the animal kingdom, having evolved to meet the metabolic needs of each organism.

a. Would you expect a carnivore, like a lion, or an herbivore, such as a gazelle, to have a longer digestive tract? Why?

b. As humans, we are unable to digest the sugar cellulose found in plant cell walls because we lack the enzyme cellulase. If humans cannot digest cellulose, why do we eat plants?

c. Herbivores, such as cattle, have evolved with a solution to this problem. How do cows derive nutrition from a grass-based diet?

d. Despite their sharp canines, giant pandas feed exclusively on bamboo. How does this diet affect their activity rates?

10 In Greek mythology, the sea god Proteus had the ability to change the shape of his body. Proteus syndrome, a rare genetic disease that causes an overgrowth of tissues, is named after this Greek god. Babies appear normal at birth but soon develop abnormal tissue growth. Joseph Merrick, also known as the "Elephant Man," lived from 1862 to 1890 and is suspected of having had this syndrome.

a. While this disorder involves multiple systems, based on Merrick's appearance, what are some of the tissue types impacted by this disease?

b. Proteus syndrome results from a mutation to the *AKT1* gene, which codes for a protein involved in cell growth and division, differentiation, and cell survival. Since patients with Proteus syndrome have tissue that divides excessively, what other disease are they more prone to developing?

For more, visit digital.wwnorton.com/bionow2 for access to:

Body (Re)Building

Could engineered human tissues, brought to life in the lab, replace failing organs in people?

After reading this chapter you should be able to:

- Create a flowchart depicting the movement of blood through a cardiovascular system.

- Describe the function of the different components of blood, and compare the three types of blood vessels.

- Diagram the elements of the respiratory system, and show how air moves through it.

- Explain how gases are exchanged in the lungs, and why.

- Describe how the structure of a nephron relates to its function in the urinary system.

- Distinguish between the central and peripheral nervous systems.

- Identify the sensory systems active in humans.

CHAPTER
24

CIRCULATORY, RESPIRATORY, URINARY, AND NERVOUS SYSTEMS

DRIVING QUESTION

As waiting lists for organ transplants grow, what alternative options are currently available or under development?

- HS-LS1.A: Structure and Function
- HS-ETS1.A: Defining and Delimiting Engineering Problems

Standing over the operating table, covered head to toe in teal scrubs, physician Jeffrey Lawson lifts a long, white tube out of a bath of clear liquid. Carefully, slowly, he threads the tube through the unconscious patient's upper arm and stitches each side of the tube to an exposed blood vessel.

Bioengineer Laura Niklason stands to the side of the operating table, watching closely (**Figure 24.1**). She and Lawson created the tube being implanted in the patient. Once the operation is complete, Lawson steps back, and Niklason gives him a hug. "Congratulations to you," says Niklason happily. "We're saving the world!"

On June 5, 2013, Lawson, Niklason, and their team transplanted a laboratory-grown blood vessel into a human. It was the first such procedure in the United States and a major feat for the field of *tissue engineering*, the effort to grow or regenerate tissues or organs using engineering materials and principles. Other engineered tissues implanted into humans have included nerve grafts, bladders, and windpipes.

Lawson and Niklason imagine a future in which any type of organ can be constructed in the lab. In an ideal world, "we'll be able to grow all sorts of tissues for patients, so a surgeon can literally reach up on a shelf, pull down a tissue

Figure 24.1

A bioengineered blood vessel
In 2013, a team of doctors at Duke University was the first in the United States to implant a bioengineered blood vessel into a patient.

graft, and implant it in a patient," Niklason said in 2013, shortly after the first blood vessel transplant. "That's really going to be a revolution—being able to grow replacement parts for patients, that they don't have to wait for."

Organ shortage is a major concern in health care (**Figure 24.2**). According to the U.S. government, the number of people waiting for an organ would fill a football stadium—twice. Every day, an average of 18 people die waiting for transplants of kidneys, hearts, livers, lungs, and more.

One possible solution to this organ crisis is the Frankenstein-like approach of tissue engineering: building organs in the lab. It sounds far-fetched, but research teams have already begun to engineer complicated organs such as kidneys, lungs, even whole hearts. These are just a few of the organs that make up the major organ systems in our bodies, including the circulatory, respiratory, urinary, and nervous systems—each one critical to the healthy functioning of our bodies (see Figure 22.4 for an overview).

That healthy functioning, including growth and homeostasis (see Figure 22.5), depends on the internal transfer of transport proteins (see Figure 4.4), waste products, and other substances in the body. Distances inside the bodies of most multicellular organisms are far too great for diffusion to be an effective means of distributing these materials, so elaborate organ systems exist to transport them around the body. The **circulatory system** moves oxygen from the lungs to the heart, which then pumps oxygen-rich blood to the rest of the body. The **respiratory system** brings in oxygen and expels carbon dioxide to support cellular respiration. The **urinary system** removes excess fluid from the body, along with waste products, toxins, and other water-soluble substances that are not needed.

These three systems are interconnected in the body: Carbon dioxide collected by the circulatory system is delivered to the respiratory system for exchange with the outside environment. The circulatory system brings substances dissolved in blood to the urinary system for discharge into the environment. And a unique high-speed communication system called the **nervous system** coordinates the many muscles involved in the functioning of organs, including the heart, lungs, and bladder. The nervous system directs the rapid contractions of muscles, and processes information received by the senses, such as touch, sound, and sight. In this way, the nervous system enables an animal to detect food, find a mate, avoid predators, and respond to extremes of heat and cold.

These organ systems, along with all the other organ systems introduced in Chapter 22, govern how your body operates. If one of your organs fails, a whole system can shut down. That's why researchers are exploring how to build new organs, writing a recipe for flesh and blood.

Emergency Meeting

Niklason and Lawson met at Duke University Medical Center in 1999 while performing surgery together. Waiting in the operating room to move their patient, the two struck up a conversation. Within moments, they discovered they shared strikingly similar interests: Lawson was a surgeon who spent his days suturing blood vessels in the hospital and growing different types of blood cells in his lab, and Niklason was an anesthesiologist and bioengineer who had recently opened her own lab to build blood vessel tubes. Both were dreaming of the same goal—to create a blood vessel from scratch—and they had complementary skills. "It made us a perfect partnership," says Lawson.

That first conversation bloomed into a friendship and working partnership. Together, Lawson and Niklason started building prototypes of blood vessels, using his knowledge of blood cells and her knowledge of the engineering forces at work in blood vessels. **Blood vessels** are a critical part of the **cardiovascular system**, a closed circulatory system consisting of a muscular *heart*, a complex network of *blood vessels* that collectively form a closed loop, and *blood* that circulates through the heart and blood vessels. Almost all of our cells lie within 0.03 millimeter of a blood vessel with which they exchange materials by diffusion. Carrying

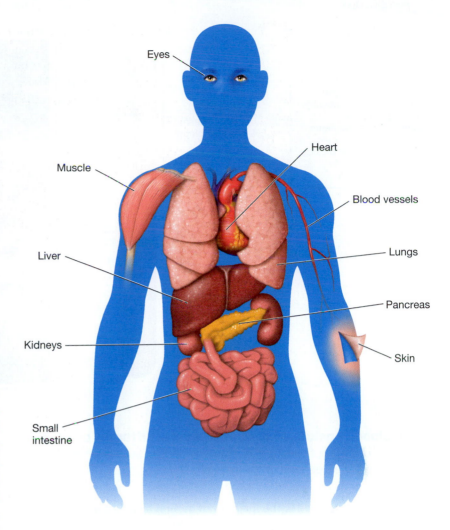

Figure 24.2

Organs needed for donation

blood so close to all the trillions of cells in our bodies requires an extensive network of vessels.

Lawson and Niklason built blood vessels created of various materials and cell types and then implanted them in rats to see whether blood would successfully flow through the tubes. **Blood** is composed of cells and cell fragments that float

JEFFREY LAWSON AND LAURA NIKLASON

Jeffrey Lawson is a physician-scientist at Duke University Medical Center in North Carolina who studies novel therapies for vascular surgery, including bioengineered blood vessels. Laura Niklason is a doctor and professor of anesthesiology and biomedical engineering at Yale University in Connecticut. Her research includes cardiac, lung, and blood vessel bioengineering.

Blood plasma is 92% water and contains dissolved gases, ions, and molecules that are critical for homeostasis, as nutrients, or as signaling molecules. Much of the carbon dioxide carried in the blood is dissolved in plasma.

Plasma
(55% of whole blood)

Cellular elements
(45% of whole blood)

Platelets

Platelets are small cell fragments. They can clump together to help stop the loss of blood if a blood vessel is damaged. Platelets release substances that stimulate plasma proteins to create a meshwork of protein strands, platelets, and blood cells to collectively form a blood clot.

Red blood cells

A mature red blood cell has no nucleus, and its cytoplasm is packed with oxygen-binding proteins called hemoglobin. Each hemoglobin molecule can carry up to four oxygen molecules. Because each human blood cell contains about 280 million hemoglobin molecules, a single one of these cells can bind over a billion molecules of oxygen.

White blood cells

Several different kinds of white blood cells help defend the body from invading organisms.

Figure 24.3

Human blood consists of fluid and several different cell types

Whole blood consists of plasma and different kinds of cells and cell fragments, three of which are shown here. Red blood cells account for about 95 percent of the cells in blood.

Q1: Where in the blood is the majority of carbon dioxide carried?

Q2: Where in the blood is the majority of oxygen carried?

Q3: What would happen if your red blood cells carried a mutation that made the hemoglobin less effective at binding to oxygen (as in sickle-cell disease)?

in a fluid known as **plasma** (**Figure 24.3**). Blood plasma itself has a low capacity for transporting dissolved oxygen, but **red blood cells** in the plasma carry significant amounts of oxygen, greatly increasing the oxygen-carrying capacity of blood.

In 2005, the duo hit upon a technique that seemed to work. They collected smooth muscle cells from the blood vessels of organ donors and grew those muscle cells on a biodegradable frame shaped like a blood vessel. The cells worked like little machines, churning out proteins that formed a three-dimensional scaffold of connective tissue called the *extracellular matrix*. Then the original, biodegradable frame dissolved, leaving behind a sturdy tube of cells and extracellular matrix.

There was still a catch. The immune system of a human body rejects cells from another person (see Chapter 25 for more on the immune system), so Lawson and Niklason had to wash away the original donor muscle cells to leave behind just the tubular extracellular matrix. "It's like we kept the mortar but the bricks all washed away," says Lawson. They tested their creation in rats. It worked. They had made a functioning blood vessel.

The vessels were first tested in humans in Poland in December 2012. The researchers found that once an engineered tube is implanted into a patient, the patient's own blood and muscle cells take up residence in the tube, filling in the cracks. "What started off as our structure now becomes your tissue," says Lawson. "And it all seals together so it doesn't leak."

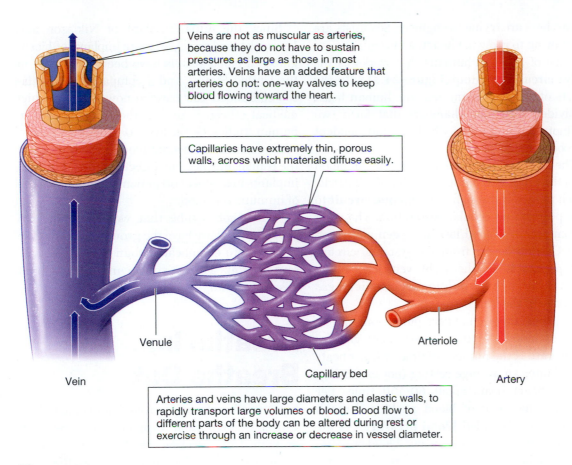

Veins are not as muscular as arteries, because they do not have to sustain pressures as large as those in most arteries. Veins have an added feature that arteries do not: one-way valves to keep blood flowing toward the heart.

Capillaries have extremely thin, porous walls, across which materials diffuse easily.

Venule

Arteriole

Vein

Capillary bed

Artery

Arteries and veins have large diameters and elastic walls, to rapidly transport large volumes of blood. Blood flow to different parts of the body can be altered during rest or exercise through an increase or decrease in vessel diameter.

Figure 24.4

Arteries, veins, and capillaries

Arteries carry blood away from the heart. Veins carry blood toward the heart. Arterioles and venules are smaller arteries and veins, respectively. The narrowest arteries and veins connect with each other in a fine network known as a capillary bed.

Q1: Why are arteries more muscular than veins?

Q2: What structural feature(s) of capillaries enable easier diffusion into and out of surrounding tissues?

Q3: Why do you think capillaries are not typically transplanted?

A critical feature of the engineered vessel is that it has the structural strength of a normal blood vessel in the body, so it can withstand the force of the blood pulsing through it. The human body has three major kinds of blood vessels through which blood flows: arteries, veins, and capillaries (**Figure 24.4**). **Arteries** are large vessels (0.1–10 mm in diameter) that transport blood away from the heart. **Veins** (0.1–2 mm) are large vessels that carry blood back to the heart. **Capillaries**, the smallest vessels, at 0.005–0.01 mm, exchange materials by diffusion with nearby cells.

The large vessels—arteries and veins—are built for mass transport of blood. Currently, Niklason and Lawson have built blood vessels with diameters of 6 mm (about the width of a pencil) and 3 mm. But they also expect to be able to make and transplant larger vessels, such as the aorta, the main artery of the body. Capillaries, however, are very small and are not typically transplanted. They are built for slower movement of blood, and their large surface area facilitates the exchange of materials with surrounding cells.

Niklason and Lawson's blood vessels are not the only tissues in the cardiovascular system that researchers are trying to engineer. Scientists are attempting to grow the **heart**, a muscular organ the size of a fist in humans that works as the body's circulatory pump (**Figure 24.5**). Like the hearts of all other mammals, the human heart is divided into four chambers that form two distinct pumping units, which are independent but coordinated.

The left atrium receives oxygenated blood from the lungs and pumps it to the left ventricle, which pumps it through the **systemic circuit** to cells performing cellular respiration. The right atrium receives blood low in oxygen and laden with CO_2 returning from the systemic circuit, and pumps it to the right ventricle, which pumps it through the **pulmonary circuit** for gas exchange in the lungs.

Together, these chambers pump some 7,000 liters of blood per day—about 1,850 gallons. **Heart rate** is the number of times a heart beats per minute; an average resting (not exercising) human heart beats about 60–100 times per minute. The force of blood pushing through blood vessels is called **blood pressure**. As the heart contracts to push blood out, blood pressure is at its highest, and when the heart relaxes after each contraction, blood pressure is at its lowest. The first pressure is referred to as **systole** and is the top number in a blood pressure reading. The second, **diastole**, is the bottom number in a blood pressure reading. The human circulatory system adjusts the heart rate and patterns of blood distribution according to the body's needs. Blood pressure is at its highest as it leaves the heart, and by the time it reaches the veins it is at its lowest.

The heart might seem like too complex an organ to replicate by means of tissue engineering techniques, but researchers have demonstrated some success using a tactic similar to the one Niklason and Lawson took with blood vessels, starting with an entire organ as a scaffold (such as a pig or rat heart, of which there is a ready supply, or a heart from an organ donor). They use detergents to strip away all the original cells that would cause an immune reaction, then repopulate the heart with cells that are a better match for the patient. Using this technique, researchers have created pumping rat hearts in a dish, though these bioengineered hearts are still too primitive to work when transplanted into an animal.

Since the initial implant of Niklason and Lawson's blood vessel, 60 more implants of bioengineered blood vessels have been performed in the United States. And a company that Niklason cofounded and Lawson now serves as chief medical officer, is testing the blood vessels in human studies. Of the first 20 patients in Poland who received engineered blood vessels, none of the transplants became infected, and additional implants in a total of 60 patients saw no evidence of immune rejection.

"We've got a tube that works to put into people, so now we have the groundwork to make things more complicated than a blood vessel," says Lawson. "There is still an unlimited amount of science to do."

Breathe In, Breathe Out

In 2010, after she had moved from Duke to Yale University, Niklason expanded her research program from creating simple blood vessels to attempting to build a lung. **Lungs** are the main organs of the respiratory system, which carries air from the nose (or mouth) to the lungs through a series of tubular passageways. These airways allow air to move between the external environment and the inside of the body—specifically, the gas exchange surfaces in the lungs.

The process of taking air into the lungs (inhaling) and expelling air from them (exhaling) is called **breathing** (**Figure 24.6**). The air we inhale is about 21 percent oxygen and contains little carbon dioxide and water vapor. The air we exhale has less oxygen (15 percent) and contains about 4 percent each of carbon dioxide and water vapor. That's because these gases are exchanged at the surface of the cells that line our lungs: Oxygen is removed from the inhaled air and sent to the bloodstream, while carbon dioxide and water vapor are removed from the bloodstream and added to the air that is exhaled. Remember that our bodies need oxygen to obtain energy via cellular respiration (see Figure 5.11).

At Yale, Niklason's team bioengineered a rat lung and implanted it in a rat. There, it briefly supported gas exchange for the animal before filling up with fluid, suggesting that more work

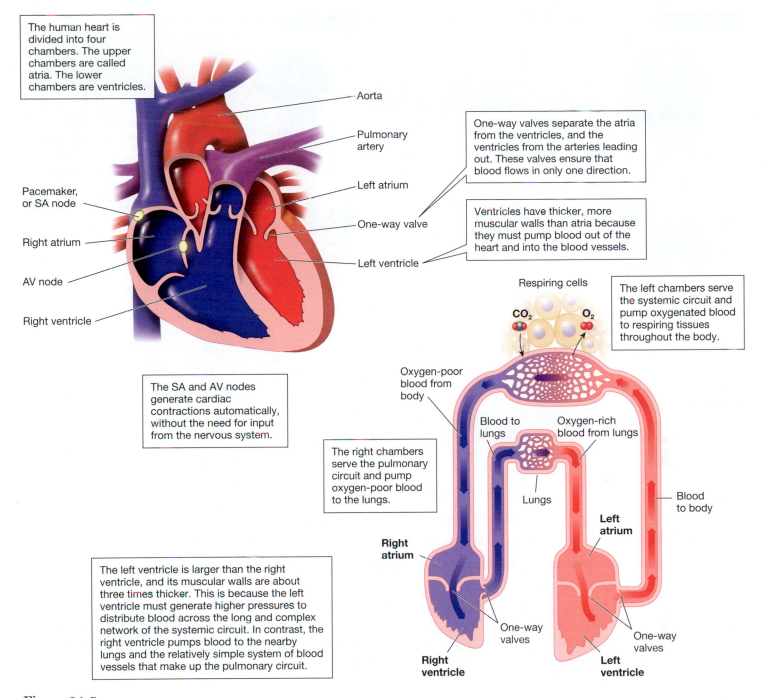

The human heart is divided into four chambers. The upper chambers are called atria. The lower chambers are ventricles.

Aorta

Pulmonary artery

Left atrium

One-way valve

Left ventricle

One-way valves separate the atria from the ventricles, and the ventricles from the arteries leading out. These valves ensure that blood flows in only one direction.

Ventricles have thicker, more muscular walls than atria because they must pump blood out of the heart and into the blood vessels.

Pacemaker, or SA node

Right atrium

AV node

Right ventricle

The SA and AV nodes generate cardiac contractions automatically, without the need for input from the nervous system.

Respiring cells

CO₂   O₂

The left chambers serve the systemic circuit and pump oxygenated blood to respiring tissues throughout the body.

Oxygen-poor blood from body

Blood to lungs

Oxygen-rich blood from lungs

The right chambers serve the pulmonary circuit and pump oxygen-poor blood to the lungs.

Lungs

Blood to body

Left atrium

Right atrium

The left ventricle is larger than the right ventricle, and its muscular walls are about three times thicker. This is because the left ventricle must generate higher pressures to distribute blood across the long and complex network of the systemic circuit. In contrast, the right ventricle pumps blood to the nearby lungs and the relatively simple system of blood vessels that make up the pulmonary circuit.

One-way valves

One-way valves

Right ventricle

Left ventricle

Figure 24.5

The human heart

The right and left sides of the heart function as two separate pumps, although the two upper chambers, or **atria** (singular "atrium"), contract in unison, as do the two lower chambers, the **ventricles**. This unified contraction begins with a signal from the **pacemaker**, or **sinoatrial (SA) node**. The signal causes both atria to contract; it also causes the **atrioventricular (AV) node** to pass the signal on to the ventricles about a tenth of a second later. The short delay allows the atria to empty completely. Note that the heart is shown from the front of the body, so the left atrium is on the right side of the diagram, and so forth.

Q1: Beginning with the left atrium, list the locations of a drop of blood as it moves through the circulatory system.

Q2: Why is the left ventricle larger than the right ventricle, and why are its walls thicker and more muscular?

Q3: Some people have an artificial pacemaker implanted in their heart. What is its function?

Inhalation

Air (rich in O₂)

Lung

Rib cage

Lung volume increases as the rib cage is pulled out and the diaphragm moves down. As a result, the pressure in the lungs drops below atmospheric pressure. Since gases will move from a region of higher pressure to a region of lower pressure, outside air rushes in and the lungs fill with oxygen-rich air.

Diaphragm contracts (downward movement)

The diaphragm is a thick sheet of muscle that forms the floor of the chest cavity.

Exhalation

Air (rich in CO₂)

Lung

Rib cage

Lung volume decreases as the rib cage is pulled in and the diaphragm moves up. This action compresses the space in the chest cavity, raising the pressure inside the lungs and forcing the air out of them.

Space inside chest decreases

Diaphragm relaxes (upward movement)

Figure 24.6

Breathing

Breathing involves two main steps: inhalation, when air is pulled into the lungs, and exhalation, when air is pushed out of the lungs. Most of the time, breathing is controlled automatically by sensory systems located in the heart and brain. If we choose, we can also control our breathing with a system of muscles, the most important of which are the rib muscles and the diaphragm.

Q1: When is air richer in oxygen—as it enters the body or as it exits the body?

Q2: The figure shows air entering through the nose. Where else can air enter the respiratory system?

Q3: Explain how the body creates a change in air pressure during breathing.

---

needs to be done. Any engineered lung needs to be highly reliable, as lungs are vital to transporting oxygen throughout the body. Although humans can live for more than a week without food and for a few days without water, a mere 4 minutes without oxygen results in irreversible brain damage. And lung tissue is notoriously bad at repairing itself, so lung transplants are often the only option in the case of lung damage from disease or trauma.

The respiratory system can be divided into two parts (**Figure 24.7**). The **upper respiratory system** includes airways in the nose, mouth, and throat. When we inhale, air enters through the nostrils and moves into each nasal cavity. Next the air enters the throat, or **pharynx**, an area where the back of the mouth and the two nasal cavities join together into a single passageway.

From the pharynx, air moves into the **larynx**, or voice box, which forms the entryway to the windpipe, or **trachea**.

The trachea is the start of the **lower respiratory system**. Within the chest, the trachea branches into two smaller tubes called **bronchi** (singular "bronchus"). Each bronchus leads to one of the paired lungs, the organs where gases are exchanged. Together, the trachea, bronchi, and lungs make up the lower respiratory system.

Inside the lungs is where the respiratory system gets intricate: The bronchi divide into **bronchioles**, a series of branching, ever-smaller tubes. The tiniest bronchioles open into the **alveoli** (singular "alveolus"), small clusters of sacs that resemble a bunch of grapes. Gases are exchanged across the moist surface

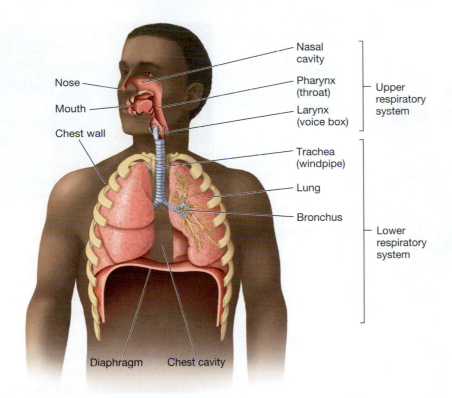

Figure 24.7

The human respiratory system

The respiratory system consists of the upper respiratory system (nose, mouth, throat, and larynx) and the lower respiratory system (the trachea, bronchi, and lungs).

Q1: Beginning with the nose, list the locations of a molecule of oxygen as it moves through the respiratory system.

Q2: From the lung, where would the oxygen molecule move?

Q3: How might an upper respiratory infection like the common cold affect your respiratory system and thus your breathing?

of the thin layer of cells that line each alveolar sac, surrounded by capillaries (**Figure 24.8**). Niklason continues to try to achieve this gas exchange in an engineered lung. So far, her rat lung transplants have managed this exchange for only about 2 hours before failing.

These organs, however, have yet to be successfully transplanted.

In 2013, Ott's lab used the scaffolding technique to engineer rat kidneys, which were then transplanted into rats. In vertebrates, **kidneys** are a set of paired organs that maintain water

Waste Not

At Massachusetts General Hospital in Boston, Harald Ott, a pioneer in the field of organ bioengineering, has engineered whole rat, pig, and human lungs by cleaning donated organs and repopulating them with new cells (**Figure 24.9**).

HARALD OTT

Harald Ott is a thoracic surgeon at Massachusetts General Hospital and heads a laboratory focused on organ engineering and regeneration at Harvard Medical School.

Trachea — Air

① The human trachea, which carries air into the lungs, divides into progressively finer tubes called bronchi, and then bronchioles.

Oxygen-poor blood pumped out from heart

Oxygen-rich blood returned to heart

② More than 150 million tiny alveoli in each human lung create a large surface area for gas exchange.

Bronchiole

Alveoli

Capillaries

Alveolar cells
Red blood cell
Capillary

③ Oxygen enters and carbon dioxide leaves through the walls of the alveoli.

④ The specialized cells of the alveolar walls are extremely thin so that the gases need diffuse only a very short distance.

CO_2

O_2

CO_2

O_2

Alveolus (air space)

Opening into neighboring alveolus

Figure 24.8

Gas exchange in the alveoli

The structure of the lungs speeds the diffusion of oxygen and carbon dioxide into and out of the body by providing a large surface area for gas exchange.

Q1: Why is a large surface area important for gas exchange?

Q2: Does carbon dioxide move into or out of the alveoli? Into or out of capillaries at the surface of the alveoli?

Q3: When a person has pneumonia, the alveoli may fill with fluids. Why would this be a problem?

Figure 24.9

A bioengineered rat lung

Researchers at Massachusetts General Hospital have bioengineered the lungs of rats, pigs, and humans. None have yet been capable of functioning within a living individual.

and solute homeostasis, serving as key components of the urinary system.

All animals must regulate the concentrations of solutes, such as sodium and calcium, in their body fluids, but terrestrial animals face the additional challenges of conserving water and retaining vital solutes. The solute composition of an animal is affected by metabolic activities within the body: as cells metabolize biomolecules, they use up chemicals dissolved in body fluids and produce new ones. This process results in waste products that must somehow be removed from the body.

Kidneys filter and regulate the composition of the blood as it moves through them. When blood leaves the kidneys and returns to the circulatory system, it is cleansed of metabolic wastes and carries water and solutes in normal amounts. The volume of blood leaving the kidneys is slightly smaller than the entering volume because some water is lost to make **urine**, the waste-carrying solution that is expelled from the body.

The blood-cleansing work of the kidneys—**filtration**—is performed by the kidney's basic functional unit, the **nephron** (**Figure 24.10**). Each human kidney has about a million of these tiny filtration units, tightly integrated with surrounding capillaries. When the kidneys fail, an individual can no longer filter waste. Approximately 100,000 individuals in the United States currently await kidney transplantation. As they wait, these patients receive dialysis three times per week, during which a machine filters their blood for them. This lifesaving technology

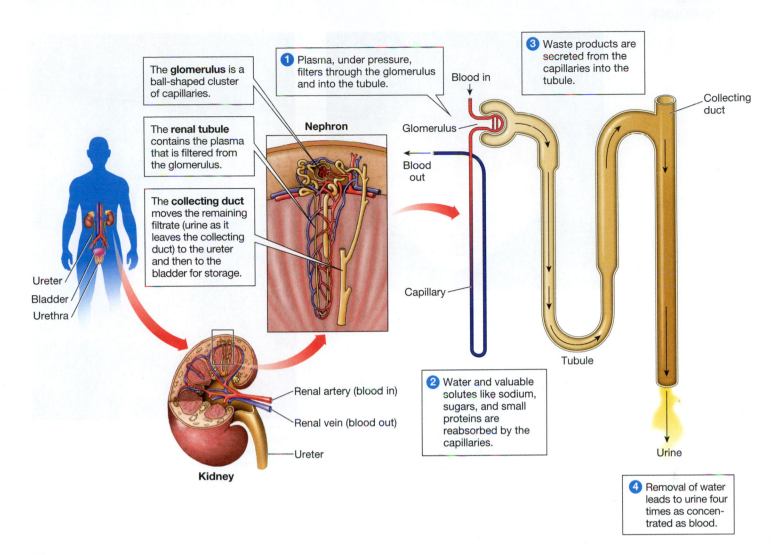

Figure 24.10

The kidney, a critical component of the human urinary system

The kidney regulates internal water content, balances solute concentrations, and removes toxic wastes. All of this work occurs within the nephron.

Q1: What is released from the kidney, and where does it go after release?

Q2: What is the difference between reabsorption and secretion?

Q3: Alcohol suppresses the kidney's ability to reabsorb water. What common consequence of drinking alcohol is related to this fact, and how might you alleviate this problem?

requires a significant time commitment, and dialysis patients have a high risk of infection.

An engineered kidney would provide a permanent solution for these patients. But an engineered kidney would need to do more than just remove waste from the blood. Filtration is only one of three parts of the kidney's job. A second important function is the **reabsorption** of water and valuable solutes such as sodium, chloride,

and sugars before they leave the kidneys. A third function of the kidney is **secretion**: the kidney actively transports excess quantities of substances such as potassium and hydrogen ions, and some medications and toxins, from the blood into the liquid passing through the kidney.

The concentrated fluid that results from the combination of filtration, reabsorption, and secretion is urine. Urine from the many

A rat kidney with all its cells removed, leaving only the extracellular matrix made of collagen.

Figure 24.11

A bioengineered kidney: before and after

The Ott lab at Massachusetts General Hospital bioengineered rat kidneys that function when transplanted. Before implanting, the kidney is flushed of its donor cells and repopulated (seeded) with recipient cells so that it will not be rejected by the recipient's immune system when transplanted.

The collagen "scaffold" has now been reseeded with cells.

collecting ducts in a kidney drains into a long tube, the *ureter*, which delivers the fluid to the urinary bladder for storage. In urination, the bladder empties through a tube called the *urethra*. In Ott's lab in Boston, the bioengineered rat kidneys successfully produced urine when transplanted into rats (**Figure 24.11**).

Coming to Your Senses

To date, the kidney is the most complex organ re-created in the lab and successfully transplanted into an animal. Meanwhile, bioengineering research in a different organ system—the nervous system—has already yielded astonishing improvements in patients' lives, including the life of navy corpsman Edward Bonfiglio Jr.

At 5:00 a.m. on August 27, 2009, Edward Bonfiglio awoke to the sound of the phone ringing. It was his son, Edward Jr., a member of the U.S. Navy serving on active duty in Afghanistan. "Dad?" Edward said, "I got shot. Don't tell Mom."

During a routine foot patrol, Edward's unit had been ambushed, and he'd been shot in his left leg. The bullet hit his sciatic nerve, a long, thick bundle of nerve fibers that runs from the lower back down the leg. Edward immediately lost all feeling and function below his left knee. Doctors found that the bullet had sliced a 5-centimeter gap in the sciatic nerve. Without that nerve intact, the leg would not function.

After Edward arrived home in the United States, doctors gave him two options: either amputate the leg, or repair the damaged nerve with a new kind of nerve graft, a bioengineered tube that could potentially reconnect the two ends of his severed nerve and bring feeling back to his leg. Edward chose the latter—to try to repair the nerve in his leg, part of his peripheral nervous system.

The nervous system of vertebrates is a communication system that transmits signals among various parts of the body. It can be divided into

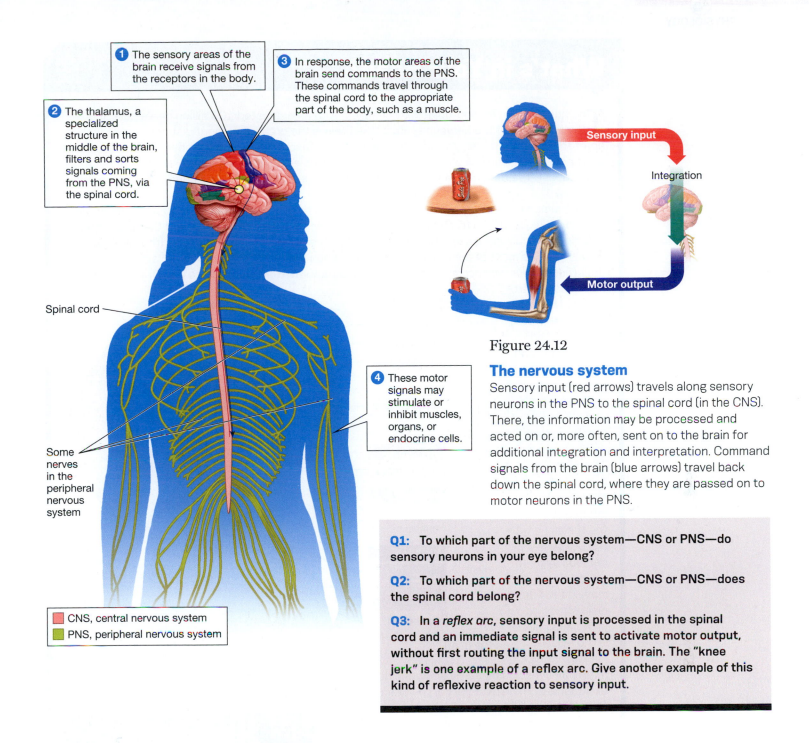

1 The sensory areas of the brain receive signals from the receptors in the body.

2 The thalamus, a specialized structure in the middle of the brain, filters and sorts signals coming from the PNS, via the spinal cord.

3 In response, the motor areas of the brain send commands to the PNS. These commands travel through the spinal cord to the appropriate part of the body, such as a muscle.

Spinal cord

Some nerves in the peripheral nervous system

4 These motor signals may stimulate or inhibit muscles, organs, or endocrine cells.

■ CNS, central nervous system
■ PNS, peripheral nervous system

Sensory input

Integration

Motor output

Figure 24.12

The nervous system

Sensory input (red arrows) travels along sensory neurons in the PNS to the spinal cord (in the CNS). There, the information may be processed and acted on or, more often, sent on to the brain for additional integration and interpretation. Command signals from the brain (blue arrows) travel back down the spinal cord, where they are passed on to motor neurons in the PNS.

Q1: To which part of the nervous system—CNS or PNS—do sensory neurons in your eye belong?

Q2: To which part of the nervous system—CNS or PNS—does the spinal cord belong?

Q3: In a *reflex arc*, sensory input is processed in the spinal cord and an immediate signal is sent to activate motor output, without first routing the input signal to the brain. The "knee jerk" is one example of a reflex arc. Give another example of this kind of reflexive reaction to sensory input.

two main units: the **central nervous system,** or **CNS,** and the **peripheral nervous system,** or **PNS (Figure 24.12)**. The CNS consists of the brain and spinal cord. The **brain** has a large capacity for processing diverse types of sensory information, and it controls and coordinates nerve signals throughout the body (see "What's in Your Head?" on page 444). The **spinal cord** is a thick central nerve cord that is continuous with the brain, acting as a filter between the brain and sensory neurons.

The PNS consists of **sensory organs**—such as the eyes and ears—plus the nerves (except for the retinal, optic, and olfactory nerves, which are part of the CNS), including the nerve in Edward Bonfiglio's leg. The PNS converts stimuli from the sensory organs into **sensory input**—signals that are received, transmitted, and processed by the CNS. Animals are constantly bombarded with sensory input from their external and internal environments. Five main classes of sensory receptors in humans receive this input; some

What's in Your Head?

The human brain is mind-bogglingly complex. With an estimated 100 billion nerve cells, it is the epicenter of the nervous system. The brain is the organ that most distinctly sets humans apart from other species, giving us our capacity to reason, feel, and remember. The brain is made up of *gray matter* (the cell bodies of neurons) and *white matter* (the branching network of neuronal projections—the dendrites and axons, winding tendrils that connect one neuron to many other neurons).

The human brain weighs about 3 pounds and has three main sections: *forebrain, midbrain,* and *hindbrain* (Figure 1). The forebrain contains the *thalamus,* a central switchboard that processes and directs incoming sensory information, and the *cerebrum,* whose outer layer (the *cerebral cortex*) handles most of the brain's actual information processing. The forebrain also contains the *hypothalamus, hippocampus,* and *amygdala,* structures that make up the *limbic system,* our emotional control center. The limbic system regulates emotions and motivations, and participates in memory formation.

The midbrain and hindbrain together make up the *brain stem.* The midbrain coordinates sensory information from the thalamus with the peripheral nervous system for simple physical movement. The main structure in the midbrain is the *reticular formation,* which regulates consciousness. Finally, the hindbrain, evolutionarily the oldest part of the brain, controls the most basic functions, including balance, heart rate, and breathing. The hindbrain consists of the *medulla oblongata, pons,* and *cerebellum.*

Neuroscience is an extremely active area of research. Scanning technologies such as diffusion MRI, which tracks the diffusion of water through white matter, and functional MRI, which detects brain activity by measuring changes in blood flow, are helping us map the brain at an unprecedented level of detail. Figure 2 is a scan from a diffusion MRI (artificially colored) depicting the elegant structure of white-matter fibers twisting through the brain.

Forebrain
Cerebrum
Thalamus
Hypothalamus

Pituitary gland

Amygdala
Hippocampus

Midbrain
Reticular formation

Hindbrain
Pons
Medulla oblongata
Cerebellum

Brain stem

Cerebral cortex

Spinal cord

Figure 1

Figure 2

Table 24.1

Different Ways to Sense the World

Receptor Type	Stimulus	Sense(s)
Chemoreceptors	Chemicals	Taste, smell
Photoreceptors	Light	Vision
Mechanoreceptors	Physical changes	Touch, hearing, proprioception (body position), balance
Thermoreceptors	Moderate heat and cold	Thermoreception (gradations of heat and cold)
Pain receptors	Injury, noxious chemicals, chemical and physical irritants	Pain, itch
Electroreceptors*	Electrical fields (especially those generated by muscle contractions of other animals)	Electrical sense
Magnetoreceptors*	Magnetic fields	Magnetic sense

*Electroreceptors and magnetoreceptors are found in many animals, including most vertebrates. They are not known to be active in humans, however.

additional categories, not known to be active in humans, are found in other animals (**Table 24.1**).

Through sensory input, the PNS gathers information from the external and internal environments and sends it to the CNS. For example, imagine placing your hand near a hot stove; the heat input to your skin is transmitted as a signal from the PNS to the CNS. The CNS integrates and processes the information and generates a signal in response, dictating a particular action, or **motor output**, such as moving your hand away from the stove. The PNS then relays that signal to the body part that will complete the action: the hand moves away from the stove.

All this action in the nervous system is conducted by specialized cells called **neurons** that transmit signals from one part of the body to another in a fraction of a second (**Figure 24.13**). The structure of a neuron reflects its unique function, and different types of neurons respond to different types of stimuli.

Neurons in the PNS are fragile and can be damaged easily. For instance, if you've ever hit your "funny bone," you've actually caused trauma to your ulnar nerve at the elbow, resulting in numbness and tingling. A more severe blow—such as trauma from a broken bone, or a bullet wound as in Edward's case—can cause loss of motor or sensory function.

Unlike lungs, kidneys, or the heart, nerves do have the ability to repair themselves, but they need help. "If you have an injury to a nerve, we can get those nerves to regenerate, but they need a pathway, like a sidewalk, to migrate on," says Christine Schmidt, a biomedical engineer at the University of Florida. As a postdoctoral fellow, Schmidt started looking for biomaterials to act as sidewalks to guide severed neurons to migrate.

The traditional approach to nerve repair is to remove a piece of nerve from somewhere else in a patient's body—such as a nerve in the leg that receives sensory input from the top of the foot—and stitch it into the damaged nerve to restore function. However, this process requires surgery to take the nerve from the leg, and feeling on the top of the foot is lost. And in some cases, using a person's own nerves is not an option. Edward's gap was 5 centimeters wide—far too big for a nerve from another part of his body to work.

CHRISTINE SCHMIDT

Christine Schmidt heads the biomedical engineering department at the University of Florida. She develops biomaterials to guide nerves to regenerate.

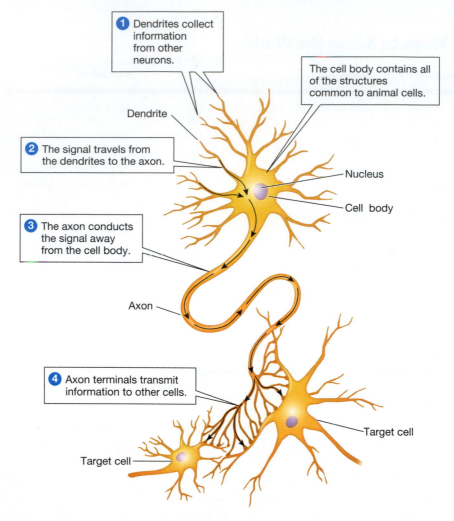

① Dendrites collect information from other neurons.

Dendrite

② The signal travels from the dendrites to the axon.

The cell body contains all of the structures common to animal cells.

Nucleus

Cell body

③ The axon conducts the signal away from the cell body.

Axon

④ Axon terminals transmit information to other cells.

Target cell

Target cell

Figure 24.13

Neurons transmit signals from one cell to another

A neuron receives information from other cells, including other neurons, through one or more **dendrites**. The neuron pictured here has a single long **axon** with branched endings. Axons carry signals away from the cell body and to another cell. Two target cells are shown here.

Q1: How do neurons look different from other cells you've learned about in this book?

Q2: What is the function of these differences?

Q3: Some people are born without the capacity to feel pain. Although it might initially sound nice not to feel pain, it is actually quite dangerous. Describe a situation in which it would be dangerous not to feel pain.

Another option is a synthetic graft, a hollow plastic tube inserted to provide a way for nerve cells to grow from one location to another. Yet these tubes can be used only in very small places and, like artificial blood vessels, can be rejected by the immune system.

"There was really a need for grafts that could provide better guidance for neurons, without using one's own nerves," says Schmidt. Schmidt's lab obtained nerves from animal cadavers and spent 3 years testing different methods of washing the nerves to remove cells that could cause immune rejection. "It was like a cooking experiment," says Schmidt—trying to find the right combination of chemicals and physical forces, such as rinsing and swirling the nerves, to retain

Have a Heart

Over 100,000 people in the United States are right now waiting for an organ transplant. For many of them, the phone call that a suitable donor organ is available may never come. On average, 21 people die each day in the U.S. while awaiting a transplant. Consider registering to be an organ donor: By donating your organs after you die, you could save or improve as many as 50 lives.

 Assessment available in smartw⊕rk**5**

Organs needed vs. organs transplanted

Waiting-list registrations and completed transplants for 2013

Heart

3,700 needed	2,531 transplanted

Lung

1,600 needed	1,923 transplanted

Liver

19,000 needed	6,455 transplanted

Pancreas

1,200 needed	256 transplanted

Kidney

108,000 needed	16,895 transplanted

Figure 24.14

Using bioengineering to repair a damaged nerve
Edward Bonfiglio chose experimental surgery over amputation, and today he can walk and run without assistance.

the nerve's architecture but remove all the original cells. Her team perfected the process in 2004 and published the results.

Right away, a company involved in nerve repair called and asked Schmidt about using her washing process on human tissue. Together, she and the company tested the process on human cadaver tissues. After a successful demonstration in those tissues, the company soon began selling off-the-shelf nerve grafts for use in hospitals. In late 2009, Edward Bonfiglio had one of these grafts implanted into his leg (**Figure 24.14**). Within months of his surgery, he wiggled his toes. It was "one of the greatest moments I had in my entire life," he later said.

There's plenty more bioengineering to be done in the nervous system, says Schmidt. For instance, no nerve grafts can currently repair injuries to the spinal cord. But she's not giving up anytime soon. "It's rewarding," says Schmidt. "It's pretty neat to take something all the way from the bench to impacting patients."

REVIEWING THE SCIENCE

- Humans and other vertebrates have a **cardiovascular system**, a closed **circulatory system** with a chambered heart that pumps blood through a complex network of **blood vessels**.

- **Blood** is composed of cells and cell fragments that float in a fluid known as **plasma**. **Red blood cells** are the main oxygen carriers in the blood.

- Large vessels are built for mass transport of blood. **Arteries** are large vessels that transport blood away from the heart. **Veins** are large vessels that carry blood back to the heart. **Capillaries**, the smallest vessels, facilitate the exchange of materials with surrounding cells.

- The human cardiovascular system has two main circuits. In the **pulmonary circuit**, oxygen-deficient blood is pumped to the lungs. In the **systemic circuit**, oxygenated blood returning from the lungs is pumped out to body tissues.

- The mammalian **heart** is composed of four chambers that make up two separate muscular pumps, each composed of an **atrium** and a **ventricle**. The left atrium and ventricle pump blood to the body; the right atrium and ventricle pump blood to the lungs.

- The human **respiratory system** carries air from the nose (or mouth) to the **lungs**, eventually reaching

clusters of tiny sacs in the lungs called **alveoli**. Gas exchange takes place in the alveoli, where oxygen diffuses into the blood and carbon dioxide diffuses out of it.

- Inhalation and exhalation are controlled by the contraction of muscles, especially those of the diaphragm and the rib cage.

- In the **urinary system** of many animals, including humans, **kidneys** regulate body water and solute concentrations. The kidney's basic unit, the **nephron**, performs three functions: **filtration**, **reabsorption**, and **secretion**. The resulting concentrated solution, **urine**, is carried by ducts to the bladder and excreted from the body.

- The vertebrate **nervous system** is divided into the **central nervous system** (**CNS**), consisting of the **brain** and **spinal cord**, and the **peripheral nervous system** (**PNS**), consisting of the sensory organs and all the remaining nerves.

- **Sensory organs** convert environmental stimuli into nerve impulses that are carried by sensory **neurons** to the CNS. All human senses rely on **sensory input** from just five types of sensory receptors.

- The CNS integrates sensory information and sends an output signal, often **motor output**, through the PNS to the appropriate body part.

THE QUESTIONS

Challenge Yourself

1 Which of the following sensory receptors is *not* active in humans?

(a) pain receptor

(b) electroreceptor

(c) mechanoreceptor

(d) chemoreceptor

2 Select the correct terms:
Kidneys conduct their primary work of filtering the blood through their basic functional unit, the (**neuron / nephron**). Besides filtration, the kidney must (**reabsorb / secrete**) water and important solutes, and (**reabsorb / secrete**) toxins and overabundant substances. Urine drains from the kidneys into the bladder via the (**urethra / ureter**) and empties from the bladder via the (**urethra / ureter**).

3 Beginning with sensory input from a mechanoreceptor in your toe, place the response events of your nervous system in the correct order by numbering them from 1 to 5.

____ a. Command signals travel from the brain to the spinal cord.

____ b. Input is sent to the brain.

____ c. A sensory signal is sent to the spinal cord.

____ d. The muscle responds to the signal.

____ e. Processing occurs in the brain.

4 How do veins and capillaries differ in both structure and function?

Try Something New

5 Which of the following consequences would most likely result from kidney failure?

(a) Heart rate would be uncontrolled.

(b) Toxin levels in the bloodstream would increase.

(c) Incorrect sensory signals would be sent to the brain.

(d) Oxygen levels would fluctuate dramatically.

6 If an organism has a greater concentration of CO_2 in its lungs than in its blood, will there be (a) net transport of CO_2 from the lung air space to the alveolar capillaries, or (b) net transport of CO_2 from the alveolar capillaries to the lung air space? Explain.

7 The number of times our hearts beat per minute is referred to as heart rate. Each heartbeat lasts a little less than 1 second and consists of a series of events called the *cardiac cycle*. The blood pressure measured in a doctor's office reflects the pressure in the arteries leading to the body from the left ventricle. A blood pressure reading of 120/80 (systole/diastole), for example, means that contraction of the left ventricle generates 120 millimeters of mercury (mm Hg) of pressure in the arteries, followed by a drop to 80 mm Hg when the ventricles relax and refill.

a. Blood vessels become less flexible with age. How might this change affect the body's ability to respond to environmental changes with changes in blood pressure?

b. How would you predict that heart rate changes with exercise in the short term, and in the long term?

c. "White coat hypertension" refers to the higher blood pressure displayed at the doctor's office than in normal daily life. How would you explain this phenomenon?

Leveling Up

8 **What do *you* think?** In the highly competitive world of endurance sports such as marathon running and bicycle racing, any means of improving performance offers a significant advantage. Many cyclists have admitted to injecting the hormone erythropoietin (EPO), also known as blood doping. Made naturally in the kidneys, EPO increases the oxygen-carrying capacity of blood by stimulating red blood cell production. In addition to increasing the production of red blood cells, EPO stimulates the growth of capillaries that carry oxygen to tissues.

A synthetic form of EPO developed by drug companies is used to treat patients with anemia, kidney damage, and malaria. Athletes engaging in "blood doping" subject themselves to many health risks. An excess of red blood cells can make the blood so thick that it clots or fails to flow easily through the heart. Nearly two dozen endurance athletes are thought to have died of heart attacks caused by doping with EPO. Because EPO is a naturally occurring hormone, identifying synthetic EPO in the blood or urine samples of athletes has been difficult.

a. What is EPO, and why does it offer a performance advantage in sports, especially endurance events such as cycling and rowing? How could taking EPO kill a person?

b. Some cyclists increase their red blood cell counts by training at high altitudes. The low oxygen content of mountain air triggers the natural release of EPO. Other athletes have accomplished the same thing by spending time in special low-oxygen tents. Do you think either of these approaches is more acceptable than injecting EPO or cells engineered to express EPO? Where would you draw the line, and why?

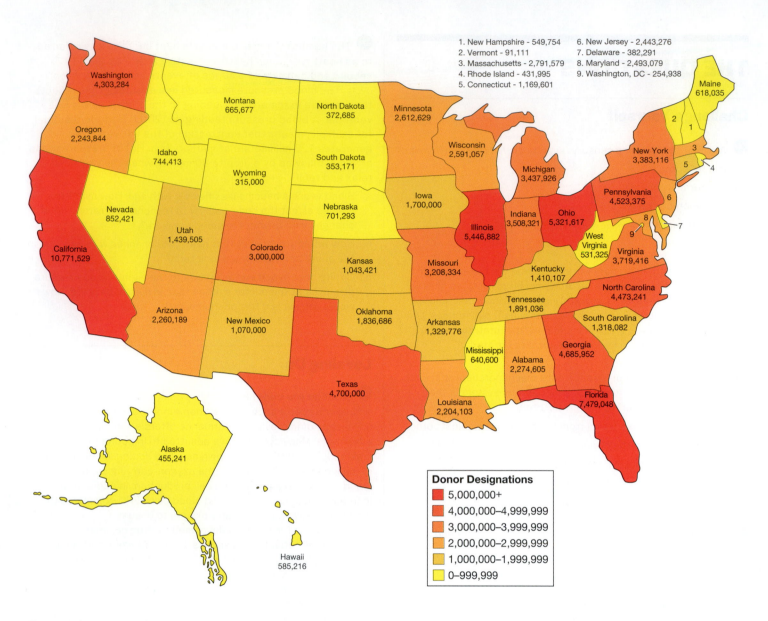

Donor Designations

■ (red)	5,000,000+
■ (orange-red)	4,000,000–4,999,999
■ (orange)	3,000,000–3,999,999
■ (light orange)	2,000,000–2,999,999
■ (gold)	1,000,000–1,999,999
■ (yellow)	0–999,999

1. New Hampshire - 549,754
2. Vermont - 91,111
3. Massachusetts - 2,791,579
4. Rhode Island - 431,995
5. Connecticut - 1,169,601
6. New Jersey - 2,443,276
7. Delaware - 382,291
8. Maryland - 2,493,079
9. Washington, DC - 254,938

9 Life choices As of 2016, almost half of American adults have identified themselves as organ donors in the event of their death. Unfortunately, many more people are in need of an organ than can be helped, and this number continues to rise. Every day, about 80 people receive organ transplants but another 18 people die, waiting for a transplant that never came. Refer to the two maps above to answer the first three questions here.

a. Which state has the highest number of organ donors? The lowest number?

b. Which state has the highest percentage of organ donors? The lowest?

c. Where does your state fall, in terms of the percentage of organ donors? Does this surprise you? Why or why not?

d. Are you an organ donor? Why or why not?

e. Suggest one way that a state (yours or another) could increase the number of its residents who identify themselves as organ donors.

Thinking Multidimensionally

10 Free divers, people who deep dive without oxygen tanks, can hold their breath underwater for sustained periods of time, with the current record holder beating 24 minutes! To accomplish this incredible feat, they enter into a meditation-like state in which heart and breathing rates slow down. Some free divers breathe pure oxygen prior to a dive, while others hyperventilate before going underwater.

a. Explain how hyperventilating impacts the amount of gases in the blood and the blood's pH.

b. If the body is deprived of oxygen, explain the impact this has on cellular respiration and energy production.

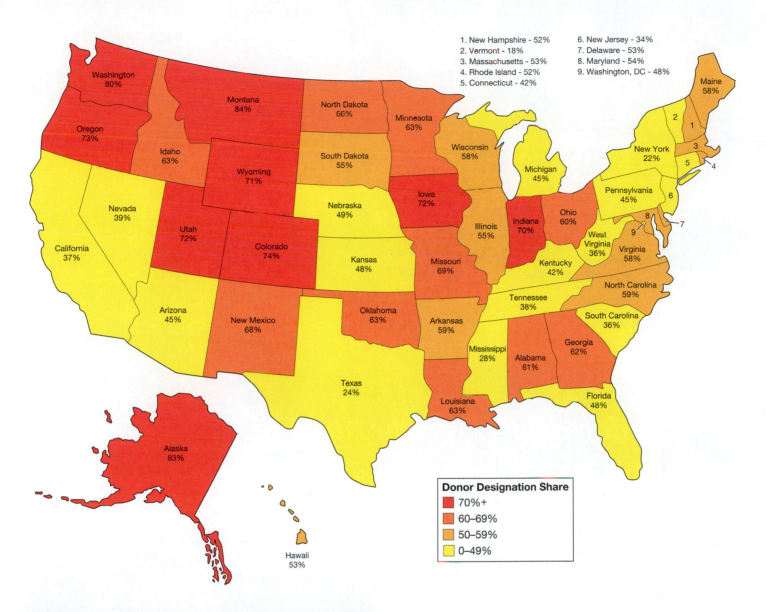

1. New Hampshire - 52%
2. Vermont - 18%
3. Massachusetts - 53%
4. Rhode Island - 52%
5. Connecticut - 42%
6. New Jersey - 34%
7. Delaware - 53%
8. Maryland - 54%
9. Washington, DC - 48%

Washington 80%
Montana 84%
North Dakota 66%
Minnesota 63%
Maine 58%
Oregon 73%
Idaho 63%
Wyoming 71%
South Dakota 55%
Wisconsin 58%
Michigan 45%
New York 22%
Nevada 39%
Nebraska 49%
Iowa 72%
Pennsylvania 45%
California 37%
Utah 72%
Colorado 74%
Kansas 48%
Illinois 55%
Indiana 70%
Ohio 60%
West Virginia 36%
Virginia 58%
Arizona 45%
New Mexico 68%
Oklahoma 63%
Missouri 69%
Kentucky 42%
North Carolina 59%
Arkansas 59%
Tennessee 38%
South Carolina 36%
Mississippi 28%
Alabama 61%
Georgia 62%
Texas 24%
Louisiana 63%
Florida 48%
Alaska 83%
Hawaii 53%

Donor Designation Share
- 70%+
- 60–69%
- 50–59%
- 0–49%

scaffolding where they transplanted human cardiac cells that beat like real heart tissue. Next, red dye was added to the vessels of the engineered leaves to mimic human blood flow through capillaries.

a. Compare and contrast the role and structure of the vascular tissue in plants to the vascular system in humans.

b. While using engineered plants as substitutes for human organ transplants is still in the distant future, what would be the drawbacks and benefits of using plants instead of human and/ or animal organ donors?

11 Using spinach plants, scientists transformed the leaves into working human heart muscle in 2017. By dissolving all the cellular material inside the spinach, this left behind a cellulose

For more, visit digital.wwnorton.com/bionow2 for access to:

Testing the Iceman

A Dutch daredevil claims he can fend off disease with his mind. Two skeptical scientists take the case.

After reading this chapter you should be able to:

- Explain how cells communicate with each other via the endocrine system.
- Describe different ways a hormone can act on a target cell.
- Identify the immune system's first, second, and third lines of defense.
- Compare and contrast the role of white blood cells in the innate and adaptive immune systems.
- Diagram the processes of inflammation and blood clotting.
- Distinguish between a primary and a secondary adaptive immune response.
- Create a flowchart depicting the sequence of events as a vertebrate immune system responds to a pathogen.

CHAPTER
25

ENDOCRINE
AND IMMUNE
SYSTEMS

DRIVING QUESTION

How can the Iceman control the body's stress response, and what effect does this have on his immune system?

• HS-LS1.A: Structure and Function

The scantily clad young men lie on the ground, looking up toward the sky. With sunglasses on and hands propped behind their heads, they look as if they're tanning at the beach—but there are no piña coladas or warm sand here. Instead, these 18 men, wearing only swim trunks, are lying on cold, white snow in the mountains of Poland. And lying with them is the Iceman.

Wim Hof, a Dutch daredevil known as the "Iceman," who holds numerous world records for cold exposure, breathes deeply, leading the youths in an exercise. Over 4 days, he will train them to tolerate extreme cold. During his rigorous program, they will swim in near-freezing water every day and climb a snow-covered mountain in just shorts (**Figure 25.1**). Hof claims that exposure to the cold, combined with meditation and breathing exercises, will enable the men to fend off illness and disease.

Matthijs Kox stands to the side of the Iceman's trainees, taking notes. Kox, a researcher in intensive-care medicine at Radboud University Medical Center in the Netherlands, first met Hof in 2010, when the Iceman was visiting another laboratory at the university. A team in the physiology department was measuring Hof's ability to regulate his core temperature while standing in an ice bath (**Figure 25.2**). The scientists were surprised to find that rather than decreasing as expected, Hof's core temperature actually increased, and his metabolism climbed. While standing in the ice bath talking to his examiners, Hof mentioned that he could also consciously modulate his autonomic nervous system and immune system.

It was an unbelievable claim. The autonomic nervous system operates body functions that humans cannot voluntarily control, such as heartbeat and blood pressure. The **immune system**—a remarkable defense system that protects us against most infectious agents—has also long been known to be involuntary.

Figure 25.1

"Iceman" Wim Hof trains volunteers under extreme conditions

Figure 25.2

Hof's vital signs are monitored while he is immersed in ice

But Hof had a history of doing the unbelievable. He claimed the Guinness World Record for longest ice bath by staying immersed in ice for 1 hour, 52 minutes, and 42 seconds. He climbed part of Mount Everest wearing nothing but shorts. He ran a marathon through the snow at 4°F (20°C), again wearing only shorts.

Hof's testers in the physiology unit told him that a Radboud University professor named Peter Pickkers had a way to measure a person's immune response. So, Hof hoofed it to Pickkers's office, shook his hand, and said, "I can modulate my immune system. I heard you can measure it. Will you measure mine?"

WIM HOF

Wim Hof, better known as the "Iceman," is a Dutch celebrity who holds numerous world records for withstanding extreme cold.

Hormonal Changes

Pickkers was skeptical of Hof's claim, which had a whiff of pseudoscience. But Hof was an interesting character, so Pickkers went online and watched videos of his feats. "There were remarkable things I did not know of—things that, if you had asked me beforehand, I would have said, 'That's not possible. It's not possible to run half a marathon barefoot in the snow,'" says Pickkers. "But he did that."

Pickkers raised the idea of testing Hof to Kox, who was one of Pickkers's PhD students at the time, studying how the brain and immune system interact. Pickkers and Kox discussed the possibility at length, and they decided to give Hof a chance to document his claim. But they would do it while adhering strictly to the principles of the scientific process. "You can imagine some people wondered what we were doing with this guy," says Kox. "So we really focused on doing this in a very sound, precise manner, with no doubt about the scientific integrity of the project."

Hof claimed that the regimen for consciously controlling his immune system required three components: cold exposure, meditation, and breathing exercises. So, the team tested Hof's blood before and after an 80-minute full-body ice bath while Hof performed breathing and meditation exercises. Each time the scientists took blood, they went back to the lab and exposed the blood cells to molecules of endotoxin, a substance found in the cell walls of bacteria that activates an immune response in the human body. They wanted to see how Hof's immune cells in the blood would react to the endotoxin. After the regimen of ice, breathing, and meditation, Hof's cells had a far more subdued immune system response, showing very low levels of proteins associated with activation of the immune system, compared to similar cells before the regimen. The cause of that subdued immune response was unclear, but the researchers suspected that stress hormones played a role.

Hormones are signaling molecules produced by certain cells that tell other cells what to do under specific situations or at certain times in the life cycle of the individual. Hormones are produced by specialized secretory cells of the **endocrine system** (**Figure 25.3**).

Secretory cells are often organized into discrete organs called **endocrine glands**. Major

Figure 25.3

The endocrine system is composed of hormone-secreting cells

The endocrine system consists of cells organized into ductless glands, plus scattered endocrine cells embedded in other tissues or organs. These cells all release hormones directly into the circulatory system.

Q1: What organ coordinates the endocrine system?

Q2: How does an endocrine gland differ from an exocrine gland?

Q3: How do male and female endocrine systems differ?

endocrine glands are located throughout the human body. Unlike exocrine glands, such as tear ducts, endocrine glands do not have ducts or tubes that deliver secretions from the gland to the site of action. Instead, endocrine glands release hormones into body fluids such as blood, which

Figure 25.4

Hormones enable cells to communicate with one another

Hormones released by endocrine cells travel through the circulatory system to produce a response in target cells often located at a distance in the body.

> **Q1:** How do hormones travel to target cells?
>
> **Q2:** Distinguish between an endocrine cell and a target cell.
>
> **Q3:** Why is a hormone called a signaling molecule?

Circulating hormones

Some hormones exert their effects by binding to a cell surface receptor on the target cell.

Target cell A

Endocrine cells

Blood vessel

Target cell B

Some hormones cross the plasma membrane and act on receptors inside the target cell.

carries these chemical messengers throughout the body (**Figure 25.4**). In a subsequent experiment, Kox measured the levels of a stress hormone called cortisol in Hof's blood. After the ice, breathing, and meditation regimen, Hof's blood contained far higher levels of cortisol than before.

In the human body, most hormones can travel only as fast as the blood moves, which means they take several seconds or more to arrive at their target cells. Hormones coordinate functions that take place over timescales ranging from seconds (such as quickly increasing one's heartbeat in reaction to fear) to months (such as preparing a uterus to contract during the birthing process; see Figure 22.11).

Typically, hormones become greatly diluted after they are released into the circulatory system. They must therefore be able to exercise their effects at very low concentrations, as vitamins do. Hormones are effective in small amounts because they bind to their targets with great specificity. Cortisol, for example, binds to a very specific receptor present on the surface of almost every cell in the body.

Cortisol, adrenaline (also called epinephrine), and noradrenaline (norepinephrine) are three hormones produced by the **adrenal glands**, a pair of endocrine glands that sit atop the kidneys. The release of these hormones launches a number of rapid physiological responses, including boosting blood glucose levels to increase energy levels in times of stress.

When a single hormone molecule binds to its receptor, it sets in motion a chain of events that may ultimately activate thousands of protein molecules in the target cell (**Figure 25.5**). When cortisol binds to its receptor, it initiates a pathway that results in the regulation of genes involved in development, metabolism, and immune response. This signal amplification—from a single hormone molecule to the activation of many proteins and genes—means that just a few hormone molecules can have a substantial impact on target cells. Through its effects on many cells, a hormone can exert a profound influence on the body as a whole.

Some hormone-secreting cells are not organized into distinct glands like the adrenal glands, but are instead embedded as single cells or clusters of cells within other specialized tissues and organs. For example, the main role of the kidneys is to filter blood, yet some cells in the kidneys produce hormones that stimulate red blood cell production. Altogether, the endocrine glands and the endocrine cells embedded in other organs, such as the kidneys, make up the endocrine system.

PETER PICKKERS AND MATTHIJS KOX

Peter Pickkers (left) is a professor of experimental intensive-care medicine who studies the innate immune system. Matthijs Kox (right) is a researcher in intensive-care medicine. Both work at Radboud University Medical Center in the Netherlands.

A hydrophobic hormone can cross the plasma membrane.

Plasma membrane receptor

A hydrophilic hormone must bind to a membrane receptor because it cannot cross the plasma membrane.

Plasma membrane

Intracellular receptor

Cytoplasm

Effect: Change in membrane transport

Effect: Changes in gene expression

Nucleus

Gene 1   DNA   Gene 2

Effect: Change in metabolism

mRNA

Effect: Change in cytoskeleton

Protein 1   Protein 2

Figure 25.5

Hormonal signals are amplified within the cell

Hormones are effective at low concentrations because of their specificity and because tiny amounts of a hormone can generate a large internal signal within a target cell.

Q1: Describe the two ways that a hormone outside a cell can exert its effect on a cell.

Q2: Within the cell, how does a hormone bring about a change in cell activity?

Q3: It takes very little of the hormone cortisol to have large effects throughout the body. Explain why.

Brain-Body Connection

Testing Hof's cells alone wasn't enough for Kox and Pickkers. They wanted to measure his entire body's immune response, and wanted to know if the breathing and meditation techniques made any difference. After the ice bath, they asked Hof to perform his meditation and breathing techniques while they injected him directly with the endotoxin. In previous experiments, healthy volunteers injected with endotoxin had experienced fever, headaches, and shivering, accompanied by high levels of signaling proteins, called **cytokines**, that immune system cells use to communicate when an invader is present.

At various times before and after the injection, Kox measured Hof's blood levels for hormones and cytokines. Kox then compared Hof's results to those of a control group of 112 healthy volunteers who had previously taken the same test. To the scientists' surprise, as soon as Hof began practicing his breathing techniques,

his cortisol and adrenaline levels skyrocketed. And unlike the other volunteers, Hof reported almost no flu-like symptoms. Topping it off, the concentration of cytokines in his blood—indicative of an immune response—was less than half that of the control group. Hof appeared to have suppressed his immune system—voluntarily.

How was that possible? Kox considered the possibilities. First, a tiny region at the base of the vertebrate brain, the **hypothalamus**, coordinates the endocrine system and integrates it with the nervous system (refer back to **Figure 25.3**). The hypothalamus contains neurons, which interact with the brain, and endocrine cells, which produce hormones. It is a literal brain-body connection.

One well-known part of that connection involves the adrenal glands. In response to stress messages from the brain, the adrenal glands release adrenaline into the blood. If a man sees a rattlesnake in front of him, for example, he is likely to jump back or at least freeze in place, his heart racing. This quick response is due to the connection between the nervous system and the adrenal glands (**Figure 25.6**): The nervous system processes visual information (*Snake!*) and transmits an alarm signal to the adrenal glands within a fraction of a second. The adrenal glands kick in right away, pouring adrenaline and noradrenaline into the blood. Adrenaline stimulates glycogen breakdown in liver and skeletal muscle cells, causing glucose to be released into the bloodstream. It also speeds up the heartbeat and increases the force with which the heart contracts, so that glucose is delivered throughout the body more rapidly. In this way, glucose becomes available to fuel a rapid response to a stressful situation.

Within just a few seconds, then, these hormones increase the pumping of blood and trigger the release of glucose, all of which support the next move: fight or flight. In the case of an encounter with a snake, that may mean either arming oneself with a stout stick or running away. It turns out, however, that adrenaline and cortisol play another role, aside from triggering glucose delivery: research has shown that the hormones also subdue the activity of immune system cells.

Hof appeared to be able to consciously activate his nervous system (in the absence of real stress) to prompt the release of cortisol and adrenaline, thereby suppressing his immune response to the endotoxin voluntarily. It was a "remarkable" finding, says Pickkers, but he wasn't ready to jump to conclusions. The case study of a single individual is weak scientific evidence for any phenomenon. Perhaps Hof was simply an outlier: "Everyone can play a little baseball, but there is only one Derek Jeter," says Pickkers. Perhaps Hof had a unique genetic mutation or another factor that

Figure 25.6

Adrenal hormones produce a rapid response to stress or fear

The adrenal glands produce adrenaline (epinephrine) and noradrenaline (norepinephrine), which trigger the rapid release and delivery of stored energy.

Q1: Describe an event (other than the one illustrated in the figure) that might cause the release of adrenaline.

Q2: What organs does adrenaline affect?

Q3: What do you think would happen if your adrenal glands were constantly releasing adrenaline?

enabled him to control his autonomic nervous and immune systems.

But Hof claimed that he was not an outlier, that he could teach his technique to anyone. "I'm sure everybody is able to do this," Hof told Pickkers. Pickkers challenged him to demonstrate that he could do this. For scientific validation, Hof needed to teach his method to a group of healthy volunteers so that Pickkers could then compare that group's immune responses to those of an untrained control group of volunteers. In this controlled way, Hof might produce stronger scientific evidence for his claim.

Innate Defenders

If Hof was right—if it really was possible to voluntarily control the immune system—the discovery would do more than change our understanding of the immune system; it would offer hope to people with autoimmune diseases, individuals in whom the immune system is overactive.

When healthy, the immune system protects animals from most infectious agents, called **pathogens**. Human pathogens include viruses, bacteria, and protists, as well as some fungi and multicellular animals such as parasitic worms. A well-known example is human immunodeficiency virus or HIV, the virus responsible for AIDS (see "What Makes HIV so Deadly?" on page 460).

Pathogens infect animals only if they can find a way into the body. An animal's first line of defense against pathogens is its **external defenses**, which reduce the likelihood that a harmful organism or virus will gain access to internal tissues. Linings that separate the "outside" from the "inside" of the body—the skin and the linings of the lungs, for example—act as a physical barrier to keep out most pathogens. Other external defenses include chemical agents (such as enzymes) and chemical environments (such as acidic conditions) that keep invaders from attaching to or growing on body surfaces (**Figure 25.7**).

Although external defenses do a good job of keeping out most pathogens, the body is still vulnerable. Mucous membranes, in the nasal cavity and other parts of the body, are a common point of entry for pathogens. Wounds, in the form of cuts, abrasions, and punctures, are common, and many pathogens will take advantage of breaks in the skin to gain entry to their hosts.

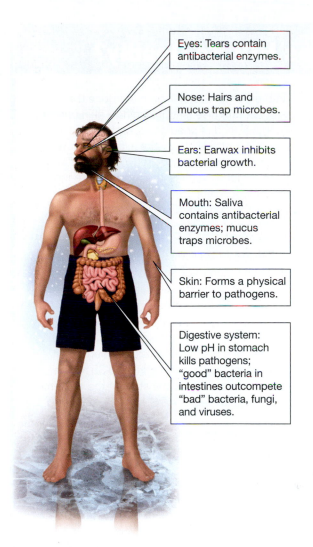

Eyes: Tears contain antibacterial enzymes.

Nose: Hairs and mucus trap microbes.

Ears: Earwax inhibits bacterial growth.

Mouth: Saliva contains antibacterial enzymes; mucus traps microbes.

Skin: Forms a physical barrier to pathogens.

Digestive system: Low pH in stomach kills pathogens; "good" bacteria in intestines outcompete "bad" bacteria, fungi, and viruses.

Figure 25.7

The immune system's first line of defense is to prevent the entry of pathogens
Our skin and the linings of our respiratory and digestive systems form physical and chemical barriers against pathogens.

Q1: What is the main physical barrier that animals use to keep out pathogens?

Q2: Give an example of a chemical defense within the digestive system.

Q3: Explain why rubbing your eyes and nose during flu and cold season is not recommended.

Once inside, pathogens confront a second line of defense: the cells and defensive proteins of the **innate immune system**. To mount an internal defense that kills, disables, or isolates invading

What Makes HIV So Deadly?

In the early 1980s, doctors in the United States began to notice that gay men were dying of a variety of rare diseases, including a skin cancer called Kaposi's sarcoma, an unusual kind of pneumonia, and other infections that most people ordinarily shake off. By the mid-1980s it was clear that patients with the syndrome—named acquired immunodeficiency syndrome, or AIDS—had broken immune systems, the result of infection by a virus called the human immunodeficiency virus, or HIV.

In North America and Europe, the number of new cases rapidly increased, claiming the lives of tens of thousands of people each year. Initially, most new cases were limited to gay men, intravenous drug users, and people who had received blood transfusions. The common denominator was contact with the blood or body fluids of others: couples during sex, drug users when sharing used needles, and surgical and hemophilia patients who received blood transfusions contaminated with HIV.

In time, safe-sex education and clean-needle programs reduced the rates of infections among gay men and blood transfusion patients, but the virus spread to other populations. Globally, an estimated 35 million people have died of AIDS-related illnesses since the start of the epidemic. In just one year, 2015, 1.1 million people died of AIDS, and 36.7 million were living with HIV.

Inside the bloodstream, HIV enters immune system cells and reproduces inside them, eventually killing so many immune cells that the body's defenses are crippled. In the short term, remaining immune system cells track down HIV-infected cells and destroy them. Because the immune cells do such a good job of killing HIV-infected cells in the blood, most people with HIV have about a decade of normal health before they become ill, even without any treatment.

Over time, however, the HIV particles in the body evolve. As HIV evolves, the immune system cells no longer recognize and kill the virus. The population of HIV particles increases and begins destroying immune system cells faster than they can multiply. The body no longer has the immune system cells it needs to fight off infections by bacteria, yeasts, and other viruses. Once the immune system collapses, a person is vulnerable to almost any infection.

So far, there is no effective vaccine or cure for HIV. But a variety of new drugs enable people with AIDS to live years longer with fewer symptoms than they used to. "HIV cocktails," as the standard mixture of therapeutic drugs are called, prevent the viral genetic material from replicating and prevent the virus from merging with plasma membranes and entering cells. But the drugs can cost hundreds or thousands of dollars a month. Because treatment is so costly, only one in five AIDS patients in Africa and Asia receives effective treatment. For now, the best way to slow the spread of the disease remains safe-sex education, the free availability of condoms, and clean-needle programs.

pathogens, the body first must recognize that an invader is present. Although a person is not consciously aware of it, a healthy body can distinguish foreign invaders (nonself) from its own cells (self). If the internal defenders fail to tell self from nonself, they mistakenly attack the body's own cells, leading to autoimmune diseases such as rheumatoid arthritis (in which immune cells attack the lining of the membranes that surround joints) or type 1 diabetes (in which immune cells attack the pancreas, which makes insulin).

If there was a way, as Hof claims, to subdue the immune system at will, people with autoimmune diseases might have another avenue, aside from expensive drug therapies, by which to control their rebellious immune systems.

Team Effort

Despite Hof's personal achievement, Pickkers and Kox didn't really think he would be able to teach others to voluntarily control their innate immune systems. "We thought it would be a negative result," says Pickkers. If nothing else, Hof had been performing his technique for 30 years, so even if he could teach it, Pickkers doubted he could teach a novice enough to influence his or her own immune system in just a few days. Hof disagreed, arguing that a short training regimen would be sufficient to impart the ability. Kox didn't think most of the volunteers would even make it through the training.

It was no easy study for the participants. Over 4 days, 18 healthy, young, male volunteers were taken into the mountains of Poland and exposed to the cold in various ways: standing in the snow barefoot for 30 minutes, lying in the snow bare-chested for 20 minutes, swimming in ice-cold water each day for several minutes, hiking up a snowy mountain in nothing but shorts and shoes. Hof also taught them his meditation and breathing techniques, including deep inhalations and exhalations.

Contrary to Kox's expectation, all 18 participants completed the training, and 12 of them were then randomly selected to come back to the lab for the final part of the experiment: exposure to the endotoxin, to test whether they could consciously regulate their immune response. Their results would then be compared to those of 12 healthy controls who had also been exposed to the endotoxin but had not received Hof's training. Now the scientists would finally find out whether Hof could teach someone to control the activity of the innate immune system.

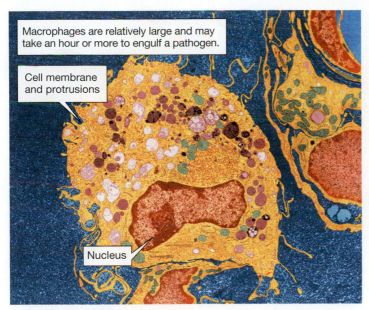

Macrophages are relatively large and may take an hour or more to engulf a pathogen.

Cell membrane and protrusions

Nucleus

Neutrophils attack pathogens with antimicrobial chemicals.

Nucleus

Cell membrane and protrusions

Figure 25.8

Phagocytes destroy pathogens by engulfing them

Phagocytes are a kind of white blood cell, a family of defense cells found in body fluids including blood, where they intercept invading pathogens. Two different kinds of phagocytes—a macrophage and a neutrophil—are seen in these colorized transmission electron micrographs (TEMs).

Q1: Place these terms in order from most to least inclusive: neutrophil, white blood cell, phagocyte, innate immune system.

Q2: Compare and contrast macrophages and neutrophils.

Q3: Why would it be a problem if your innate immune system identified the insulin-producing cells in your pancreas as "nonself"?

The innate immune system reacts to cells or molecules that do not belong in the body by activating defense cells and proteins to eliminate the unwelcome guests. A suite of pathogen-recognizing cells called **phagocytes**, a type of white blood cell, mark and destroy foreign invaders by engulfing and digesting them (**Figure 25.8**).

This immune response is said to be innate (inherent) because the necessary components are constantly at the ready for deployment against an invading pathogen. The innate response can be local, occurring at the point of entry, or global, involving the whole body. Like the external defense system, innate immunity is indiscriminate as to which foreign invaders it repels, so it is considered a **nonspecific response**. The innate immune system is an ancient defense mechanism found in both invertebrates and vertebrates.

In addition to defending against invaders, the innate immune system plays two other critical roles. First, it responds to tissue damage from a pathogen invasion or wound by mounting an immediate and coordinated sequence of events known as **inflammation** (**Figure 25.9**). Cytokines are a clear marker of inflammation, and thus a good way to measure the action of the immune system. A second role of the innate immune system is clotting blood to close a wound. Sealing an open wound reduces blood loss and restores the integrity of external defense barriers (**Figure 25.10**).

To test the innate immune response of the 24 study participants (the 12 trained volunteers and the 12 untrained controls), the scientists injected each participant with endotoxin and monitored them for 6 hours. Hof visited his trainees during the experiment, coaching them through his breathing techniques.

Skin   Bacteria

1 The skin is torn.

Cytokines

2 Damaged cells release cytokines, histamines, and other chemical signals.

Histamine

Blood vessel   Neutrophil

3 Histamines dilate blood vessels and make them "leaky" so that macrophages can migrate toward the wound.

Macrophage

4 Macrophages, neutrophils, and other white blood cells engulf bacteria and cellular debris.

Platelets

5 Platelets from blood plasma enter the damaged area to help heal the wound.

Figure 25.9

The inflammatory response acts against invading pathogens

Inflammation occurs when the innate immune system swings into action after cellular damage is detected, cleaning up damaged tissues and preventing the spread of pathogens. Inflammation can occur anywhere inside the body. Here we see an inflammatory response following a puncture wound to the skin.

Q1: What is the role of white blood cells in inflammation?

Q2: What would happen if histamines were not produced during inflammation?

Q3: Why is inflammation called a "nonspecific" immune response?

Figure 25.10

Blood clots help prevent pathogens that may be present in a wound from spreading

Sticky cell fragments, or **platelets** (shown here in light blue), and clotting proteins (yellow) form a gel-like mesh that traps blood cells, creating a blood clot that seals broken skin. Clotting can begin as quickly as 15 seconds after tissue damage occurs. Growth of new tissue eventually repairs the wound more permanently.

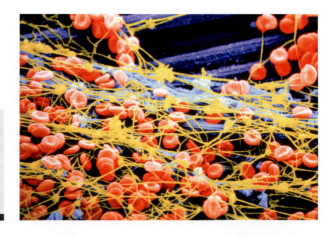

Q1: Why is blood clotting an important immune response?

Q2: How are the inflammatory response and blood clotting similar?

Q3: Some people have a genetic disorder in which their blood cannot clot. Why would this be a problem?

The results were clear: after being injected, and while performing the breathing techniques, the trainees showed higher adrenaline levels than the controls—higher even than the adrenaline produced by a person's first bungee jump. "They produced more adrenaline just lying in bed than somebody standing in front of an abyss going to jump in fear for the first time," says Hof. "That means direct control of your hormone system, and your hormones have a direct relationship to the immune system."

In addition, the trainees had fewer flu-like symptoms and lower fevers, and their cytokines—the signaling proteins of the immune system and markers of inflammation—were at less than half the level of the control group. "We were very surprised," says Kox. "It was impressive that these guys could do all this cold exposure training, but I still thought the chances were slim they'd be able to modulate their immune systems. But the results were so convincing."

While Hof's trainees' results may point the way to identifying new, less expensive treatments for autoimmune diseases, don't try this at home! We have an innate immune system for a reason—to protect against pathogens—so voluntarily subduing it without the guidance of a doctor is not a good idea.

Adapting to the Enemy

Kox and Pickkers's study did not address the human immune system's third line of defense. In contrast to the broad responses of external defenses and the innate immune system, the more complex **adaptive immune system** is tailored against specific invaders. There is no evidence, yet, that Wim Hof is able to control his adaptive immune response.

Adaptive immunity goes beyond simply recognizing something as nonself. Instead, specialized defense cells are trained to recognize only one strain of pathogen and to activate a **specific response**. The adaptive immune system is based in the lymphatic system, which is itself an important part of both the immune system and the circulatory system. It relies on two main weapon systems: antibody-mediated immunity and cell-mediated immunity (**Figure 25.11**).

Antibody-mediated immunity uses powerful Y-shaped proteins called **antibodies** to recognize and attack invaders. Antibodies recognize nonself markers—**antigens**—on the pathogen and mark the pathogen for destruction. **B cells**, specialized lymphocytes created and matured in the bone marrow, produce thousands of antibodies per second aimed specifically at the pathogen that has been recognized.

Cell-mediated immunity recognizes cells that have been infected by a pathogen such as a virus, as well as cancer cells (for more on cancer, see "Cancer: Uncontrolled Cell Division" on page 101 of Chapter 6). **T cells**, lymphocytes created in the bone marrow and matured in the thymus, recognize markers on the surface of a cell that has been infected. The T cells then kill the infected cells so that thay cannot spread the disease to other cells.

Compared to innate immunity, adaptive immunity is slow to mobilize. However, it is the most sophisticated and effective of animal defense systems because of the amazing selectivity with which the adaptive immune system attacks a particular invader.

Adaptive immunity occurs in two stages. The very first time a person is exposed to a particular pathogen, the **primary immune response** is activated. This response takes time—more than 2 weeks sometimes—to reach full steam. Because of that slow start, and because pathogens multiply so rapidly, people infected with an aggressive pathogen for the first time can sometimes lose the race, becoming ill and dying. Therefore, any pathogen that is new to humans is particularly dangerous, and the nonspecific response of innate immunity may be more effective in combating it.

A distinctive feature of the adaptive immune system is **immune memory**, the capacity of this defense system to remember a first encounter with a specific pathogen and to mobilize a speedy and targeted response to future infection by the same strain. This "memory" is what enables us to become immune to attacks by the same strain after we suffer the disease a first time. Once you've had measles, for example, you never get sick from the measles virus again, because the adaptive immune system recognizes the virus and quickly eradicates it the next time. Keep in mind that this means you are not born with immune memory. Each individual must

Lymphatic ducts and associated organs

Adenoids

Tonsils

Lymph nodes

Lymphocytes move around the body through the lymphatic and circulatory systems.

Thymus

They develop in the thymus…

Spleen

Lymphatic ducts

Appendix

…and in the marrow of the bones.

Origin and maturation of lymphocytes

Bone marrow

Stem cell

Immature lymphocytes (white blood cells)

To thymus via blood

Thymus

Antigen-recognizing proteins

B cell (antibody-mediated immunity)

T cell (cell-mediated immunity)

To lymph, spleen, and other organs

Figure 25.11

Adaptive immunity resides in the lymphatic system

(Left) The **lymphatic system** consists of lymphatic ducts, lymph nodes, and associated organs. (Right) **Lymphocytes** originate from stem cells in bone marrow. B cells mature in the bone marrow; T cells mature in the thymus. Lymphocytes circulate in the lymphatic and circulatory systems and accumulate in lymph nodes and other organs, such as the spleen, appendix, and tonsils.

Q1: Why are B and T cells so named?

Q2: In what way is this immune system "adaptive"?

Q3: Why is the adaptive immune response considered the third layer of the immune system?

build one over time by being exposed to various pathogens.

Because of immune memory, the adaptive immune system produces a faster, more dramatic response to pathogens when it encounters them a second time. The second encounter, when the adaptive immune system is poised and ready to respond, is the **secondary immune response**.

We acquire immunity in two ways: either actively or passively. **Active immunity** to a particular pathogen develops when our own bodies produce antibodies against that pathogen; they are not received from an outside source.

This happens naturally when we're exposed to certain pathogens, such as the measles virus. We can also acquire active immunity to certain diseases through vaccination, which introduces antigens to the body in a harmless form.

Passive immunity develops when antibodies that were not made by our own bodies are received. A human fetus acquires antibodies from exchanges between its blood and its mother's blood. This antibody sharing continues after birth: mother's milk is rich in antibodies because the mother's immune system has encountered many antigens and made many antibodies in her lifetime. Thanks

Driven by Hormones

Hormones regulate not only the classical fight-or-flight response, but also the sleep and wake cycles of all people. These signaling molecules in the blood also become elevated or depressed during periods of stress (cramming for a test, anyone?) and exercise. What are your levels of melatonin, adrenaline, and cortisol right now?

Assessment available in smartwork5

Hormones in the human body

Melatonin

Regulates sleep timing, blood pressure, and more. Levels fluctuate over a 24-hour cycle, peaking at night.

Serum melatonin (pg/mL)

60 50 40 30 20 10

Stress hormone that quickens the heartbeat, among other effects. Levels increase as exercise intensity goes up.

Stress hormone that regulates homeostasis in the body. Levels drop during prolonged high-intensity exercise.

Adrenaline

(Epinephrine)

Exercise intensity
(in watts)

100
150   200
50   250
0

200
400
600
800
1000

Plasma epinephrine (pg/mL)

Cortisol

Exercise time
(in minutes)

90
60   120
30   150
0

4
8
12

Cortisol (µg/dL)

Figure 25.12

Hof continues to train recruits to use his novel methods

to that antibody-rich milk, a nursing baby receives passive immunity to a broad range of pathogens. Passive immunity produces no memory cells, so it wears off as the received antibodies degrade, usually within a few weeks or months.

The Iceman Cometh

In 2014, Pickkers and Kox published the results of their experiment in the *Proceedings of the National Academy of Sciences*, one of the world's most respected and cited peer-reviewed journals. They are conducting follow-up studies to determine whether one or more of the three parts of Hof's technique—cold exposure, breathing, and meditation—is primarily responsible for the adrenaline release and subsequent immune suppression, and exactly how these effects come about. Kox suspects that the breathing techniques are the main factor, since Hof's breathing appears to trigger the release of hormones, but he cannot yet be sure.

"It needs to be studied a whole lot more," agrees Hof, who is eager to continue putting his method under the magnifying glass of the scientific process (**Figure 25.12**). "By meticulous experiments and measurements—not speculation—we want to show this works. I'm very thankful to the professors at Radboud who dared to go into this."

Whether Hof's technique can help individuals with autoimmune disorders is still up for debate. Although the training worked for young, physically fit men, the team does not know whether it will work for older people with autoimmune diseases who already have compromised organ systems. "We would not advise people to do this [right now]," says Pickkers, before there are additional studies. "We have to be careful there are no unwanted side effects or risks."

But he is optimistic about the future and still sounds surprised about how well the training worked. "We confirmed that, indeed, using the techniques of Wim Hof, humans are able to modulate their autonomic nervous system and influence their immune response," says Pickkers. "It is remarkable."

mild. The **secondary immune response** is a faster, stronger response to a pathogen that has been encountered one or more times, creating immune memory.

- **Active immunity** can be acquired through natural exposure to a pathogen or through a vaccine. **Passive immunity** comes from receiving antibodies that were not made by our own bodies, such as when a fetus acquires antibodies from its mother.

REVIEWING THE SCIENCE

- A **hormone** is a signaling molecule distributed through the body by the circulatory system. Because hormones move only as quickly as the blood moves, they tend to coordinate functions that are slower and longer-lasting than those under the influence of the nervous system.

- A single hormone may affect many different kinds of target cells, potentially triggering a different response in each. Hormones act on target cells either by moving through the plasma membrane to the cell's interior or by acting on receptors embedded in the plasma membrane.

- The **endocrine system** is made up of the glands and specialized cells that produce hormones. The **hypothalamus**, located in the forebrain, coordinates the endocrine system and integrates it with the nervous system. The **adrenal glands**, located in the kidneys, produce hormones responsible for the fight-or-flight response.

- The vertebrate **immune system** possesses three layers of defenses against **pathogens**.
  - The first layer consists of **external defenses**: physical and chemical barriers, including the skin and the linings of the respiratory and digestive systems.

  - The second line of defense is the **innate immune system**. Several types of blood cells and molecules produce the **nonspecific responses** of the innate immune system, including **phagocytes** such as macrophages and neutrophils, which engulf and destroy pathogens. Tissue damage stimulates **inflammation** and blood clotting.

  - The third line of defense is the **adaptive immune system**, providing long-term defenses in the form of **specific responses** to pathogens and parasites. These responses are mediated by powerful proteins called **antibodies**, or by cells.

- The **lymphatic system** provides the primary sites for adaptive immunity. White blood cells called **lymphocytes** confer specific immunity. Immature lymphocytes differentiate into **B cells** in the bone marrow and **T cells** in the thymus. Each lymphocyte has special membrane proteins that bind to only a specific **antigen** of a specific pathogen.

- The **primary immune response** from the adaptive immune system is relatively slow and

THE QUESTIONS

Challenge Yourself

1 Identify whether each of the following is characteristic of either antibody-mediated (A) or cell-mediated (C) immunity.

____ a. The immune response relies on Y-shaped proteins to identify pathogens.

____ b. B cells produce proteins specific to a pathogen.

____ c. Lymphocytes that matured in the thymus identify infected cells.

____ d. Antigens on the pathogen allow it to be identified as nonself.

____ e. Infected cells are destroyed so that an infection cannot spread to other cells.

2 Select the correct terms:

The first time you are exposed to a pathogen, the (**primary / secondary**) immune response is activated. The (**primary / secondary**) immune response to a pathogen is stronger and more rapid. You acquire (**active / passive**) immunity to a pathogen when your own body creates the antibodies against that pathogen. (**Active / Passive**) immunity comes from the antibodies produced by another person, such as your mother when you were in utero or nursing. The immunity conferred by vaccines is an example of (**active / passive**) immunity.

3 Beginning with a perceived threat (for example, a spider), identify the correct order of events in the stress response by numbering them from 1 to 5.

____ a. Target cells amplify the hormonal signal to produce a response.

____ b. The liver breaks down glycogen to glucose, and the heart increases its rate and the force of its contractions.

____ c. Adrenaline reaches target cells in the liver and heart.

____ d. The hypothalamus signals the adrenal glands that a threat is present.

____ e. The adrenal glands release adrenaline into the bloodstream.

Try Something New

4 Wim Hof and his trainees had increased levels of the stress hormone adrenaline and decreased immune function during the experiments described in this chapter. How might these changes negatively affect the endocrine and immune systems over the long term?

5 Which of the following is *not* true of a B cell?

(a) It is a kind of lymphocyte.

(b) It is produced in the bone marrow.

(c) It matures in the thymus.

(d) It is part of the adaptive immune response.

(e) All of the above are true of B cells.

6 Increased body temperature (fever) is part of the body's innate immune response. Fever is uncomfortable and can be dangerous if very high. It is often treated with over-the-counter medicines like acetaminophen, ibuprofen, naproxen, or aspirin. What are possible negative effects of this treatment?

Leveling Up

7 **Life choices** While clotting is an important component of the innate immune response, it can also be dangerous. For example, a blood clot could block an artery to the heart or brain, leading to a heart attack or stroke. Aspirin reduces blood clotting by interfering with the body's production of a lipid called thromboxane A2. This lipid normally helps platelets clump together (see Figure 25.10), so aspirin, by inhibiting its production, reduces clotting and "thins the blood." Some doctors may prescribe a daily dose of aspirin for patients who are at risk of heart attack or stroke. Review the U.S. Preventive Services Task Force 2016 recommendations on daily aspirin therapy (https://www.uspreventiveservicestaskforce .org/Page/Document/RecommendationStatementFinal/aspirin-to -prevent-cardiovascular-disease-and-cancer), and then answer the following questions.

a. Do you fall into one of the categories for which daily aspirin therapy is recommended? If yes, which one? If no, is there an aspirin therapy category that you think you'll be in eventually?

b. How strong is the evidence supporting aspirin therapy in the category you identified in the previous question, if any? (See the "Grade" column in the task force recommendations.)

c. With this information in hand, do you plan to take aspirin daily at some point in your life?

d. Will you speak with your doctor before taking aspirin daily? Why or why not?

e. Do you know anyone who has had a heart attack or stroke? Do they take aspirin daily?

8 **Doing science** Pickkers and Kox say that their next experiment will attempt to determine whether one or more of Hof's techniques—cold exposure, meditation, and breathing exercises—is primarily responsible for the increased adrenaline release and decreased immune response. Kox predicts that the breathing exercises will prove to be most important. (You can view a video of cold-exposure training at https://www.youtube.com /watch?v=ziXm9oWJm6A.) Imagine that it is your responsibility to design the next experiment for Pickkers and Kox. Include answers to the following questions in the description of your experimental design.

a. What are your experimental hypotheses?

b. Give at least one prediction for each hypothesis.

c. Identify your control group and treatment group(s). How many subjects will be in each group? Justify your sample size.

d. Give a detailed description of the treatment for each group.

Thinking Multidimensionally

9 If you're one of the millions of allergy sufferers, you know the discomfort of watery eyes and nose, difficulty breathing, and itchy rashes that can follow exposure to allergens. To understand how seemingly harmless substances like dust, pollen, dander, and so on can trigger an allergic response, analyze the mast cell image below.

a. What are the Y-shaped proteins on the surface of the mast cell, and what is their role in an allergic response?

b. When an antigen (such as pollen) enters the body, what gets released from the mast cell and how does this chemical cause the allergy symptoms?

c. How do antihistamines stop an allergic response?

d. What is thought to be the cause of allergies?

e. People who suffer from extreme allergies to peanuts, shellfish, or bee stings can go into anaphylactic shock and possibly die from exposure to these allergens. Draw an outline of a person, and diagram what happens inside the body during an extreme allergic reaction. Provide labels to contrast how extreme allergies differ from mild allergic reactions. In a written explanation, describe the role an EpiPen plays to counteract the symptoms.

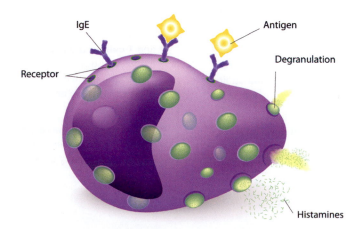

MAST CELL

IgE · Antigen · Degranulation · Receptor · Histamines

lockjaw and body rigidity. Is rust really the culprit? It turns out that tetanus is a bacterium commonly found in soil, dust, and animal feces that reproduces in oxygen-free environments.

a. What do the bacteria release to cause muscle contractions?

b. Draw an image showing how the chemical a tetanus bacterium releases impacts a neuron and then a muscle cell.

c. Every 10 years or after receiving a deep puncture wound, doctors recommend getting a tetanus booster shot. What is contained in the tetanus vaccine, and how does it work?

d. Is a vaccination considered active or passive immunity?

10 You've probably heard that stepping on a rusty nail can cause tetanus, which leads to painful muscle contractions, including

For more, visit digital.wwnorton.com/bionow2 for access to:

Amber Waves of Grain

We've been growing the same domesticated crops for thousands of years. To survive the future, we're going to need new ones.

After reading this chapter you should be able to:

- Identify the structure and explain the function of plant tissues, organs, and organ systems.

- Give an example of how plants use chemical means to survive, grow, and reproduce.

- Compare and contrast how plants and animals grow.

- Explain how a plant obtains a given nutrient.

- Diagram the alternation of generations in a plant life cycle.

- Explain the variety of ways in which plants are pollinated, and how this relates to the structure of flowers.

- Describe how plants disperse their seeds and how fruits have evolved to assist this effort.

PHYSIOLOGY

CHAPTER

26

PLANT
PHYSIOLOGY

DRIVING QUESTION

What are the process and requirements to creating a new food crop?

- HS-LS4.D: Biodiversity and Humans
- HS-LS1.A: Structure and Function
- HS-ETS1.B: Developing Possible Solutions

Lee DeHaan wanders through a field of grain, golden knee-high stalks brushing against his faded jeans. Tall, with an angular face and goatee, DeHaan looks the part of a farmer, with a floppy hat and tanned arms to boot. But though he grew up on a corn farm in Minnesota, DeHaan is now an *agronomist*, a researcher studying the science of producing and using plants for food, energy, land preservation, and more. In these fields around his office at the Land Institute in Salina, Kansas, DeHaan is domesticating a new crop with the potential to transform agriculture—a grain bred from humble prairie grass (**Figure 26.1**).

The way we plant and harvest crops is unsustainable. Nearly 70 percent of the freshwater used by humans is put toward irrigation. Farmland is losing productivity because of deforestation, overgrazing, and poor agricultural practices that lead to erosion and pollution. Staple crop production worldwide has leveled off. All this does not bode well when our population is projected to grow by another 3 billion people over the next half century.

We need new crops. The common crops grown in the United States today—corn, wheat, and soybeans—were domesticated by our ancestors thousands of years ago to produce high yields and be easily harvested and replanted. Yet these crops are failing to meet our sustainability needs. They require large amounts of water, fertilizer, and pesticides, and they are vulnerable to weather changes, pests, and diseases. They are inefficient and delicate; we need food crops that are hardy and resilient.

Unfortunately, humans stopped domesticating new crop plants long ago. So, despite Earth's rich diversity of plants—over 300,000 species, including more than 50,000 species of edible plants—we rely on just 30 of them to provide 95 percent of our food (see "Food Banks" on page 296 in Chapter 16).

More than 250,000 of Earth's plant species are *flowering plants* (also called *angiosperms*). Worldwide, people get over 80 percent of their calories from flowering plants such as grasses (wheat, rice, corn), legumes (peas, beans, peanuts), potatoes, and sweet potatoes. But as climate change affects agriculture—increasing temperatures and severe weather, and causing more disease—and as the global population grows, we're going to need tough, plentiful crops. That's no easy requirement to meet.

Perfecting Plants

In 2001, as a young and ambitious plant breeder, DeHaan joined a research team at the Land Institute, a 600-acre research center devoted to developing sustainable alternatives in agriculture. At the time, he was the young scientist on the team, so his bosses handed him an ambitious long-term project: to create a new type of grain by domesticating a wild grass that grows year after year.

Plants can be grouped into three categories on the basis of their life cycle: annuals, biennials, and perennials.

- **Annuals** complete their entire life cycle in one year. In flowering plants, an annual has one year to grow from a seed into a mature plant, produce flowers, and make the seeds that will start the next generation. Annual crops must be replanted every year.

- **Biennials** grow and mature for a year but do not initially reproduce. Reproduction takes place in the second year of growth. After the second year, biennial crops must be replanted.

Figure 26.1

Harvest at the Land Institute

Researchers gather ripe stalks of a prairie grass that holds hope for feeding our growing population.

- **Perennials** live three years or more, and sometimes for hundreds or even thousands of years. Perennials, which once made up much of the natural grasslands that dominated Earth, are alive year-round and are efficient at nutrient cycling and water management.

Wheat, corn, rice, soybeans—these are all annuals. Our ancestors saw the advantages of breeding annuals: compared with perennials, annuals produce more seed, and replanting them every season speeds the process of domestication. But perennials have advantages too, and it was those advantages that DeHaan wanted to tap into by breeding a perennial as a food crop. First, perennials do not waste energy by regrowing **roots** each year, as annuals do. Instead, they grow long, deep roots that anchor in the soil and remain there year after year. These roots enable perennials to absorb water and nutrients more efficiently than annuals do. The roots also outcompete weeds, so less weed killer is needed to grow perennials. Deep roots also hold carbon in the soil, acting as a *carbon sink*.

Flowering plants can also be classified as either monocots or dicots (**Figure 26.2**).

LEE DEHAAN

Lee DeHaan is an agronomist at the Land Institute in Salina, Kansas, where he leads a new crop domestication program.

Grain-producing plants such as wheat, rice, and corn are **monocots**, with fibrous, branching roots. Other plants, including soybeans and oak trees, are **dicots**, which grow straight, thick taproots.

Roots are part of the plant body, which is relatively simple in its organization, compared to the bodies of vertebrate animals (**Figure 26.3**). Plant bodies are made of three basic tissue types: dermal, ground, and vascular.

- **Dermal tissues**, which form the outermost layer of the plant, protect the plant from the outside environment and control the flow of materials into and out of the plant.

- **Ground tissues**, which form the intermediate layer, make up the bulk of the plant body and perform a wide range of functions, including physical support, wound repair, and photosynthesis.

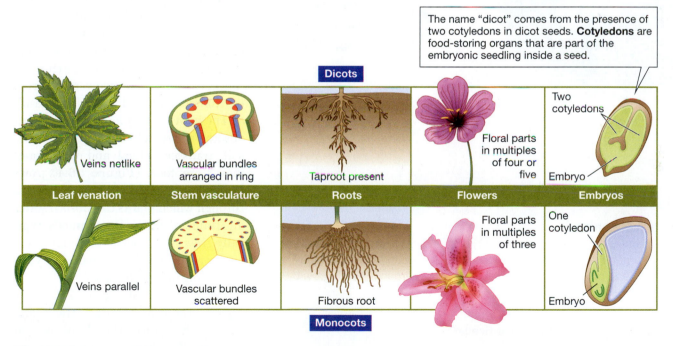

The name "dicot" comes from the presence of two cotyledons in dicot seeds. **Cotyledons** are food-storing organs that are part of the embryonic seedling inside a seed.

Dicots

| Leaf venation | Stem vasculature | Roots | Flowers | Embryos |

Veins netlike — Vascular bundles arranged in ring — Taproot present — Floral parts in multiples of four or five — Two cotyledons / Embryo

Veins parallel — Vascular bundles scattered — Fibrous root — Floral parts in multiples of three — One cotyledon / Embryo

Monocots

Figure 26.2

Monocots and dicots

Flowering plants have traditionally been classified into two main groups—*dicots* and *monocots*—on the basis of their external form and internal structure. The dicots are the larger of these two informal categories and encompass about 175,000 species, including beans, squash, oak trees, and roses. Monocots include all the grasses, members of the lily family, palm trees, and banana plants.

Figure 26.3

How plants are organized

Plants have three basic tissue types (dermal, ground, and vascular) which make up three types of organs (roots, stems, and leaves). Belowground, plants grow by extending old roots and producing new lateral roots. Aboveground, plants grow by adding new bud-stem-leaf units. ▶️

Q1: What is the function of the vascular tissue?

Q2: A plant organ is green if the cells within it contain chloroplasts. In the figure, which plant organ does not contain chloroplasts, and why do you think that is?

Q3: Which tissue type has chloroplast-containing cells? Why?

• **Vascular tissues**, which are found in or near the center of the plant body, contain stacks of long cells forming continuous tubes that run throughout the plant body, linking all organs of the root and shoot systems. The vascular tissue known as **phloem** ships sugars from the leaves, where they are produced, to living cells in every part of the plant; **xylem**, in contrast, transports water and minerals, absorbed from the soil, upward from the roots and outward to the leaves.

DeHaan's attempt to domesticate intermediate wheatgrass (*Thinopyrum intermedium*)—a wild, flowering perennial plant—started as a side project. Widely used for hay and pasture, intermediate wheatgrass grows wild across the western United States and Canada. It has tall, thin shoots that grow green in the fall and turn golden brown in the spring and summer, and long, deep roots that stretch belowground.

The body of a flowering plant, whether intermediate wheatgrass, corn, or roses, can be divided into those two basic organ systems: the belowground *root system* and the aboveground *shoot system*. These two organ systems are specialized for life in two very different environments: roots in soil, shoots in air.

The **root system** anchors the plant, absorbs water and nutrients from the soil, transports food and water, and may store food (**Figure 26.4**). *Root hairs* greatly increase the surface area through which plants can absorb water and mineral nutrients. Annuals typically grow short roots. Perennials grow longer roots, which last year after year, depending on the species.

Roots are one of plants' three basic organs. The other two are stems and leaves, which form the **shoot system** (**Figure 26.5**). **Stems** provide the plant with structural support, transport food and water, and hold leaves up to intercept light. Although cells in the stems of many plants can perform photosynthesis, most sugars are produced by

Figure 26.4

The root system

Roots produce numerous outgrowths from dermal cells called *root hairs*, which aid in water and nutrient absorption. Plant roots have a region of active cell division, protected by the root cap, and a region of cell elongation, in which cells increase in size and complete their development.

Q1: How do root hairs increase the amount of water and nutrients that a plant can absorb?

Q2: How do roots make a plant more stable?

Q3: Given question 2, which do you predict would be more stable in harsh weather conditions: annuals or perennials?

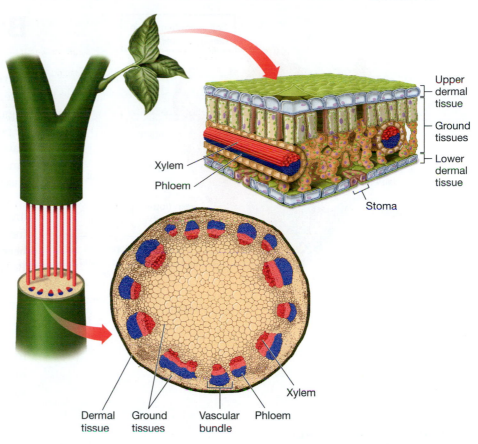

Figure 26.5

The shoot system

Leaves produce the majority of the plant's food through photosynthesis. Stems provide structural support and may perform a limited amount of photosynthesis.

Q1: Which part of the shoot system—the bud, stem, or leaf—produces flowers?

Q2: Find the stoma in the figure. How is its location on the leaf important for its function?

Q3: What nutrients does phloem move from the leaves to other parts of the plant?

photosynthesis in the **leaves**. Leaves provide a broad, sunlight-capturing surface to maximize energy capture. Wheat and intermediate wheatgrass have long, pointy leaves that grow from the stems. At the tip of each shoot, and the base of many leaves, is a bud. Under the right conditions, buds produce new shoots or flowers.

Flowers house the structures that produce male and female gametes and, in many species, also facilitate the delivery of the sperm-bearing pollen to the female reproductive organs

Pollen contains male gametes (sperm).

Petal

Stamen — Anther

Filament

Sepal

Stigma

Style

Carpel

Ovary (future fruit)

Each ovule encloses an embryo sac containing a single egg (the female gamete). The eggs, when fertilized, become seeds.

The ovary develops into a fruit, which is fleshy and juicy in some species.

Ovule (future seed)

Figure 26.6

Four whorls make a flower

The various parts of a flower are arranged in concentric rings, or whorls. From the outermost whorl inward, a typical flower consists of four whorls: sepals, petals, stamens, and carpels. (All of the petals in a flower are collectively known as the *corolla*. The collective term for all of the sepals is the *calyx*.)

(**Figure 26.6**). The ovary of the flower, following fertilization, develops into fruit, which helps disperse seeds in a highly effective ways.

The outer layer of leaves is made up of dermal tissues, which include pores known as **stomata** (singular "stoma") that control gas exchange. Plants open stomata to let in carbon dioxide needed for photosynthesis. When open, stomata release oxygen and lose water through evaporation. Most plants open their stomata in the daytime and close them at night, conserving water when photosynthesis is not an option. A plant experiencing water stress because of an inadequate water supply will close its stomata to conserve water, no matter the time of day. Intermediate wheatgrass and other deep-rooted perennials are less likely to experience water stress, because of their extensive underground network of roots to extract water from the ground. Yet they do often become dormant and close their stomata during the hot summer.

Breeding Begins

Intermediate wheatgrass first came to the attention of agronomists in 1983, when the Rodale Institute, a Pennsylvania research facility that studies organic agriculture, evaluated close to 100 species of perennial grasses for traits including seed size, flavor, and harvestability. Intermediate wheatgrass emerged the winner: it is hardy and currently not susceptible to any major pest or disease.

Disease resistance is important because crops are at the mercy of pathogens. About 20 years ago, for example, a fungal disease called "Fusarium head blight" wiped out Minnesota's barley crop. Because plants are under attack by pathogens and predators, they have developed a rich variety of mechanisms to deter attacks (**Figure 26.7**, left). In addition to physical defenses such as thorns, many plants contain chemical substances that are toxic to herbivores. Nicotine, the addictive chemical in cigarette smoke, protects tobacco plants from insect predators. Caffeine protects the leaves and seeds of coffee and tea plants from potential predators. Plants also circulate and secrete antimicrobial chemicals to kill pathogens that could infect them.

In addition to using chemicals to protect themselves, plants produce chemicals called **hormones** to coordinate internal activities necessary for growth and reproduction (**Figure 26.7**, right). Hormones are active at very low concentrations (see Chapter 25 for more on hormones).

After intermediate wheatgrass proved to be resistant to pests and produced better seeds than other perennial grasses do, the Rodale Institute began collecting strains of the plant from all around the world. Over 12 years, researchers bred the strains, selected the best 20, and then identified the top 14 individual plants.

At the Land Institute, DeHaan received seed from the 14 Rodale plants and began breeding them, tinkering with the possibilities of intermediate wheatgrass. In his first year, DeHaan planted some 3,000 plants in an effort to create a large group of diverse individuals. Then, he watched the grass grow.

Plants can increase in length by **primary growth**, in which the plant's shoots and roots lengthen. They can also grow in thickness, called **secondary growth**. Secondary growth includes the thickening of both stems and roots.

Unlike most animals, which grow early in their lives and then level off in a pattern

1 A tough outer surface protects the plant from water loss and heat and cold stress, and serves to block pathogens from entry. Thorns and other defenses keep predators away.

2 Plants deploy nonspecific chemical defenses, which involve a range of broadly targeted antipathogen and antiherbivore chemicals, including hormones that signal other parts of the plant to manufacture chemical weapons and have them "battle ready."

3 Plants also have specific chemical defenses. For example, they have a number of genes that respond to complementary genes in a pathogen. If a specific pathogen is detected, protective defenses including toxic chemicals are unleashed.

1 **Auxins** are necessary for cell division and the formation of organs such as roots. They are involved in *phototropism*, the growth of shoots toward the light, and in *gravitropism*, the growth of roots toward the ground and the growth of shoots away from the ground.

2 **Cytokinins** are necessary for cell division, and they promote shoot formation. The levels of both cytokinins and auxins decline sharply just before plants drop their leaves in the fall season.

3 **Gibberellins** bring about stem growth through both cell elongation and cell division. They also stimulate seeds to germinate.

4 **Abscisic acid (ABA)** mediates adaptive responses to drought, cold, heat, and other stresses.

5 **Ethylene** stimulates the ripening of some fruits, including apples, bananas, avocadoes, and tomatoes. Ethylene activates enzymes that convert starches into sugars, resulting in a sweeter fruit.

Figure 26.7

Plants use chemicals to survive, grow, and reproduce

Plants use chemicals to protect themselves from both living and nonliving threats in their environment. And like animals, plants use hormones to coordinate internal activities necessary for growth and reproduction.

Q1: Give an example of how plants use chemicals to defend themselves.

Q2: Of the plant hormones introduced in this figure, which promote growth?

Q3: What is the first line of plant defense? Is this a chemical defense?

termed **determinate growth**, most plants grow throughout their lives—a pattern called **indeterminate growth**. Also unlike most animals, plants are able to grow through **modular growth**, the repeated addition of "modules" of budstem-leaf units aboveground or new lateral roots belowground.

Indeterminate and modular growth habits give plants great flexibility to respond to changing environmental conditions, such as high or low levels of sunlight, water, or nutrients. Plants tend to add many new parts when conditions are favorable and few new parts when conditions are not. This flexibility also enables plants to replace damaged tissues and organs. In fact, plants are so flexible in their development that most living cells in the adult plant body can generate whole new plants.

When it was time to harvest the intermediate wheatgrass, DeHaan and his team went around their field and recorded traits of the plants, such as their heights and how early they flowered. Then the team plucked the heads off the plants and placed them in bar-coded bags. Back in the lab, technicians analyzed the seeds from each plant, recording size and weight—the larger the seed, the more grain can be produced for food—and noting how easily the seeds shed their outer husks, because the less sticky a grain's husk is, the easier it is to process that grain.

DeHaan's first experimental crop grew plants of many sizes and traits. "It's like scratching off lottery tickets," DeHaan told National Public Radio in 2009. "Maybe there's something amazing in there. We'll see." As he recorded traits, DeHaan found that most of his best performers—those with larger seeds and deeper roots—came from a few families, so DeHaan went back into the field and uprooted those specific plants. These were brought into the warm greenhouse for the winter and bred with each other. Their offspring were then planted the following fall, and the process repeated.

But the more DeHaan bred these particular plants, the more he restricted their gene pool, creating a genetic bottleneck by excluding genes that had been in the original population of 3,000 plants. He began to worry that he might have lost genes needed to improve certain traits. One particular trait he struggled to identify and retain in his crop was the ability to mature early in the season.

Many plants, especially species native to temperate regions, perceive the seasons by sensing the length of the day. This is possible because day length varies with the season: days are shorter in winter and longer in summer. This sensing of the duration of light and dark in a 24-hour cycle is known as **photoperiodism**. Plants use night length to sense when conditions are favorable for flowering and seed germination. The dormancy of buds through fall and winter, as well as their regrowth in spring, is also influenced by photoperiodism, as well as changes in temperature and precipitation. DeHaan wanted to identify wheatgrass that flowered early in the spring, so from the U.S. Department of Agriculture (USDA) he obtained hundreds of wild collections of wheatgrass that mature early, and those that flowered early he crossed with his own plants.

By 2010, DeHaan had something good: he had doubled both the seed yield and the weight of seeds of his intermediate wheatgrass. "It was succeeding a lot faster than we thought," says DeHaan. "It was really easy to grow." Unfortunately, "intermediate wheatgrass" doesn't sound like a tasty grain one might like in cereal or bread. So DeHaan and his team renamed the newly domesticated grain "Kernza," a combination of the word "kernel" and the name of a native tribe of the region, Kanza (which also inspired the state name "Kansas").

Kernza® perennial grain is different from wild intermediate wheatgrass in several ways. First, the seeds are larger. "When I started working with it, the typical seed weighed 3.5 milligrams," DeHaan said in 2010. "Now, our best seeds are 10 milligrams." That's progress, but there's more to be done: an average wheat seed weighs 35 milligrams (**Figure 26.8**).

Wheat seeds are larger and heavier.

Kernza seeds are smaller and lighter.

Figure 26.8

Wheat and Kernza seeds, side by side

Kernza growers hope to eventually breed plants with seeds as large as those of wheat. They still have a long way to go.

Figure 26.9

Kernza roots grow deeper than wheat roots

Kernza's longer roots (right) make the plant better able to collect water from surrounding soil, and to store carbon and nitrogen that would otherwise be lost to the air or waterways.

Then there are the roots. Kernza's roots grow down to 10 feet (**Figure 26.9**). The deep roots make the plant very efficient at sucking water out of the soil. That makes it more resistant to climate change than annual crops are, and the plant has already been shown to fare better in drought conditions than traditional wheat does. Furthermore, the long roots hold soil together, preventing erosion, which carries away fertile soil, fertilizers, and pesticides with it. Major crop-producing countries all around the world lose tens of billions of tons of topsoil every year to erosion.

Kernza's long roots have additional benefits. A recent study found that within 2 years of planting, Kernza beat annual wheat at accessing groundwater, accumulating carbon in the soil, and absorbing nitrogen fertilizer. To grow, plants need CO_2, water, and mineral nutrients, especially nitrogen, phosphorus, and potassium. Most of the dry weight of a plant comes from CO_2 absorbed from the air and converted into carbohydrates by photosynthesis. Plants require **macronutrients** (nitrogen, phosphorus, potassium, calcium, sulfur, and magnesium) in relatively large amounts and **micronutrients** (including iron, zinc, and copper) in small amounts. Carbon, oxygen, and hydrogen are obtained from air or water; the rest of the nutrients that plants need must be obtained from soil. In agriculture, farmers add most of the nutrients to the soil as fertilizer, but fertilizer can run off into water, polluting it. Perennial plants make better use of fertilizer than do annuals, by efficiently retaining and recycling it. In the same study mentioned earlier, Kernza reduced the amount of nitrogen leached into the nearby ecosystem by up to 99 percent compared with wheat.

Crop Collaboration

Large seeds and deep roots are good traits, but they aren't enough to make Kernza a successful crop. Other specialty grain crops yield about 1,000 pounds per acre. In Kansas, Kernza was yielding less than 300 pounds per acre. So DeHaan sought help. Around 2007, DeHaan took his Kernza plants north to the University of Minnesota, where he had once been a student. He began working with agronomist Donald Wyse and other wheat breeders at the university. Planting in Minnesota turned out to be a boon: Kernza had larger yields there because intermediate wheatgrass is a cool-season grass and performs better in a cooler place.

DONALD WYSE

Donald Wyse develops and promotes winter annual and perennial crops at the University of Minnesota, where he is a professor of agronomy and plant genetics.

Wyse was an ideal collaborator. He had spent years breeding perennial and winter annual crops for agriculture in Minnesota, including perennial flaxseed and perennial sunflower. One of the great values of a perennial, in addition to its having deep roots that prevent erosion and requiring less fertilizer and pesticide, says Wyse, is that its roots and ground tissues are active during a longer period of the year, absorbing more solar energy and thus increasing an area's net primary productivity (see Chapter 21 for a discussion of NPP). This, he says, is "high-efficiency agriculture."

In Minnesota, for example, the roots of corn and soybeans are active for just 2½ months a year, usually after the heaviest rains. That means all the sunlight from the other 9½ months of the year is wasted. But perennials can perform photosynthesis from the moment the snow melts in the spring until the first snowfall of the winter, says Wyse. With perennials, he says, "we're harvesting more energy."

In 2011, DeHaan, Wyse, and the Minnesota team planted a field of more than 2,000 intermediate-wheatgrass plants from 69 families and then measured traits like biomass, grain yield, and seed shape. They've continued this process every year, sometimes pairing complementary strengths and weaknesses. In the summer of 2014, they planted 14,000 seedlings.

The domestication continues, and it is not easy. DeHaan estimates it will be another 10 years before Kernza is ready to compete economically with wheat. One of the challenges of breeding intermediate wheatgrass has to do with how it reproduces. Many plants, such as dandelions and poplar trees, can reproduce asexually, when a parent plant forms a genetically identical clone. Some crops, including potatoes, apples, and grapes, are also propagated as clones. But staple crops, including grains, reproduce sexually.

The overall principle of sexual reproduction in plants is similar to that in animals: A haploid male gamete (*sperm*) fuses with a haploid female gamete (*egg*) to give rise to a diploid cell, the *zygote*, which undergoes mitosis to create a multicellular diploid *embryo*. In time, the embryo develops into an individual offspring, which represents the next generation (**Figure 26.10**).

Plant and animal life cycles differ in one key respect: In animals, meiosis creates gametes and nothing but gametes. In plants, meiosis generates haploid cells called **spores**. Each spore undergoes mitotic divisions to create a haploid, multicellular **gametophyte**. Specialized cells in the gametophyte differentiate to produce sperm or egg cells.

In an animal, gametes are the only haploid cells, and animals have no such thing as a gametophyte. However, gametophytes are part of the life cycle of every plant. In plants like mosses, the gametophytes are a major multicellular phase of the life cycle. In flowering plants like Kernza, gametophytes are reduced to several cells contained in flowers.

Once fertilized, the zygote develops into an embryo inside the ovule, which is contained within the ovary, at the base of the flower. The outer layers of the ovule harden into a protective seed coat. Each seed contains the ingredients for growing a young plant of the next generation: a mature embryo, a food source, and the seed coat.

The ovary surrounding the seed forms a *fruit*, yet another plant organ. Once a fruit is formed, the embryo enters dormancy, and the seed is dispersed from its parent (**Figure 26.11**). Seeds are often dispersed by wind, by attaching to the fur of animals, or by being eaten (and excreted) by animals. Dormancy ends when the embryo is stimulated by favorable conditions to start growing again. Then the seed germinates, and a seedling grows into a plant that will mature and produce flowers and fruits, the beginning of another generation. This adult plant is a **sporophyte**, analogous to an individual animal in that both are multicellular, diploid organisms.

Annual wheat plants *self-pollinate*: they reproduce by fertilizing themselves with their own pollen to create a zygote before their flowers even open. But intermediate wheatgrass, and many other plants, do not self-pollinate. They require a mate. Given that plants cannot travel to find mates, how does sperm-containing pollen from one plant reach the eggs of another plant?

In some species, like grasses and pine trees, pollen is transported by wind. Wind can carry pollen long distances, but most pollen blown by wind lands in inhospitable places (such as parking lots or lakes), not on the flower of another plant of the same species. Many flowering plants

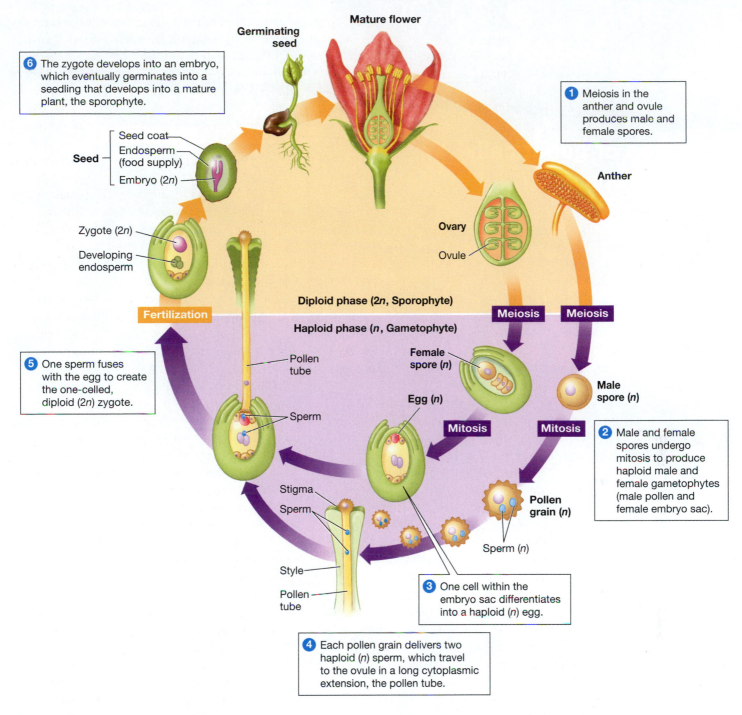

Mature flower

Germinating seed

6 The zygote develops into an embryo, which eventually germinates into a seedling that develops into a mature plant, the sporophyte.

1 Meiosis in the anther and ovule produces male and female spores.

Anther

Seed
- Seed coat
- Endosperm (food supply)
- Embryo (2n)

Zygote (2n)

Developing endosperm

Ovary

Ovule

Diploid phase (2n, Sporophyte)

Fertilization

Meiosis **Meiosis**

Haploid phase (n, Gametophyte)

Pollen tube

Female spore (n)

5 One sperm fuses with the egg to create the one-celled, diploid (2n) zygote.

Sperm

Egg (n)

Male spore (n)

Mitosis **Mitosis**

2 Male and female spores undergo mitosis to produce haploid male and female gametophytes (male pollen and female embryo sac).

Stigma

Sperm

Pollen grain (n)

Sperm (n)

Style

Pollen tube

3 One cell within the embryo sac differentiates into a haploid (n) egg.

4 Each pollen grain delivers two haploid (n) sperm, which travel to the ovule in a long cytoplasmic extension, the pollen tube.

Figure 26.10

From generation to generation

The life cycle of flowering plants is marked by **alternation of generations**. Haploid (n) stages of the life cycle are shown in purple; diploid (2n) stages, in orange.

Q1: Are eggs and sperm haploid or diploid? Are they part of the sporophyte or gametophyte generation?

Q2: Why is the plant life cycle, but not the animal life cycle, referred to as an "alternation of generations"?

Q3: How does asexual reproduction differ from the plant life cycle diagrammed here?

GM Crops Take Root

Since 1996, the total land area used to grow genetically modified (GM) crops has expanded dramatically. Most of these crops, including the varieties depicted here, are enhanced with the insertion of a gene that imparts insect resistance or herbicide tolerance. Broad scientific consensus has concluded that food derived from GM crops poses no greater risk to health than food from nonmodified crops.

Assessment available in smartwork5

Global area of genetically modified crops

Area shown in million hectares

Soybean   Maize   Cotton   Canola

1998
14
3
3
8

2000
26
6
3
10

2002
36
7
3
12

2004
48
9
5
19

2006
58
13
6
25

2008
65
15
7
37

2010
74
20
8
46

2012
80
24
9
56

By wind

Milkweed

Dandelion

Maple

By water

Lotus

Cattail

Coconut

By animals

Sandbur

Beggar-ticks

Blackberry

By humans

(artificial selection)

Bean

Wheat

Cherry

Figure 26.11

Plants spread their seeds in a variety of ways

Many plants spread their seeds via the wind or water, but others use modified seeds and fruit to attract animals or attach to their fur. Humans have artificially selected the fruit and seeds of many species to make them more edible.

Q1: What kinds of seeds would you expect to be contained within a sugary fruit: seeds spread by water or by animals? Why?

Q2: What kinds of seeds would you expect to weigh less: those spread by air or those artificially selected by humans? Why?

Q3: How does the height of a tree affect its ability to spread its seeds widely?

sidestep this problem by having their pollen transported by animals such as insects, birds, and even mammals (**Figure 26.12**). Plants attract these **pollinators** with brightly colored, sweet-smelling flowers filled with sugary nectar. It is a mutualistic relationship, in which both species benefit.

DeHaan and his team need to mate specific plants together in order to breed for specific traits. So, instead of relying on wind or pollinators, they put a bag over the flowering stalks of two plants they want to breed so that pollen from one plant passes directly to the second plant in the bag, and no others (**Figure 26.13**). "We call them 'plant condoms,'" says DeHaan with a laugh. "The bags keep crosses we don't want to happen from happening."

And because an intermediate-wheatgrass plant can breed with only another plant—rather than with itself, as annual wheat can do—the offspring are genetically diverse; no child is *exactly* like its parent. In fact, a child can be very different from its parents, because there is a great deal of variation in the genome of intermediate wheatgrass. Each gene may have six or more versions, or alleles, in a single plant, says DeHaan. So, when breeding new plants, the team can never match the exact genotype and phenotype of a parent but rather must constantly juggle traits to try to capture the right set of traits in a single line of plants. For example, DeHaan is currently working to grow shorter plants, because taller plants tend to tip over, which makes them difficult to harvest.

The distinctive colors, shapes, and smells of flowers often attract highly specific pollinators that are efficient pollen dispersal agents.

Honeybees carry the pollen that dusts their bodies from flower to flower as they search for food.

Birds have good color vision and favor red flowers with long floral tubes.

Figure 26.12

Bribing animals to do the work

Pollinators provide stationary plants with a way of transporting sperm to eggs. The spectacular colors, shapes, and odors of flowers, in combination with food rewards such as nectar, lure pollinating animals into visiting several flowers of the same species, incidentally transferring pollen in the process.

Q1: Many crops, including apples, peppers, and tomatoes, depend on insect pollinators. What would happen if no pollinator visited a plant?

Q2: Although they look very different, flower petals are actually modified leaves. Do flowers still perform the main function of leaves? Explain your reasoning.

Q3: Some flowers look like an insect. How might this resemblance attract pollinators?

Perennial Pancakes

Early on, DeHaan estimated it would take some 50 years for a new plant to be domesticated to match the yield of wheat, but now USDA and University of Minnesota researchers are

Figure 26.13

Breeding the next generation of Kernza plants

The flowering stalks of Kernza plants are enclosed in bags to prevent the pollen from spreading to any other plants. Scientists identify plants with the traits they want to pass on to the next generation and selectively breed them in this way.

Q1: What traits are scientists selecting for in each new generation of Kernza?

Q2: What would happen if the Kernza plants were allowed to freely pollinate, rather than being selectively pollinated by hand?

Q3: After choosing two plants to breed, researchers put the plant heads together under a bag as shown in the photo. Why do they do this before the plants begin to pollinate?

supplementing his hands-on breeding work with gene sequencing efforts. These scientists have been associating genetic markers with particular traits, such as seed size. Once they have enough traits mapped on the genome, they will no longer have to wait for a plant to mature to observe its phenotype; instead they will simply sample the DNA of young seedlings and predict the plant's traits on the basis of genetic markers.

"Within a very short period of time, we'll be able to identify key genes in the domestication process, and at a very low cost," says Wyse. With these tools, Wyse thinks the breeders won't take too long to match the yield of wheat. "It may only take half a decade to see dramatic improvements in these new crops," says Wyse. "We think crops of this type are going to have great value."

But there's more to a food crop than how well it grows and can be harvested. It also has to taste good. DeHaan has used Kernza flour to make cookies, cakes, scones, bread, and more (**Figure 26.14**). He and other Land Institute staff have enjoyed them, though at least one food scientist says the flour will need some flavor improvement to go mainstream. A team at the University of Minnesota is currently experimenting with Kernza's food traits, including taste, texture, and ability to rise.

Both beer brewers and sustainable food companies have expressed an interest in using the new grain. DeHaan and Wyse also imagine that the crop by-product could be used as hay or to produce biofuel. "We'd love this to be a dual-use crop," says DeHaan. Farmers, for example, could harvest the grain for a food company and then use the "residue" plant material in the field, typically the leftover bottom of the stalk, to make biofuel. "Those two things together could result in an economically viable crop," he says.

Figure 26.14

Kernza flour

Scientists are finding uses for Kernza in traditional baking, and also using it to make alcohol, animal feedstock, and hopefully in the future, even biofuel.

REVIEWING THE SCIENCE

- Plant bodies are made of three basic tissue types: **Dermal tissues** control the flow of materials into and out of the plant and protect the plant from attack. **Ground tissues** make up the bulk of the plant body and participate in support, wound repair, and photosynthesis. **Phloem** and **xylem** are the two types of **vascular tissue**, transporting sugars and water, respectively, through the entire plant.

- The plant body contains two basic organ systems: the belowground **root system** and the aboveground **shoot system**. Plant organs include roots, stems, leaves, flowers, and fruits. **Roots** absorb water and mineral nutrients, anchor the plant, and store food. **Leaves** produce the majority of the plant's food through photosynthesis. **Stems** support the plant and may perform a limited amount of photosynthesis.

- Plants defend themselves against the external environment with a tough outer covering, physical weapons such as thorns, and an arsenal of toxic chemicals.

- Like animals, plants use **hormones**, chemicals produced at low concentrations, to coordinate growth and reproduction.

- Plants have **indeterminate growth**: they can grow throughout their lives. Plants grow by adding repeating bud-stem-leaf units aboveground and root units below; a pattern called **modular growth**.

- Aboveground, plants increase in length by **primary growth**, the lengthening of shoots and roots. Plants increase in thickness by **secondary growth**, the thickening of stems and roots.

- To grow, plants need CO_2, water, and mineral nutrients. **Macronutrients** (especially nitrogen, phosphorus, and potassium) are required in large quantities.

- Many plants can reproduce asexually, but most can also reproduce sexually. Meiosis in plants produces a single-celled **spore**, which divides through mitosis to create a haploid, multicellular **gametophyte**. Cells in the gametophyte differentiate into egg or sperm. The fertilization of the egg by sperm generates a diploid zygote, which gives rise to the diploid, multicellular **sporophyte**. This **alternation of generations** (gametophyte and sporophyte) is a hallmark of plant life cycles.

- Fertilization produces a zygote, which divides and develops into an embryo. The embryo is located within the ovule, which hardens into a protective seed coat. Each seed contains the ingredients for growing a young plant of the next generation: a mature embryo, a food source, and the seed coat.

- In flowering plants, male and female reproductive parts are contained in flowers. Flowers attract animal **pollinators**, which provide immobile plants with a way of transporting sperm-containing pollen to eggs.

THE QUESTIONS

Challenge Yourself

1 Which of the following is *not* a plant organ?

(a) stem

(b) fruit

(c) vascular bundles

(d) root

(e) flower

2 Which *micro*nutrient do plants need to absorb from the environment to survive and grow?

(a) carbon

(b) sugar

(c) potassium

(d) nitrogen

(e) zinc

3 Select the correct terms:

(**Annual / Biennial**) plants complete their life cycle in one year, whereas (**biennial / perennial**) plants live for three or more years. Wheat is a(n) (**annual / perennial**), whereas Kernza is a(n) (**annual / perennial**). Kernza is also a (**monocot / dicot**), which has a (**fibrous root / taproot**).

4 Beginning with the stage following spores, place the following steps of the plant life cycle in the correct order by numbering them from 1 to 4.

____ a. Zygote develops into an embryo.

____ b. Gametophyte develops.

____ c. Seed germinates.

____ d. Egg is fertilized.

Try Something New

5 Some plants are pollinated by bats, which are active at night. What would you predict about how bat-pollinated flowers differ from bird-pollinated flowers? (*Hint:* Remember that birds are active during the day.)

6 Ethylene is a plant hormone that causes fruits to ripen by converting starches into sugars. It also signals flowers to open, seeds to germinate, and leaves to shed. You may have observed that placing a ripe banana or apple next to another fruit causes it to ripen more quickly. Furthermore, if you place fruit in a paper bag (rather than exposed on the counter), it ripens more quickly. How would you explain this?

7 Several of the plants domesticated by humans have highly modified organs. For example, potatoes are modified stems, and carrots and sweet potatoes are modified roots. From which plant organ do you think onions and tomatoes are modified? Why?

Leveling Up

8 **Looking at data** The debate over genetically modified (GM) crops and genetically modified organisms (GMOs) has been impassioned but not always informed. Genetically modified crops are those with DNA altered through biotechnology, in order to increase productivity or resistance to disease.

More than 93 percent of the soybeans harvested in the United States, fed almost exclusively to domesticated animals,

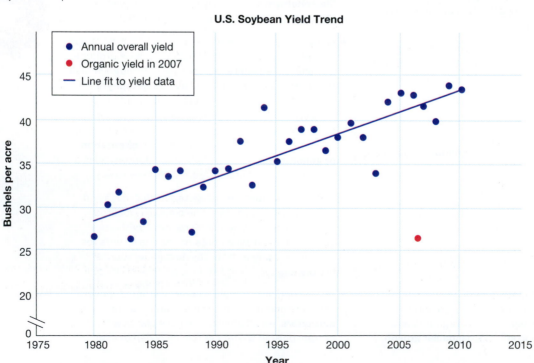

U.S. Soybean Yield Trend

are genetically modified for pesticide and herbicide resistance. Examine the accompanying graph, which shows the yield in bushels per acre of soybeans from 1980 to 2010. The red circle is the yield for organic soybeans in 2007, which is based on information from 1,331 organic farms.

a. What does the x-axis represent? The y-axis? What does the data line on the graph represent?

b. Describe the trend in soybean productivity from 1980 to 2010.

c. The yield for organic soybeans in 2007 is only 66 percent of the average yield for that year. What year of conventional soybean production does this production match?

d. Critics say that the widespread growth and consumption of GM crops may have unexpected health and environmental costs. Proponents argue that genetic modification is the only way to feed a rapidly growing population. What do you think?

9 **Doing science** Although you are an expert on Kernza® perennial grain, you have been asked to contribute your scientific expertise to developing another perennial crop, *Silphium integrifolium* (sometimes referred to as a "perennial sunflower"), that could replace annual crops grown as sources of cooking oil.

a. Read the Land Institute's overview of its research on *Silphium*: http://landinstitute.org/our-work/perennial-crops/silphium.

b. The *Silphium* researchers say that they aim to identify critical ways in which *Silphium* responds to changes in the environment each year, but that they will need to follow up such observations with experiments. Are the scientists conducting observational or experimental research on *Silphium*? Explain your answer.

c. Propose a hypothesis about an environmental variable that you believe might increase oil production in *Silphium*, and then identify a testable prediction from that hypothesis. Design an experiment to test the prediction, identifying your independent and dependent variables, the control group, and the treatment conditions. Create a graph to show the results you expect to find (1) if your hypothesis is supported and (2) if your hypothesis is not supported.

Thinking Multidimensionally

10 Historically, plants have been used to treat illness and even today still continue to be a source of prescribed and over-the-counter medications. In 1826, the opiate morphine was isolated from the milky sap of poppy plants and used to relieve pain after war injuries and surgery.

a. What illegal street drug is also derived from poppy plants, and how is it created?

b. Acetylsalicylic acid, commercial name aspirin, is derived from the herb meadowsweet. It is used to relieve pain and acts as a blood thinner. Explain how aspirin is able to block pain perception and prevent the blood from clotting.

c. While not medicinal, caffeine's origins can be traced to cacao, coffee, and tea plants. Why did these three plants evolve this central nervous system stimulant?

11 The first flower to pop through the snow in the Northeast belongs to a plant called skunk cabbage. Using a form of accelerated cellular respiration, these plants actually generate heat up to 90°F to melt the snow around them.

a. In what plant organ does skunk cabbage store a large supply of sugar to power thermogenesis, and why did it evolve this heat-making ability?

b. Named skunk cabbage for the rotting-flesh odor the flower releases, what pollinator is the skunk cabbage is trying to attract?

c. Make an argument for or against the claim that skunk cabbage is warm-blooded.

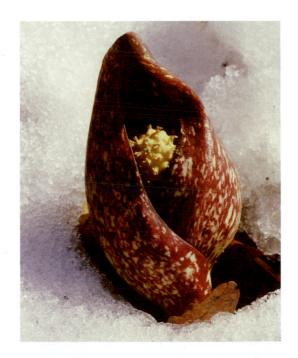

For more, visit digital.wwnorton.com/bionow2 for access to:

Glossary

ABA See **abscisic acid**.

abiotic Nonliving. Compare *biotic*.

abscisic acid (ABA) A plant hormone that mediates adaptive responses to drought, cold, heat, and other stresses.

absorption The uptake of mineral ions and small molecules by cells lining the cavity of the digestive tract.

acid A chemical compound that loses hydrogen ions (H^+) in aqueous surroundings. Compare *base* (definition 1).

actin filament One of the two kinds of protein filaments, made of the protein actin, that is found in sarcomeres. The sliding of *myosin filaments* against actin filaments enables sarcomeres to contract.

active immunity Immunity to a particular pathogen that is conferred by antibodies made by the body itself. Compare *passive immunity*.

active site The location within an enzyme where substrates are bound.

active transport The movement of a substance in response to an input of energy. Compare *passive transport*.

adaptation 1. An evolutionary process by which a population becomes better matched to its environment over time. 2. See **adaptive trait**.

adaptive immune system The immune system's third line of defense against pathogens, which mounts responses against specific invaders via antibodies (antibody-mediated immunity) and phagocytes and other specialized cells (cell-mediated immunity). Compare *external defenses* and *innate immune system*.

adaptive radiation The expansion of a group of organisms to take on new ecological roles and to form new species and higher taxonomic groups.

adaptive trait Also called an *adaptation*. A feature that gives an individual improved function in a competitive environment.

adenine (A) One of the four nucleotides that make up DNA. The other three are thymine (T), guanine (G), and cytosine (C).

ADP Adenosine diphosphate.

adrenal glands The paired glands, located atop the kidneys, that release the hormones cortisol, adrenaline (epinephrine), and noradrenaline (norepinephrine), which launch a number of rapid physiological responses, including boosting blood glucose levels.

aerobic Requiring oxygen. Compare *anaerobic*.

algae One of two main groups of protists, whose members are photosynthetic and may or may not be motile. Compare *protozoans*.

alleles Different versions of a given gene.

allopatric speciation The formation of new species from geographically isolated populations. Compare *sympatric speciation*.

alternation of generations The life cycle of flowering plants, in which haploid stages (gametophytes) alternate with diploid stages (sporophytes).

alveoli (sing. alveolus) Small clusters of minute sacs resembling a bunch of grapes where gases are exchanged in the lungs.

amino acid Any of a class of small molecules that are the building blocks of proteins.

amniocentesis A prenatal genetic screening technique in which amniotic fluid is extracted from the pregnancy sac that surrounds a fetus by means of a needle that is inserted through the abdomen into the uterus. Compare *chorionic villus sampling*.

anabolism Metabolic pathways that create complex biomolecules from smaller organic compounds. Compare *catabolism*.

anaerobic Not requiring oxygen. Compare *aerobic*.

analogous trait A feature that is shared across species because of convergent evolution, not because of modification by descent from a recent common ancestor. Compare *homologous trait*.

analytical study An observational study that looks for patterns in the information collected and addresses how or why those patterns came to exist. Compare *descriptive study*.

anatomy The study of the structures that make up a complex multicellular body. Compare *physiology*.

anchorage dependence The phenomenon in which cells stop dividing when detached from their surroundings.

androgens Any of the hormones produced in the testes that stimulate cells to develop characteristics of maleness, such as beard growth and the production of sperm. Compare *estrogens* and *progestogens*.

angiogenesis The formation of new blood vessels.

angiosperms Flowering plants. One of two main groups of seed-bearing plants, characterized by seeds enclosed in an ovary and by flowers. Compare *gymnosperms*.

Animalia The animals. One of the six kingdoms of life, in the domain Eukarya, encompassing all animals, including humans, birds, and dinosaurs.

annual A plant that completes its life cycle in one year. Compare *biennial* and *perennial*.

antibody Any of various Y-shaped proteins that recognize invaders by their *antigens* and attack them in the antibody-mediated response of the adaptive immune system.

antibody-mediated immunity A response mounted by the adaptive immune system in which specific invaders are attacked by antibodies. Compare *cell-mediated immunity*.

anticodon A unique sequence of three nitrogenous bases at one end of a tRNA molecule that binds to the corresponding *codon* on an mRNA molecule.

antigen Any of various nonself markers found on pathogens that mark them for destruction by *antibodies*.

anus The muscle-lined opening through which solid waste is eliminated from the body.

appendicular skeleton The part of the skeleton that has to do with motion. It is made up of the arms, legs, and pelvis. Compare *axial skeleton*.

applied research Research in which scientific knowledge is applied to human issues and often commercial applications. Compare *basic research*.

Archaea One of the three domains of life (compare *Bacteria* and *Eukarya*) and also one of the six kingdoms of life. The domain and kingdom Archaea consists of single-celled organisms best known for living in extremely harsh environments.

artery Any of the large vessels that transport blood away from the heart. Compare *capillary* and *vein*.

artificial selection The process by which individuals display specific traits through selective breeding. Compare *natural selection*.

ascomycetes Sac fungi, one of three main groups of fungi. Compare *basidiomycetes* and *zygomycetes*.

asexual reproduction The process by which clones, offspring that are genetically identical to the parent, are generated. Compare *sexual reproduction*.

atom The smallest unit of an element that retains the element's distinctive properties. Atoms are the building blocks of all matter.

atomic mass number The sum of the number of protons and the number of neutrons in an atom's nucleus. Compare *atomic number*.

atomic number The number of protons in an atom's nucleus. Compare *atomic mass number*.

ATP Adenosine triphosphate, a small, energy-rich organic molecule that is used to store energy and to move it from one part of a cell to another. Every living cell uses ATP.

atrioventricular (AV) node The part of the heart that sends a signal from the atria to the ventricles telling the ventricles to contract. Compare *sinoatrial node*.

atrium (pl. atria) Either of the two upper chambers of the heart. Compare *ventricle*.

autosome Any chromosome that is not one of the *sex chromosomes*.

autotroph A metabolic producer, an organism that makes food on its own. *Chemoautotrophs* acquire their energy from inorganic chemicals in their environment. *Photoautotrophs* absorb the energy of sunlight and take in carbon dioxide to conduct photosynthesis. Compare *heterotroph*.

auxin Any of a group of plant hormones that are necessary for cell division and promote root formation. Compare *cytokinin* and *gibberellin*.

AV node See **atrioventricular node**.

axial skeleton The part of the skeleton that supports and protects the long axis of the body. It includes the skull, the ribs, and a long, bony spinal column. Compare *appendicular skeleton*.

axon The part of a neuron that sends information to other cells. Compare *dendrite*.

B

B cell A type of lymphocyte that matures in the bone marrow. B cells are involved in antibody-mediated immunity. Compare *T cell*.

Bacteria One of the three domains of life (compare *Archaea* and *Eukarya*) and also one of the six kingdoms of life. The domain and kingdom Bacteria includes familiar disease-causing bacteria.

base 1. A chemical compound that accepts hydrogen ions (H^+) in aqueous surroundings. Compare *acid*. 2. The nitrogen-containing component of nucleotides. The four nitrogenous bases of DNA are adenine, cytosine, guanine, and thymine. In RNA, thymine is replaced by uracil.

base pair Also called *nucleotide pair*. Two nucleotides that form one rung of the DNA ladder.

base-pairing rules The rules that govern the pairing of nucleotides in DNA. Adenine (A) on one strand can pair only with thymine (T) on the other strand, and cytosine (C) can pair only with guanine (G).

basic research Research that is intended to expand the fundamental knowledge base of science. Compare *applied research*.

basidiomycetes Club fungi, one of three main groups of fungi. Compare *ascomycetes* and *zygomycetes*.

behavioral trait A characteristic of an individual's behavior, such as shyness or extroversion. Compare *biochemical trait* and *physical trait*.

benign Referring to a tumor, confined to one site.

bias A prejudice or opinion for or against something.

biennial A plant that takes two years to complete its life cycle, growing and maturing in the first year and reproducing in the second year. Compare *annual* and *perennial*.

bilateral symmetry An animal body plan in which the body can be divided by just one plane passing vertically from the top to the bottom of the animal into two halves that mirror each other. Compare *radial symmetry*.

bile A fluid, produced by the liver, that helps digest fats by creating a coating that enables the fat globules to interact with water molecules to partially dissolve them.

binary fission A type of cell division in which a cell simply divides into two equal halves, resulting in daughter cells that are genetically identical to each other and to the parent cell.

biochemical trait A characteristic due to specific chemical processes of an individual, such as the level of a particular enzyme. Compare *behavioral trait* and *physical trait*.

biodiversity The variety of all the world's life-forms, as well as their interactions with each other and the ecosystems they inhabit.

biogeography The geographic locations where organisms or the fossils of a particular species are found.

biological hierarchy A way to visualize the breadth and scope of life, from the smallest structures to the broadest interactions between living and nonliving systems that we can comprehend.

biological species concept The idea that a species is defined as a group of natural populations that can interbreed to produce fertile offspring and cannot breed with other such groups.

biome A large region of the world defined by shared physical characteristics (especially climate) and a distinctive community of organisms.

biomolecule Also called *macromolecule*. A large organic molecule that is critical for living cells. Four major classes of

biomolecules are proteins, carbohydrates, nucleic acids, and lipids.

biosphere All the world's living organisms and the physical spaces where they live.

biotic Living. Compare *abiotic*.

bipedal Walking upright on two legs.

blood The cells, cell fragments, and plasma that circulate through the heart and blood vessels.

blood pressure The force of blood pushing from the heart through blood vessels.

blood vessel A vessel (for example, vein, capillary, artery) that transports blood throughout the body.

boreal forest A terrestrial biome that is found in northern or high-latitude regions, has cold, dry winters and mild summers, and is dominated by coniferous trees.

brain The part of the central nervous system that controls and coordinates nerve signals throughout the body and has a large capacity for processing diverse types of sensory information.

breathing The process of taking air into the lungs (inhaling) and expelling air from them (exhaling).

bronchioles A series of ever-smaller tubes that branch from the bronchi and open into the alveoli.

bronchus (pl. bronchi) Either of two small tubes that the trachea branches into in the chest and that lead to the lungs.

bryophytes A group of nonflowering plants that includes liverworts and mosses.

C

Calvin cycle See **light-independent reactions**.

Cambrian explosion The burst of evolutionary activity, occurring about 540 million years ago, that resulted in a dramatic increase in the diversity of life. Most of the major living animal groups first appear in the fossil record during this time.

camouflage Any type of coloration or appearance that makes an organism hard to find or hard to catch. Compare *mimicry* and *warning coloration*.

cancer cell Also called *malignant cell*. A tumor cell gains anchorage independence and starts invading other tissues.

capillary Any of the tiny blood vessels that exchange materials by diffusion with nearby cells. Compare *artery* and *vein*.

capsule An additional protective layer that surrounds the cell wall in prokaryotes.

carbohydrate Any of a major class of biomolecules, including sugars and starches, built of repeating units of carbon, hydrogen, and oxygen.

carbon cycle The movement of carbon within biotic communities, between living organisms and their physical surroundings, and within the abiotic world.

carbon dioxide (CO_2) The most abundant and consequential of the greenhouse gases.

carbon fixation See **light-independent reactions**.

carbon sink A natural or artificial reservoir that absorbs more carbon than it releases. Compare *carbon source*.

carbon source A natural or artificial reservoir that releases more carbon than it absorbs. Compare *carbon sink*.

cardiac muscle The specialized muscle that helps produce the coordinated contractions known as heartbeats. It has a banded appearance and branched muscle fibers, and its contractions are entirely involuntary. Compare *skeletal muscle* and *smooth muscle*.

cardiovascular system A closed circulatory system consisting of a muscular heart, a complex network of blood vessels that collectively form a closed loop, and blood that circulates through the heart and blood vessels.

carnivore An animal (or, rarely, plant) that eats other animals. Compare *herbivore* and *omnivore*.

carrying capacity The maximum population size that can be sustained in a given environment.

cartilage A dense tissue of the skeleton that combines strength with flexibility. It is found almost everywhere that two bones meet and prevents them from grinding together.

catabolism Metabolic pathways that release chemical energy in the process of breaking down complex biomolecules. Compare *anabolism*.

causation A statistical relation indicating that a change in one aspect of the natural world causes a change in another aspect. Compare *correlation*.

cell The smallest and most basic unit of life—a microscopic, self-contained unit enclosed by a water-repelling membrane.

cell cycle The sequence of events that make up the life of a typical eukaryotic cell, from the moment of its origin to the time it divides to produce two daughter cells.

cell division The final stage of the cell cycle. Cell division includes the transfer of DNA from the parent cell to the daughter cells.

cell-mediated immunity A response mounted by the adaptive immune system in which specific invaders are attacked by phagocytes and other specialized cells. Compare *antibody-mediated immunity*.

cell theory One of the unifying principles of biology; a theory stating that every living organism is composed of one or more cells, and that all cells living today came from a preexisting cell.

cellular respiration The reciprocal process to *photosynthesis*, in which sugars are broken down into energy usable by the cell.

cellulose A chemical substance that provides structural strength in plant cells. Compare *lignin*.

central nervous system (CNS) One of two main parts of the nervous system, consisting of the brain and spinal cord. Compare *peripheral nervous system*.

centromere The central region of a chromosome that attaches sister chromatids together.

cervix The lower portion of the uterus, which narrows and connects the uterus to the vagina.

chaparral A terrestrial shrubland biome characterized by cool, rainy winters and hot, dry summers, and dominated by drought-resistant plants.

chemical bond A force that holds two atoms together.

chemical compound A molecule that contains atoms from two or more different elements.

chemical reaction The process of breaking existing chemical bonds and creating new ones.

chemoautotroph See **autotroph**.

chemoheterotroph See **heterotroph**.

chlorophyll A green pigment that is specialized for absorbing light energy.

chloroplast An organelle of plant cells and some protist cells that captures energy from sunlight and uses it to manufacture food molecules via photosynthesis.

chordates A large phylum that encompasses all animals with backbones, such as fishes, birds, and mammals.

chorionic villus sampling (CVS) A prenatal genetic screening technique in which cells are extracted by gentle suction from the villi (a cluster of cells that attaches the pregnancy sac to the wall of the uterus). Ultrasound is used to guide the narrow, flexible suction tube through a woman's vagina and into her uterus. Compare *amniocentesis*.

chromatin fiber A chromosome in relaxed form, made up of DNA and nucleosomes.

chromosomal abnormality Any change in the chromosome number or structure, compared to what is typical for a species.

chromosome A DNA double helix wrapped around spools of proteins.

chromosome theory of inheritance The theory that genes are located on chromosomes, and that these chromosomes are the basis for all inheritance.

circulatory system The organ system that moves oxygen from the respiratory system to the heart, which then pumps oxygen-rich blood to the rest of the body.

citizen science Research that is assisted by members of the public, who participate by collecting and sometimes analyzing data in cooperation with professional scientists.

citric acid cycle See **Krebs cycle**.

clade A branch of an evolutionary tree, consisting of an ancestor and all its descendants.

class The unit of classification in the Linnaean hierarchy above order and below phylum.

climate The prevailing *weather* of a specific place over relatively long periods of time (30 years or more).

climate change A large-scale and long-term alteration in Earth's climate, including such phenomena as *global warming*, change in rainfall patterns, and increased frequency of violent storms.

climax community A mature community whose species composition remains stable over long periods of time.

CNS See **central nervous system**.

codominance An interaction between two alleles of a gene that causes a heterozygote to display a phenotype that clearly displays the effects of both alleles. Compare *incomplete dominance*.

codon A unique sequence of three mRNA bases that either specifies a particular amino acid during translation or signals the ribosomes where to start or stop translation. Compare *anticodon*.

coevolution The tandem evolution of two species that results because interaction between the two so strongly influences their survival.

collagen A tough but pliable protein that, in addition to being the main component of cartilage, is found in a great variety of tissues.

colon Also called *large intestine*. The final portion of the digestive system, where water and nutrients are absorbed before the remaining solid waste is expelled.

commensalism A species interaction in one species benefits at no cost to the other species. Compare *competition, mutualism*, and *predation*.

common ancestor An organism from which many species have evolved.

common descent The sharing of a common ancestor by two or more different species.

community The populations of different species that live and interact with one another in a particular place.

compact bone One of the two major types of bone tissue. It forms the hard, white outer region of bones. Compare *spongy bone*.

competition A species interaction in which both species may be harmed. Compare *commensalism, mutualism*, and *predation*.

competitive exclusion principle The idea that different species that use the same resource can coexist only if one of the species adapts to using other resources.

complementary base-pairing The relationship between two nucleic acid strands in which each purine on one strand hydrogen-bonds with a specific pyrimidine on the opposite strand. A pairs with T or U, and C pairs with G.

complex trait A genetic trait whose pattern of inheritance cannot be predicted by Mendel's laws of inheritance.

condensation The transition from gas to liquid. Compare *evaporation*.

conjugation The physical process of transferring genetic material through direct contact.

connective tissue A tissue that binds and supports tissues and organs.

consumer An organism that obtains energy by eating all or parts of other organisms or their remains. Compare *producer*.

contraceptive Any means of preventing pregnancy.

control group The group of subjects in an experiment that is maintained under a standard set of conditions with no change in the independent variable. Compare *treatment group*.

controlled experiment An experiment that measures the value of a dependent variable for two groups of subjects that are comparable in all respects except that one group (the treatment group) is exposed to a change in the independent variable and the other group (the control group) is not.

convection cell A large and consistent atmospheric circulation pattern in which warm, moist air rises and cool, dry air sinks. Earth has four giant convection cells.

convergent evolution Evolution that results in organisms that have different genetics but appear very much alike.

corpus luteum The cells of the ruptured follicle that remain behind in the ovary after ovulation to produce the hormone progesterone.

correlation A statistical relation indicating that two or more aspects of the natural world behave in an interrelated manner: if one shows a particular value, we can predict a particular value for the other aspect. Compare *causation*.

cotyledon A food-storing organ that is part of the tiny, embryonic seedling that lies inside a plant seed.

covalent bond The sharing of electrons between two atoms. Compare *hydrogen bond* and *ionic bond*.

credentials Evidence of qualifications and competence to be recognized as an authority on a subject. Such evidence would include education and accomplishments.

CRISPR Clustered regularly interspaced short palindromic repeats. An RNA sequence that guides molecular machinery to a DNA sequence through complementary base-pairing.

cross See **genetic cross**.

crossing-over The physical exchange of chromosomal segments between homologous chromosomes. Compare *genetic recombination*.

CVS See **chorionic villus sampling**.

cyclical fluctuation A relatively predictable pattern of change in population size that occurs when at least one of two species is strongly influenced by the other. Compare *irregular fluctuation*.

cytokine Any of a group of signaling proteins that immune system cells use to communicate when an invader is present.

cytokinesis Division of the cytoplasm, the second step of mitotic division, resulting in two self-contained daughter cells. Compare *mitosis*.

cytokinin Any of a group of plant hormones that are necessary for cell division and promote shoot formation. Compare *auxin* and *gibberellin*.

cytosine (C) One of the four nucleotides that make up DNA. The other three are adenine (A), thymine (T), and guanine (G).

cytoskeleton The network of protein cylinders and filaments that forms the framework of a cell.

D

data (sing. datum) Information collected in a scientific study.

decomposer A scavenger that dissolves the dead bodies of other organisms to consume them. Compare *detritivore*.

deletion A mutation in which a base is deleted from the DNA sequence of a gene. Compare *insertion* and *substitution*.

dendrite The part of a neuron that receives information from other cells. Compare *axon*.

density-dependent population change A change in population size that occurs when birth and death rates change as the population density changes. Compare *density-independent population change*.

density-independent population change A change in population size that occurs when populations are held in check by factors that are not related to the density of the population. Compare *density-dependent population change*.

deoxyribonucleic acid See **DNA**.

dependent variable Any variable that responds, or could potentially respond, to changes in an *independent variable*.

dermal tissue One of three types of plant tissue. Forming the outermost layer of the plant, dermal tissue protects the plant from the outside environment and controls the flow of materials into and out of the plant. Compare *ground tissue* and *vascular tissue*.

descriptive study An observational study that reports information about what is found in nature. Compare *analytical study*.

desert A terrestrial biome characterized by a lack of moisture and temperature fluctuation from very hot during the day to very cold at night, and dominated by succulent plants such as cacti.

determinate growth The general growth pattern of animals, in which they grow early in their lives and then level off. Compare *indeterminate growth*.

detritivore A scavenger that mechanically breaks apart the dead bodies of other organisms to consume them. Compare *decomposer*.

diastole The second pressure, or bottom number, in a blood pressure reading—the pressure at the point the heart relaxes after a contraction. Compare *systole*.

dicot Any plant that has two cotyledons in each of its seeds. Dicots are characterized by parallel veins in the leaves, scattered vascular bundles, fibrous roots, and floral parts in multiples of three. Examples include beans, squash, oak trees, and roses. Compare *monocot*.

diffusion The movement of a substance from a region of higher concentration to a region of lower concentration.

digestion The chemical breakdown of food.

digestive system The organ system that breaks down food in the mouth, stomach, and small intestine and absorbs nutrients in the small and large intestine.

dihybrid cross A controlled mating experiment involving organisms that are heterozygous for two traits.

diploid Possessing a double set of genetic information, represented by $2n$. Somatic cells are diploid. Compare *haploid*.

directional selection The most common pattern of natural selection, in which individuals at one extreme of an inherited phenotypic trait have an advantage over other individuals in the population. Compare *disruptive selection* and *stabilizing selection*.

disruptive selection The least common pattern of natural selection, in which individuals with either extreme of an inherited trait have an advantage over individuals with an intermediate phenotype. Compare *directional selection* and *stabilizing selection*.

DNA Deoxyribonucleic acid, the genetic code of life, consisting of two parallel strands of nucleotides twisted into a double helix. DNA is the genetic material that transfers information from parents to offspring.

DNA polymerase The enzyme that builds new strands of DNA in DNA replication.

DNA replication The duplication of a DNA molecule.

DNA sequence similarity The degree to which the sequences of two different DNA molecules are the same—a measure of how closely related two DNA molecules are to each other.

domain The highest hierarchical level in the organization of life, describing the most basic and ancient divisions among living organisms. The three domains of life are Bacteria, Archaea, and Eukarya.

dominant allele An allele that prevents a second allele from affecting the phenotype when the two alleles are paired together. Compare *recessive allele*.

dominant genetic disorder A genetic disorder that is inherited as a dominant trait on an autosome. Compare *recessive genetic disorder*.

double helix The spiral formed by two complementary strands of nucleotides that is the backbone of DNA.

down-regulation The slowing down of gene expression. Compare *up-regulation*.

E

Earth equivalent The number of planet Earths needed to provide the resources we use and absorb the wastes we produce.

ecological community An association of species that live in the same area.

ecological footprint The area of biologically productive land and water that an individual or a population requires to produce the resources it consumes and to absorb the waste it produces.

ecological isolation The condition in which closely related species in the same territory are reproductively isolated by slight differences in habitat. Compare *geographic isolation* and *reproductive isolation*.

ecological niche The set of conditions and resources that a population needs in order to survive and reproduce in its habitat.

ecology The scientific study of interactions between organisms and their environment, where the environment of an organism includes both biotic factors (other living organisms) and abiotic (nonliving) factors.

ecosystem A particular physical environment and all the communities in it.

ecosystem process Any of four processes—nutrient cycling, energy flow, water cycling, and succession—that link the biotic and abiotic worlds in an ecosystem.

egg Also called *ovum*. The female gamete. Compare *sperm*.

electron A negatively charged particle found outside the nucleus of an atom. Compare *neutron* and *proton*.

electron transport chain An elaborate chain of chemical events in which electrons and protons (H^+) are handed over to other molecules that ultimately generates ATP and NADPH.

element A pure substance that has distinctive physical and chemical properties, and that cannot be broken down into other substances by ordinary chemical methods.

elimination The removal from the body of solid waste, consisting mostly of indigestible material and bacteria that inhabit the digestive tract.

embryo The earliest stage of development of an individual after fertilization, up to 2 months of age in humans. Compare *fetus*.

embryonic development The process by which an embryo develops. Common patterns of embryonic development across species provide evidence of evolution.

endocrine gland A gland that releases hormones into body fluids, such as the bloodstream, for transport to target cells throughout the body.

endocrine system The organ system, consisting of a number of glands and secretory tissues that produce and secrete hormones, that works closely with the nervous system to regulate all other organ systems.

endocytosis The process by which materials are transported into a cell via vesicles. Compare *exocytosis*.

endoplasmic reticulum (ER) An extensive and interconnected network of sacs made of a single membrane that is continuous with the outer membrane of the nuclear envelope. See also **rough ER** and **smooth ER**.

endoskeleton An internal skeleton. Compare *exoskeleton*.

energy The capacity of any object to do work, which is the capacity to bring about a change in a defined system.

energy capture The trapping and storing of solar energy by the producers at the base of an ecosystem's energy pyramid.

energy carrier A molecule that can store and deliver usable energy.

energy pyramid A pyramid-shaped representation of the amount of energy available to organisms in a food chain.

enzyme Any of a class of small molecules that speed up chemical reactions.

epistasis A form of inheritance in which the phenotypic effect of the alleles of one gene depends on the presence of alleles for another, independently inherited gene.

epithelial tissue A tissue that covers organs and lines body cavities.

ER See **endoplasmic reticulum**.

esophagus The food tube that connects the pharynx to the stomach.

essential amino acid Any of the eight amino acids that can be obtained only from food.

estradiol The primary estrogen. Compare *progesterone* and *testosterone*.

estrogens Any of the hormones produced in the ovaries that play a role in determining female characteristics such as wide hips, a voice that is pitched higher than that of males, and the development of breast tissues. Compare *progestogens* and *androgens*.

estuary An aquatic biome characterized by shallow depth and high productivity. It is the tidal ecosystem where a river flows into the ocean and consequently has a constant ebb and flow of fresh and salt water. It is dominated by grasses and sedges.

ethylene A plant hormone that stimulates the ripening of some fruits, including apples, bananas, avocadoes, and tomatoes.

Eukarya One of the three domains of life, including all the living organisms that do not fit into the domains *Archaea* or *Bacteria*, from amoebas to plants to fungi to animals.

eukaryote An organism that belongs to the Eukarya. Animals, plants, fungi, and protists are all eukaryotes. Compare *prokaryote*.

eutherians One of three main groups of mammals, whose members have a placenta and produce offspring that are born in a well-developed state. Compare *marsupials* and *monotremes*.

evaporation The transition from liquid to gas. Compare *condensation*.

evolution A change in the overall inherited characteristics of a group of organisms over multiple generations.

evolutionary tree A model of evolutionary relationships among groups of organisms that is based on similarities and differences in their DNA, physical features, biochemical characteristics, or some combination of these. It maps the relationships between ancestral groups and their descendants, and it clusters the most closely related groups on neighboring branches.

exocytosis The process by which materials are exported out of a cell via vesicles. Compare *endocytosis*.

exon A stretch of DNA that carries instructions for building a protein. Compare *intron*.

exoskeleton An external skeleton. Compare *endoskeleton*.

experiment A repeatable manipulation of one or more aspects of the natural world.

experimental group See **treatment group**.

exploitative competition Competition between species in which the two species indirectly compete for shared resources, such as food. Compare *interference competition*.

exponential growth A pattern of population growth in which the population increases by a constant proportion over a constant time interval, such as one year. Exponential growth occurs when there are no constraints on resources and is represented by a J-shaped curve. Compare *logistic growth*.

external defenses The immune system's first line of defense against pathogens, consisting of physical and chemical barriers that reduce the likelihood of harmful organisms or viruses gaining access to internal tissues. Compare *adaptive immune system* and *innate immune system*.

F

F_1 generation The first generation of offspring in a series of genetic crosses. Compare *F_2 generation* and *P generation*.

F_2 generation The second generation of offspring in a series of genetic crosses. Compare *F_1 generation* and *P generation*.

facilitated diffusion Diffusion that requires transport proteins. Compare *simple diffusion*.

fact A direct and repeatable observation of any aspect of the natural world. Compare *theory*.

fallopian tube See **oviduct**.

falsifiable Able to be refuted.

family The unit of classification in the Linnaean hierarchy above genus and below order.

fat-soluble vitamin A vitamin that cannot dissolve in water and therefore tends to accumulate in body fat because it cannot be so easily excreted in urine. Compare *water-soluble vitamin*.

feces The solid waste produced by digestion.

feedback loop The steps of a process that either decrease (*negative feedback*) or increase (*positive feedback*) the output of that process.

fermentation A metabolic pathway by which most anaerobic organisms extract energy from organic molecules. It begins with glycolysis and is followed by a special set of reactions whose only role is to help perpetuate glycolysis. Fermentation enables organisms to generate ATP anaerobically.

ferns A group of vascular, nonflowering plants that reproduce via spores.

fertilization The fusion of two gametes, resulting in a zygote.

fetus The second stage of development of an individual, from 2 months to birth in humans. Compare *embryo*.

filtration The blood-cleansing work of the kidneys.

flagellum (pl. flagella) A long, whiplike structure that assists bacteria in locomotion.

flower A structure in angiosperms that enhances sexual reproduction by bringing male gametes (sperm cells) to the female gametes (egg cells) in highly efficient ways, by attracting animal pollinators through scent, shape, and color.

follicle-stimulating hormone (FSH) A hormone, produced by the pituitary gland, that launches each menstrual cycle by stimulating the growth of ovarian follicles.

food chain A single direct line of who eats whom among species in a community. Compare *food web*.

food web The way in which various and often overlapping *food chains* of a community are connected.

fossil The mineralized remains or impression of a formerly living organism.

founder effect A form of genetic drift that occurs when a small group of individuals establishes a new population isolated from its original, larger population.

freshwater biome An aquatic biome whose character is defined by the terrestrial biomes that it borders or through which its water flows. Lakes, rivers, and wetlands are all part of the freshwater biome.

fruiting body In fungi, the structure resulting from mating that releases offspring as sexual spores.

FSH See **follicle-stimulating hormone**.

Fungi The fungi. One of the six kingdoms of life, in the domain Eukarya, distinguished by their modes of reproduction. Fungi are absorptive heterotrophs.

G

G_0 phase A nondividing state of the cell.

G_1 phase "Gap 1," the first phase in the life of a newborn cell.

G_2 phase "Gap 2," the phase of the cell cycle between the S phase and cell division.

gallbladder An organ of the digestive system that stores bile made by the liver and dispenses the bile into the small intestine as needed.

gamete A sex cell. Male gametes are sperm; female gametes are eggs. Compare *somatic cell*.

gametophyte A haploid, multicellular structure produced by mitotic division of a plant's spore. Cells in the gametophyte differentiate into egg or sperm. Compare *Sporophyte*.

gene The basic unit of information, consisting of a stretch of DNA, that codes for a distinct genetic characteristic.

gene expression The process by which genes are transcribed into RNA and then translated to make proteins.

gene flow The exchange of alleles between populations.

gene regulation The changing of the genes that are expressed in response to internal signals or external cues that allows organisms to adapt to their surroundings by producing different proteins as needed.

gene therapy A genetic engineering technique for correcting defective genes responsible for disease development.

genetic bottleneck A form of genetic drift that occurs when a drop in the size of a population causes a loss of genetic variation.

genetic carrier An individual who has only one copy of a recessive allele for a particular disease and therefore can pass on the disorder allele but does not have the disease.

genetic code The information specified by each of the 64 possible codons.

genetic cross A controlled mating experiment performed to examine how a particular trait may be inherited.

genetic disorder A disease caused by an inherited mutation, passed down from a parent to a child.

genetic divergence The presence of differences in the DNA sequences of genes.

genetic drift A change in allele frequencies produced by random differences in survival and reproduction among the individuals in a population.

genetic engineering The permanent introduction of one or more genes into a cell, tissue, or organism.

genetic modification (GM) Altering the genes of an organism for a specific purpose.

genetic recombination The exchange of DNA between homologous chromosomes brought about by *crossing-over*, contributing to variation in gametes.

genetic trait Any inherited characteristic of an organism that can be observed or detected in some manner.

genome The complete set of genes of an organism.

genotype The allelic makeup of a specific individual with respect to a specific genetic trait. Compare *phenotype*.

genus (pl. genera) The unit of classification in the Linnaean hierarchy above species and below family.

geographic isolation The condition in which populations are separated by physical barriers. Compare *ecological isolation* and *reproductive isolation*.

gibberellin Any of a group of plant hormones that bring about stem growth through both cell elongation and cell division. Compare *auxin* and *cytokinin*.

global warming A significant increase in the average surface temperature of Earth over decades or more. Compare *climate change*.

glycolysis The first of three stages of cellular respiration. During glycolysis, sugars (mainly glucose) are split to make the three-carbon compound pyruvate. For each glucose molecule that is split, two molecules of ATP and two molecules of NADH are released. Compare *Krebs cycle* and *oxidative phosphorylation*.

GM See **genetic modification**.

Golgi apparatus A collection of flattened membranes that packages and directs proteins and lipids produced by the ER to their final destinations either inside or outside the cell.

grassland A terrestrial biome characterized by low moisture levels (but not as low as in deserts), and dominated by grasses and herbaceous plants.

greenhouse effect The process by which greenhouse gases let in sunlight and trap heat.

greenhouse gas A gas in Earth's atmosphere that absorbs heat that radiates away from Earth's surface. Examples include carbon dioxide (CO_2), water vapor (H_2O), methane (CH_4), and nitrous oxide (N_2O).

ground tissue One of three types of plant tissue. Forming the intermediate layer of the plant, ground tissue makes up the bulk of the plant body and performs a wide range of functions, including support, wound repair, and photosynthesis. Compare *dermal tissue* and *vascular tissue*.

guanine (G) One of the four nucleotides that make up DNA. The other three are adenine (A), thymine (T), and cytosine (C).

gymnosperms One of two main groups of seed-bearing plants, characterized by naked seeds. Compare *angiosperms*.

H

halophile A prokaryote, usually archaean, that can live in extremely salty environments.

haploid Possessing a single set of genetic information, represented by $2n$. Gametes are haploid. Compare *diploid*.

heart A muscular organ the size of a fist that works as the body's circulatory pump.

heart rate The number of times a heart beats per minute.

herbivore An animal that eats plants. Compare *carnivore* and *omnivore*.

herd immunity Protection against disease that is brought about by vaccination of a critical portion of a population.

hermaphrodite An individual that produces both functional testes and functional ovaries and is therefore both male and female.

heterotroph A metabolic consumer, an organism that obtains energy by taking it from other sources. *Chemoheterotrophs* consume organic molecules as a source of energy and carbon. *Photoheterotrophs* absorb the energy of sunlight but require an organic source of carbon. Compare *autotroph*.

heterozygous Carrying two different alleles for a given phenotype (*Bb*). Compare *homozygous*.

histone protein One of a class of specific proteins that, together with DNA, form nucleosomes.

homeostasis The process of maintaining constant internal conditions.

homeostatic pathway The sequence of steps that reestablishes homeostasis if there is any departure from the genetically determined normal state of a particular internal characteristic.

hominids The ape family, which includes humans and chimpanzees. All hominids are capable of tool use, symbolic language, and deliberate acts of deception. Compare *hominins*.

hominins The "human" branch of the *hominids*, including modern humans and extinct relatives such as Neanderthals.

homologous pair A pair of chromosomes consisting of one chromosome received from the father and one from the mother.

homologous trait A feature that is similar across species because of common descent. Homologous traits may begin to look different from one another over time. Compare *analogous trait*.

homozygous Carrying two copies of the same allele (such as *BB* or *bb*) for a particular gene. Compare *heterozygous*.

horizontal gene transfer The transfer of genes on plasmids from one bacterium to another.

hormone A signaling molecule that coordinates internal activities necessary for growth and reproduction.

host An organism in which a *parasite* lives.

human microbiome The complete collection of microbes that live in and on our cells and bodies.

hydrogen bond The weak electrical attraction between a hydrogen atom with a partial positive charge and a neighboring atom with a partial negative charge. Compare *covalent bond* and *ionic bond*.

hydrologic cycle The movement of water as it circulates from the land to the sky and back again.

hydrophilic Literally, "water-loving." Soluble in water. Compare *hydrophobic*.

hydrophobic Literally, "water-fearing." Excluded from water. Compare *hydrophilic*.

hypertonic Describing a fluid that has a solute concentration higher than that of the cell it surrounds. Compare *hypotonic* and *isotonic*.

hyphae (sing. hypha) The fine, branching threads of fungi that absorb nutrients from the environment.

hypothalamus (pl. hypothalami) A small organ at the base of the vertebrate brain that coordinates the endocrine system and integrates it with the nervous system.

hypothesis (pl. hypotheses) An informed, logical, and plausible explanation for observations of the natural world.

hypotonic Describing a fluid that has a solute concentration lower than that of the cell it surrounds. Compare *hypertonic* and *isotonic*.

I

immune memory The capacity of the adaptive immune system to remember a first encounter with a specific pathogen and to mobilize a speedy and targeted response to future infection by the same strain.

immune system The organ system that defends the body from invaders such as viruses, bacteria, and fungi.

in vitro fertilization (IVF) Fertilization of an egg by a sperm in a petri dish, followed by implantation of one or more embryos into a woman's uterus.

incomplete dominance An interaction between two alleles of a gene in which neither one can exert its full effect, causing a heterozygote to display an intermediate phenotype. Compare *codominance*.

independent assortment The random distribution of the homologous chromosomes into daughter cells during meiosis I.

independent variable The variable that is manipulated by the researcher in a scientific experiment. Compare *dependent variable*.

indeterminate growth The general growth pattern of plants, in which they grow throughout their lives. Compare *determinate growth*.

induced fit The way an enzyme changes shape when molecules bind to its active site.

inflammation The immediate and coordinated sequence of events mounted by cytokines in the innate immune system in response to tissue damage from a pathogen invasion or wound.

ingestion The taking in of food, the first stage in the processing of food by the digestive system.

innate immune system The immune system's second line of defense against pathogens, consisting of cells and proteins that recognize the presence of an invader and mount an internal defense to kill, disable, or isolate it. Compare *adaptive immune system* and *external defenses*.

insertion A mutation in which a base is inserted into the DNA sequence of a gene. Compare *deletion* and *substitution*.

integumentary system The largest organ system in the human body, covering and protecting the surface of the body.

interference competition Competition between species in which one organism directly excludes another from the use of a resource. Compare *exploitative competition*.

intermediate fossil A fossil that displays physical characteristics in between those of two known fossils in a family tree.

interphase The longest stage of the cell cycle. Most cells spend 90 percent or more of their life span in interphase.

intron A stretch of DNA that does not code for anything. Compare *exon*.

invariant trait A trait that is the same in all individuals of a species. Compare *variable trait*.

ion An atom that has lost or gained electrons and therefore is either negatively or positively charged.

ionic bond The chemical attraction between a negatively charged ion and a positively charged ion. Compare *covalent bond* and *hydrogen bond*.

irregular fluctuation An unpredictable pattern of change in population size. Compare *cyclical fluctuation*.

isotonic Describing a fluid that has a solute concentration equal to that of the cell it surrounds. Compare *hypertonic* and *hypotonic*.

isotopes Two or more forms of an element that have the same number of protons but different numbers of neutrons.

IVF See **in vitro fertilization**.

J

J-shaped growth curve The type of graphical curve that represents exponential growth. Compare *S-shaped growth curve*.

joint A junction in the skeletal system that lets the skeleton move in specific ways.

K

karyotype A depiction showing all the chromosomes of a particular individual or species arranged in homologous pairs.

keystone species A species that has a disproportionately large effect on a community, relative to the species' abundance.

kidneys The paired organs that maintain water and solute homeostasis.

kingdom The second-highest hierarchical level in the organization of life; the unit of classification in the Linnaean hierarchy above phylum and below domain. The six kingdoms of life are Bacteria, Archaea, Protista, Plantae, Fungi, and Animalia.

Krebs cycle Also called *citric acid cycle*. The second of three stages of cellular respiration. In this sequence of enzyme-driven reactions, the pyruvate made in glycolysis is broken down, releasing CO_2 and producing large amounts of energy carriers, including ATP, NADH, and $FADH_2$. Compare *glycolysis* and *oxidative phosphorylation*.

L

large intestine See **colon**.

larynx Also called *voice box*. The breathing and sound-producing structure that forms the entryway to the trachea.

law of independent assortment The law, proposed by Gregor Mendel, stating (in modern terms) that when gametes form, the two alleles of any given gene segregate during meiosis independently of any two alleles of other genes. Compare *law of segregation*.

law of segregation The law, proposed by Gregor Mendel, stating (in modern terms) that the two alleles of a gene are separated during meiosis and end up in different gametes. Compare *law of independent assortment*.

leaf A structure in plants that produces the majority of the plant's food through photosynthesis.

LH See **luteinizing hormone**.

lichen A mutualistic association between a photosynthetic microbe (usually a green alga or cyanobacterium) and a fungus.

ligament A specialized, flexible band of tissue that joins bone to bone. Compare *tendon*.

light-independent reactions Also called the *Calvin cycle* or *carbon fixation*. The second of two principal stages of photosynthesis, in which a series of enzyme-catalyzed chemical reactions converts carbon dioxide (CO_2) into sugar, using energy delivered by ATP and electrons and hydrogen ions donated by NADPH. Compare *light reactions*.

light reactions The first of two principal stages of photosynthesis, in which chlorophyll molecules absorb energy from sunlight and use that energy for the splitting of water, which in turn produces oxygen gas (O_2) as a by-product that is released into the atmosphere. Compare *light-independent reactions*.

lignin A strengthening substance that links together *cellulose* fibers in plant cells to create a rigid network.

lineage A single line of descent.

Linnaean hierarchy A system of biological classification devised by the Swedish naturalist Carolus Linnaeus in the eighteenth century.

lipid Any of a major class of biomolecules built of fatty acids and insoluble in water.

liposome A sphere formed by a phospholipid bilayer.

liver A large organ of the digestive system that produces bile, stores glycogen, and detoxifies dangerous chemicals in the body.

locus (pl. loci) The physical location of a gene on a chromosome.

logistic growth A pattern of population growth in which the population grows nearly exponentially at first but then stabilizes at the maximum population size that can be supported indefinitely by the environment. Logistic growth is represented by an S-shaped curve. Compare *exponential growth*.

lower respiratory system The part of the respiratory system made up of the trachea, bronchi, and lungs. Compare *upper respiratory system*.

lungs The paired organs in which gases (oxygen and carbon dioxide) are exchanged.

luteinizing hormone (LH) A hormone, produced by the pituitary gland, that triggers ovulation.

lymphatic system The network of ducts, lymph nodes, and associated organs that are the primary sites for action by the adaptive immune system.

lymphocyte A type of white blood cell that confers specific immunity as part of the adaptive immune system. Immature lymphocytes differentiate into B cells and T cells.

lysosome An organelle in animal cells that acts as a garbage or recycling center. Compare *vacuole*.

M

macromolecule See **biomolecule**.

macronutrient Any of nine nutrients—carbon, oxygen, hydrogen, nitrogen, phosphorus, potassium, calcium, sulfur, and magnesium—that plants need in relatively large amounts. Compare *micronutrient*.

malignant cell See **cancer cell**.

mammals A large class of animals that have body hair, sweat glands, and milk produced by mammary glands.

marine biome An aquatic biome characterized by salt water and encompassing both the coastal regions of all continents and the open ocean.

marrow A tissue found in the cavities of hollow bones that, depending on the type of bone, stores fat or produces blood cells.

marsupials One of three main groups of mammals, whose members have a simple placenta and produce offspring that complete development in their mother's pouch. Compare *eutherians* and *monotremes*.

mass extinction A period of time during which a great number of Earth's species goes extinct. The fossil record shows that there have been five mass extinctions in the history of Earth.

matter Anything that has mass and occupies a volume of space.

meiosis A specialized type of cell division that kicks off sexual reproduction. It occurs in two stages: meiosis I and meiosis II, each involving one round of nuclear division followed by cytokinesis. Compare *mitosis*.

meiosis I The first stage of meiosis, in which the chromosome set is reduced by the separation of each homologous pair into two different daughter cells. Each homologous chromosome lines up with its partner and then separates to the two ends of the cells. Compare *meiosis II*.

meiosis II The second stage of meiosis, in which sister chromatids are separated into two new daughter cells. Compare *meiosis I*.

Mendelian trait A trait that is controlled by a single gene and unaffected by environmental conditions.

menstrual cycle The process in human females by which individual eggs mature and are released in a hormone-driven sequence of events approximately monthly.

messenger RNA (mRNA) A type of RNA that is complementary to a DNA template strand. Compare *ribosomal RNA* and *transfer RNA*.

meta-analysis Work that combines results from different studies.

metabolic heat The heat released as a by-product of chemical reactions within a cell, typically during cellular respiration.

metabolic pathway Any of various chains of linked events that produce key biological molecules in a cell, including important chemical building blocks like amino acids and nucleotides.

metabolism All the chemical reactions that occur inside living cells, including those that release and store energy.

metastasis The spread of a disease from one organ to another.

methanogen An anaerobic archaean that feeds on hydrogen and produces methane gas as a by-product of its metabolism.

microbe A microscopic, single-celled organism.

micronutrient Any of a variety of nutrients (including iron, zinc, and copper) that plants need in relatively small amounts. Compare *macronutrient*.

mimicry Coloration of a nonpoisonous animal that resembles the coloration of a toxic species. Compare *camouflage* and *warning coloration*.

mineral Any of various small, inorganic nutrients needed by the human body for critical biological function, but only in small amounts. Compare *vitamin*.

mitochondrial-DNA inheritance The passing down of DNA from the mitochondria in an egg cell to a new generation. Mitochondrial DNA passes virtually unchanged from mother to child, so it can be tracked from one generation, or one species, to another. Sequencing of mitochondrial DNA can determine how related an individual is to its female ancestors on its mother's side. Compare *nuclear-DNA inheritance*.

mitochondrion (pl. mitochondria) An organelle that is a tiny power plant fueling cellular activities. Mitochondria are the main source of energy in eukaryotic cells.

mitosis Division of the nucleus, the first step of mitotic division. Mitosis is divided into four main phases: prophase, metaphase, anaphase, and telophase. Compare *cytokinesis* and *meiosis*.

mitotic division A type of cell division that generates two genetically identical daughter cells from a single parent cell in eukaryotes. It consists of two steps: mitosis and cytokinesis.

modular growth The way plants grow by repeatedly adding the same module of bud-stem-leaf unit aboveground or new lateral roots belowground.

molecule An association of atoms held together by chemical bonds.

monocot Any plant that has one cotyledon in each of its seeds. Monocots are characterized by netlike veins in the leaves, vascular bundles arranged in rings, taproots, and floral parts in multiples of four or five. Examples include all the grasses, members of the lily family, palm trees, and banana plants. Compare *dicot*.

monomer A small molecule that is the repeating unit of a *polymer*. For example, amino acids are the monomers that make up protein polymers.

monotremes One of three main groups of mammals, whose members lack a placenta and lay eggs. Compare *eutherians* and *marsupials*.

morphology An organism's physical characteristics.

most recent common ancestor The most immediate ancestor that two lineages share.

motor output A particular action of the body.

mRNA See **messenger RNA**.

muscle fiber A long, narrow cell that can span the length of an entire muscle because it is made up of several muscle cells that fused together during development.

muscle tissue A tissue that generates force by contracting.

muscular system The organ system that, working closely with the skeletal system, produces the force that moves structures within the body.

mutation A random change to the sequence of bases in an organism's DNA.

mutualism A species interaction in which both species benefit. Compare *commensalism*, *competition*, and *predation*.

mycelium (pl. mycelia) The entire bundle of hyphae that composes the main body of a fungus.

mycorrhizal fungi Fungi that form mutualistic associations with the root systems of plants, that help the plants absorb more water and nutrients from the soil.

myofibril Any of the cylindrical structures packed inside of muscle fibers containing proteins that contract by bracing against each other.

myosin filament One of the two kinds of protein filaments, made of the protein myosin, that is found in sarcomeres. The sliding of myosin filaments against *actin filaments* enables sarcomeres to contract.

N

natural selection The process by which individuals with advantageous genetic characteristics for a particular environment survive and reproduce at a higher rate than do individuals with other, less useful characteristics. Compare *artificial selection*.

negative feedback The steps of a process that decrease its output. Compare *positive feedback*.

nephron The basic functional unit of the kidney.

nervous system The organ system that directs the rapid contractions of muscles and processes information received by the senses, such as touch, sound, and sight.

nervous tissue A tissue that communicates and processes information.

net primary productivity (NPP) The energy acquired through photosynthesis that is available for the growth and reproduction of producers within an ecosystem. It is the amount of energy captured by photosynthetic organisms, minus the amount they expend on cellular respiration and other maintenance processes.

neuron A specialized cell of the nervous system that transmits signals from one part of the body to another in a fraction of a second.

neutron An electrically neutral particle found in the nucleus of an atom. Compare *electron* and *proton*.

nitrogen fixation The process, carried out by bacteria, of taking nitrogen gas from the air and converting it to ammonia, making it available for plants.

node The point on an evolutionary tree indicating the moment in time when an ancestral group split, or diverged, into two separate lineages. The node represents the most recent common ancestor of the two lineages in question.

noncoding DNA DNA that does not code for any kind of functional RNA.

nonspecific response A response mounted by the immune system against pathogens that is indiscriminate as to the invaders it repels. External defenses and the innate immune system are both nonspecific responses. Compare *specific response*.

notochord In chordates, a flexible yet rigid rod along the center of the body that is critical for development.

NPP See **net primary productivity**.

nuclear-DNA inheritance The passing down of DNA from the nucleus in an egg or sperm cell to a new generation. Sequencing of nuclear DNA can determine how related an individual is to all of its ancestors, both male and female. Compare *mitochondrial-DNA inheritance*.

nuclear envelope The boundary of a cell's nucleus, consisting of two concentric phospholipid bilayers.

nuclear pore Any of many small openings in the nuclear envelope that allow chemical messages to enter and exit the nucleus.

nucleic acid Any of a major class of biomolecules, including DNA and RNA, built of chains of nucleotides.

nucleosome DNA wrapped around histone proteins that, in multiples, form the beads-on-a-string complex in a chromatin fiber.

nucleotide The basic repeating subunit of DNA, composed of the sugar deoxyribose, a phosphate group, and one of four bases: adenine (A), cytosine (C), guanine (G), or thymine (T).

nucleotide pair See **base pair**.

nucleus (pl. nuclei) 1. The dense core of an atom, which contains protons and neutrons. 2. The control center of the eukaryotic cell, containing all of the cell's DNA and occupying up to 10 percent of the space inside the cell.

nutrient A chemical element that is required by a living organism.

nutrient cycling The process by which decomposers break down dead organisms or waste products, release the chemical elements locked in the biological material, and return them to the environment.

O

observation A description, measurement, or record of any object or phenomenon.

omnivore An animal that eats both animals and plants. Compare *carnivore* and *herbivore*.

oogenesis The series of cell divisions in human females that results in an egg. Compare *spermatogenesis*.

opposable Able to be placed opposite other digits of the hand or foot. For example, opposable thumbs can be placed opposite each of the other four fingers.

oral cavity The mouth.

order The unit of classification in the Linnaean hierarchy above family and below class.

organ A collection of different types of tissues that form a functional unit with a distinctive shape and location in the body.

organ system A network of organs that work in a closely coordinated manner to perform a distinct set of functions in the body.

organelle Any of the membrane-enclosed subcellular compartments found in eukaryotic cells.

organic molecule A molecule that includes at least one carbon-hydrogen bond.

organism An individual living thing composed of interdependent parts.

origin of replication A DNA sequence where DNA replication is initiated.

osmosis A form of simple diffusion in which water moves in and out of cells (and compartments inside cells).

osteocyte A specialized bone cell that surrounds itself with a hard, nonliving mineral matrix composed largely of calcium and phosphate.

ovary Either of a pair of female reproductive organs that produce eggs and estrogens in vertebrates. Compare *testis*.

oviduct Also called *fallopian tube*. The tube through which an egg travels from the ovary to the uterus.

ovule The egg-bearing structure in plants.

ovum (pl. ova) See **egg**.

oxidative phosphorylation The third of three stages of cellular respiration. During this process, the chemical energy of NADH and $FADH_2$ is converted into the chemical energy of ATP, while electrons and hydrogen atoms removed from NADH and $FADH_2$ are handed over to molecular O_2, creating water (H_2O). A large amount of ATP is generated. Compare *glycolysis* and *Krebs cycle*.

oxytocin A hormone—secreted by the fetus and, later in the birth process, by the mother's pituitary gland—that stimulates the uterine muscles and causes the placenta to secrete prostaglandins, which reinforce contractions.

P

P generation The first set of parents in a series of genetic crosses. Compare *F₁ generation* and *F₂ generation*.

pacemaker See **sinoatrial node**.

pancreas A gland that produces insulin and secretes fluids that aid in the digestion of food.

parasite An organism that lives in or on another species and harms it by stealing resources. For example, some parasites suck blood or live off the food in our intestines. Compare *host*.

passive immunity Immunity to a particular pathogen that is conferred by antibodies not made by the body, but received from an outside source. Compare *active immunity*.

passive transport The movement of a substance without the addition of energy. Compare *active transport*.

pathogen An infectious agent.

PCR See **polymerase chain reaction**.

pedigree A chart similar to a family tree that shows genetic relationships among family members over two or more generations of a family's medical history.

peer-reviewed publication The publishing of original research only after it has passed the scrutiny of experts who have no direct involvement in the research under review, or a scientific journal that follows this standard.

penis The male reproductive organ that introduces sperm into a female or hermaphrodite sexual partner. The penis is also involved in urination in mammals.

perennial A plant that lives three or more years. Compare *annual* and *biennial*.

peripheral nervous system (**PNS**) One of two main parts of the nervous system, consisting of the sensory nerves plus all the nerves that are not part of the *central nervous system*.

PGD See **preimplantation genetic diagnosis**.

pH scale A logarithmic scale that indicates the concentration of hydrogen ions. The scale goes from 0 to 14, with 0 representing an extremely high concentration of free H⁺ ions and 14 representing the lowest concentration.

phagocyte A type of white blood cell that functions as part of the innate immune system to mark and destroy foreign invaders by engulfing and digesting them.

phagocytosis Literally, "cellular eating." A large-scale version of endocytosis in which particles considerably larger than biomolecules are ingested. Compare *pinocytosis*.

pharynx Also called *throat*. An area where the back of the mouth and the two nasal cavities join together into a single passageway.

phenotype The physical, biochemical, or behavioral expression of a particular version of a trait. Compare *genotype*.

phloem One of two types of vascular tissue in plants. Phloem transports sugars from the leaves, where they are produced, to living cells in every part of the plant. Compare *xylem*.

phospholipid An organic molecule with a hydrophilic head and a hydrophobic tail.

phospholipid bilayer A double layer of phospholipids in which the heads face out and the tails face in. Plasma membranes are phospholipid bilayers.

photoautotroph See **autotroph**.

photoheterotroph See **heterotroph**.

photoperiodism The ability of plants to sense the duration of light and dark in a 24-hour cycle.

photosynthesis The process by which organisms capture energy from the sun and use it to create sugars from carbon dioxide and water, thereby transforming light energy into chemical energy stored in the covalent bonds of sugar molecules. Compare *cellular respiration*.

phylum (**pl. phyla**) The unit of classification in the Linnaean hierarchy above class and below kingdom.

physical trait An anatomical or physiological characteristic of an individual, such as the shape of an animal's head. Compare *behavioral trait* and *biochemical trait*.

physiology The science that focuses on the functions of anatomical structures. Compare *anatomy*.

pili (**sing. pilus**) Short, hairlike projections that cover the surface of many bacteria.

pinocytosis Literally, "cellular drinking." A large-scale version of endocytosis in which fluids are ingested. Compare *phagocytosis*.

Plantae The plants. One of the six kingdoms of life, in the domain Eukarya, encompassing all plants, which are multicellular photosynthetic autotrophs.

plasma The fluid portion of the blood.

plasma membrane A barrier consisting of a phospholipid bilayer that separates a cell from its external environment.

platelet A type of sticky cell fragment found in circulating blood that helps form blood clots.

pleiotropy The pattern of inheritance in which a single gene influences multiple different traits. Compare *polygenic trait*.

PNS See **peripheral nervous system**.

point mutation A mutation in which only a single base is altered.

polar molecule A molecule whose electrical charge is shared unevenly, with some regions being electrically negative and others electrically positive.

pollen In plants, a microscopic structure containing sperm cells that can be lofted into the air in massive quantities.

pollinator An animal that transports pollen from one plant to another.

polygenic trait A genetic trait that is governed by the action of more than one gene. Compare *pleiotropy*.

polymer A long strand of repeating units of small molecules called *monomers*. For example, proteins are polymers made up of amino acid monomers.

polymerase chain reaction (PCR) A technique for replicating DNA that can produce millions of copies of a DNA sequence in just a few hours from a small initial amount of DNA.

polyploidy The condition in which an individual's somatic cells have more than two sets of chromosomes.

population A group of individuals of the same species living and interacting in a shared environment.

population density The number of individuals per unit of area.

population doubling time The time it takes a population to double in size, as a measure of how fast a population is growing.

population ecology The study of the number of organisms in a particular place.

population size The total number of individuals in a population.

positive feedback The steps of a process that increase its output. Compare *negative feedback*.

postzygotic barrier A barrier that prevents a zygote from developing into a healthy and fertile individual—that is, a reproductive barrier that acts after a zygote exists. Compare *prezygotic barrier*.

predation A species interaction in which one species benefits and the other species is harmed. Compare *commensalism*, *competition*, and *mutualism*.

predator A consumer that eats either plants or animals. Compare *prey*.

preimplantation genetic diagnosis (PGD) The removal of one or two cells from an embryo developing in a petri dish, usually 3 days after fertilization, followed by testing for genetic disorders. One or more embryos that are free of disorders are then implanted into a woman's uterus.

prey An animal that is eaten by a *predator*.

prezygotic barrier A barrier that prevents a male gamete and a female gamete from fusing to form a zygote—that is, a reproductive barrier that acts before a zygote exists. Compare *postzygotic barrier*.

primary consumer An organism that eats producers. Compare *secondary consumer*, *tertiary consumer*, and *quaternary consumer*.

primary growth In plants, an increase in length by the division of cells at the tips of shoots, stems, and roots. Compare *secondary growth*.

primary immune response The slow response mounted by the adaptive immune system against an invading pathogen the very first time a person is exposed to that pathogen. Compare *secondary immune response*.

primary literature Scientific literature in which research is first published. Compare *secondary literature*.

primary oocyte An immature egg cell.

primary succession Succession that occurs in a newly created habitat, usually from bare rock or sand. Compare *secondary succession*.

primates The order of mammals to which humans belong. All primates have flexible shoulder and elbow joints, five functional fingers and toes, opposable thumbs, flat nails, and brains that are large in relation to the body.

primer A short stretch of RNA or DNA that is complementary-base-paired to a DNA template to provide a 3'-hydroxyl group for polymerase to initiate DNA replication or the polymerase chain reaction (PCR), respectively.

process of science See **scientific method**.

producer An organism at the bottom of a food chain that uses energy from the sun to produce its own food. Compare *consumer*.

product A substance that results from a chemical reaction. Compare *reactant*.

progesterone The primary progestogen. Compare *estradiol* and *testosterone*.

progestogens Any of the hormones produced in the ovaries that have a number of functions in the female body, including thickening the lining of the uterus and increasing the blood supply to it to create a suitable environment for a developing fetus. Compare *estrogens* and *androgens*.

prokaryote An organism that belongs to either the Bacteria or the Archaea. Compare *eukaryote*.

promoter A segment of DNA near the beginning of a gene that RNA polymerase recognizes and binds to begin transcription. Compare *terminator*.

protein Any of a major class of biomolecules built of amino acids.

Protista The protists. One of the six kingdoms of life, in the domain Eukarya; a diverse group that is composed of mainly single-celled, microscopic organisms grouped together simply because they are not plants, animals, or fungi, and that includes amoebas and algae.

proton A positively charged particle found in the nucleus of an atom. Compare *electron* and *neutron*.

protozoans One of two main groups of protists, whose members are nonphotosynthetic and motile. Compare *algae*.

pseudoscience Scientific-sounding statements, beliefs, or practices that are not actually based on the scientific method.

pulmonary circuit The circuit in the heart, consisting of the two chambers on the right side, that receives blood low in oxygen and pumps it to the lungs for gas exchange. Compare *systemic circuit*.

Punnett square A grid-like diagram showing all possible ways that two alleles can be brought together through fertilization.

Q

quaternary consumer An organism that eats *tertiary consumers*. Compare *primary consumer* and *secondary consumer*.

quorum sensing A system of cell-to-cell communication used by prokaryotes that enables them to sense and respond to other bacteria in the area in accordance with the density of the population.

R

radial symmetry An animal body plan in which the body can be sliced symmetrically along any number of vertical planes that pass through the center of the animal. Compare *bilateral symmetry*.

reabsorption The removal of valuable solutes such as sodium, chloride, and sugars from the fluid filtered by the kidneys so that they don't exit the body in the urine.

reactant A substance that undergoes change in a chemical reaction. Compare *product*.

receptor-mediated endocytosis A form of specific endocytosis in which receptor proteins embedded in the membrane recognize specific surface characteristics of substances that will be incorporated into the cell.

receptor protein A site where a molecule from another cell can bind.

recessive allele An allele that has no effect on the phenotype when paired with a *dominant allele*.

recessive genetic disorder A genetic disorder that is inherited as a recessive trait on an autosome. Compare *dominant genetic disorder*.

red blood cell A cell in the blood that greatly increases the oxygen-carrying capacity of blood.

relative species abundance How common one species is when compared to another.

reproduction The making of a new individual like oneself.

reproductive barrier A barrier that prevents two species from interbreeding, making them reproductively isolated.

reproductive isolation The condition in which barriers prevent populations from interbreeding. Compare *ecological isolation* and *geographic isolation*.

reproductive system The organ system that generates gametes and may also support fertilization and prenatal development.

respiratory system The organ system that brings in oxygen and expels carbon dioxide to support cellular respiration.

ribonucleic acid See **RNA**.

ribosomal RNA (rRNA) A type of RNA that is an important component of ribosomes. Compare *messenger RNA* and *transfer RNA*.

ribosome The site of protein synthesis (translation) in the cytoplasm. Ribosomes are embedded in the rough endoplasmic reticulum.

RNA Ribonucleic acid, a single-stranded nucleic acid transcribed from DNA and consisting of the ribonucleotides adenine, guanine, cytosine, and uracil.

RNA polymerase An enzyme that recognizes and binds a gene's promoter sequence and then separates the two strands of DNA during transcription.

RNA splicing Processing of mRNA in which the introns are snipped out of a pre-mRNA and the remaining pieces of mRNA—the exons—are joined to generate the mature mRNA.

root A structure in plants that absorbs water and mineral nutrients, anchors the plant, and stores food.

root system One of two plant organ systems. It anchors the plant, absorbs water and nutrients from the soil, transports food and water, and may store food. Compare *shoot system*.

rough ER A part of the endoplasmic reticulum, having a knobby appearance because of embedded ribosomes, where proteins are assembled. Compare *smooth ER*.

rRNA See **ribosomal RNA**.

rubisco The enzyme that catalyzes the first step in the light-independent reactions of photosynthesis, fixing a carbon molecule from CO_2.

S

S phase The "synthesis" phase of the cell cycle, in which preparations for cell division begin. A critical event during this phase is the replication of all the cell's DNA molecules.

S-shaped growth curve The type of graphical curve that represents logistic growth. Compare *J-shaped growth curve*.

SA node See **sinoatrial node**.

saliva A fluid secreted into the oral cavity to aid in the digestion of food.

sarcomere Any of the contractile units of the muscular system that make up each myofibril.

scavenger An animal that eats dead or dying animals and plants. Scavengers are categorized as either *decomposers* or *detritivores*.

science A body of knowledge about the natural world, and an evidence-based process for acquiring that knowledge.

scientific claim A statement about how the world works that can be tested using the scientific method.

scientific literacy An understanding of the basics of science and the scientific process.

scientific method Also called *process of science*. The practices that produce scientific knowledge.

scientific name The unique two-word Latin name, consisting of the genus and species names, that is assigned to a species in the Linnaean hierarchy.

secondary consumer An organism that eats *primary consumers*. Compare *tertiary consumer* and *quaternary consumer*.

secondary growth In plants, an increase in thickness in either stems or roots. Compare *primary growth*.

secondary immune response The rapid response mounted by the adaptive immune system against an invading pathogen the second and subsequent times a person is exposed to that pathogen. Compare *primary immune response*.

secondary literature Scientific literature that summarizes and synthesizes an area of research. Compare *primary literature*.

secondary succession Succession that occurs after a disturbance in a community. Compare *primary succession*.

secretion The active transport by the kidneys of excess quantities of substances such as potassium and hydrogen ions, and some toxins, from the blood into the liquid passing through the kidney and out of the body in the urine.

seed In plants, the embryo and a supply of stored food, all encased in a protective covering.

selective breeding The process by which humans allow only individuals with certain inherited characteristics to mate.

selective permeability The quality of plasma membranes by which some substances are allowed to cross the membrane at all times, others are excluded at all times, and still others can pass through the membrane when they are aided by transport proteins.

semiconservative replication The mode of replication by which DNA is duplicated, where one "old" strand (the template strand) is retained (conserved) in each new double helix.

sense To perceive the world through a sensory system such as sight, touch, or smell.

sensory input Signals that are received, transmitted, and processed by the central nervous system.

sensory organ An organ of the body, such as the eyes or ears, that receives sensory input.

serotype See **viral strain**.

set point The genetically determined normal state of any physical or chemical characteristic of the body's internal environment.

sex chromosome One of the two chromosomes (X and Y) that determine gender. Compare *autosome*.

sex-linked Found solely on the X or Y chromosome. See also **X-linked** and **Y-linked**.

sexual dimorphism A distinct difference in appearance between the males and females of a species.

sexual reproduction The process by which genetic information from two individuals is combined to produce offspring. It has two steps: cell division through meiosis, followed by fertilization. Compare *asexual reproduction*.

sexual selection Natural selection in which a trait increases an individual's chance of mating even if it decreases the individual's chance of survival.

shared derived trait A unique feature common to all members of a group that originated in the group's most recent common ancestor and then were passed down in the group.

shoot system One of two plant organ systems, consisting of stems and leaves. Stems provide the plant with structural support, transport food and water, and hold leaves up to intercept light so that they can perform photosynthesis. Compare *root system*.

simple diffusion Diffusion in which substances such as the small, uncharged molecules of water, oxygen, or carbon dioxide, slip between the large molecules in the phospholipid bilayer without much hindrance. Compare *facilitated diffusion*.

sinoatrial (SA) node Also called *pacemaker*. The part of the heart that sends a signal telling the atria to contract. Compare *atrioventricular node*.

sister chromatids The two identical DNA molecules produced by the replication of a chromosome.

skeletal muscle The specialized muscle that is associated with the skeleton. It has a banded appearance, and its contractions are mainly voluntary. Compare *cardiac muscle* and *smooth muscle*.

skeletal system The organ system, consisting of bones, cartilage, and ligaments, that provides an internal framework to support the body of vertebrates.

small intestine The highly coiled tube, specialized for absorption, into which food moves from the stomach during digestion.

smooth ER A part of the endoplasmic reticulum, having a smooth appearance, where lipids and hormones are manufactured. Compare *rough ER*.

smooth muscle The specialized muscle found in the walls of the digestive system and blood vessels. It has no visible bands, and its contractions are entirely involuntary. Compare *cardiac muscle* and *skeletal muscle*.

soluble Able to mix completely with water.

solute A dissolved substance, such as sugar in water. Compare *solvent*.

solution Any combination of a solute and a solvent.

solvent The fluid, such as water, into which a substance has dissolved. Compare *solute*.

somatic cell A non-sex cell. Compare *gamete*.

speciation The process by which one species splits to form two species or more.

species 1. Members of a group that can mate with one another to produce fertile offspring. 2. The smallest unit of classification in the Linnaean hierarchy.

species interaction Any of four ecological interactions—mutualism, commensalism, predation, and competition—that occur between different species.

species richness The total number of different species that live in an ecological community.

specific response A response mounted by the adaptive immune system against a specific strain of pathogen. Compare *nonspecific response*.

sperm The male gamete. Compare *egg*.

spermatogenesis The series of cell divisions in human males that results in sperm. Compare *oogenesis*.

spinal cord A thick central nerve cord that is continuous with the brain, acting as a filter between the brain and the sensory neurons.

spongy bone One of the two major types of bone tissue. Honeycombed with numerous tiny cavities, it lies inside the *compact bone*. Spongy bone is most abundant at the knobby ends of our long bones.

spore 1. In fungi, a reproductive structure that can survive for long periods of time in a dormant state and will sprout under favorable conditions to produce the body of the organism. 2. In plants, a haploid cell produced by meiosis that divides mitotically to produce the gametophyte.

sporophyte A diploid, multicellular individual that arises from the zygote in plants. Compare *gametophyte*.

sporulation The formation of thick-walled dormant structures called spores.

stabilizing selection The pattern of natural selection in which individuals with intermediate values of an inherited phenotypic trait have an advantage over other individuals in the population. Compare *directional selection* and *disruptive selection*.

start codon The codon AUG; the point on an mRNA strand at which the ribosomes begin translation. Compare *stop codon*.

statistics A branch of mathematics that estimates the reliability of data.

stem A structure in plants that provides structural support, transports food and water, and holds leaves up to intercept light.

stoma (pl. **stomata**) An air pore in the leaf of a plant that controls gas exchange.

stomach The organ of the digestive system, located between the esophagus and intestines, in which most digestion occurs, through mechanical and chemical means.

stop codon The codon UAA, UAG, or UGA; the point on an mRNA strand at which the ribosomes end translation. Compare *start codon*.

substitution A mutation in which one base is substituted for another in the DNA sequence of a gene. Compare *deletion* and *insertion*.

substrate A molecule that will react to form a new product.

succession The process by which the species in a community change over time.

sustainable Able to be continued indefinitely without causing serious damage to the environment.

swallowing reflex The reaction of the digestive system when food comes into contact with the pharynx, in which the epiglottis seals off the entry into the trachea and food is pushed into the esophagus.

sympatric speciation The formation of new species in the absence of geographic isolation. Compare *allopatric speciation*.

synovial sac The cavity inside a joint that is filled with a lubricating fluid that reduces friction between two bony surfaces.

systemic circuit The circuit in the heart, consisting of the two chambers on the left side, that receives oxygenated blood from the lungs and pumps it to the body. Compare *pulmonary circuit*.

systole The first pressure, or top number, in a blood pressure reading—the pressure at the point the heart contracts to push blood out. Compare *diastole*.

T

T cell A type of lymphocyte that matures in the thymus. T cells are involved in cell-mediated immunity. Compare *B cell*.

temperate forest A terrestrial biome characterized by snowy winters and moist, warm summers and dominated by trees and shrubs adapted to relatively rich soil.

template strand The strand of DNA that is used as a template to make a new strand of DNA.

tendon A specialized, flexible band of tissue, rich in collagen, that joins muscle to bone. Compare *ligament*.

terminator A segment of DNA that, when reached by RNA polymerase, stops transcription. Compare *promoter*.

tertiary consumer An organism that eats *secondary consumers*. Compare *primary consumer* and *quaternary consumer*.

testis (pl. testes) Either of a pair of male reproductive organs that produce sperm and androgens in vertebrates. Compare *ovary*.

testosterone The primary androgen. Compare *estradiol* and *progesterone*.

theory A hypothesis, or a group of related hypotheses, that has received substantial confirmation through diverse lines of investigation by independent researchers. Compare *fact*.

thermophile A prokaryote, usually archaean, that can live in extremely hot environments, such as geysers, hot springs, and hydrothermal vents.

throat See **pharynx**.

thymine (T) One of the four nucleotides that make up DNA. The other three are adenine (A), guanine (G), and cytosine (C).

tissue A group of cells that function in an integrated manner to perform a unique set of tasks in the body.

trachea Also called *windpipe*. The structure that connects the pharynx to the bronchi.

transcription The synthesis of RNA based on a DNA template. Compare *translation*.

transfer RNA (tRNA) A type of RNA that facilitates translation by delivering specific amino acids to the ribosomes as codons are read off of an mRNA. Compare *messenger RNA* and *ribosomal RNA*.

translation The process by which ribosomes convert the information in mRNA into proteins. Compare *transcription*.

transpiration The process of plants absorbing water through their roots and releasing this water through their leaves into the atmosphere.

transport protein A protein that acts like a gate, channel, or pump that allows molecules to move into and out of a cell.

treatment group Also called *experimental group*. The group of subjects in an experiment that is maintained under the same standard set of conditions as the control group but is subjected to manipulation of the independent variable. Compare *control group*.

trimester Any of the three defined stages of human pregnancy. Each trimester is about 3 months long.

tRNA See **transfer RNA**.

trophic level Each level of the energy pyramid, corresponding to a step in a food chain.

tropical forest A terrestrial biome characterized by warm temperatures, about 12 hours of daylight, and seasonally heavy or year-round rains, and containing a rich diversity of organisms.

tumor A cell mass that results from runaway cell division.

tundra A terrestrial biome containing permafrost that is found at the poles and on mountaintops. Tundra is dominated by low-growing flowering plants and a lack of trees.

U

up-regulation The speeding up of gene expression. Compare *down-regulation*.

upper respiratory system The part of the respiratory system that includes airways in the nose, mouth, and throat. Compare *lower respiratory system*.

urinary system The organ system that removes excess fluid from the body, along with waste products, toxins, and other water-soluble substances that are not needed.

urine The waste-carrying watery solution that is expelled from our bodies.

uterus The female reproductive organ in which a fertilized egg implants and develops until birth.

V

vacuole An organelle in plant cells that acts as a garbage or recycling center and that stores water. Compare *lysosome*.

vagina The female reproductive organ that connects the uterus to the external genitalia.

variable A characteristic of any object or individual organism that can change.

variable trait A trait that is different in different individuals of a species. Compare *invariant trait*.

vascular system In plants, a network of tissues that is made up of tubelike structures specialized for transporting fluids.

vascular tissue One of three types of plant tissue. Forming the innermost layer of the plant, vascular tissue (consisting of phloem and xylem) contains stacks of long cells forming continuous tubes that run throughout the plant body and transport materials throughout the plant. Compare *dermal tissue* and *ground tissue*.

vein Any of the large vessels that carry blood back to the heart. Compare *artery* and *capillary*.

ventricle Either of the two lower chambers of the heart. Compare *atrium*.

vertebrae (sing. vertebra) The strong, hollow sections of the backbone, or vertebral column.

vertebrates Chordates that possess a backbone.

vesicle A sac, formed by the bulging inward or outward of a section of the plasma membrane, that moves molecules from place to place inside a cell but also may transport substances into and out of the cell.

vestigial trait A feature that is inherited from a common ancestor but no longer used. Vestigial traits may appear as reduced or degenerated parts whose function is hard to discern.

villus (pl. villi) Any of the large number of fingerlike projections in the small intestine that are specialized for nutrient absorption.

viral strain Also called *serotype*. Any of the variant forms of a particular type of virus.

virus A small, infectious agent that can replicate only inside a living cell.

vitamin Any of various small, organic nutrients needed by the human body, but only in tiny amounts. Compare *mineral*.

voice box See **larynx**.

W

warning coloration Bright coloring of an animal that alerts a potential predator to dangerous chemicals in the animal's tissues. Compare *camouflage* and *mimicry*.

water-soluble vitamin A vitamin that can dissolve in water and therefore tends not to accumulate in body tissues because it can be easily excreted in urine. Compare *fat-soluble vitamin*.

weather Short-term atmospheric conditions, such as today's temperature, precipitation, wind, humidity, and cloud cover. Compare *climate*.

windpipe See **trachea**.

X

X-linked Found solely on the X chromosome. Compare *Y-linked*.

xylem One of two types of vascular tissue in plants. Xylem transports water and minerals, absorbed from the soil, upward from the roots and outward from the central stem to the leaves. Compare *phloem*.

Y

Y-linked Found solely on the Y chromosome. Compare *X-linked*.

yeasts Single-celled fungi that belong in the group zygomycetes and are important in the rising of bread, the brewing of beer, and the fermenting of wine.

Z

Z disc A structure, found at each end of a sarcomere, that contains a large protein that provides an anchor point for actin filaments.

zygomycetes One of three main groups of fungi, containing many species of molds. Compare *ascomycetes* and *basidiomycetes*.

zygote The single cell that results from fertilization.

Credits

Photos and Text

Front Matter

Page v top: Jonathan Mays, Wildlife Biologist, Maine Department of Inland Fisheries and Wildlife/USFWS; bottom: Alexandra Grablewski/DigitalVision/Getty Images; **p. vi** top: Peter Zurek/Shutterstock; center: (puzzle) George Diebold/Getty Images; (cell) Mopic/Shutterstock; bottom: Michael Melford/National Geographic/Getty Images; **p. vii** top: Tsuneo Yamashita/Getty Images; bottom: (background) Iaroslav Neliubov/Shutterstock; (dogs) Eric Isselee/Shutterstock; **p. viii** top: (background) bitterfly/Thinkstock/Getty Images Plus; (toy) Gaby Kooijman/Thinkstock/Getty Images Plus; center: Mike Kemp/Rubberball/Getty Images; bottom: (background) shironosov/iStock/Getty Images Plus; (tobacco) joannawnuk/Shutterstock; **p. ix** top: Andrey Nekrasov/Alamy; center: (background) muss/Shutterstock; (IV bag) Thinkstock/Getty Images; (VSRA) Science Source; bottom: (lizard) MattiaATH/Shutterstock; (diver) CyberEak/Shutterstock; **p. x** top: Lou Linwei/Alamy; center: Elke Van de Velde/Photodisc/Getty Images Plus; bottom: JimmyWrangles/iStock/Getty Images Plus; **p. xi** top: (karyotype) Leonard Lessin/Science Source/Getty Images; (background) thomaslenne/iStock/Getty Images Plus; (table) Piotr Adamowicz/iStock/Getty Images Plus (scientist) LattaPictures/iStock/Getty Images Plus; (left skull) Leemage/Universal Images Group/Getty Images; (right skull) Sabena Jane Blackbird/Alamy Stock Photo; bottom: (background) Franz Lanting/Offset/Shutterstock; (smoke) Shutterstock; (charcoal edges) Anan Kaewkhammul/Shutterstock; (pre-burned edges) Sylwia Brataniec/Shutterstock; **p. xii** top: Paul Starosta/Corbis Documentary/Getty Images; center: Holly Kuchera/Shutterstock; bottom: SECCHI Disk; (water) TFoxFoto/Shutterstock; **p. xiii** top: SCIEPRO/Science Photo Library/Getty Images; center: (sky) Arcady/Shutterstock; (girl) Alex Mares-Manton/Getty Images; bottom: Vince Michaels/Getty Images; **p. xiv** top: Henny Boogert; bottom: Adam Gault/OJO Images/Getty Images; **p. xv**: (Houtman) Will Prouty; (Scudellari): Sarah Der Photography; (Malone) courtesy Cindy Malone; (Brewer): Rebecca Brewer.

Chapter 1

Pages 2–3: Jonathan Mays, Wildlife Biologist, Maine Department of Inland Fisheries and Wildlife/USFWS; **p. 4:** Gerrit Vyn; **p. 6** (left to right): Judith Collins/Alamy, Scott Camazine/Science Source, AS Food studio/Shutterstock, Lev Kropotov/Shutterstock, Rosa Jay/Shutterstock; **p. 7** top: Photo by Dr. Kimberli Miller, courtesy USGS National Wildlife Health Center; bottom: Stephen Alvarez/National Geographic Creative; **p. 8** left: Photo courtesy Ryan von Linden. Photo used with permission from New York State Department of Environmental Conservation. All rights reserved; right top: Kevin Wenner/Pennsylvania Game Commission; right bottom: David S. Blehert, U.S. Geological Survey—National Wildlife Health Center; **p. 9** top: Image Courtesy of The Advertising Archives; bottom: photograph by Steve Grodsky; **p. 10** (all) Amy Smotherman Burgess/Knoxville News Sentinel/ZUMAPRESS.com; **p. 13** top left: Photo used with permission from New York State Department of Environmental Conservation; center left: Greg Turner/Pennsylvania Game Commission; bottom left: Photo courtesy Ryan von Linden. Photo used with permission from New York State Department of Environmental Conservation. All rights reserved; top right: Deborah Springer; center right: Design Pics Inc/Alamy; bottom right: Dr. Mary Hausbeck; **p. 15** top (left to right): Valeriy Vladimirovich Kirsanov/Shutterstock, Valeriy Vladimirovich Kirsanov/Shutterstock, Rosa Jay/Shutterstock, Vitalii Hulai/Shutterstock; bottom right: nico99/Shutterstock; bottom left: ASA-Carlos Asanuma/Getty Images; **p. 17:** Lindsey Heffernan, PA Game Commission/Patty Stevens, USGS; **p. 19** top left: SPL/Science Source; bottom left: Wally Eberhart/Visuals Unlimited, Inc.; right: NASA Ames/SETI Institute/JPL-Caltech.

Chapter 2

Pages 20–21: Alexandra Grablewski/Digital Vision/Getty Images; **p. 22:** Dr. Anna K. Eaton; **p. 23** top left: Gohil JR, Shah BA, Parekh AN. Diphtheria WebmedCentral INFECTIOUS DISEASES 2011;2(12): WMC002594 doi: 10.9754/journal.wmc.2011.002594. CC BY 3.0; center: Science Source/Colorization by Mary Madsen; top right: South West News Service; bottom left: AP Photo; bottom right: CDC; **p. 24:** Rachel Torres/Alamy Stock Photo; **p. 25** left: Dr. Anna K. Eaton; right: Ryan Collerd/The New York Times/Redux Pictures; **p. 28:** Graph (Fig. 2.6): "The Real Cause of Increasing Autism Prevalence?" by J. Emory Parker. Reprinted by permission of the author; **p. 31:** Republished with permission of Elsevier Science and Technology Journals, from "Ileal-Lymphoid-Nodular, Hyperplasia, Non-specific Colitis, and Pervasive Developmental Disorder in Children." Wakefield, AJ, et al. *Lancet*, v351(9103) 1998; permission conveyed through Copyright Clearance Center, Inc. **p. 34:** Mara008/Shutterstock; **p. 35:** Dr. Anna K. Eaton; **p. 37:** Chart courtesy Tyler Vigen, tylervigen.com.

Chapter 3

Pages 38–39: Peter Zurek/Shutterstock; **p. 40:** David McNew/Reuters/Newscom; **p. 41:** Time & Life Pictures/Getty Images; **p. 43** top: Scripps Institution of Oceanography, UC San Diego; bottom: Courtesy of Henderson James Cleaves III; **p. 45:** Scripps Institution of Oceanography, UC San Diego; **p. 46:** Photolukacs/Shutterstock; **p. 47:** Kerstin Langenberger/imagebroker/Alamy; **p. 48** top: NASA, ESA, and D. Jewitt (UCLA); bottom: NASA Ames Research Center/SETI/Peter Jenniskens; **p. 49** top: Helen Sessions/Alamy; center: Shcherbinator/Shutterstock; bottom: Helen Sessions/Alamy; **p. 52** top: Africa Studio/Shutterstock; center: Now Foods; bottom: Kateryna Kon/Shutterstock; **p. 53** top left: Deep OV/Shutterstock; top center: designelements/Shutterstock; center right: Photo Melon/Shutterstock; center left: iofoto/Shutterstock; top right: Andrey Starostin/Shutterstock; center: Abramova Elena/Shutterstock; center bottom right: tryton2011/Shutterstock; bottom right: Jacqui Hurst/Corbis Documentary via Getty Images; **p. 57** top left: Paramount Television/Kobal/ REX /Shutterstock; bottom left: guentermanaus/Shutterstock; top right: inomasa/Shutterstock; bottom right: Carol Farneti-Foster/Oxford Scientific/Getty Images.

Chapter 4

Pages 58–59 (puzzle): George Diebold/Getty Images, (cell) Mopic/Shutterstock; **p. 60** top left: A. Barry Dowsett/Science Source; top right: Biophoto Associates/Science Source/Colorization by Mary Martin; bottom left: EMBL/Johanna Höög; bottom right: CNRI/Science Source; **p. 61:** Jessica Rinaldi/Reuters/Newscom; **p. 62:** Photos courtesy J. Craig Venter Institute. From Gibson, DG et al., "Creation of a Bacterial Cell Controlled by a Chemically Synthesized Genome." *Science*. 2010 Jul 2; 329(5987):52–6. doi: 10.1126/science.1190719; **p. 63:** Courtesy of Neal Devaraj; **p. 65:** Photo Researchers, Inc./Composition by: Eric Cohen/Science Source; **p. 66:** Roger Harris/Science Source; **p. 67** top left: Don W. Fawcett/Science Source; top right: SPL/Science Source; bottom left and right: Dennis

p. 227: blackboard1965/Shutterstock; **p. 229:** Karen Haydock.

Chapter 13

Pages 230–31 top: MattiaATH/Shutterstock; bottom: CyberEak/Shutterstock; **p. 232** bottom: Joseph T. Collins/Science Source; top: Anthony Herrel of the University of Antwerp; **p. 233** bottom: Thomas Hoogendyk, Coherent Bytes, LLC; top left; bottom left; bottom right: Anthony Herrel of the University of Antwerp; **p. 234** left: Photo by Luke Mahler; right: Photo by Marc C. Perkins; **p. 235** top: Juniors Bildarchiv GmbH/Alamy; bottom: Cyril Ruoso/Biosphoto/Getty Images; **p. 236** left: Thomas & Pat Leeson/Science Source; center: Larry Geddis/Alamy; right: Gerald C. Kelley/Science Source; **p. 238** left; center; right: Courtesy of Carlos Prada; **p. 238** bottom: Mary Leavines/LSU Daily Reveille; **p. 239** all: Todd C LaJeunesse; **p. 240:** Gregory Basco/Visuals Unlimited, Inc.; **p. 241:** Sohns/imageBROKER/Alamy; **p. 243** top; bottom left and right: Simon E. Coppard; **p. 244** (clockwise from far left): Boule/Shutterstock, Martina_L/Shutterstock, Eric Isselee/Shutterstock, Blinka/Shutterstock, IrinaK/Shutterstock, Mike Truchon/Shutterstock, BOONROONG/Shutterstock, oksana2010/Shutterstock, StockPhotoAstur/Shutterstock, Sakdinon Kadchiangsaen/Shutterstock; **p. 245** left: Robert Sivinski; right: Doug Sokell/Visuals Unlimited; **p. 246** left: gallimaufry/Shutterstock; right: Sinclair Stammers/Science Source; **p. 247** left: Eric Isselee/Shutterstock; top right: Yuri Kevhiev/Alamy Stock Photo; bottom right: Coprid/Shutterstock.

Chapter 14

Pages 248–49: Lou Linwei/Alamy; **p. 250:** Carlos Munoz/Getty Images; **p. 251** top: MichaelTaylor3d/Shutterstock; center: Eye of Science/Science Source; bottom: Ray Simons/Science Source; **p. 252:** Lou Linwei/Alamy; **p. 253:** Royal Saskatchewan Museum (RSM/R.C. McKellar); **p. 255:** Image courtesy of Dr. Xu Xing. Reprinted by permission from Macmillan Publishers Ltd: Nature 475:465–470 Xing, X. et al. "An Archaeopteryx-like Theropod from China and the Origin of Avialae," 2011; **p. 258** top left: Roland Birke/Getty Images; top center: Rainer Fuhrmann/Shutterstock; top right: Dr. Peter Siver/Visuals Unlimited; center left: Robert and Jean Pollock/Science Source; center: Chris Mattison/FLPA/Science Source; bottom left: Dennis Flaherty/Science Source; bottom right: Schmitz Olaf/Getty Images; **p. 259** top left: Andrew Syred/Science Source; top center: Fletcher & Baylis/Science Source; top right: Ingeborg Knol/imageBROKER/Corbis; center left: Borut Furlan/Getty Images; center: Stuart Wilson/Science Source; center right: Andrew J. Martinez/Science Source; bottom left: Andrew J. Martinez/Science Source; bottom center: Marius Dobilas/Shutterstock; bottom right: Martin Shields/Alamy; **p. 260** 1st image in each row: Hans Kim/Shutterstock; row 1 (left to right): Science Source (Colorization by: Eric Cohen), BSIP/Science Source, Georges Antoni/Hemis/Alamy, Gary Meszaros/Science Source, Picture Partners/Science Source; row 2 (left to right): Gary Meszaros/Science Source, Irochka_T/iStock/Getty Images Plus, Anthony Mercieca/Science Source, James DeBoer/Shutterstock; row 3 (left to right): Valeriy Vladimirovich Kirsanov/Shutterstock, FLPA/Alamy, James Hager/Robert Harding/Alamy; row 4 (left to right): P. Wegner/Arco Images/Alamy, Bernd Rohrschneider/FLPA/Science Source; row 5 (left to right): Image Source/Alamy Stock Photo, Eric Isselee/Shutterstock; row 6 Sabena Jane Blackbird/Alamy Stock Photo; **p. 261:** The Royal Belgian Institute of Natural Sciences, Brussels; **p. 262:** upper left and right: The Royal Belgian Institute of Natural Sciences, Brussels; top: Emily Willoughby/Stocktrek Images/Alamy; center top: Masato Hattori; center: Natural History Museum, London; center bottom: Masato Hattori; bottom: Walt Anderson/Visuals Unlimited/GettyImages; **p. 267** left: Terry Whittaker/Alamy Stock Photo; right: Corey Ford/Stocktrek Images/National Geographic Creative.

Chapter 15

Pages 268–69: Elke Van de Velde/Photodisc/Getty Images Plus; **p. 270:** image courtesy Allen Nutman, University of Wollongong. A.P. Nutman et al. "Rapid Emergence of Life Shown by Discovery of 3,700-Million-Year-Old Microbial Structures," Nature. Published online August 31, 2016. doi: 10.1038/nature19355; **p. 271:** photo by Amanda Ward/Dunn Lab; **p. 272** top left: Gillian Conquest, Extreme Citizen Science (ExCiteS) research group, University College London, top right: Photograph courtesy of Gardenista; bottom: Jim Davis/The Boston Globe via Getty Images; **p. 273** all: Dunn Lab; **p. 274** top left: Eye of Science/Science Source; center left: David McCarthy/Science Source; bottom left: Michael Abbey/Science Source; top right: Scimat/Science Source; bottom right: Eye of Science/Science Source; **p. 275** top left: David Henderson/caia image/Alamy Stock Photo; top left inset: Eye of Science/Science Source; top right: Jose Arcos Aguilar/Shutterstock; top right inset: image courtesy Luis R. Comolli. Comolli, L. R., Baker, B., Downing, K.H., Siegerist, C. E., and Banfield, J. (2008). "Three-Dimensional Analysis of the Structure and Ecology of a Novel, Ultra-Small Archaeon." ISME J, 3(2), 159–167; bottom: IFE, URI-IAO, UW, Lost City Science Party; NOAA/OAR/OER; The Lost City 2005 Expedition; bottom inset: MichaelTaylor3d/Shutterstock; **p. 277:** Photo courtesy of Lauren Nichols; **p. 278** top left: Dr. Ron Dengler/Visuals Unlimited/Getty Images; top right: Dennis Kunkel Microscopy/Science Source; bottom left: Eye of Science/Science Source; bottom right: CDC/James Archer, Illustrators: Alissa Eckert and Jennifer Oosthuizen; **p. 279:** Universal History Archive/REX/Shutterstock; **p. 280** background: Michael Heim/EyeEm/Getty Images; top right; top left and bottom: Dennis Kunkel Microscopy/Science Source; **p. 281:** photo by Lauren Nichols, Your Wild Life/Dunn Lab; **p. 283** top: Joel Sartore, National Geographic Photo Ark/Nat Geo Creative; bottom: Caroline Dahl.

Chapter 16

Pages 284–85: JimmyWrangles/iStock/Getty Images Plus; **p. 286** left: photo courtesy of Kumar Paudel; right: Jacob Phelps; **p. 287** top: Wim van Egmond/Science Photo Library via Science Source; bottom: Plant-Success; **p. 290:** Leon Neal/AFP/Getty Images; **p. 291** left: divedog/Shutterstock; right: Casther/Shutterstock; **p. 292:** Lance Cheung/USDA; **p. 293:** Carroll & Carroll/AgStock/Design Pics Inc/Alamy Stock Photo, inset: Shawn Poynter/The New York Times/Redux; **p. 295** top left: Stefano Rellandini/Reuters/Newscom; right: Laszlo Podor/Moment Open/Getty images; bottom left: Stefano Rellandini/Reuters/Newscom; **p. 296** left: Le Do/Shutterstock; center: Suzifoo/iStock/Getty Images Plus; right: Egor Rodynchenko/Shutterstock; bottom (left to right): Amazing snapshot/Shutterstock, xpixel/Shutterstock, kzww/Shutterstock, tarapong srichaiyos/Shutterstock, Scott Sinklier/Alamy Stock Photo, Vitaly Korovin/Shutterstock; **p. 297** top: Glass and Nature/Shutterstock; bottom: Frank Kohn/USFWS; **p. 299:** Designua/Shutterstock.

Chapter 17

Pages 300–301: (background) thomaslenne/iStock/Getty Images Plus; (table) Piotr Adamowicz/iStock/Getty Images Plus; (karyotype) Leonard Lessin/Science Source/Getty Images; (scientist) LattaPictures/iStock/Getty Images Plus; (left skull) Leemage/Universal Images Group/Getty Images; (right skull) Sabena Jane Blackbird/Alamy Stock Photo; **p. 302:** Dr. Mirjana Roksandic; **p. 306:** Edward Westmacott/Shutterstock; **p. 307** left: Steve Bloom Images/Alamy; center: John W Banagan/Getty Images; right: Dave Watts/Nature Picture Library/Alamy Stock Photo; **p. 311:** John Reader/Science Source; **p. 313:** Frank Vinken; **p. 318** top: Sabena Jane Blackbird/Alamy Stock Photo; bottom: Silvana Condemi; **p. 319:** Mark Thiessen/National Geographic Creative; **p. 321** left: T. Michael Keesey; top right: Stan Osolinski/Getty Images; bottom right: gritsalak karalak/Shutterstock.

Chapter 18

Pages 322–23: (background) Franz Lanting/Offset/Shutterstock; (smoke) Shutterstock; (charcoal edges) Anan Kaewkhammul/Shutterstock; (pre-burned edges) Sylwia Brataniec/Shutterstock; **p. 324** top and bottom: Jennifer Balch; **p. 325** left: NASA; right: Fotos593/Shutterstock; **p. 326** top: Paul Lefebvre, Woods Hole Research Center; bottom: Dr. Michael T. Coe; **p. 327** top left: David Greedy/Getty Images; top right: John SommersII/Reuters/Newscom; center: Greenpeace/Ges.oek.Forschung; bottom left: James Steinberg/Science Source; bottom right: AP Photo/Nick Ut; **p. 330** top: Hervé Collart/Sygma via Getty Images; bottom: Ricky Rogers/Reuters/Newscom; **p. 331:** (leaves) Valentyn Volkov/Shutterstock, (tree section 1) My Life Graphic/Shutterstock, (tree section 2) andersphoto/Shutterstock; **p. 337:** Andrew Orlemann/Shutterstock; **p. 339:** 19th era 2/Alamy Stock Photo.

Chapter 19

Pages 340–41: Paul Starosta/Corbis Documentary/Getty Images; **p. 343** top: Estefan Radovicz/Agencia o Dia/AGENCIA ESTADO/Xinhua/Alamy Live News; bottom: photo courtesy of Dr. Matthew DeGennaro; **p. 344:** CDC; **p. 345:** CDC/Cynthia Goldsmith; **p. 346:** 7th Son Studio/Shutterstock; **p. 349** left: Dyakova Yulia/Shutterstock; right: Mark Newman/Tom

Stack & Associates; **p. 350:** Tom Brakefield/Getty Images; **p. 351** top: Alan G. Nelson/Dembinsky Photo Associates; bottom: photo courtesy of Dr. Margareth Capurro; **p. 352:** Angel Valentin/The New York Times/Redux; **p. 353:** (mosquito) Antagain/E+/Getty Images Plus, (animal silhouettes) nikiteev_konstantin/Shutterstock; **p. 355** left: Graph: "Oxitec Approach Suppresses Mosquitoes in Piracicaba, Brazil." Appeared in "No GM Mosquitoes Didn't Start the Zika Outbreak," Discovermagazine.com, January 31, 2016. Reprinted by permission of Oxitec Ltd; top right: RNGS/RTR/Newscom; bottom right: Paul Darrow/Polaris/Newscom.

Chapter 20

Pages 356–57: Holly Kuchera/Shutterstock; **p. 358** top: photo by Dr. William Ripple; bottom: Pat & Chuck Blackley/Alamy; **p. 359:** Dr. William Ripple; **p. 361:** Don Johnston_IH/Alamy Stock Photo, inset: Kevin Ebi/Alamy Stock Photo; **p. 362:** Ksanawo/Shutterstock; **p. 363** left: Martin Harvey/Corbis via Getty Images; right: Luc Novovitch/Alamy Stock Photo; **p. 364:** Daniel Stahler/NPS; **p. 365** top: Mark Conlin/Oxford Scientific/Getty Images; bottom: David Fleetham/Visuals Unlimited/Getty Images; **p. 366:** (wolves) rachisan alexandra/Shutterstock, (deer) KatarinaF/Shutterstock, (trees) Vertyr/Shutterstock; **p. 367:** PhotoStock-Israel/Alamy; **p. 368** top: Joe McDonald/Visuals Unlimited; center left: Nancy Nehring/Getty Images; center right: CHAINFOTO24/Shutterstock; bottom: Gerry Bishop/Visuals Unlimited; **p. 369:** Norbert Rosing/National Geographic Creative; **p. 371:** Anne-Marie Kalus; **p. 372:** Adam Burton/Alamy; **p. 373** left: Stan Osolinski/Dembinsky Photo Associates; center: Howard Garrett/Dembinsky Walt Anderson/Visuals Unlimited; **p. 375** bottom right: George Gentry/U.S. Fish and Wildlife Service.

Chapter 21

Pages 376–77: (water) TFoxFoto/Shutterstock; SECCHI Disk; **p. 378:** Gary Corbett/Alamy; **p. 379** top: Jacques Descloitres, MODIS Rapid Response Team, NASA/GSFC, inset: D.P. Wilson/FLPA/Science Source; bottom left: Tania Wong; bottom right: Dr. Boris Worm; **p. 382:** NASA image by Jesse Allen & Robert Simmon; **p. 383:** Chris Linder/Visuals Unlimited; **p. 384** top left: Bernd Zoller/imagebroker/Alamy; top right: John E Marriott/Alamy; center: Ken Lucas/Visuals Unlimited; bottom left: Pixtal Images/Media Bakery; bottom right: Design Pics Inc/Photolibrary/Getty Images; **p. 385** top left: Michael P. Gadomski/Earth Sciences/Animals Animals; top right: Mark De Fraeye/SPL/Science Source; center: Willard Clay/Dembinsky Photo Associates; bottom left: Mark Goodreau/Alamy; bottom right: Jim Zipp/National Audubon Society Collection/Science Source; **p. 387:** Abigail McQuatters-Gollop; **p. 391:** Map courtesy Dr. Danielle Medek. From Danielle E. Medek, Joel Schwartz, and Samuel S. Myers Estimated Effects of Future Atmospheric CO2 Concentrations on Protein Intake and the Risk of Protein Deficiency by Country and Region. Environmental Health Perspectives, August 2017, Volume 125, Issue 8 DOI:10.1289/EHP4.

Chapter 22

Pages 392–93: SCIEPRO/Science Photo Library/Getty Images; **p. 394:** Spies 2015; **p. 399** left: Reinhard Dirscherl/agefotostock; center: Matteo photos/Shutterstock; right: cbpix/Alamy; **p. 400** left: Clouds Hill Imaging Ltd/Getty Images; right: Dennis Kunkel Microscopy/Science Source; **p. 402:** Dr. Tina Kold Jensen; **p. 403:** Lukiyanova Natalia/frenta/Shutterstock; **p. 404:** Dr. Niels Jorgensen; **p. 406** left: MedicalRF.com/Corbis; center: Joo Lee/Corbis Documentary via Getty Images; right: Anatomical Travelogue/Science Source; **p. 411** left (left to right): pattara puttiwong/Shutterstock; Butterfly Hunter/Shutterstock; Daniel Prudek/Shutterstock; right: JMx Images/Shutterstock.

Chapter 23

Pages 412–13: (sky) Arcady/Shutterstock; Alex Mares-Manton/Getty Images; **p. 414:** BSIP/Science Source; **p. 415** top left: Gtranquillity/Shutterstock; top center: Juliya Shangarey/Shutterstock; top right: Media Bakery; bottom left: Nik Merkulov/Shutterstock; bottom center: D7INAMI7S/Shutterstock; bottom right: Shutterstock; **p. 421:** James L. Amos/Science Source; **p. 424:** Philippe Autier; **p. 426** top (left to right): jeehyun/Shutterstock, Maks Narodenko/Shutterstock, Petr Malyshev/Shutterstock, Nattika/Shutterstock; bottom (left to right): Levent Konuk/Shutterstock, Binh Thanh Bui/Shutterstock, Viktor1/Shutterstock, (pill bottle) HeinzTeh/Shutterstock; **p. 427:** Chloe Johnson/Alamy; **p. 429:** Mediscan/Alamy Stock Photo.

Chapter 24

Pages 430–31: Vince Michaels/Getty Images; **p. 432** top and bottom right: Shawn Rocco/Duke Medicine; **p. 433:** Laura Niklason; **p. 434** top and center: Dennis Kunkel/Science Source; bottom: Richard Kessel & Dr. Randy Kardon/Visuals Unlimited; **pp. 439, 440, 442** (both) The Ott Lab/Massachusetts General Hospital, Boston MA; **p. 444:** Alfred Pasieka/Science Source; **p. 445:** Christine E. Schmidt, Ph.D.; **p. 448:** Scott Lewis Photography, LLC; **p. 451:** Matthew Burgos, Worcester Polytechnic Institute.

Chapter 25

Pages 452–53: Henny Boogert; **p. 454** top: Matthijs Kox, PhD; **p. 454** center: Henny Boogert; bottom: Enahm at Innerfire; **p. 456:** Marcel Rekers; **p. 461** left: Don W. Fawcett/Science Source; right: MedImage/Science Source; **p. 462:** Volker Steger/Science Source; **p. 466:** Henny Boogert; **p. 468:** Designua/Shutterstock; **p. 469:** CDC.

Chapter 26

Pages 470–71: Adam Gault/OJO Images/Getty Images; **pp. 472, 473:** courtesy of The Land Institute; **p. 478** left: Dragomir Radovanovic/Shutterstock; right: Photograph by Kathleen Bauer, GoodStuffNW.com; **p. 479** top: courtesy of The Land Institute; bottom: University of Minnesota, College of Food, Agricultural and Natural Resources Science, Department of Agronomy and Plant Genetics; **p. 482** (left to right): Vasilius/Shutterstock, bergamont/Shutterstock, natu/Shutterstock, Madlen/

Shutterstock; **p. 484** left: Stephen Ausmus/USDA ARS; center: Anthony Mercieca/Science Source; right: courtesy of The Land Institute; **p. 485:** courtesy of The Land Institute; **p. 487** top: Jean-Loup Charmet/Science Source; bottom: Ray Coleman/Science Source.

Infographics

Chapter 1

Page 15, "Bug Zappers": Data from Justin G. Boyles, Paul M. Cryan, Gary F. McCracken, and Thomas H. Kunz, "Economic Importance of Bats in Agriculture," *Science*, April 1, 2001, www.sciencemag.org/content/332/6025/41.

Chapter 2

Page 34, "Safety in Numbers": Data from Caroline L. Trotter and Martin C. J. Maiden, "Meningococcal Vaccines and Herd Immunity: Lessons Learned from Serogroup C Conjugate Vaccination Programs," Expert Review of Vaccines, July 2009, www.ncbi.nlm.nih.gov/pubmed/19538112.

Chapter 3

Page 54, "What's It All Made Of?": Earth's Atmosphere data from Wikipedia, en.wikipedia.org/wiki/Abundance_of_the_chemical_elements. The Universe data from WebElements, www.webelements.com/periodicity/abundance_universe/. The Human Body data from Wikipedia, en.wikipedia.org/wiki/Chemical_makeup_of_the_human_body. Earth's Crust data from Wikipedia, en.wikipedia.org/wiki/Abundance_of_elements_in_Earth%27s_crust.

Chapter 4

Page 73, "Sizing Up Life": Data from Genetic Science Learning Center at the University of Utah, "Cell Size and Scale," learn.genetics.utah.edu/content/cells/scale/.

Chapter 5

Page 92, "Making Way for Renewables": Data from REN21, "Renewables 2016 Global Status Report," www.ren21.net/status-of-renewables/global-status-report/.

Chapter 6

Page 112, "Cancer's Big 10": Data from Centers for Disease Control and Prevention, "United States Cancer Statistics: 2013 Top Ten Cancers," nccd.cdc.gov/uscs/toptencancers.aspx.

Chapter 7

Page 130, "Does Bigger Mean Better?": Data from National Center for Biotechnology Information, U.S. National Library of Medicine, "Genome Information by Organism," www.ncbi.nlm.nih.gov/genome/browse/.

Chapter 8

Page 143, "Genetic Diseases Affecting Americans": Data from Cold Spring Harbor Laboratory, DNA Learning Center, "Your Genes, Your Health," www.ygyh.org.

Chapter 9

Page 168, "The Meteoric Rise of CRISPR": PubMed search results data from National Center for Biotechnology Information, U.S. National Library of Medicine, www.ncbi.nlm.nih.gov/pubmed/?term=crispr. Timeline data from Broad Institute, "CRISPR Timeline," www.broadinstitute.org/what-broad/areas-focus/project-spotlight/crispr-timeline, and from CRISPR Update, "CRISPR Timeline," www.crisprupdate.com/crispr-timeline/.

Chapter 10

Page 185, "The Deadly Price of a Pandemic": Data from U.S. Department of Health and Human Services, "Pandemic Flu History," www.flu.gov/pandemic/history/index.html, and from World Health Organization, "Global Influenza Programme," www.who.int/influenza/en/.

Chapter 11

Page 206, "Watching Evolution Happen": Data from Oliver Tenaillon, Alejandra Rodríguez-Verdugo, Rebecca L. Gaut, Pamela McDonald, Albert F. Bennett, Anthony D. Long, and Brandon S. Gaut, "The Molecular Diversity of Adaptive Convergence," Science, January 27, 2012, www.ncbi.nlm.nih.gov/pubmed/22282810.

Chapter 12

Page 227, "Race against Resistance": Antibacterial drugs data from Infectious Diseases Society of America, "Bad Bugs, No Drugs: As Antibiotic Discovery Stagnates. . . A Public Health Crisis Brews," www.idsociety.org/uploadedfiles/idsa/policy_and_advocacy/current_topics_and_issues/antimicrobial_resistance/10x20/images/bad%20bugs%20no%20drugs.pdf, and from Dalia Deak, Kevin Outterson, John H. Powers, and Aaron S. Kesselheim, Annals of Internal Medicine, September 6, 2016, "Progress in the Fight Against Multidrug-Resistant Bacteria? A Review of U.S. Food and Drug Administration–Approved Antibiotics, 2010–2015," annals.org/aim/article/2526197/progress-fight-against-multidrug-resistant-bacteria-review-u-s-food/. Timeline data from Centers for Disease Control and Prevention, "Antibiotic Resistance Threats in the United States, 2013," www.cdc.gov/drugresistance/threat-report-2013/pdf/ar-threats-2013-508.pdf.

Chapter 13

Page 244, "On the Diversity of Species": Data from A.D. Chapman, "Numbers of Living Species in Australia and the World, 2nd Edition, A Report for the Australian Biological Resources Study, September 2009," www.environment.gov.au/node/13875.

Chapter 14

Page 264, "The Sixth Extinction": Data from International Union for Conservation of Nature and Natural Resources, The IUCN Red List of Threatened Species, Summary Statistics, www.iucnredlist.org/about/summary-statistics.

Chapter 15

Page 280, "The Bugs in Your Belly Button": Data from Robert R. Dunn, Julie Horvath-Roth, Sarah P. Council, Holly Menninger, and Matthew Fitzpatrick, Your Wild Life, November 25, 2013, "The Belly Buttons Will be Revealed, Slowly," yourwildlife.org/2013/11/the-belly-buttons-will-be-revealed-slowly/.

Chapter 16

Page 296, "Food Banks": Data from Food and Agriculture Organization of the United Nations, "Biodiversity for Food Security and Nutrition: 30 Years of the Commission on Genetic Resources for Food and Agriculture," www.fao.org/resources/infographics/infographics-details/en/c/174199/, and "Commission on Genetic Resources for Food and Agriculture: Plant Genetic Resources," www.fao.org/nr/cgrfa/cthemes/plants/en/.

Chapter 17

Page 309, "Hereditary Heirlooms": Data from National Geographic, "Genes Are Us. And Them," ngm.nationalgeographic.com/2013/07/125-explore/shared-genes, and from Stefan Lovgren, National Geographic News, August 31, 2005, "Chimps, Humans 96 Percent the Same, Gene Study Finds," news.nationalgeographic.com/news/2005/08/0831_050831_chimp_genes.html.

Chapter 18

Page 331, "Forest Devastation": Deforestation and forest degradation data from G. Kissinger, M. Herold, and V. De Sy, "Drivers of Deforestation and Forest Degradation: A Synthesis Report for REDD+ Policymakers," Lexeme Consulting, Vancouver, Canada, August 2012, www.gov.uk/government/uploads/system/uploads/attachment_data/file/65505/6316-drivers-deforestation-report.pdf. Top 5 country data from Food and Agriculture Organization of the United Nations, "Change in Extent of Forest and Other Wooded Land 1990–2005," www.fao.org/forestry/32033/en/.

Chapter 19

Page 353, "World's Deadliest Animals": Data collected from multiple sources at Bill Gates, GatesNotes, April 25, 2014, "The Deadliest Animal in the World," www.gatesnotes.com/Health/Most-Lethal-Animal-Mosquito-Week/.

Chapter 20

Page 366, "Cause and Effect": Data from William J. Ripple and Robert L. Beschta, "Restoring Yellowstone's Aspen with Wolves," Biological Conservation 138 (2007), www.cof.orst.edu/leopold/papers/Restoring%20Yellowstone%20aspen%20with%20wolves.pdf.

Chapter 21

Page 388, "Productive Plants": Data from Oak Ridge National Laboratory Distributed Active Archive Center, "NPP and the Global Carbon Cycle," daac.ornl.gov/NPP/html_docs/npp_ccycle.html.

Chapter 22

Page 403, "Preventing Pregnancy": Data from James Trussell, "Contraceptive Failure in the United States," Contraception, May 2011; 83(5): 397–404, www.ncbi.nlm.nih.gov/pmc/articles/PMC3638209/.

Chapter 23

Page 426, "Nutritional Needs": Multivitamin data from One A Day® Men's Pro Edge and Women's Pro Edge vitamins. RDA data from Jennifer J. Otten, Jennifer Pitzi Hellwig, and Linda D. Meyers (editors), Dietary Reference Intakes: The Essential Guide to Nutrient Requirements, The National Academies Press, Washington, DC, 2006, www.nap.edu/catalog/11537.html; www.nal.usda.gov/fnic/DRI/Essential_Guide/DRIEssentialGuideNutReq.pdf. Food data from National Institutes of Health, Office of Dietary Supplements, "Dietary Supplements Fact Sheets," ods.od.nih.gov/factsheets/list-all/.

Chapter 24

Page 447, "Have a Heart": Waiting list registration data from United Network for Organ Sharing, "Data Slides for Spring Regional Meetings: 2013 Data," www.unos.org/docs/DataSlides_Spring_2014.pdf. Completed transplants data from U.S. Department of Health and Human Services, Organ Procurement and Transplantation Network, optn.transplant.hrsa.gov.

Chapter 25

Page 465, "Driven by Hormones": Melatonin data from B. Selmaoui and Y. Touitou, "Reproducibility of the Circadian Rhythms of Serum Cortisol and Melatonin in Healthy Subjects: A Study of Three Different 24-h Cycles over Six Weeks," Life Sci., 73 (2003): 3339–3349. Epinephrine data from University of Mississippi Medical Center, "Physical Exercise—Epinephrine," www.umc.edu/Education/Schools/Medicine/Basic_Science/Physiology_and_Biophysics/Core_Facilities(Physiology)/Physical_Exercise_-_Epinephrine.aspx. Cortisol data from J. H. Wilmore, D. L. Costill, and W. L. Kenney, Physiology of Sport and Exercise, Human Kinetics, Champaign, IL, 2007.

Chapter 26

Page 482, "GM Crops Take Root": Data from Clive James, "Global Status of Commercialized Biotech/GM Crops: 2012," ISAAA Brief No. 44, ISAAA, Ithaca, NY, 2012, www.isaaa.org/resources/publications/briefs/44/download/isaaa-brief-44-2012.pdf.

Index

Page numbers set in *italic* type refer to materials in figures or tables.

A

Abert's squirrel, *236*
abiotic factors, 325, 336–37, 378, 380–82
abscisic acid (ABA), *477*
absorption of nutrients, *417*, 418–19
absorptive heterotrophs, 294, *295*
abstinence for pregnancy prevention, 404
achondroplasia, 146–47
acid–base chemistry, 49, 51
Acidithiobacillus, *278*
acids, defined, 49
acquired immunodeficiency syndrome (AIDS), 459, 460
actin filaments, 423, *425*
active immunity, 464
active sites, 87
active transport, 64
adaptation. *See also* evolution; speciation
    characteristics of, 234
    defined, 195, 232–33
    Indohyus adaptation to aquatic life, 198–99, *200*
    sea fans (*Eunicea flexuosa*), 238, 239–40
    whales, adaptation to aquatic life, 192, 193, 198–99
adaptive immune system, 463–64, 466
adaptive radiation, 265
adaptive traits
    natural selection, 199–200, 215
    overview, 199–200, 232
    Podarcis sicula lizards, 232, 233–34
    sea fans, 238
ADA-SCID (bubble boy disease), 150
Addison's disease, 118, 124
adenine (A), 61, 157, *158*, 178
adipose tissue, 416
ADP (adenosine diphosphate), 84
adrenal glands, 456, 457–58
adrenaline (epinephrine), 456, 457–58, 463, 465, 466

adult-onset lactose intolerance, 145
Aedes aegypti, *342*, 343, 348, 350–51
Aedes albopictus, 343
aerobic organisms (aerobes), 277
aerobic processes, 90
African hare, *367*
African seed crackers, 216, 218, *219*
Agrobacterium ("agrobacteria"), 176, *177*, 181, 184
agronomist, defined, 472
AIDS (acquired immunodeficiency syndrome), 459, 460
alcohol, effect on semen quality, 404
alcoholic fermentation, *91*
Aldrich, Robert Anderson, 137, 139, 142
alfalfa, 175
algae
    coevolution with corals, 239
    comparison to terrestrial plants, *291*
    photosynthesis, 85, 239, 291, *336*
    symbiosis with sea fans, 239, 240
alleles
    allele frequencies, 214, 224
    black-fur allele (*B*), 120
    brown-fur allele (*b*), 120
    codominance and codominant traits, 126
    defined, 118, 214
    dominant alleles, 119–20, 146
    genetic diversity and, 119
    incomplete dominance, 126
    recessive alleles, 120, 142, 145–46
allopatric speciation, 236, *237*, 238, *243*
"Almahata Sitta" meteorites, 48
alternation of generations, 480, *481*
alveoli (singular "alveolus"), 438–39, *440*
Amazon rainforest
    carbon sink and carbon source, 337
    conversion to grasslands (savannas), 324, 329, 332, *334*
    deforestation, 324, *331*, 335

effect of droughts, 324, 326, 330, 332
effect of population and agriculture, 324, *331*
effects of global warming, 324, 328–29
experimental burning of plots, 324, 325–26, 330, 332, 335–36, 337–38
hydrologic cycle, 332–35
overview, 324
projection of rainforest in 2030, *334*
rainforest ecology, *325*
trees, ecological role, 328–29, 335
wildfires in, 324, 326, 329, 330, 335, 337, 338
Ambulocetus, 196, *197*
amino acids
    defined, 41
    essential amino acids, 414
    genetic code, 181, *182*
    left- and right-handed forms, 48
    in meteorite fragments, 48
    Miller experiments, 41, 44, 45, 47–48, 49
    in proteins, 42, 48, 51
    protein synthesis, 174, *180*, 181, 183
amniocentesis, 148
Amoeba proteus, 73
amoebas, *73*, 287, 288
amphibians, *244*, *252*, 257, 305
amygdala, 444
anabolism, 83, 90. *See also* photosynthesis
anaerobic organisms (anaerobes), 84–85, 90, *91*, 277
anaerobic processes, 90
analogous traits, 218
analytical methods, 8–9, *10*
anaphase I in meiosis, *107*, *109*, *110*
anaphase II in meiosis, *107*
anaphase in mitosis, 102, *103*
anatomy, defined, 395